CW00722642

AN INTRODUCTION TO HIGHER MATHEMATICS

The self-taught mathematician Hua Loo-Keng (1910–1985) has been credited with inspiring generations of mathematicians, while his papers on number theory are regarded as some of the most significant contributions made to the subject during the first half of the twentieth century. *An Introduction to Higher Mathematics* is based on lectures given by Hua at the University of Science and Technology of China from 1958. The course reflects Hua's instinctive technique, using the simplest tools to tackle even the most difficult problems, and contains both pure and applied mathematics, emphasising the interdependent relationships between different branches of the discipline. With hundreds of diagrams, examples and exercises, this is both a wide-ranging reference text for university mathematics and a testament to the teaching of one of the most eminent mathematicians of his generation.

THE CAMBRIDGE CHINA LIBRARY

The Cambridge China Library is a series of new English translations of books by Chinese scholars that have not previously been available in the west. Covering a wide range of subjects in the arts and humanities, the social sciences and the history of science, the series aims to foster intellectual debate and to promote closer cross-cultural understanding by bringing important works of Chinese scholarship to the attention of western readers.

AN INTRODUCTION TO HIGHER MATHEMATICS

VOLUME II

HUA LOO-KENG

Translated by

PETER SHIU

CAMBRIDGE UNIVERSITY PRESS
Cambridge, New York, Melbourne, Madrid, Cape Town,
Singapore, São Paulo, Delhi, Mexico City

Cambridge University Press
The Edinburgh Building, Cambridge CB2 8RU, UK

Published in the United States of America by Cambridge University Press, New York

www.cambridge.org
Information on this title: www.cambridge.org/9781107020009

Originally published by Higher Education Press as An Introduction to Advanced Mathematics, volume III (9787040258394) and An Introduction to Advanced Mathematics, volume IV (9787040258400) in 2009

This updated edition is published by Cambridge University Press with the permission of Higher Education Press under the China Book International programme.
For more information on the China Book International programme, please visit
http://www.cbi.gov.cn/wisework/content/10005.html

First published 2012

Printed in the United Kingdom at the University Press, Cambridge

A catalogue record for this publication is available from the British Library

Library of Congress Cataloguing in Publication data
Hua, Luogeng, 1910–1985.
[Gao deng shu xue yin lun. English]
Introduction to higher mathematics / Hua Loo-keng ; translated by Peter Shiu. – English translation ed.
p. cm.
Includes index.
ISBN 978-1-107-01998-0 (v. 1 : hardback) – ISBN 978-1-107-02000-9 (v. 2 : hardback) –
ISBN 978-1-107-02001-6 (2-vol. set)
1. Mathematics. I. Title.
QA37.3.H8313 2012
510 – dc23 2011049197

ISBN 978-1-107-02000-9 Hardback

Only available as a two-volume set:

ISBN 978-1-107-02001-6 2-volume set

NOT FOR SALE IN THE PEOPLE'S REPUBLIC OF CHINA (EXCLUDING HONG KONG SAR, MACAU SAR AND TAIWAN)

Contents

Foreword by Professor H. Halberstam — *page* xv
Preface — xix
Translator's note — xxi
Introduction — xxiii

1 The geometry of the complex plane — 1
1.1 The complex plane — 1
1.2 The geometry of the complex plane — 3
1.3 The bilinear transformation (Möbius transformation) — 5
1.4 Groups and subgroups — 7
1.5 The Riemann sphere — 10
1.6 The cross-ratio — 12
1.7 Corresponding circles — 14
1.8 Pencils of circles — 16
1.9 Bundles of circles — 18
1.10 Hermitian matrices — 21
1.11 Types of transformations — 25
1.12 The general linear group — 27
1.13 The fundamental theorem of projective geometry — 30

2 Non-Euclidean geometry — 32
2.1 Euclidean geometry (parabolic geometry) — 32
2.2 Spherical geometry (elliptic geometry) — 33
2.3 Some properties of elliptic geometry — 37
2.4 Hyperbolic geometry (Lobachevskian geometry) — 37
2.5 Distance — 39
2.6 Triangles — 42
2.7 Axiom of parallels — 44
2.8 Types of non-Euclidean motions — 45

3 Definitions and examples of analytic and harmonic functions — 47
3.1 Complex functions — 47
3.2 Conformal transformations — 48

3.3 Cauchy–Riemann equations 52
3.4 Analytic functions 55
3.5 The power function 58
3.6 The Joukowsky transform 59
3.7 The logarithm function 60
3.8 The trigonometric functions 61
3.9 The general power function 63
3.10 The fundamental theorem of conformal transformation 64

4 Harmonic functions 66
4.1 A mean-value theorem 66
4.2 Poisson's integral formula 67
4.3 Singular integrals 71
4.4 Dirichlet problem 72
4.5 Dirichlet problem on the upper half-plane 73
4.6 Expansions of harmonic functions 75
4.7 Neumann problem 77
4.8 Maximum and minimum value theorems 78
4.9 Sequences of harmonic functions 80
4.10 Schwarz's lemma 80
4.11 Liouville's theorem 83
4.12 Uniqueness of conformal transformations 84
4.13 Endomorphisms 85
4.14 Dirichlet problem in a simply connected region 85
4.15 Cauchy's integral formula in a simply connected region 87

5 Point set theory and preparations for topology 89
5.1 Convergence 89
5.2 Compact sets 90
5.3 The Cantor–Hilbert diagonal method 91
5.4 Types of point sets 92
5.5 Mappings or transformations 93
5.6 Uniform continuity 94
5.7 Topological mappings 96
5.8 Curves 97
5.9 Connectedness 98
5.10 Special examples of Jordan's theorem 99
5.11 The connectivity index 102

6 Analytic functions 105
6.1 Definition of an analytic function 105
6.2 Certain geometric concepts 107

6.3 Cauchy's theorem 108
6.4 The derivative of an analytic function 112
6.5 Taylor series 114
6.6 Weierstrass' double series theorem 115
6.7 Analytic functions defined by integrals 119
6.8 Laurent series 121
6.9 Zeros and poles 124
6.10 Isolated singularities 125
6.11 Being analytic at infinity 127
6.12 Cauchy's inequality 129
6.13 Analytic continuations 131
6.14 Multi-valued functions 133
6.15 The location of singularities 135

7 The residue and its application to the evaluation of definite integrals 138
7.1 The residue 138
7.2 Integrals of rational functions along a circle 139
7.3 Evaluation of certain integrals over the reals 140
7.4 Certain integrals involving sin and cos 142
7.5 The integral $\int_0^\infty x^{a-1}Q(x)dx$ 145
7.6 The Γ function 147
7.7 Cauchy principal value 150
7.8 An integral related to a moment problem 151
7.9 Counting zeros and poles 152
7.10 The roots of an algebraic equation 154
7.11 Evaluation of series 155
7.12 Linear differential equations with constant coefficients 156
7.13 The Bürmann and the Lagrange formulae 158
7.14 The Poisson–Jensen formula 160

8 Maximum modulus theorem and families of functions 162
8.1 Maximum modulus theorem 162
8.2 Phragmen–Lindelöf theorem 163
8.3 Hardamard's three-circle theorem 164
8.4 Hardy's theorem on the mean value of $|f(z)|$ 165
8.5 A lemma 166
8.6 A general mean-value theorem 167
8.7 $\left(I_p(r)\right)^{1/p}$ 168
8.8 Vitali's theorem 169
8.9 Families of bounded functions 172
8.10 Normal families 174

9 Integral functions and meromorphic functions 175
9.1 Definitions 175
9.2 Weierstrass' factorisation theorem 177
9.3 The order of an integral function 178
9.4 Hadamard's factorisation theorem 181
9.5 Mittag–Leffler's theorem 182
9.6 Representations for $\cot z$ and $\sin z$ 183
9.7 The Γ function 186
9.8 The zeta function 189
9.9 The functional equation for $\zeta(s)$ 191
9.10 Convergence on the sphere 193
9.11 Normal families of meromorphic functions 194

10 Conformal transformations 197
10.1 Important basic ingredients 197
10.2 Univalent functions 199
10.3 The inverse from Taylor series 200
10.4 The image of a region 202
10.5 Sequences of univalent functions 203
10.6 The boundary and the interior 204
10.7 The Riemann mapping theorem 206
10.8 Estimating the second coefficient 208
10.9 Corollaries 210
10.10 The Koebe quarter theorem 211
10.11 Littlewood's estimate 213
10.12 Star regions 215
10.13 Real coefficients 217
10.14 Mapping a triangle onto the upper half-plane 218
10.15 The Schwarz reflection principle 220
10.16 Mapping a quadrilateral onto the upper half-plane 222
10.17 Mapping polygons onto the upper half-plane – the Schwarz–Christoffel method 225
10.18 Continuation 227
10.19 More on polygonal mappings 230

11 Summability methods 232
11.1 Cesàro summability 232
11.2 Hölder summability 235
11.3 Two lemmas related to means 236
11.4 The equivalence of (C, k) and (H, k) summabilities 238
11.5 (C, α) summability 240
11.6 Abel summability 242
11.7 Introduction to general summability 243

11.8 Borel summability 244
11.9 Hardy–Littlewood theorems 247
11.10 Tauber's theorem 251
11.11 Asymptotic properties on the circle of convergence 252
11.12 A Hardy–Littlewood theorem 254
11.13 Littlewood's Tauberian theorem 259
11.14 Being analytic and being convergent 262
11.15 The Borel polygon 266

12 Harmonic functions satisfying various types of boundary conditions 270
12.1 Introduction 270
12.2 Poisson's equation 272
12.3 Doubly harmonic equation 275
12.4 Formula for a doubly harmonic function in the unit disc 277
12.5 The background to Cauchy's integral 278
12.6 Cauchy's integral 281
12.7 Sokhotsky's formula 282
12.8 The Hilbert–Privalov problem 286
12.9 Continuation 289
12.10 The Riemann–Hilbert problem 290
12.11 The uniqueness of the solution to a mixed boundary value problem 291
12.12 The Keldysh–Sedov formula 293
12.13 The Keldysh–Sedov formula in other regions 296
12.14 A mixed form of partial differential equations 299

13 Weierstrass' elliptic function theory 303
13.1 Modules 303
13.2 Periodic functions 305
13.3 Expansions of periodic integral functions 306
13.4 The fundamental region 308
13.5 General properties of an elliptic function 309
13.6 Algebraic relationships 311
13.7 Two types of theories for elliptic functions 312
13.8 Weierstrass' γ function 313
13.9 The algebraic relationship between $\gamma(z)$ and $\gamma'(z)$ 315
13.10 The function $\zeta(z)$ 316
13.11 The function $\sigma(z)$ 318
13.12 General representations for elliptic functions 319
13.13 Addition formulae 322
13.14 The integral of an elliptic function 323
13.15 The field of algebraic functions 325
13.16 The inverse problem 326
13.17 Modular substitutions 327

13.18 The fundamental region 329
13.19 The net of the fundamental region 333
13.20 The structure of the modular group 334
13.21 The definition and properties of modular functions 335
13.22 The function $J(\tau)$ 338
13.23 Solutions to $g_2(w, w') = a$, $g_3(w, w') = b$ 341
13.24 Any modular function is a rational function of $J(\tau)$ 341
13.25 Appendix: Important formulae (Weierstrass) 342

14 Jacobian elliptic function theory 346
14.1 The theta functions $\theta(z, q)$ 346
14.2 The zeros of, and infinite products for, θ functions 348
14.3 $G = \prod_{n=1}^{\infty}(1 - q^{2n})$ 350
14.4 Using θ functions to represent elliptic functions 353
14.5 Various quadratic relationships between θ functions 355
14.6 Formulae for sums and differences 356
14.7 Differential equations satisfied by ratios of θ functions 359
14.8 Jacobian elliptic functions 360
14.9 Periodicity 362
14.10 Being analytic 363
14.11 The relationship between Weierstrass and Jacobian functions 364
14.12 Addition formulae 365
14.13 Representations of K, K' as functions of k, k' 365
14.14 Various formulae associated with the Jacobian elliptic functions 367
14.15 Notes 368
14.16 Important formulae (Jacobi) 369

15 Linear systems and determinants (review) 374
15.1 Linear systems 374
15.2 Method of eliminations 375
15.3 Geometric interpretation of the method 377
15.4 Mechanical interpretation of the method 378
15.5 Economic equilibrium 379
15.6 Linear regression analysis 380
15.7 Determinants 382
15.8 The Vandermonde determinant 385
15.9 Symmetric functions 390
15.10 The fundamental theorem on symmetric functions 393
15.11 Common roots for two algebraic equations 395
15.12 Intersections of algebraic curves 397
15.13 Power series associated with determinants 397
15.14 Power series expansion for the Wronskian 400

16 Equivalence of matrices 403
16.1 Notations 403
16.2 Rank 405
16.3 Elementary operations 406
16.4 Equivalence 409
16.5 The n-dimensional vector space 411
16.6 Transformations in a vector space 412
16.7 Length, angle, area and volume 414
16.8 The determinant of a transformation (Jacobian) 416
16.9 Implicit function theorem 417
16.10 The Jacobian of a complex transformation 419
16.11 Functional dependence 421
16.12 Algebraic considerations 426
16.13 The Wronskian 429

17 Functions, sequences and series of square matrices 431
17.1 Similarity of square matrices 431
17.2 Powers of a square matrix 433
17.3 The limit of powers of a square matrix 434
17.4 Power series 436
17.5 Examples of power series 437
17.6 Method of successive substitutions 439
17.7 The exponential function 441
17.8 The derivative of a square matrix of a single variable 442
17.9 Power series for the Jordan normal form 443
17.10 A number raised to a matrix power 445
17.11 The matrix e^X for some special X 445
17.12 The relationship between e^X and X 448

18 Difference and differential equations with constant coefficients 449
18.1 Difference equations 449
18.2 Linear difference equations with constant coefficients – the method of generating functions 452
18.3 Second method – order reduction method 455
18.4 Third method – Laplace transform method 456
18.5 Fourth method – matrix method 456
18.6 Linear differential equations with constant coefficients 457
18.7 The motion of a particle over the surface of the Earth 458
18.8 Oscillations 461
18.9 Absolute values of matrices 463
18.10 Existence and uniqueness problems for a linear differential equation 464
18.11 Iterated integrals 468
18.12 Solutions with full rank 470

18.13 Inhomogeneous equations 471
18.14 Asymptotic expansions 473
18.15 A functional equation 474
18.16 Solutions to the differential equation $dX/dt = AX + XB$ 476

19 Asymptotic properties of solutions 479
19.1 Difference equations with constant coefficients 479
19.2 Generalised similarity 482
19.3 Ordinary linear differential equations with constant coefficients 483
19.4 Introduction to Lyapunov's method 485
19.5 Stability 488
19.6 Lyapunov transformation 490
19.7 Differential equations with periodic coefficients 491
19.8 Lyapunov equivalence 493
19.9 Difference and differential equations with coefficients which are asymptotically constant 494

20 Quadratic forms 495
20.1 Completing squares 495
20.2 Completing the square on pieces 499
20.3 Affine types of the affine geometry of quadratic surfaces 500
20.4 Projective geometry 504
20.5 Projective types of quadratic surfaces 507
20.6 Positive definite forms 508
20.7 Finding the least value from completion of squares 510
20.8 The Hessian 512
20.9 Types of second order partial differential equations with constant coefficients 513
20.10 Hermitian forms 515
20.11 The real shape of a Hermitian form 516

21 Orthogonal groups corresponding to quadratic forms 518
21.1 Orthogonal groups 518
21.2 Square roots of positive definite forms for distance functions 522
21.3 Normed spaces 523
21.4 The Gram–Schmidt process 525
21.5 Orthogonal projections 528
21.6 Unitary spaces 531
21.7 Introduction to inner product function spaces 533
21.8 Eigenvalues 536
21.9 Eigenvalues of integral equations 539
21.10 Orthogonal types of symmetric matrices 540
21.11 Euclidean types of quadratic surfaces 541

21.12 Pairs of square matrices 543
21.13 Orthogonal types of skew-symmetric matrices 546
21.14 Symplectic groups and symplectic types 547
21.15 Various types 548
21.16 Particle oscillations 549

22 Volumes 552
22.1 Volume elements in m-dimensional manifolds 552
22.2 Dirichlet's integral 556
22.3 Normal distribution integrals 559
22.4 Normal Parent's distribution 561
22.5 Determinants of matrix transformations 563
22.6 Integration elements in a unitary group 566
22.7 Integration elements in a unitary group (continuation) 568
22.8 Volume elements for real orthogonal matrices 571
22.9 The total volume of the real orthogonal group 572

23 Non-negative square matrices 575
23.1 Similarity of non-negative matrices 575
23.2 Standard forms 576
23.3 Proof of the fundamental theorem 577
23.4 Another form of the fundamental theorem 580
23.5 Arithmetic operations on standard forms 582
23.6 Large and small matrices 583
23.7 Strongly indecomposable matrices 587
23.8 Markov chains 588
23.9 Continuous stochastic transition processes 590

Index 593

Foreword

I welcome this opportunity to join Professor Wang and Dr Shiu in launching this classic text book into contemporary mathematics literature; although written half a century ago, it demonstrates in its range and depth of material the ambition with which Hua set out to educate his people at a crucial time in their history, in mathematics and its applications. With the erosion of standards in the teaching of these areas in many parts of the English-speaking world, Hua's books have as much to offer now as they did when they were written.

In his fine biography[1] of Hua, Professor Wang often refers to him as 'Teacher Hua', and I think that is how Hua would have liked best to be remembered. He was a fluent speaker and prolific writer, always ready to report on new and exciting topics, and he disliked having to repeat talks he had given before – he was a born communicator. Hua had a commanding presence, a sunny temperament and an ever optimistic outlook, and teaching was his passion; the joy of making new advances was partly to tell others about them. The 'glittering prizes' of a successful academic career held little attraction for him; he wanted only to lead his students to success and to advance the progress of his people by all the means at his disposal. I see him as one of the architects of modern China.

A few biographical remarks may not be out of place. Hua was born into humble circumstances in Jintan, once a village but now a flourishing city proudly displaying tokens of his fame. He received little beyond an elementary education, but his love of mathematics surfaced early and was nurtured by a sympathetic teacher. Despite health problems and a severe physical handicap, Hua found his way somehow to a position as assistant in a university library and, having studied mathematics there on his own, was by 1932 teaching mathematics to others; not only that, but he had also begun to publish. When two famous mathematicians[2] visited China, Hua came to their notice, and before long he was on his way to Cambridge, England, into the charmed circle of G. H. Hardy and the bright young people at his feet. Hardy was impressed by Hua's talent, and told him that he could easily gain a Ph.D. within the space of two years, but Hua politely declined; although he did not say so, the reason was that he could not afford the registration fee.

[1] *Hua Loo-Keng*, by Wang Yuan (in Chinese, 1994); English translation by Peter Shiu (Springer Verlag, Singapore, 1998).
[2] Jaques Hadamard of France and Norbert Weiner of the USA.

In 1937 Japan invaded China, and the following year Hua returned to China by a circuitous route, not to his home, but to Kunming, to the South West Associated University, an amalgam of several universities displaced by the advancing Japanese forces. There Hua and his family lived the next nine years in wretched conditions, often hungry and lacking adequate library facilities. Nevertheless Hua kept up the spirit of a group of young mathematicians with his lectures, and he even managed to keep up with developments in Europe and the USA. It was probably in these years of scarcity that Hua developed a taste for simplicity, both in exposition and in proof. In these years Hua also widened the range of his own mathematical interests, partly in the course of supporting the work of his colleagues, and soon he was making original contributions in areas that had been new to him.

World War II came to an end in 1945, and with it the Japanese occupation, but in China there ensued a bitter civil war. Despite this, Hua visited Russia in 1946 at the invitation of I. M. Vinogradov, who had been impressed by Hua's contribution to additive number theory – Hua liked to tell how he narrowly escaped being awarded a Stalin Prize! Later that same year, Hua was invited to visit the Institute of Advanced Studies in Princeton; he was well received there and his researches prospered. When he accepted a professorial position at the University of Illinois at Urbana-Champaign a few years later, he seemed well launched on an academic career in America. However, he had no sooner seemed well settled there and had begun to build a research group when a call came in 1950 to return home, and he did not hesitate. At the age of 40, at the height of his powers, he set sail for China – the new China of Mao.

Hua threw himself enthusiastically into the effort to repair his devastated country: to organize mathematical activity at all levels, to support young mathematicians, and to teach wherever he was needed. For example, he travelled with a team of assistants throughout the country to give a hand with start-up industries where workers would run into problems; in later years he liked to describe some of these problem-solving episodes.[3]

Civil wars are untidy and often violent affairs, and the national prominence at which Hua had arrived by now exposed him to hostilities from those in power and even to personal danger. He was harassed frequently, his home was searched, and confiscated manuscripts were never returned; his most successful students were pressed to bear witness against him. Hua survived, sometimes thanks to the protection of Premier Zhou En-Lai, but, although he never ceased to do mathematics, often in partnership with his one-time student and friend Wang Yuan, the prime time of his invention was over.

The Cultural Revolution came to an end in 1976 and Hua was allowed to travel abroad again. Not long after that, Peter Shiu and I met him for the first time in Birmingham in the English Midlands, where Hua was visiting. And so began the last phase of his life, when he toured the places he had spent time in and in which he had made friends in his youth, collecting honours and lecturing on his educational adventures in China. We met once more in 1984, in the USA, when I found him very tired but determined to continue work in the

[3] Hua, L. K. and Wang Yuan, *Popularizing Mathematical Methods in the People's Republic of China* (Birkhauser, Boston 1989).

service of his country. He died in Japan after giving yet another talk to an admiring audience in Tokyo on 12 June 1985, in his 75th year, exhausted by unceasing work in the service of mathematics and of his people.

H. Halberstam
Urbana-Champaign, August 2011

Preface

The material in this book is based on my notes when I gave a course of lectures on advanced mathematics in 1963 at the University of Science and Technology of China. The main part of the course was on complex function theory, and the material was intended to be published separately as Part I of Volume Two[1] to my *Introduction to Higher Mathematics*.

At the end of the lecture course the material was thoroughly revised and much expanded, but unfortunately the manuscripts for the revised material are now irretrievably lost. The content here is thus based on the original lithographic material left at the University of Science and Technology. Considering that the first volume of *Introduction to Higher Mathematics* was published some fifteen years ago, readers must have hoped that the second volume would have been published some time ago. If this part is to be rewritten completely, the nature of my work load and the condition of my health would mean that the project would have to be further delayed. Fortunately the basic material still reflects much of the author's opinion and attitude, and in any case one cannot afford to be too fussy when faced with time constraints. I now prcscnt thc matcrial as it is in this volume, and it is my hope that, after studying the material, readers will offer me their valuable suggestions so that there will be a revised edition.

Hua Loo-Keng
Beijing Hospital, 19 January, 1978

My original intention was to write six or seven volumes to *Introduction to Higher Mathemtics*. The first volume, in two parts,[2] was published in 1963. However, the drafts for most of the rest for the project were lost, leaving only scattered bits and pieces. Nevertheless, Part One of Volume II managed to see the light of day in 1981. There was some yearning that I would return to the original plan for the project, but after the realisation that the drafts were irretrievably lost, and my assessment that time was not on my side, I reluctantly had to accept that there was no real prospect for me to rewrite the lost material. Nevertheless, I

[1] Translator's footnote: In this English edition, Volume I comprises Parts I and II of the original Volume One, published in 1962 in Chinese. Thus Chapters 1–14 in this current edition comprise what the author refers to here as 'Part I of Volume Two'.

[2] Translator's footnote: In this English edition, 'Volumes I and II' are what were Parts One and Two of Volume I in the first edition in Chinese, and Volume III in this edition is what the author referred to in this preface as his 'Part One of Volume II'.

have decided to revise whatever was left of my notes for the lecture course given in 1962 at the University of Science and Technology in China, and to publish them piecemeal (with the same, or perhaps a different, title).

Fortunately, this part of the project can be considered to be reasonably independent of the other parts. When lecturing or studying the material, the reader should bear in mind the following: The teacher ought to be aware of the level of knowledge of the students, and should not advance too fast, while the student should not feel that we are only repeating rather simple material. The main reason is that we are essentially taking the second step in the '$1, 2, 3; n; \infty$' approach, as in the study of matrices, for example. Thus, in the preparation volumes, we spoke of one, two and three variables in one, two and three dimensions, whereas we are now dealing with n variables in an n-dimensional space. There were twelve chapters altogether in the original draft to this volume, and I recall that differential geometry in n dimensions was discussed in the final three chapters. Although the material is now lost, nevertheless the reader may wish to develop it, using as a model what has been presented on the differential geometry of the curve in space given in the first volume. One has to use different types of matrices corresponding to the underlying orthogonal group in order to derive the differential properties associated with curves in an n-dimensional space. Actually this is an excellent exercise, and, when completed satisfactorily, one can then replace the orthogonal group by other groups and study the properties of the differential invariants.

Time not being on my side, I regret that I cannot do much revision to the manuscript, and I beg the diligent reader to make the necessary corrections.

Hua Loo-Keng
11 October 1981

I should have added that, during the year when I wrote the material at the University of Science and Technology, comrade Gong Sheng had given me a lot of assistance, and he was even more helpful later in the search for the lost material.

Gong Sheng's diligence and attitude went beyond his responsibility as a comrade, and I would like to record my gratitude to him and to thank him.

Hua Loo-Keng
9 September 1983

Translator's note

In the *Biographical Memoir* (Vol. 81, 2002) of the (American) National Academy of Sciences, Professor Halberstam wrote:

> If many Chinese mathematicians nowadays are making distinguished contributions at the frontiers of science and if mathematics in China enjoys high popularity in public esteem, that is due in large measure to the leadership Hua gave his country, as scholar and teacher, for 50 years.

In 1979 the author and Halberstam were introduced to each other, and they immediately became good friends. As gifts for each other, *Sieve Methods* from Halberstam, and *Introduction to Number Theory* from the author, were exchanged, and I was asked by both of them to have the latter translated into English. I was overwhelmed with the honour of being asked, and also with fear because my Chinese was no longer fluent, not to mention the depth of the mathematics involved. The author's only piece of advice to me was to read the whole book thoroughly first. It was sound advice because I regained my confidence in the language, and also managed to learn the mathematics. Years later Professor Wang Yuan persuaded me that I could also take on the translation of his fine biography of Teacher Hua Loo-Keng, and I learned much more about the man, of course.

In fact, Hua had mentioned to me the existence of the titles to the present volumes, indicating that he would also like these books to be translated into English one day. The invitation from Higher Education Press (China) to take on the task therefore came as no surprise. Once again I was overwhelmed by the honour, and this time the fear of my inadequacy lies in the breadth of the mathematics involved. My memory of the man, and his sense of duty, left me no doubt that I should take on the task. It only remains for me to thank HEP for their efficiency, and also Cambridge University Press for providing me with their expertise.

Peter Shiu
Sheffield, August 2011

Hua Loo-Keng and *An Introduction to Higher Mathematics*

WANG YUAN
Institute of Mathematics, Chinese Academy of Sciences

I

The University of Science and Technology of China was established in September 1958. Following the policy of 'pooling the entire strength of the Chinese Academy of Sciences for the development of the University and combining the institutes of the Academy with the departments at the University', Hua Loo-Keng and other distinguished scholars from the Academy, such as Wu Youxun, Yan Jici and Qian Xuesen, took up positions and gave lectures there.

Hua Loo-Keng became Head of the Applied Mathematics Department (later renamed the Mathematics Department). He then initiated the 'single-dragon method' for mathematical instructions. Hua had always emphasised the interdependent relationships between all branches of mathematics, and considered the separation of such foundation subjects as calculus, algebra and complex function theory into compartments for teaching to be artificial. He therefore decided to put all such foundation materials together for a single course, which he taught for some three to four years.

Frankly speaking, the production of teaching materials for the 'single-dragon method' could only be undertaken by a mathematician who had a complete command of the whole subject together with a true understanding of the relationships between the various branches. On the other hand, there is no doubt that Hua Loo-Keng was admirably suitable for such an undertaking. This is because Hua himself had made significant contributions in several areas of mathematics and he well understood the subtle interrelationships across the various subjects.

We note the following special points from the four volumes that have already been published.[1]

First, for a university foundation course on mathematics, it is important to deal with the basics. Hua Loo-Keng had often said the following:

> I love to lecture in such a way that the new material seems to be already familiar, while the old material is reviewed in a new setting. It should appear that we are just introducing some new stuff

[1] The new English language edition comprises two volumes: Volume I contains the first two and Volume II the second two previously published volumes. In this Introduction all references to volume and chapter numbers refer to the original, four-volume, work.

into rather familiar territories. This is because, speaking generally, when the new is compared to the old there is not much that is really new or very different; often it is just a new line of thought with the same material being dressed up in new clothes only. If we get used to reviewing and revising in this way it becomes easy to discover and understand the genuinely innovative ideas ... The 'difference' and 'similarity' between 'numbers' and 'shapes', and the 'abstract' and the 'concrete' etc., are over and over again mere transitions of essentially the same objects, but perhaps set in a new environment.

Secondly, given a sufficient command of the mathematics, when it comes to the exposition of any topic, we can almost always consider it as a part of a more general scheme.

Thirdly, Hua Loo-Keng was a very hard-working mathematician who would never dismiss the easy material as trivial. He invested much of his time and effort practising 'shadow boxing', frequently using basic mathematics to deliver an interesting or pertinent application. The reader should feel that the mathematics here is really alive, and yet never so esoteric that it has become out of reach.

Fourthly, Hua Loo-Keng's prevailing attitude towards mathematics is to employ a 'direct method'. Without any hesitation he would use the simplest and the most elementary tools to tackle even the most difficult problems. He would never start from the abstract, and would always consider a concrete example as a first approach to the problem, before following it up with more general conclusions. In writing these books, he kept faith with this attitude. The proofs of theorems are not lengthy, most of the time never more than one or two pages, so that the reader can absorb them with ease.

II

In Volumes One and Two, the materials covered are calculus, advanced calculus and analysis. The theory of real numbers is presented in the first chapter. Hua uses the infinite decimal system to define the real numbers; although the presentation is deliberately on the descriptive side, the treatment is still quite rigorous. After this, the ϵ-N approach is given via Cauchy sequences. In the supplement to the first chapter, besides writing about the use of the binary system in computers, Hua also shows that a necessary and sufficient condition for a number to be rational is that there is periodicity in its decimal expansion. He then goes on to consider rational approximations to real numbers, and the use of 'continued fractions'. This is the usual material for 'elementary number theory', but he now applies the theory to calculations, first involving the solar and lunar calendars and then various astronomical phenomena, such as eclipses and conjunctions.

The concept of a 'limit' is often the first difficulty to be encountered by students in their transition from school to university mathematics. Here Hua goes back to talking about limits for numbers again in Chapter 4. The notions of upper and lower limits are introduced, and the concept of continuity and the ϵ-δ theory are then discussed. There will be yet more on limits, of course, but the reader will have had its gentler introduction.

The second chapter is on vector algebra, and the main material is concerned with various geometric formulae in Euclidean space; the 'supplement' includes material on spherical trigonometry and the application of the vector method to Newtonian mechanics.

After the consideration of continuity it is natural to deal with the applications of the differential and integral calculus. In Chapter 10 the Euler summation formula is introduced:

Let $\phi(x)$ have a continuous derivative in $[a, b]$. Then

$$\sum_{a<n\leq b} \phi(n) = \int_a^b \phi(x)dx + \int_a^b \left(x - [x] - \frac{1}{2}\right)\phi'(x)dx + \left(a - [a] - \frac{1}{2}\right)\phi(a) - \left(b - [b] - \frac{1}{2}\right)\phi(b),$$

where $[x]$ denotes the integer part of x.

Hua first applies Euler's formula to deduce Stirling's formula, and then goes on to use the summation formula to derive the various approximation formulae for integrals, such as the trapezium and Simpson rules, together with estimates for the error terms. This is different from the usual approach and, apart from seeing the generality of the Euler summation formula, the reader can see immediately how the dominating term for a sum can be given the integral representation.

In Chapter 13 the author deals with sequences of functions, leading to a deeper consideration of limit. This involves the notion of uniform convergence and the associated criteria for its application to infinite products, differentiation under the integral sign, and the interchange of orders for integration and summation. Besides the various results associated with the differential and integral calculus, the chapter also includes materials on the contraction mapping principle and the theorem of Cauchy–Kovalevskaya.

In the supplement for Chapter 15 the author discusses various methods in the calculation of area and volume and their practical applications. Such methods were first found in texts on geography and mining, using elementary geometry to effect the required calculations which are usually rather complicated. Here, in only a dozen pages, the author sets out the basic ideas and develops a theory to obtain the analytic results.

Differential geometry is the subject matter in Chapters 14 and 18. Having dealt with the differential calculus, the new notion of differential geometry should no longer be that formidable. This includes the usual results in the subject such as the first and second differential forms of Gauss, curvature, tensor and the formulae of Gauss and Codazzi.

Chapter 19 is on Fourier series and covers the usual material in the subject.

Chapter 20 is on systems of ordinary differential equations. Here the author introduces the formulae associated with the orbits for artificial satellites, including the three laws governing their motions, and a discussion of the many-bodies problem. Some of the materials are taken from interesting exercises considered by the author soon after the launching of the first artificial satellite from the USSR.

III

The third volume is mainly concerned with 'complex variable theory', but much else is also included. The author starts with the geometry of the complex plane, from where the Möbius transformation is introduced, together with discussions on the linear group, the

Riemann sphere, cross-ratios, harmonic points, etc. Finally the fundamental theorem (von Staudt) of projective geometry is established:

A one-to-one continuous transformation in the (complex) projective space into itself which maps the harmonic points to themselves must be a generalised linear transformation.

This important theorem has relevance in the study of the space of matrices and is related to some of the research work by the author in matrix geometry.

The second chapter is on non-Euclidean geometry. The author introduces us to parabolic (Euclidean) geometry, spherical (elliptic) geometry and hyperbolic (Lobachevsky) geometry. Here the reader will see the definitions for the various different 'metrics'.

The third chapter is on analytic functions and harmonic functions. The author leads us to the important Riemann mapping theorem:

Let D be a simply connected region with more than one boundary point, and let z_0 be an interior point with a certain specified direction. Then there is a unique conformal transformation mapping D bijectively onto the unit circle with z_0 being mapped to the origin, with the specified direction along the positive real axis.

Hua then writes that, with this theorem, a problem in a simply connected region can now be transformed to one in the unit circle, and thus if we can understand clearly what is happening in the unit circle then there is hope for a solution to the more general situation. This is because, within the unit circle, there is now the useful tool of power series representation for a complex function. Throughout the text we often see the author emphasising such points.

Cartan has proved the following:

There are six types of irreducible homogeneous bounded symmetric domains for analytic mappings. Two of these are exceptional and the other four are the so-called classical domains.

Classical domains may be regarded as the higher dimensional analogues of the unit disc and other domains in the complex plane. Therefore the importance of the unit disc is easily recognised. Hua established the theory of harmonic analysis on classical domains (an orthogonal system for each of the four classical domains), and consequently he obtained the Cauchy kernel, the Poisson kernel, etc. for the classical domains. Hua's idea on this research lies in his deep understanding of the theory of functions on the unit disc.

By applying Riemann's theorem many important results in complex function theory become much easier to establish and understand.

In Chapter 5 the author introduces the distance function and uses it to define convergence. This then extends the notion of convergence, reflecting on what has been said about new material appearing to be familiar already.

In this volume, besides the theory of complex functions, the author also gives us plenty of other material. In Chapter 11 we have various summability methods applied to some divergent series; they include the methods of Cesàro, Hölder, Borel and Abel. Indeed there are even Tauberian theorems and other related material which is usually considered only in treatises on the theory of Fourier series.

Chapter 12 deals with problems arising from solutions to differential equations, such as the Riemann–Hilbert problems and mixed types of differential equations.

The elliptic functions of Weierstrass and Jacobi together with other more recent results in number theory are dealt with in Chapters 13 and 14; in fact it is rare that such materials are given in a university course.

IV

The main material in Volume Four is concerned with the algebra of matrices, but it goes well beyond that in scope.

Chapter 4 is on systems of difference equations and ordinary differential equations, while Chapter 5 deals with the convergence of solutions. The use of matrix theory to solve ordinary differential equations includes a discussion of Lyapunov's method.

Chapter 8 has *Volume* as the title. Here the author derives the formula for the volume of an m-dimensional manifold, and deals with the evaluation in the general case associated with an orthogonal group.

Chapter 9 is on non-negative matrices, and would have included the author's work on their applications to economics. This is the unfinished volume. The author had indicated that there would have been three more chapters on differential geometry in n-dimensional space, making use of materials on differential geometry and orthogonal groups given in the second volume.

V

The author wrote in the Prefaces to Volumes One and Two that the books were:

. . . written in a bit of hurry [and] done without much editing . . .

The reader may discover that there are materials not found in other texts, or that the treatment of some material is slightly different, but not that much . . .

Thinking in a vacuum there may be plenty of errors . . .

When just copying down materials which are familiar, mistakes are sometimes unavoidable already. More worrying is the material that is being written for the first time. . . .

In particular there must be places where the more advanced material has been somewhat 'dumbed-down', or a genuinely complicated argument over-simplified. We invite readers to put us back on the right path.

Such words are, of course, a reflection of the integrity and high standard that Hua Loo-Keng set himself. In fact, much detail was given in both Volumes One and Two. However, in Volumes Three and Four, there are more than several places at which the author wrote 'Similarly it is not difficult to prove that . . . ' or words to that effect. This is acceptable when writing for someone who is as knowledgeable as the author. However, when it comes to students coming across such material for the first time, or even for myself when I was working closely with him as his assistant, it was not at all easy. For this reason, when studying the material, special attention must be given to such places where one needs to put in the extra work.

There are well over a thousand pages in these four volumes. As teaching material there is obviously too much for an ordinary student and, as the author himself had pointed out, teachers need to make a judicious selection. However, for the teachers themselves, I believe the books as a whole are excellent. Indeed, capable students should be able to learn much from any of the chapters under suitable guidance from their teachers.

Apart from the rewriting of Theorem 7 in Section 3 of Chapter 4 in Volume One,[2] there had only been minor corrections of misprints during each reprinting of the books. Operating in this way, the spirit given to the original edition is naturally preserved. On the other hand, much revision could also be given, but an informed discussion is required for a decision. Perhaps this is not yet the moment.

Take the following example. The famous Bieberbach conjecture:

The coefficients of the normalised univalent function $f(z) = z + a_2 z^2 + \cdots\ |z| < 1$, *satisfy the inequality* $|a_n| \le n$, $n = 2, 3, \ldots$.

This conjecture was mentioned in Chapter 10 of Volume Three, in which the partial results of Littlewood, Nevanlinna, Dieudonné and Rogossinski were given. However, the Bieberbach conjecture was proved by L. de Branges in 1985. Whether such material should be included in a new edition is perhaps worth considering.

More to the point is the following. Hua Loo-Keng had worked tirelessly on the project: he produced six or seven manuscripts, and yet he had never spoken to any of his assistants concerning their overall plans, nor did anyone ask him about the project. Looking back at it now, it appears that material on abstract algebra, algebraic topology, Lebesgue theory and much else might also have been included in the project.

Back in the 1950s Hua spoke to me several times on the teacher and student relationship between Dirichlet and Dedekind. In the nineteenth century Dirichlet wrote a certain text (*Lectures on the Theory of Numbers*), and in each subsequent edition Dedekind wrote a 'Supplement' updating the text, so much so that the supplements eventually took up more space than the original text.

Hua Loo-Keng encouraged us to revise his texts continually, be they modifications or addition of materials. For example, there were several editions of his *Introduction to Number Theory* during his lifetime. Xiao Wenjie (P. Shiu) and myself wrote some supplements to the latest edition that Hua himself approved and much appreciated. However, for *Introduction to Higher Mathematics*, the area being covered is so enormous that I simply do not have the required knowledge and stamina to write the supplements.

Indeed, with the passage of time, those university students who attended his lectures are themselves rather old, retired and also lack the energy to take on the task. If there are to be supplements, we shall have to wait for the next generation, or perhaps even another generation, of scholars. However, I am confident that the young mathematicians in our country will keep the fire in the stove first lit by Hua Loo-Keng.

[2] See Theorem 4.21.

The year 2010 will be the centenary of the birth of Hua Loo-Keng, and also the twenty-fifth anniversary of his death. We thank Higher Education Press for the work done for the new edition of these books. At the same time they are also preparing to have an English edition for these books, which is very far sighted of them because it will be a very worthwhile undertaking.

Thinking back, it has been some fifty years since I was a student and an assistant. I was in the fortunate position to help Teacher Hua Loo-Keng with his lectures and the writing of Volumes One and Two of these books at the University of Science and Technology. The vivid memory of the time spent together will be treasured by me forever. With the new printing I take the opportunity to wish sincerely that the publication of these books will help to develop and expand mathematical knowledge in our country, and to nurture the talented ones so that new and important contributions will be forthcoming.

1

The geometry of the complex plane

1.1 The complex plane

A complex number can be written in the form

$$z = x + iy, \tag{1.1}$$

where x and y are real numbers; they are called the real and the imaginary parts of z, respectively, and we shall write $x = \Re z$ and $y = \Im z$. Two complex numbers are equal if and only if their real parts and their imaginary parts are separately the same. The complex number $x - iy$ is called the conjugate of z, and is denoted by $\bar{z}$.

The following are the two operational rules for the arithmetic of complex numbers: Given any two complex numbers $z_1 = x_1 + iy_1$, $z_2 = x_2 + iy_2$, their sum and their product are given by

$$z_1 + z_2 = x_1 + x_2 + i(y_1 + y_2), \quad z_1 z_2 = x_1 x_2 - y_1 y_2 + i(x_1 y_2 + x_2 y_1).$$

Take a system of orthogonal coordinates for the plane. Corresponding to the complex number (1.1), there is a point P with coordinates (x, y), and we call this point the representation of z on the plane. This correspondence then sets up a one-to-one relationship between the set of complex numbers and the set of points on the plane. The real numbers then correspond to the x-axis, and for this reason we sometimes call it the real axis; similarly the y-axis is called the imaginary axis. Henceforth we shall not distinguish between a complex number and the point representing it on the plane; for example, when we say the point $1 + i$, we mean the point (x, y) on the plane with $x = 1$, $y = 1$.

Take a line segment from the origin O to the point P, and call it the vector $\overrightarrow{OP}$. Then each complex number corresponds to a vector from the origin, and vice versa. Moreover, the sum of complex numbers corresponds to the sum of the vectors. The length of the vector $\overrightarrow{OP}$ is

$$\rho = \sqrt{x^2 + y^2} = \sqrt{z\bar{z}},$$

and is called the absolute value, or the modulus, of z, which is denoted by $|z|$. The angle θ that the vector $\overrightarrow{OP}$ makes with the x-axis is called the argument of z, and we write $\theta = \arg z$; the angle is measured in the anti-clockwise direction with respect to the positive real axis; clearly,

$$z = x + iy = \rho(\cos\theta + i\sin\theta) = \rho e^{i\theta} = |z|e^{i\arg z}. \tag{1.2}$$

It is not difficult to show that, under such a representation, the product of two complex numbers $z_1 = \rho_1 e^{i\theta_1}$ and $z_2 = \rho_2 e^{i\theta_2}$ is $z_1 z_2 = \rho_1\rho_2 e^{i(\theta_1+\theta_2)}$. Note, however, that the angle is not uniquely specified, because if (ρ, θ) represents z, then so does $(\rho, \theta + 2\pi)$. More generally, we may set

$$\phi = \operatorname{Arg} z = \begin{cases} \arctan\frac{y}{x} + 2k\pi \text{ (if } z \text{ is in the first or fourth quadrant)}, \\ \arctan\frac{y}{x} + (2k+1)\pi \text{ (if } z \text{ is in the second or third quadrant)}, \end{cases}$$

where k is any integer, and arctan is the branch of the inverse tangent function which satisfies

$$-\frac{\pi}{2} < \arctan\frac{y}{x} \leq \frac{\pi}{2}.$$

The argument ϕ which satisfies

$$-\pi < \phi = \arg z \leq \pi$$

is called the principal value of the argument, and is denoted by $\arg z$.

The plane with such a set up to represent the complex numbers is called the Argand plane, or the Gauss plane, or simply the complex plane.

Let $c = \beta + i\gamma$ be the centre of a circle with radius ρ, so that the equation for the circle is

$$|z - c| = \rho, \tag{1.3}$$

that is $(z - c)(\overline{z - c}) = \rho^2$, or

$$z\bar{z} - \bar{c}z - c\bar{z} + c\bar{c} = \rho^2.$$

More generally, let us consider

$$\alpha z\bar{z} - \bar{c}z - c\bar{z} + c\bar{c} + \delta = 0, \tag{1.4}$$

where α, δ are real numbers. When $\alpha = 0$, equation (1.4) represents the straight line

$$\beta x + \gamma y - \frac{\delta}{2} = 0,$$

and indeed is the equation for a general straight line. If $\alpha \neq 0$ then we may rewrite (1.4) in the form

$$\left(z - \frac{c}{\alpha}\right)\left(\overline{z - \frac{c}{\alpha}}\right) = \frac{c\bar{c}}{\alpha^2} - \frac{\delta}{\alpha}.$$

If $c\bar{c} > \delta\alpha$ then (1.4) represents a circle with centre c/α and radius $\sqrt{(c\bar{c}/\alpha^2) - (\delta/\alpha)}$. When $c\bar{c} = \delta\alpha$, the circle degenerates to the single point c/α, which we may call a point circle. When $c\bar{c} < \delta\alpha$, there is no real locus corresponding to (1.4), so that it represents only an imaginary circle.

Note. We may use the coefficients in (1.4) to form the Hermitian matrix

$$H = \begin{pmatrix} \alpha & -\bar{c} \\ -c & \delta \end{pmatrix}.$$

A Hermitian matrix now represents a circle, and conversely different Hermitian matrices may represent the same or different circles. It is not difficult to prove that the condition for two Hermitian matrices to represent the same circle is that they should differ by a real factor.

If the determinant of H is positive then it represents an imaginary circle; if it is negative then H represents a real circle; if it is zero then H represents a point circle.

1.2 The geometry of the complex plane

Consider the substitution (or transformation)

$$w = az + b, \quad a \neq 0, \tag{1.5}$$

where a, b are complex constants. Given a complex number z, we have a unique complex number w. Solving the equation to give

$$z = \frac{1}{a}w - \frac{b}{a}, \tag{1.6}$$

we see that, given any w there is also a unique z, so that the substitution (1.5) is a transformation of the complex plane onto itself. Also, if

$$z = a_1 z_1 + b_1, \quad a_1 \neq 0,$$

then

$$w = a(a_1 z_1 + b_1) + b = a a_1 z_1 + a b_1 + b.$$

This shows that two successive substitutions of the form (1.5) comprise just another substitution of the form (1.5), and this brings us to the notion of a 'group'. Thus the set of substitutions (1.5) is called the group of integral linear transformations, or just linear transformations.

We now study the 'geometry' under this group of transformations.

First, any point can be transformed into any other point; that is, given any two points z_0 and w_0, we can find a substitution of the form (1.5) which transforms z_0 to w_0. Clearly

$$w = z + (w_0 - z_0)$$

already does the job admirably. Such a property is called being reachable, so that under the group of (integral) linear substitutions the whole complex plane forms a reachable set.

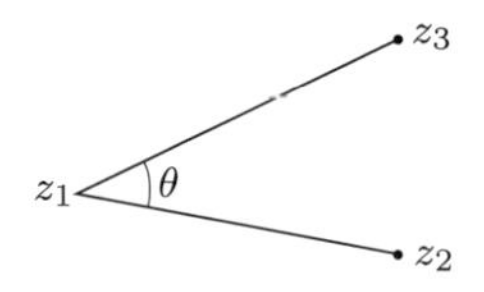

Figure 1.1.

Next, any two points can be transformed into any two other points; for example, the substitution

$$w = \frac{w_1 - w_2}{z_1 - z_2} z + \frac{z_1 w_2 - z_2 w_1}{z_1 - z_2} \tag{1.7}$$

transforms z_1 to w_1, and z_2 to w_2.

Consider now the problem for three points – we cannot expect to be able to transform any three points into any three other points. A suitable condition for three points z_1, z_2, z_3 to be transformed to w_1, w_2, w_3, respectively, is that

$$\frac{w_2 - w_1}{w_3 - w_1} = \frac{z_2 - z_1}{z_3 - z_1}. \tag{1.8}$$

Clearly we may apply (1.7) to transform z_1, z_2 to w_1, w_2. Now, by (1.8), we have

$$\begin{aligned} \frac{w_1 - w_2}{z_1 - z_2} z_3 + \frac{z_1 w_2 - z_2 w_1}{z_1 - z_2} &= w_1\left(\frac{z_3 - z_2}{z_1 - z_2}\right) + w_2\left(\frac{z_1 - z_3}{z_1 - z_2}\right) \\ &= w_1\left(\frac{w_3 - w_2}{w_1 - w_2}\right) + w_2\left(\frac{w_1 - w_3}{w_1 - w_2}\right) = w_3, \end{aligned}$$

so that z_1, z_2, z_3 can be transformed to w_1, w_2, w_3, respectively. Conversely, if

$$w_i = az_i + b,$$

then

$$\begin{vmatrix} w_1 & z_1 & 1 \\ w_2 & z_2 & 1 \\ w_3 & z_3 & 1 \end{vmatrix} = 0,$$

which is equivalent to (1.8). Collecting both parts together, we may state the following: Three points define a certain ratio $(z_2 - z_1)/(z_3 - z_1)$, and a necessary and sufficient condition for three points to be transformed into another three points is that the values for the two ratios concerned should be the same.

What is the geometric interpretation of the value for this ratio? Mark the three points z_1, z_2, z_3 on the complex plane, and let

$$\frac{z_2 - z_1}{z_3 - z_1} = \rho e^{i\theta}.$$

Here ρ is the ratio of the two sides of the triangle formed by the three points with the intersection at z_1, and θ is the angle at z_1 subtended by the two sides (see Fig. 1.1). Thus if the two triangles concerned are similar, then there is a linear substitution which transforms one to the other.

In the above we have taken points as the material objects for our geometry. Now, let us take straight lines and circles as our objects for consideration. It is clear that the substitution (1.5) transforms straight lines into straight lines, and circles into circles.

A general straight line can be written in the form

$$\Im(cz + d) = 0, \tag{1.9}$$

and the substitution $w = cz + d$ transforms the straight line (1.9) into the line $\Im w = 0$, that is the real axis. Therefore any straight line on the complex plane can be transformed into the real axis by a linear substitution, so that the set of straight lines form a reachable set under the group of (integral) linear substitutions.

We leave it to the reader to show that a necessary and sufficient condition for two straight lines to be transformed into another two straight lines is that the corresponding angles of intersections for the lines should be the same.

A general circle has the form

$$(z - c)(\overline{z - c}) = \rho^2,$$

and the substitution $z - c = \rho w$ can be applied to transform it to $w\bar{w} = 1$, the unit circle with centre the origin. Thus, any (real) circle can be transformed into the unit circle, so that, under the group of (integral) linear substitutions, the (real) circles form a reachable set.

We leave it to the reader to find the necessary and sufficient condition for two circles to be transformed into another two circles. Separate the cases depending on whether the two circles intersect or not. If they intersect, we examine their angle of intersection; if they do not intersect, then what should we be looking for?

1.3 The bilinear transformation (Möbius transformation)

Consider now a more general group of transformations than the (integral) linear ones: the group of transformations having the form

$$w = \frac{az + b}{cz + d}, \quad ad - bc \neq 0, \tag{1.10}$$

in which the special case $c = 0$ is that of an integral linear transformation. We call this the group of bilinear transformations, and an individual transformation of the form (1.10) is also known as a linear fractional transformation, or a Möbius transformation. Obviously the transformation (1.10) has the inverse

$$z = \frac{dw - b}{-cw + a}. \tag{1.11}$$

Also, it is not difficult to verify that two successive transformations of the form (1.10) comprise just another such transformation – we call it the product of the two bilinear transformations.

We need, however, to take note of the following point: The mapping (1.10) does not give a one-to-one transformation of the complex plane onto itself, for it is clear that there is no point w which corresponds to the point $z = -d/c$, and similarly there is no point z which corresponds to the point $w = a/c$.

If we are to insist on a one-to-one correspondence, then we need to extend our complex plane by the introduction of an extra point, called the point at infinity, which is denoted by ∞. With the addition of such a point, we can then let $w = \infty$ be the corresponding point to $z = -d/c$, and $z = \infty$ corresponds to $w = a/c$.

The plane with the addition of the point at infinity is called the function theory plane, or the one dimensional complex projective space. It is sometimes also called the extended complex plane, or just the projective plane. The study of the geometry on the projective plane is concerned with the properties associated with transformations forming various groups.

A more precise definition can be given as follows: Consider the set of non-zero ordered pairs of complex numbers (z_1, z_2). Two such pairs (z_1, z_2) and (w_1, w_2) are said to be equivalent, denoted by

$$(z_1, z_2) \sim (w_1, w_2),$$

if there is a non-zero complex number p such that $z_1 = pw_1$, $z_2 = pw_2$. The equivalence relation $\sim$ enjoys the following three properties:

(i) $(z_1, z_2) \sim (z_1, z_2)$;
(ii) if $(z_1, z_2) \sim (w_1, w_2)$, then $(w_1, w_2) \sim (z_1, z_2)$;
(iii) if $(z_1, z_2) \sim (w_1, w_2)$ and $(w_1, w_2) \sim (u_1, u_2)$ then $(z_1, z_2) \sim (u_1, u_2)$.

Such an equivalence relation partitions the set of non-zero ordered pairs of complex numbers into equivalence classes: pairs which are equivalent form a class, and pairs belonging to different classes are not equivalent. Each class can then be identified with, or defines, a point in the one dimensional projective space. More specifically, if $z_2 \neq 0$ then we assign the class to which (z_1, z_2) belongs the number $z = z_1/z_2$, just an ordinary complex number on the complex plane, and we call (z_1, z_2) its homogeneous coordinates; we should clarify by saying that any pair (pz_1, pz_2) in the same equivalence class also represents the same point z. If $z_2 = 0$, then the corresponding class for which $(z_1, 0) \sim (1, 0)$ is identified with the unique point at infinity.

Under the system of homogeneous coordinates, a bilinear mapping can be written as

$$\begin{cases} w_1 = az_1 + bz_2, \\ w_2 = cz_1 + dz_2, \end{cases} \qquad \begin{vmatrix} a & b \\ c & d \end{vmatrix} \neq 0, \tag{1.12}$$

and, on using matrices, can also be written as

$$(w_1, w_2) = (z_1, z_2) \begin{pmatrix} a & b \\ c & d \end{pmatrix}.$$

Note, however, that not every matrix corresponds to a mapping.

From the equivalence relation, corresponding to any non-zero p, the matrix

$$\begin{pmatrix} ap & bp \\ cp & dp \end{pmatrix}$$

also represents the same mapping; this can also be seen clearly from (1.10) because the introduction of the non-zero factor p to the numerator and denominator of a fraction does not alter the value of the fraction.

The matrix

$$\begin{pmatrix} a & b \\ c & d \end{pmatrix}$$

is now called the matrix of the transformation (1.10), even though, as we have just explained, the matrix with a non-zero factor p also represents the same transformation.

The product of the transformations corresponding to the two matrices

$$\begin{pmatrix} a & b \\ c & d \end{pmatrix}, \quad \begin{pmatrix} a_1 & b_1 \\ c_1 & d_1 \end{pmatrix}$$

is a transformation with a matrix that is the product of the two matrices:

$$\begin{pmatrix} a & b \\ c & d \end{pmatrix}\begin{pmatrix} a_1 & b_1 \\ c_1 & d_1 \end{pmatrix} = \begin{pmatrix} aa_1 + b_1c & ac_1 + cd_1 \\ ba_1 + db_1 & c_1b + dd_1 \end{pmatrix}.$$

1.4 Groups and subgroups

Definition 1.1 A set of bilinear transformations satisfying the following three conditions is said to form a group:

(i) It includes the unit or identity transformation: $w = z$.

(ii) Each transformation in the set has an inverse transformation: if $w_1 = (az + b)/(cz + d)$ is in the set, then so is $w = (dz - b)/(-cz + a)$.

(iii) The product of any two transformations in the set also belongs to the set: If $w_1 = (az + b)/(cz + d)$ and $w_2 = (a_1z + b_1)/(c_1z + d_1)$ are in the set then so is

$$w = \frac{a(a_1z + b_1)/(c_1z + d_1) + b}{c(a_1z + b_1)/(c_1z + d_1) + d} = \frac{(aa_1 + bc_1)z + (ab_1 + bd_1)}{(ca_1 + dc_1)z + (cb_1 + dd_1)}.$$

Definition 1.2 A subset of a group which forms a group is called a subgroup.

Example 1.3 The set of all bilinear transformations form a group.

Example 1.4 The set of all integral linear transformations form a group, a subgroup of the group of bilinear transformations.

Example 1.5 The set of transformations having the form $w = z + a$ forms a group, called the group of plane translations. It is a subgroup of the group of integral linear transformations. Another subgroup of the group of integral linear transformations has the form

$$w = az, \quad a \neq 0.$$

This group also has two important subgroups: The group of mappings $w = kz$, with k positive, is called the magnification group, and the group of mappings $w = e^{i\theta} z$, with θ real, is called the rotation group.

Example 1.6 The set of mappings (1.10), with a, b, c, d real, also forms a group, called the real group. Those mappings satisfying the additional condition that $ad - bc = 1$ form a subgroup of the real group.

Example 1.7 The set of mappings

$$w = \frac{az - b}{\bar{b}z + \bar{a}}, \quad a\bar{a} + b\bar{b} = 1,$$

also forms a group, called the unitary group.

Example 1.8 The set of mappings

$$w = az + b, \quad |a| = 1,$$

forms a group. If we write this in terms of real numbers, say

$$w = u + iv, \quad z = x + iy, \quad a = e^{i\theta}, \quad b = p + iq,$$

then

$$u = x \cos\theta - y \sin\theta + p, \quad v = x \sin\theta + y \cos\theta + q.$$

This is the mapping corresponding to what we studied in rigid motions; it represents a rotation followed by a translation, and for this reason the group is also known as the group of rigid motions.

Example 1.9 The set of mappings

$$w = e^{i\theta} \frac{z - a}{1 - \bar{a}z}$$

also forms a group, called the non-Euclidean motions, or Lobachevsky group.

Example 1.10 The set of mappings

$$w = z + \omega k, \quad k = 0, \pm 1, \pm 2, \ldots,$$

forms a group, called the periodic group with period ω. It is a subgroup of the doubly periodic group

$$w = z + \omega\ell + \omega' k, \quad \ell, k = 0, \pm 1, \pm 2, \ldots.$$

Example 1.11 The set of mappings (1.10), with a, b, c, d being integers satisfying $ad - bc = 1$, also forms a group, called the modular group.

Example 1.12 A complex number having the form $a = \alpha_1 + i\alpha_2$, where α_1, α_2 are integers, is called a complex integer, or a Gaussian integer. The set of mappings (1.10), with a, b, c, d being complex integers satisfying $ad - bc = 1$, also forms a group.

Definition 1.13 Let G be a group of transformations. Suppose that we can select certain transformations from G with the property that every member of G can be expressed as a product of these transformations and their inverses. Then we call these transformations the generators of the group.

Example 1.14 The group in Example 1.10 can be generated by the single transformation $w = z + \omega$. The doubly periodic group in that example has

$$w = z + \omega, \qquad w = z + \omega'$$

as generators.

Example 1.15 The group of integral linear transformations can be generated by the following three sets of transformations:

(i) $w = z + b$ (translation),
(ii) $w = e^{i\theta}z$ (rotation),
(iii) $w = kz$ (magnification), $k > 0$.

A magnification is also called an affine mapping. Generators have the following usefulness: Suppose that a certain property is invariant under translation, rotation and affine mapping. Then the property is invariant under the whole group of integral linear transformations. We may take, for example, the angle between two straight lines. Conversely, the distance between two points is invariant under translation and rotation, but not under an affine map, so that distance is not an invariant property associated with the group of integral linear transformations.

Example 1.16 Let us examine the group of bilinear transformations again. The transformation (1.10) can be considered as the product of the transformations

$$w = \frac{a}{c} + z', \quad z' = \frac{1}{z''}, \quad z'' = \frac{c(cz + d)}{bc - ad}.$$

Consequently this group can be considered as that of the integral linear group with the addition of the transformation $w = 1/z$. If $c = 0$ then the transformation itself is already an integral linear transformation. Therefore the group of bilinear transformations can be generated by translation, affine mapping, rotation, together with $w = 1/z$.

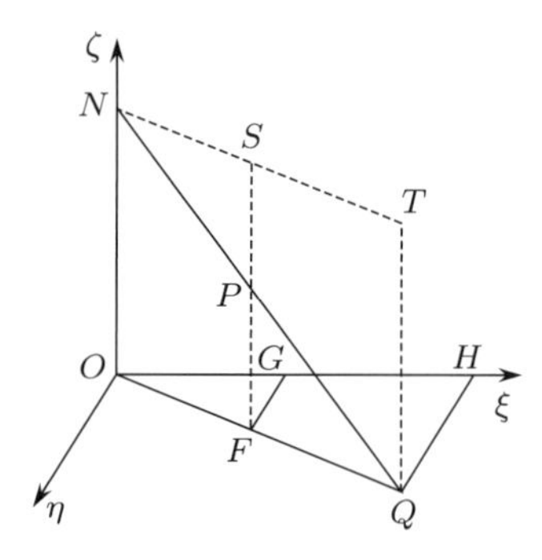

Figure 1.2.

1.5 The Riemann sphere

In order to be even clearer when speaking about the point at infinity, we now introduce the projection of the surface of a sphere onto the plane, thus establishing a relationship between such a surface and the plane.

Take a sphere with radius 1, with centre the origin O. Let a variable point on the surface of the sphere be $P(\xi, \eta, \zeta)$, so that the equation for the surface is

$$\xi^2 + \eta^2 + \zeta^2 = 1. \tag{1.13}$$

From the point $N(0, 0, 1)$, project the point $P(\xi, \eta, \zeta)$ onto the $\xi\eta$-plane with the image $Q(x, y)$. Let T be the point that completes the rectangle $ONTQ$, and let S and F be points on NT and OQ, respectively, such that SPF is parallel to NO. Next, let G and H be points on the ξ-axis chosen so that FG and QH are parallel to the η-axis (see Fig. 1.2). Then $OG = \xi$, $FG = \eta$, $PF = \zeta$, $OH = x$, $QH = y$. The triangles NSP and NTQ being similar, we see that

$$(1 - \zeta) : 1 = SP : TQ = NS : NT = OF : OQ = \eta : y = \xi : x,$$

so that

$$x = \frac{\xi}{1 - \zeta}, \quad y = \frac{\eta}{1 - \zeta}. \tag{1.14}$$

From (1.13) it follows that

$$1 + x^2 + y^2 = \frac{(1 - \zeta)^2 + \xi^2 + \eta^2}{(1 - \zeta)^2} = \frac{1 - 2\zeta + 1}{(1 - \zeta)^2} = \frac{2}{1 - \zeta},$$

and therefore

$$\xi = \frac{2x}{r}, \quad \eta = \frac{2y}{r}, \quad \zeta = 1 - \frac{2}{r}, \quad r = 1 + x^2 + y^2. \tag{1.15}$$

Writing $z = x + iy$, formulae (1.14) and (1.15) give the one-to-one relationship between the points on the surface of the sphere and the points on the plane, together with the point $z = \infty$, which corresponds to N.

A circle on the plane with the equation

$$\alpha(x^2 + y^2) + 2\beta x + 2\gamma y + \delta = 0 \tag{1.16}$$

is then transformed to

$$\alpha\left(\frac{2}{1-\zeta} - 1\right) + 2\beta\frac{\xi}{1-\zeta} + 2\gamma\frac{\eta}{1-\zeta} + \delta = 0,$$

that is

$$(\alpha - \delta)\zeta + 2\beta\xi + 2\gamma\eta + \alpha + \delta = 0. \tag{1.17}$$

This is the equation for a plane which intersects the unit sphere to form a circle. Conversely, given a plane (1.17), the image under the projection map will also deliver a circle on the plane, or rather a real circle, a point circle or an imaginary circle, depending on whether the plane (1.17) intersects, is tangential to or does not intersect the unit sphere, respectively.

The proof for such a claim is very easy by geometric consideration. If (1.16) is a real circle, then there must be an image path on the unit sphere, and it is given by the intersection of the sphere and the plane (1.17), so that there must be such an intersection. The same argument applies when the plane is tangential to, or when it fails to intersect, the unit sphere. In particular, even an imaginary circle has a real representation in three dimensional space, namely a plane which does not intersect the unit sphere. Note, however, that there is a unique exception, namely the plane corresponding to the imaginary circle $z\bar{z} + 1 = 0$ via (1.17) is the ridiculous equation $2 = 0$.

We now study the properties associated with the great circles on the unit sphere. Each great circle intersects the equator at two points that lie on the extreme ends of a diameter of the unit circle on the plane; we call these the antipodal points. Therefore each great circle corresponds to a circle which intersects the unit circle at antipodal points, and conversely. From such a consideration we now have an easy proof of the following theorem in plane geometry.

Theorem 1.17 *Suppose that the two circles A, B both intersect a circle Γ at antipodal points. Then A, B must intersect at two points. Moreover, the circle C passing through these two points also intersects Γ at antipodal points.*

Proof. We may take Γ to be the unit circle. The circles A, B are now mapped onto great circles A_1, B_1 on the unit sphere. Being great circles, A_1, B_1 must intersect at two points, and indeed such points are antipodal points of the sphere. The first part of the theorem follows. If C_1 is the image of C on the unit sphere then C_1 is also a great circle, so that the last part of the theorem also follows. □

1.6 The cross-ratio

We now consider the geometry associated with the group of bilinear transformations.

Since the integral linear transformations form a subgroup, and in that group any two points can be transformed into any other two points, it follows that projective space forms a reachable set in that any two points can be transformed into any two other points.

We now prove that any three points can be transformed into any three other points. From the property mentioned above, it does not matter if we assume that two of the three chosen points are 0 and ∞. Let the remaining one be a, and apply the transformation $w = z/a$. Therefore any three given points can be transformed to $0, 1, \infty$.

Since the bilinear transformation which maps $0, 1, \infty$ to themselves is the identity transformation, it follows that any four points cannot be mapped into any other four points.

Definition 1.18 By the *cross-ratio* for the four points z_1, z_2, z_3, z_4 we mean the number

$$(z_1, z_2, z_3, z_4) = \frac{z_3 - z_1}{z_2 - z_3} \Bigg/ \frac{z_4 - z_1}{z_2 - z_4}.$$

When $(z_1, z_2, z_3, z_4) = -1$ we say that the four points z_1, z_2, z_3, z_4 form a harmonic sequence.

Theorem 1.19 *The cross-ratio is invariant under a bilinear transformation.*

Proof. Let $w_i = (az_i + b)/(cz_i + d)$. Then

$$w_i - w_j = \frac{(ad - bc)(z_i - z_j)}{(cz_i + d)(cz_j + d)},$$

so that $(w_1, w_2, w_3, w_4) = (z_1, z_2, z_3, z_4)$. □

Theorem 1.20 *A necessary and sufficient condition for four points w_1, w_2, w_3, w_4 to be mapped to z_1, z_2, z_3, z_4, respectively, is that they should have the same cross-ratio.*

Proof. The necessity part follows immediately from the previous theorem.

Next, take any four points, and transform three of them to $0, \infty, 1$. Suppose that the four points have been transformed to $0, \infty, z, 1$, which has the cross-ratio z. This means that any four points with cross-ratio z can be mapped to $0, \infty, z, 1$. The theorem is proved. □

Let us examine the problem again in a more concrete manner. The bilinear transformation

$$w = (z_1, z_2, z_3, z) = \frac{z_3 - z_1}{z_2 - z_3} \Bigg/ \frac{z - z_1}{z_2 - z} = \frac{z_3 - z_1}{z_2 - z_3} \frac{z_2 - z}{z - z_1}$$

maps z_2, z_1, z_3 to $0, \infty, 1$, respectively. Therefore

$$(w_1, w_2, w_3, w) = (z_1, z_2, z_3, z) \tag{1.18}$$

defines a bilinear transformation which maps $z = z_1, z_2, z_3$ to $w = w_1, w_2, w_3$.

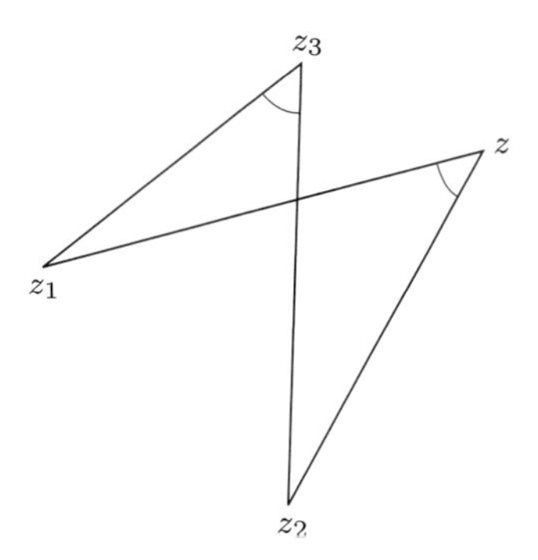

Figure 1.3.

From the invariance of the cross-ratio, the bilinear transformation is the unique one with such a property. Therefore the transformation represented by (1.18) is the general form, and the unique one which transforms three given points to three other given points.

Consider the angle associated with the cross-ratio (see Fig. 1.3):

$$\arg(z_1, z_2, z_3, z) = \arg \frac{z_3 - z_1}{z_2 - z_3} - \arg \frac{z - z_1}{z_2 - z} = \angle z_1 z z_2 - \angle z_1 z_3 z_2.$$

If the cross-ratio is real then $\angle z_1 z z_2 = \angle z_1 z_3 z_2$, and we find that z lies on the circle through the three points z_1, z_2, z_3. The converse is also true.

If (z_1, z_2, z_3, z) is real then (w_1, w_2, w_3, w), as defined by (1.18), is also real. Consequently, as z describes the circle specified by z_1, z_2, z_3, its image w describes the circle specified by w_1, w_2, w_3. Note, however, that z_1, z_2, z_3 may lie on a straight line, in which case we need to consider such a line also to be a 'circle', with a radius ∞.

Therefore a bilinear transformation maps circles into circles. Take, in particular, $z_1 = 1$, $z_2 = -1$, $z_3 = -i$, so that the circle through z_1, z_2, z_3 is the unit circle. Take $w_1 = \infty$, $w_2 = 0$, $w_3 = 1$, so that the circle through w_1, w_2, w_3 is the real axis. The corresponding transformation is

$$w = (w_1, w_2, w_3, w) = (z_1, z_2, z_3, z) = i\frac{1+z}{1-z}, \tag{1.19}$$

and it maps the unit circle onto the real axis, with the interior of the circle $|z| < 1$ being mapped onto the upper half-plane $\Im w > 0$.

If we set $z_1 = -1$, $z_2 = 1$, $z_3 = -i$ instead, then

$$w = -i\frac{1-z}{1+z},$$

and now the exterior of the circle $|z| > 1$ is mapped onto the upper half-plane $\Im w > 0$.

Again, if we set $z_1 = 0$, $z_2 = 1$, $z_3 = \infty$, then

$$w = \frac{z-1}{z} = 1 - \frac{1}{z}.$$

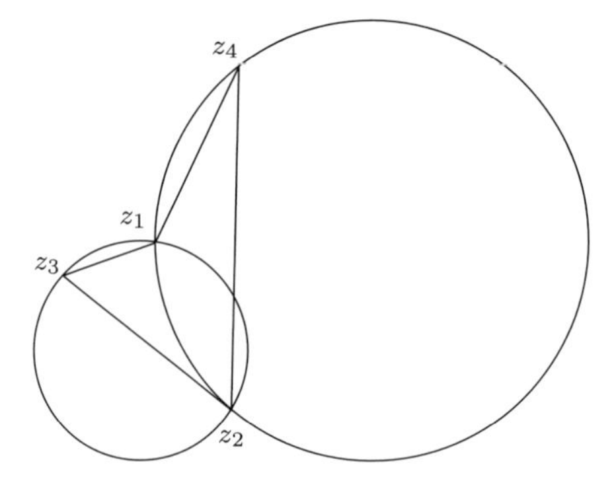

Figure 1.4.

This transformation maps the upper half-plane onto itself. If we set $z_1 = 1, z_2 = 0, z_3 = \infty$, then

$$w = \frac{z}{z-1},$$

which maps the upper half-plane onto the lower half-plane.

The reader should try to write down all the transformations corresponding to z_1, z_2, z_3 taking $0, 1, \infty$ in some order, and to identify those which map the upper-half plane onto itself, and those which map the upper half-plane onto the lower half-plane.

1.7 Corresponding circles

Let us now take the circles on the plane as our objects in the study of geometry. It has been shown in Section 1.6 that, under the group of bilinear transformations, all the circles on the plane (including straight lines) form a reachable set. We consider the problem of the correspondence between the circles.

Theorem 1.21 *The angle of intersection between two circles remains the same under a bilinear transformation.*

Proof. Let z_1, z_2 be the two points of intersection for the two circles (see Fig. 1.4). Take z_3, z_4 from each of the two circles. The angle associated with the cross-ratio of the four points is given by

$$\arg(z_1, z_2, z_3, z_4) = \angle z_3 z_2 z_4 - \angle z_3 z_1 z_4.$$

As z_3 and z_4 approach z_1, the expression here delivers the angle of intersection between the two circles. The proof of the theorem now follows from the invariance of the cross-ratio under a bilinear transformation. □

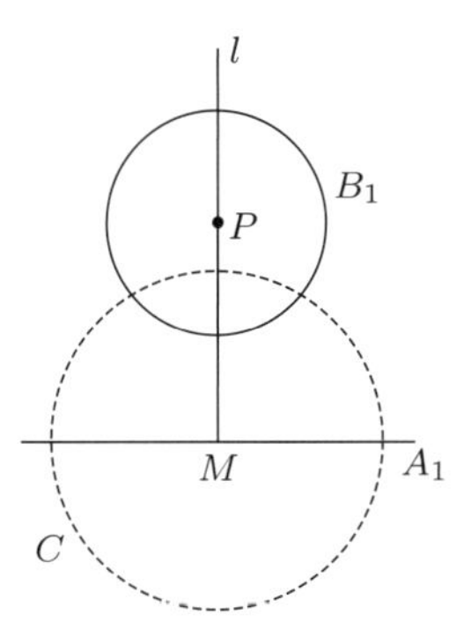

Figure 1.5.

Theorem 1.22 *Any two tangential circles can be mapped into any two other tangential circles. If two pairs of intersecting circles have the same angle of intersection, then one pair can be mapped into the other pair.*

If one of the points of intersection is mapped to ∞ *then:* (i) *tangential circles are mapped into parallel lines,* (ii) *intersecting circles are mapped into intersecting straight lines.*

Proof. Any two parallel lines can be mapped into the lines $y = 0$ and $y = 1$. Therefore the first part of the theorem follows. Any pair of straight lines intersecting at an angle θ with each other can be mapped into the lines $y = 0$ and $x \sin\theta - y \cos\theta = 0$. Therefore the second part of the theorem follows. □

Thus, for a pair of intersecting circles, their angle of intersection is the only invariance under the group of bilinear transformations.

We examine again the situation when there is no intersection. Suppose that A, B are two circles that do not intersect. First, transform A into a straight line A_1, while B is mapped into a circle B_1 (see Fig. 1.5), and, of course, A_1 and B_1 do not intersect. Drop a perpendicular ℓ from the centre P of B_1 onto A_1 at the point M, and construct the circle C, centre at M, orthogonal to B_1. Since C cuts ℓ orthogonally, by Theorem 1.21 there is a bilinear transformation which maps them onto (orthogonal) straight lines; under such a transformation A_1, B_1 are mapped into two circles A_2, B_2 which are orthogonal to two orthogonal lines, and therefore A_2, B_2 are concentric circles. We have therefore proved the following theorem.

Theorem 1.23 *Any two non-intersecting circles can be transformed into two concentric circles.*

What is the invariant associated with two non-intersecting circles? A straight line through the centre of two concentric circles cuts the circles at four points, and the cross-ratio of these four points is the same regardless of the chosen line. We can describe this invariant in another way: The only invariant associated with two non-intersecting circles is the cross-ratio of the four points of intersection obtained from any circle cutting both circles orthogonally.

Note 1. The centres and radii of the two circles

$$\alpha(x^2 + y^2) + 2\beta x + 2\gamma y + \delta = 0,$$
$$\alpha_1(x^2 + y^2) + 2\beta_1 x + 2\gamma_1 y + \delta_1 = 0,$$

are $(-\beta/\alpha, -\gamma/\alpha)$, $(-\beta_1/\alpha_1, -\gamma_1/\alpha_1)$ and $\sqrt{(\beta^2+\gamma^2-\alpha\delta)/\alpha^2}$, $\sqrt{(\beta_1^2+\gamma_1^2-\alpha_1\delta_1)/\alpha_1^2}$, respectively. Therefore the condition for the circles to intersect orthogonally is that

$$\frac{\beta^2 + \gamma^2 - \alpha\delta}{\alpha^2} + \frac{\beta_1^2 + \gamma_1^2 - \alpha_1\delta_1}{\alpha_1^2} = \left(\frac{\beta}{\alpha} - \frac{\beta_1}{\alpha_1}\right) + \left(\frac{\gamma}{\alpha} - \frac{\gamma_1}{\alpha_1}\right),$$

which simplifies to

$$\alpha_1\delta + \alpha\delta_1 = 2\beta\beta_1 + 2\gamma\gamma_1. \tag{1.20}$$

The derivation of the condition here is under the assumption that $\alpha \neq 0, \alpha_1 \neq 0$, but it is not difficult to show that it is still valid if either $\alpha = 0$ or $\alpha_1 = 0$ when the corresponding circle is a straight line.

Note 2. Let $Q = -2\beta\beta_1 - 2\gamma\gamma_1 + \alpha\delta_1 + \delta\alpha_1$, so that condition (1.20) becomes $Q = 0$. When $\alpha = \alpha_1$, $\beta = \beta_1$, $\gamma = \gamma_1$, $\delta = \delta_1$, so that $Q = \alpha\delta - \beta^2 - \gamma^2$, and now $Q > 0, = 0, < 0$ means that we have an imaginary circle, a point circle and a real circle, respectively.

1.8 Pencils of circles

Definition 1.24 Let A, B be two given circles. The set of circles orthogonal to both A, B is called the conjugate pencil of A, B.

Theorem 1.25 *A pencil of circles has one single parameter.*

Proof. Suppose first that A, B intersect with each other. We may assume without loss that they are straight lines intersecting at the origin, O. A circle orthogonal to both lines must be a circle with centre the origin. We may now take the radius of the circle as the parameter for the pencil of circles.

Next, suppose that A, B are tangential to each other. We may assume without loss that they are parallel straight lines, from which it follows that the pencil is the set of straight lines orthogonal to these parallel lines, and only one parameter will do.

Finally, suppose that A, B do not intersect. We may assume without loss that A, B are concentric, from which it follows that the pencil is the set of straight lines through the common centre, and again only one parameter will do. □

Definition 1.26 The pencils associated with three cases in the proof of Theorem 1.25 are classified as hyperbolic, parabolic and elliptic, respectively.

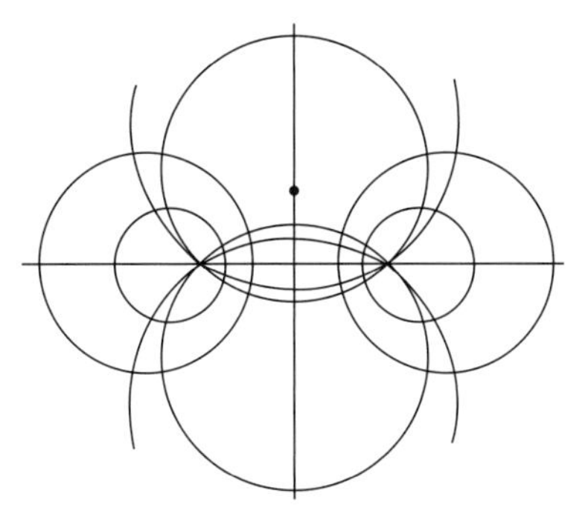

Figure 1.6.

The standard form for a pencil has already been identified in the proof of Theorem 1.25.

Any hyperbolic pencil can be transformed into the set of 'circles with centre the origin', including the two point circles, one at the origin, the other at ∞; they are called the limiting points P, Q of the pencil. At the same time we also deduce that any two circles from a hyperbolic pencil do not intersect.

Any parabolic pencil can be transformed into the set of 'straight lines orthogonal to the x-axis', with ∞ as their tangential point. Thus, members of a parabolic pencil share a common tangent point, called the common point.

Any elliptic pencil can be transformed into the set of 'straight lines through the origin'; any such line passes through 0 and ∞. Therefore circles from an elliptic pencil pass through two fixed points.

Therefore any pencil of circles can be transformed into one of the above three types. Because an elliptic pencil has two common points, a parabolic pencil has one common point, and a hyperbolic pencil has no common point, there is no transformation which can map a pencil from one type to another.

It is also clear that any pencil can be uniquely identified by two of its members.

Since 'circles with centre the origin' and 'straight lines through the origin' are orthogonal to each other, we may say that the corresponding pencils are orthogonal to each other, in the sense that any member of one pencil is orthogonal to every member of the other pencil. Thus an elliptic pencil is orthogonal to a hyperbolic pencil. At the same time, a parabolic pencil can also be orthogonal to another parabolic pencil. This provides us with the notion of pencils being conjugates to each other (see Fig. 1.6), and explains why we chose the conjugate pencil in the initial definition. To rephrase the above, every elliptic pencil has exactly one conjugate pencil, which is hyperbolic; conversely, the unique conjugate of any given hyperbolic pencil is always an elliptic pencil. The parabolic pencils are paired off into pairs of conjugates.

From the standard representation viewpoint we see that given a point on the plane (apart from the limiting points and the common point), there is one and only one circle from the pencil which passes through the point.

Note 1. The restriction being imposed on the parameters also identifies the pencil concerned. The parameter for a hyperbolic pencil runs over the points on a ray (for example, $(0, \infty)$); the parameter for a parabolic pencil runs over the points on a line; the parameter for an elliptic pencil runs over the points on a circle. It can also be deduced from this that there is no bilinear transformation which can map a pencil from one type to a pencil in another type.

Note 2. Let the circle $\alpha(x^2 + y^2) + 2\beta x + 2\gamma y + \delta = 0$ be represented by the vector $(\alpha, \beta, \gamma, \delta)$. Then the pencil of circles defined by the two real circles $(\alpha, \beta, \gamma, \delta)$ and $(\alpha_1, \beta_1, \gamma_1, \delta_1)$ is the set of circles given by

$$\lambda(\alpha, \beta, \gamma, \delta) + \mu(\alpha_1, \beta_1, \gamma_1, \delta_1),$$

where λ, μ are real constants, not both zero.

The types of pencils can now be investigated algebraically by considering the quadratic form

$$(\lambda\alpha + \mu\alpha_1)(\lambda\delta + \mu\delta_1) - (\lambda\beta + \mu\beta_1)^2 - (\lambda\gamma + \mu\gamma_1)^2.$$

Such a quadratic form may be (i) indefinite (hyperbolic pencil), (ii) degenerate (parabolic pencil), (iii) negative definite (elliptic pencil).

1.9 Bundles of circles

Let us now examine the situation when there are three circles A, B, C, and suppose that there is no common intersection point.

If A, B do not intersect then we use a bilinear transformation to map A, B into concentric circles A_1, B_1; and suppose that C is transformed into C_1. The straight line through the centres of A_1, B_1 and C_1 cuts these circles orthogonally. Therefore, in this case, there is a circle which is orthogonal to all three circles A, B, C.

Next, if A, B are tangential to each other then we use a bilinear transformation to map A, B into parallel lines A_1, B_1; and suppose that C is transformed into C_1. There is a straight line through C_1 which is orthogonal to A_1, B_1. Therefore, in this case, there is a circle which is orthogonal to all three circles A, B, C.

Finally, if A, B has a point of intersection then we use a bilinear transformation to map A, B into straight lines intersecting at the origin O; and suppose that C is transformed into C_1. (i) The point O lies outside C_1, in which case there is a circle with centre O which is orthogonal to C_1, and therefore there is a circle which is simultaneously orthogonal to A, B, C (see Fig. 1.7); (ii) the point O lies inside C_1, in which case there is a circle Γ with centre O which intersects C_1 at antipodal points for Γ (see Fig. 1.8). We have therefore proved the following theorem.

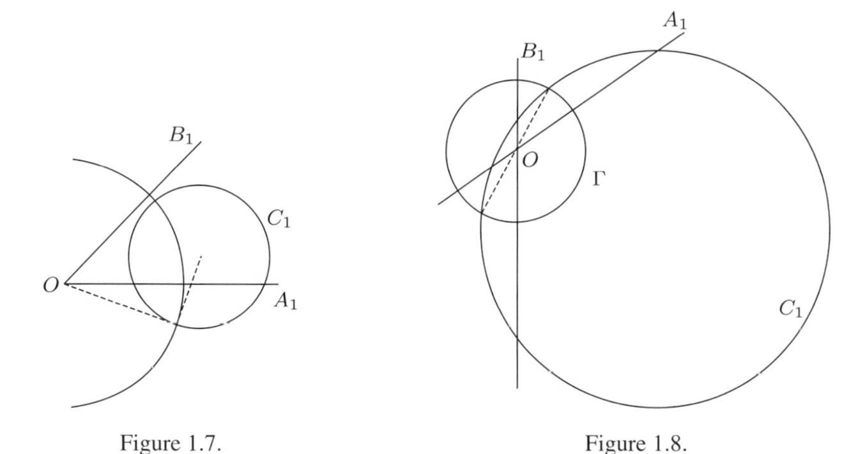

Figure 1.7. Figure 1.8.

Theorem 1.27 *Let there be three circles on the projective plane. Then one of the following cases must be valid:* (i) *there is a common point of intersection;* (ii) *there is a common orthogonal circle;* (iii) *there is a bilinear transformation which maps the three circles into circles which intersects a circle* Γ *at pairs of antipodal points for* Γ.

Definition 1.28 A set of circles orthogonal to a fixed circle is called a hyperbolic bundle of circles. A set of circles passing through a fixed point is called a parabolic bundle of circles. A set of circles with the property that, after a bilinear transformation they intersect a circle at antipodal points, is called an elliptic bundle of circles.

There are thus three types of bundles of circles, and there is no bilinear transformation which can map a bundle from one type to a bundle in another type. This is because any two circles in an elliptic bundle must intersect at two distinct points; two circles in a parabolic bundle may intersect at two distinct points, or be tangential at a point, but there cannot be two circles in the bundle with no common points; there are two circles in a hyperbolic bundle which do not intersect.

Let us examine the situation concerning the parameters associated with a bundle.

We may assume without loss that the circle to which members of a hyperbolic bundle are orthogonal is the x-axis. The circles in the hyperbolic bundle then have centres on the x-axis, and they form a two-parameter bundle: the position of the centre, which takes a parameter in $(-\infty, \infty)$, and the radius, which takes another parameter in $(0, \infty)$; we may consider the region in which the parameters vary to be a half-plane.

Next, we may assume without loss that ∞ is the common point for members of a parabolic bundle. The bundle is then the set of all straight lines, so that there are again two parameters for the bundle; we may consider the region in which the parameters vary to be the whole plane.

Next, we may assume without loss that the unit circle is the fixed circle associated with an elliptic bundle, so that members of the bundle pass through the antipodal points of

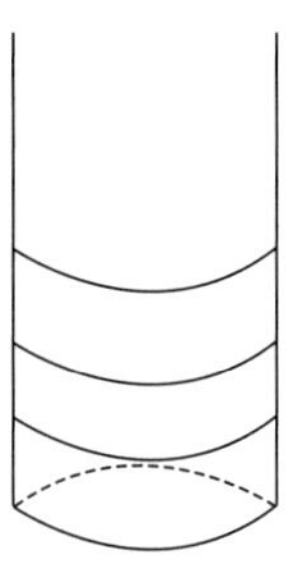

Figure 1.9.

the unit circle. A diameter of the unit circle can be specified by the angle it makes with the x-axis, which accounts for one parameter, and there is still the radius of the circle in the bundle, so that there are two parameters; we may consider the region in which the parameters vary to be the surface of a cylinder, as shown in Fig. 1.9.

Theorem 1.29 *If a bundle of circles contains two circles A, B, then it must also contain all the circles in the pencil defined by A, B.*

Proof. (1) If it is a parabolic bundle then there is a common point P, and there is no problem here. Any two circles A, B must have the common point P, and circles in the pencil also pass through P.

(2) If there is a circle in a hyperbolic bundle which is orthogonal to A, B then it must also be orthogonal to the circles in the pencil defined by A, B.

(3) The case for an elliptic bundle can be deduced from the following theorem from elementary geometry. If A, B both intersect Γ at the antipodal points of the Γ, then so does any circle which passes through the intersections of A, B. Indeed this theorem has already been established in Section 1.5. □

It is not difficult to prove the following theorem.

Theorem 1.30 *Let P be any point on the plane. Then there are infinitely many circles in a bundle passing through P, and such circles form a pencil (P not being the common point for a parabolic bundle).*

Theorem 1.31 *Take three circles A, B, C in a bundle, not all belonging to the same pencil, and let D be a fourth circle in the bundle. Then D can be obtained from successive pencils arising from A, B, C.*

Proof. We can select a point P on D which is not the limiting point or the common point of the pencils (A, B), (A, C), and not lying on A. We can construct a circle E belonging to the pencil (A, B) passing through P, and similarly a circle F belonging to the pencil (A, C). Since A, B, C do not belong to the same pencil, the circles E, F are distinct. By Theorem 1.30 the circle D belongs to the pencil (E, F). □

As a consequence we have the following theorem.

Theorem 1.32 *A bundle is uniquely specified by three of its members not belonging to the same pencil.*

In the discussion the hyperbolic and parabolic bundles have the property that they are invariant under a bilinear transformation. However, the definition for an elliptic bundle is not that it is invariant under a bilinear transformation. We are now in a position to amend the definition in order to cater for a new situation. A bundle may be defined as the set of circles obtainable from three circles not belonging to the same pencil. The definitions for hyperbolic and parabolic bundles remain the same, and we may redefine an elliptic bundle to be one which is neither hyperbolic nor parabolic.

Exercise 1.33 Let

$$\alpha_i(x^2+y^2)+2\beta_i x+2\gamma_i y+\delta_i=0, \quad i=1,2,3,$$

represent three circles. The totality of circles represented by

$$\sum_{i=1}^{3}\lambda_i(\alpha_i,\beta_i,\gamma_i,\delta_i)=0$$

is said to form a bundle of circles. Try and examine the various types of bundles from here.

1.10 Hermitian matrices

We now use Hermitian matrices as a method to deal with pencils of circles, thereby giving the simplest geometric background in the study of Hermitian matrices.

(1) A circle

$$\alpha z\bar{z}+h\bar{z}+\bar{h}z+\delta=0, \quad h=\beta+\imath\gamma, \tag{1.21}$$

that is

$$\alpha(x^2+y^2)+2\beta x+2\gamma y+\delta=0,$$

corresponds to the Hermitian matrix

$$H=\begin{pmatrix}\alpha & h\\ \bar{h} & \delta\end{pmatrix}. \tag{1.22}$$

However, two Hermitian matrices which differ by a real constant represent the same circle; that is, if there is a real $\lambda\neq 0$ such that $H=\lambda H_1$, then H, H_1 represent the same circle. There should be no confusion if we sometimes call H a circle in the following.

(2) The circle H being an imaginary one, a point one or a real one corresponds to the value of the determinant of H:

$$\det(H)=\alpha\delta-|h|^2 \quad \text{being} \ >0,\ =0 \quad \text{or} \quad <0. \tag{1.23}$$

The condition can be rewritten as follows: Let

$$F = \begin{pmatrix} 0 & 1 \\ -1 & 0 \end{pmatrix}$$

so that

$$\mathrm{tr}(FHF\bar{H}) = \mathrm{tr}\begin{pmatrix} \bar{h} & \delta \\ -\alpha & -h \end{pmatrix}\begin{pmatrix} h & \delta \\ -\alpha & -\bar{h} \end{pmatrix} = 2(|h|^2 - \alpha\delta),$$

and hence

$$\det(H) = -\frac{1}{2}\mathrm{tr}(FHF\bar{H}).$$

Therefore the circle H being an imaginary one, a point one or a real one corresponds to the trace

$$\mathrm{tr}(FHF\bar{H}) \quad \text{being} \; < 0, \; = 0 \quad \text{or} \quad > 0. \tag{1.24}$$

(3) Under the transformation

$$z \mapsto \frac{az+b}{cz+d}, \quad ad \neq bc, \tag{1.25}$$

the circle H is transformed into

$$H_1 = \bar{P}'HP, \quad \text{where} \quad P = \begin{pmatrix} a & b \\ c & d \end{pmatrix},$$

so that

$$\bar{H}_1 = \begin{pmatrix} |a|^2\alpha + \bar{a}ch + a\bar{c}\bar{h} + |c|^2\delta & \bar{a}b\alpha + \bar{a}dh + b\bar{c}\bar{h} + d\bar{c}\delta \\ a\bar{b}\alpha + c\bar{b}h + a\bar{d}\bar{h} + \bar{d}c\delta & b\bar{b}\alpha + \bar{b}dh + b\bar{d}\bar{h} + d\bar{d}\delta \end{pmatrix}. \tag{1.26}$$

The relationship between the matrices H_1 and H is the familiar one of being 'similar'. Taking determinants we have $\det H_1 = |\det P|^2 \det H$, so that there is no change in the real or imaginary nature of the circles being represented. From similarity theory concerning Hermitian matrices we know that the circle can be transformed into one represented by one of the following matrices:

$$\begin{pmatrix} 1 & 0 \\ 0 & 1 \end{pmatrix} \text{(imaginary)}, \quad \begin{pmatrix} 1 & 0 \\ 0 & 0 \end{pmatrix} \text{(point)}, \quad \begin{pmatrix} 1 & 0 \\ 0 & -1 \end{pmatrix} \text{(real)}.$$

(4) Orthogonality condition. It is not difficult to show that the condition for two circles H, H_1 to be orthogonal is that

$$\mathrm{tr}(FHFH_1) = 0; \tag{1.27}$$

thus, with

$$P = \begin{pmatrix} s & t \\ u & v \end{pmatrix}$$

so that $PFP' = (\det P)F$, we find that

$$\operatorname{tr}\big(F(\bar{P}'HP)F(P'\bar{H}_1\bar{P})\big) = \operatorname{tr}\big((\bar{P}F\bar{P}')H(PFP')\bar{H}_1\big) = |\det P|^2\operatorname{tr}(FHF\bar{H}_1).$$

(5) The totality of circles which are orthogonal to a fixed circle H_0 is said to form a bundle; they are circles H satisfying

$$\operatorname{tr}(FHF\bar{H}_0) = 0.$$

This being a linear relationship, there are two parameters. By similarity considerations we know that there are three types of bundles, corresponding to

$$H_0 = \begin{pmatrix} 1 & 0 \\ 0 & 1 \end{pmatrix}, \quad \begin{pmatrix} 1 & 0 \\ 0 & 0 \end{pmatrix}, \quad \begin{pmatrix} 1 & 0 \\ 0 & -1 \end{pmatrix},$$

with the circles in the bundles taking the form

$$\begin{pmatrix} \alpha & b \\ \bar{b} & -\alpha \end{pmatrix}, \quad \begin{pmatrix} \alpha & b \\ \bar{b} & 0 \end{pmatrix}, \quad \begin{pmatrix} \alpha & b \\ \bar{b} & \alpha \end{pmatrix}, \tag{1.28}$$

and they correspond to bundles which are elliptic, parabolic and hyperbolic.

(6) Consequently, we can find three circles H_1, H_2, H_3 in a bundle so that the bundle is represented by

$$\lambda_1 H_1 + \lambda_2 H_2 + \lambda_3 H_3.$$

According to (1.28), an elliptic bundle can be represented by

$$\alpha\begin{pmatrix} 1 & 0 \\ 0 & -1 \end{pmatrix} + \beta\begin{pmatrix} 0 & 1 \\ 1 & 0 \end{pmatrix} + \gamma\begin{pmatrix} 0 & i \\ -i & 0 \end{pmatrix},$$

while a parabolic bundle can be represented by

$$\alpha\begin{pmatrix} 1 & 0 \\ 0 & 0 \end{pmatrix} + \beta\begin{pmatrix} 0 & 1 \\ 1 & 0 \end{pmatrix} + \gamma\begin{pmatrix} 0 & i \\ -i & 0 \end{pmatrix},$$

and a hyperbolic bundle can be represented by

$$\alpha\begin{pmatrix} 1 & 0 \\ 0 & 1 \end{pmatrix} + \beta\begin{pmatrix} 0 & 1 \\ 1 & 0 \end{pmatrix} + \gamma\begin{pmatrix} 0 & i \\ -i & 0 \end{pmatrix}.$$

It is left to the reader to find the relationship between α, β, γ in order to deliver a real circle.

(7) Let H_1 and H_2 be two fixed circles. The totality of circles

$$\lambda_1 H_1 + \lambda_2 H_2$$

form a pencil. If a circle is orthogonal to both H_1 and H_2 then it is orthogonal to each member of the pencil. From

$$\operatorname{tr}(FHF\bar{H}_1) = \operatorname{tr}(FHF\bar{H}_2) = 0$$

we see that H satisfies two linear relationships, so that there is only one parameter. This means that there are two circles $H^{(1)}$ and $H^{(2)}$ such that H can be represented as

$$\mu_1 H^{(1)} + \mu_2 H^{(2)};$$

this is also a pencil.

(8) For the number of point circles in a pencil we seek the solutions to

$$\operatorname{tr}\big(F(\lambda H_1 + \mu H_2)F(\lambda \bar{H}_1 + \mu \bar{H}_2)\big) = 0.$$

Being a quadratic equation there are three cases: (i) there are two real solutions; (ii) there is one (repeated) real solution; (iii) there is no real solution. It is not difficult to show that these cases correspond to hyperbolic, parabolic and elliptic pencils, and that, using the similarity relation, they can be transformed into

(i) $\lambda \begin{pmatrix} 1 & 0 \\ 0 & 0 \end{pmatrix} + \mu \begin{pmatrix} 0 & 0 \\ 0 & -1 \end{pmatrix}$ (two point circles, hyperbolic pencil);

(ii) $\lambda \begin{pmatrix} 0 & 0 \\ 0 & 1 \end{pmatrix} + \mu \begin{pmatrix} 0 & 1 \\ 1 & 0 \end{pmatrix}$ (parallel lines, parabolic pencil);

(iii) $\lambda \begin{pmatrix} 0 & 1 \\ 1 & 0 \end{pmatrix} + \mu \begin{pmatrix} 0 & i \\ -i & 0 \end{pmatrix}$ (straight lines through origin, elliptic pencil).

Moreover, (i) and (iii) are orthogonal, while (ii) is orthogonal to

$$\lambda \begin{pmatrix} 0 & 0 \\ 0 & 1 \end{pmatrix} + \mu \begin{pmatrix} 0 & i \\ -i & 0 \end{pmatrix}.$$

It is easy to see that, given a pencil (elliptic, parabolic or hyperbolic), there is a unique pencil which is orthogonal to it.

(9) Circles in a bundle through a point (not the common point of a parabolic bundle) form a pencil. This is clear enough because 'through a point' and being orthogonal to a 'point circle' mean the same thing.

Corresponding to a bundle of circles we may also consider the quadratic form

$$\operatorname{tr}\big(F(\lambda H_1 + \mu H_2 + \nu H_3)F(\lambda \bar{H}_1 + \mu \bar{H}_2 + \nu \bar{H}_3)\big).$$

It can be seen that there are three cases: (i) there is a unique point circle; (ii) there are infinitely many point circles; (iii) there is no point circle. It is not difficult to show, under similarity conditions, that these cases correspond to the following examples.

Example 1.34 Let

$$H_1 = \begin{pmatrix} 1 & 0 \\ 0 & 0 \end{pmatrix}, \quad H_2 = \begin{pmatrix} 0 & 1 \\ 1 & 0 \end{pmatrix}, \quad H_3 = \begin{pmatrix} 0 & i \\ -i & 0 \end{pmatrix};$$

the quadratic form then becomes $-\mu^2 - \nu^2$. If $\mu^2 + \nu^2 = 0$ then $\mu = \nu = 0$, giving the point circle H_1 (parabolic).

Example 1.35 Let

$$H_1 = \begin{pmatrix} 1 & 0 \\ 0 & 1 \end{pmatrix}, \quad H_2 = \begin{pmatrix} 0 & 1 \\ 1 & 0 \end{pmatrix}, \quad H_3 = \begin{pmatrix} 0 & i \\ -i & 0 \end{pmatrix};$$

the quadratic form then becomes $-\lambda^2 + \mu^2 + \nu^2$. The equation $\lambda^2 = \mu^2 + \nu^2$ has infinitely many solutions (hyperbolic).

Example 1.36 Let

$$H_1 = \begin{pmatrix} 1 & 0 \\ 0 & -1 \end{pmatrix}, \quad H_2 = \begin{pmatrix} 0 & 1 \\ 1 & 0 \end{pmatrix}, \quad H_3 = \begin{pmatrix} 0 & i \\ -i & 0 \end{pmatrix};$$

the quadratic form then becomes $\lambda^2 + \mu^2 + \nu^2$. The equation $\lambda^2 + \mu^2 + \nu^2 = 0$ implies $\lambda = \mu = \nu = 0$, so that there is a 'no point circle' (elliptic).

These three examples also apply to (6), so that the conclusion is clear.

(10) Therefore elliptic bundles may only contain elliptic pencils (no point circle), parabolic bundles may only contain elliptic or parabolic pencils (and there must be such pencils), and hyperbolic bundles may have three types of pencils.

The reader is invited to use the method here to deduce the results obtained in the preceding sections.

1.11 Types of transformations

The fixed points (or invariant points) of the transformation

$$w = \frac{az + b}{cz + d}$$

satisfy the quadratic equation

$$cz^2 + (d - a)z - b = 0. \tag{1.29}$$

If all the coefficients in (1.29) are 0 then $w = z$, the identity transformation, and every point is a fixed point.

If $c \neq 0$ then (1.29) has the two roots

$$z_i = \frac{a - d \pm \sqrt{D}}{2c}, \qquad D = (d - a)^2 + 4bc;$$

actually there is only one fixed point if $D = 0$, and two if $D \neq 0$. In the following we separate the discussion between $c = 0$ and $c \neq 0$.

(1) If $c = 0, d - a = 0$, then there is only one fixed point, namely ∞. Such a transformation has the form

$$w = z + k, \tag{1.30}$$

which is merely a translation on the plane.

If $c = 0, d - a \neq 0$, then we can see that there are two fixed points, namely ∞ and $b/(d-a)$. Let

$$w_1 = w - \frac{b}{d-a}, \quad z_1 = z - \frac{b}{d-a},$$

so that

$$w_1 = \frac{a}{d} z_1.$$

(2) If $c \neq 0$, $D = 0$, then the single fixed point is $g = (a-d)/2c$, and the transformation (1.29) becomes

$$\frac{1}{w-g} = \frac{1}{z-g} + \frac{2c}{a+d}.$$

Letting $w_1 = 1/(w-g)$, $z_1 = 1/(z-g)$, we are back to (1.30). Therefore (1.30) is the standard form for a transformation with only one fixed point.

If $c \neq 0$, $D \neq 0$, then let

$$w' = \frac{w - z_2}{w - z_1}, \quad z' = \frac{z - z_2}{z - z_1}.$$

The fixed points have been mapped to 0 and ∞, and we have

$$w' = \rho z'. \tag{1.31}$$

Corresponding to $z = \infty$, we have $w = a/c$ and $z' = 1$, so that

$$\rho = w' = \frac{a - cz_2}{a - cz_1},$$

that is

$$\rho = \frac{a + d + \sqrt{D}}{a + d - \sqrt{D}}. \tag{1.32}$$

The above is the result of putting the two variables in the bilinear transformation $w = f(z)$ through the same transformation, that is letting $w = g(w_1)$, $z = g(z_1)$. This is based on the geometric idea that, starting with a transformation Γ on the z-plane, let the z-plane be transformed to the z_1-plane first, so that the transformation Γ on the z_1-plane gives rise to a new transformation. Putting it algebraically: If the original transformation has the matrix M, and the transformation g has the matrix P, then our new transformation has the matrix

$$M_1 = P^{-1} M P.$$

The two matrices M_1 and M are similar, and being similar is an equivalence relation. Therefore bilinear transformations can be partitioned into equivalence classes.

Definition 1.37 If a bilinear transformation has only one fixed point then we say that it is parabolic. Suppose that a bilinear transformation has two distinct fixed points. If the number ρ defined in (1.32) is real then the transformation is said to be hyperbolic. If ρ has absolute value 1 then the transformation is said to be elliptic. If ρ is a general complex number then the transformation is said be loxodromic. The number ρ is called the multiplicative factor.

What is the geometric meaning of ρ? It is in fact the cross-ratio of four points, two of which are the fixed points, and the other two are z and its image w. More precisely,

$$\rho = (z_1, z_2, z, w).$$

In particular, if $z_1 = \infty$, $z_2 = 0$, $w = \rho z$, then $(\infty, 0, z, \rho z) = \rho$.

What is the algebraic meaning of ρ? It is in fact the ratio of the two eigenvalues of the matrix M.

Note that

$$\frac{1}{\rho} = (z_2, z_1, z, w) \quad \text{and} \quad \left(\frac{1}{w}\right) = \frac{1}{\rho}\left(\frac{1}{z}\right).$$

It follows that if $\lambda \neq \rho$ or $1/\rho$, then the bilinear transformation corresponding to λ cannot be similar to that corresponding to ρ. Therefore a necessary and sufficient condition for two bilinear transformations to be similar is that their coefficients should be either the same or be the reciprocals of each other.

On the other hand, the parabolic transformation $w = z + k$ can be written as $w/k = z/k + 1$, so that any parabolic transformation can be changed to $w = z + 1$; in other words, any two parabolic transformations are similar.

Circles which are invariant under an elliptic, a parabolic or a hyperbolic transformation form hyperbolic, parabolic or elliptic pencils, respectively. The proof is not difficult if we consider the following: The translation $w = z + 1$ leaves the real axis invariant; the rotation $w = e^{i\theta} z$ leaves circles with centre the origin invariant; and the magnification $w = kz$ leaves straight lines through the origin invariant; there are no other essentially different transformations that leave circles invariant.

1.12 The general linear group

Definition 1.38 We adjoin the transformation

$$w = \bar{z} \tag{1.33}$$

to the group of bilinear transformations to form the general linear group.

Thus, for the general linear group, besides the bilinear transformations, we also have transformations of the form

$$w = \frac{a\bar{z} + b}{c\bar{z} + d}. \tag{1.34}$$

It is not difficult to show that there are no other forms of transformations in the group. Circles are associated with points having real cross-ratios, so that a general linear transformation still maps circles into circles.

If

$$w_1 = \frac{a_1\bar{w} + b_1}{c_1\bar{w} + d_1},$$

then

$$w_1 = \frac{a_1(\bar{a}z + \bar{b}) + b_1(\bar{c}z + \bar{d})}{c_1(\bar{a}z + \bar{b}) + d_1(\bar{c}z + \bar{d})} = \frac{(a_1\bar{a} + b_1\bar{c})z + a_1\bar{b} + b_1\bar{d}}{(c_1\bar{a} + d_1\bar{c})z + c_1\bar{b} + d_1\bar{d}},$$

with the matrix

$$\begin{pmatrix} a_1 & b_1 \\ c_1 & d_1 \end{pmatrix} \overline{\begin{pmatrix} a & b \\ c & d \end{pmatrix}}.$$

Thus the product of two transformations of the form (1.34) is a bilinear transformation.

Exercise 1.39 Show that any transformation (1.34) adjoined to the group of bilinear transformations only has the same effect as adjoining (1.33) to the group.

Let us now consider the fixed points of the transformation (1.34). They satisfy

$$cz\bar{z} + dz - a\bar{z} - b = 0, \tag{1.35}$$

from which it follows that $\bar{c}z\bar{z} + \bar{d}\bar{z} - \bar{a}z - \bar{b} = 0$, and hence we have the two circles

$$\begin{cases} (c + \bar{c})z\bar{z} + (d - \bar{a})z + (\bar{d} - a)\bar{z} - b - \bar{b} = 0, \\ (c - \bar{c})z\bar{z} + (d + \bar{a})z - (\bar{d} + a)\bar{z} - b + \bar{b} = 0. \end{cases} \tag{1.36}$$

Therefore, speaking generally, the two points of intersections for these circles are invariant, and hence the transformation maps the pencil defined by these circles onto itself. In the particular case $c = \bar{c}, b = \bar{b}, d = -\bar{a}$, there is only one circle in (1.36). The transformation then takes the form

$$w = \frac{-h\bar{z} - \delta}{\alpha\bar{z} + \bar{h}}, \quad \alpha, \delta \text{ real}. \tag{1.37}$$

Such a transformation has the invariant circle

$$\alpha w\bar{w} + \bar{h}w + h\bar{w} + \delta = 0. \tag{1.38}$$

On applying the transformation (1.37) twice, we have

$$u = \frac{-h\bar{w} - \delta}{\alpha\bar{w} + \bar{h}} = \frac{-h(-\bar{h}z - \delta) - \delta(\alpha z + h)}{\alpha(-\bar{h}z - \delta) + \bar{h}(\alpha z + h)} = z,$$

the identity transformation. Such a transformation is called an inversion, and the circle (1.38) is called the basic circle for the inversion; two successive applications of an inversion result in the identity transformation. The following is the geometric interpretation: Examine first the inversion $w = \bar{z}$, which is a reflection about the x-axis; points in the upper half-plane are paired with their reflections in the lower half-plane. For this reason an inversion is also called a reflection with respect to its basic circle.

The circle (1.38) can be real or imaginary. By transforming a real circle to $z\bar{z} = 1$, the inversion is reduced to

$$w = \frac{1}{\bar{z}}. \tag{1.39}$$

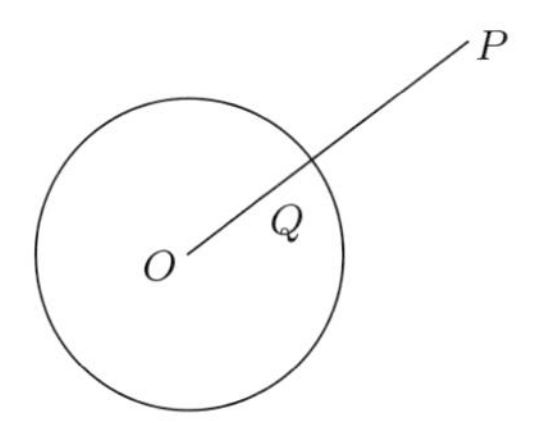

Figure 1.10.

The following is the geometric interpretation of the transformation: A point $P(z = \rho e^{i\theta})$ is mapped to $Q(w = \frac{1}{\rho} e^{i\theta})$. In other words, we take a point Q on the straight line OP such that $OP \cdot OQ = 1$ (see Fig. 1.10). The reader should be able to see the effect of an inversion with respect to a general basic circle.

The inversion with respect to an imaginary circle is

$$w = -\frac{1}{\bar{z}}. \tag{1.40}$$

Theorem 1.40 *Any bilinear transformation can be expressed as inversions with respect to four real circles.*

Proof. We need only consider the transformations

$$w = z + 1, \qquad w = \rho e^{i\theta} z.$$

(1) Parabolic form. The transformation $w = z + 1$ is the product of $w = -\bar{w}_1$, $w_1 = -\bar{z} - 1$. The former is an inversion with respect to the y-axis, and the latter is $\left(w_1 + \frac{1}{2}\right) = -\overline{\left(z_1 + \frac{1}{2}\right)}$, which is an inversion with respect to the line $x = -\frac{1}{2}$. The two lines $x = 0$ and $x = -\frac{1}{2}$ can be considered as tangential circles. Therefore, any parabolic transformation can be represented as the product of two inversions with respect to tangential circles (belonging to a parabolic pencil).

(2) Hyperbolic form. The transformation $w = \rho z$ is the product of $w = 1/\bar{w}_1$, $w_1 = 1/(\rho\bar{z})$, which are inversions with respect to the concentric circles $z\bar{z} = 1$, $z\bar{z} = 1/\rho$. Therefore, hyperbolic transformations are inversions with respect to two non-intersecting circles (from a hyperbolic pencil).

(3) Elliptic form. The transformation $w = e^{i\theta} z$ is the product of $w = e^{\frac{1}{2}i\theta}\bar{w}_1$, $w_1 = e^{-\frac{1}{2}i\theta}\bar{z}$, which are inversions with respect to two straight lines through the origin. Therefore, elliptic transformations are inversions with respect to two intersecting circles (from an elliptic pencil).

(4) Loxodromic form. The transformation $w = \rho e^{i\theta} z$ is the product of the transformations $w = \rho w_1$, $w_1 = e^{i\theta} z$. The theorem is proved. □

There is much information in the proof of Theorem 1.40. Thus it is not difficult to deduce the following theorem.

Theorem 1.41 *Elliptic, parabolic and hyperbolic transformations can be expressed as two inversions.*

Theorem 1.42 *A transformation obtained from inversions with respect to any two circles from an elliptic (or parabolic or hyperbolic) pencil is elliptic (or parabolic or hyperbolic), and the set of all such transformations forms a group with one parameter.*

Proof. First, take as an example the parabolic pencil

$$y = k, \quad k \text{ real}.$$

The transformations $w = \bar{z} + 2k_1 i$, $w = \bar{z} + 2k_2 i$ are two inversions with respect to $y = k_1$, $y = k_2$, and their product is

$$w = z + 2(k_1 - k_2)i.$$

This includes all transformations of the form $w = z + ki$, which form a subgroup, and this subgroup is the same as the 'real additive group'.

Next, consider the elliptic pencil $y = x \tan\theta$ $(0 \leq \theta \leq \pi)$. The transformation

$$w = e^{2i\theta}\bar{z}$$

is a reflection/inversion with respect to the straight line/circle $y = x\tan\theta$. The product of $w = e^{2i\theta}\bar{z}$ and $w = e^{2i\theta_1}\bar{z}$ has the form $w = e^{it}z$. Such transformations form a subgroup, and this subgroup is the same as the 'multiplicative group of numbers with unit modulus'.

Finally, consider the hyperbolic pencil $z\bar{z} = \rho^2$. The product of $w = \rho^2/\bar{z}$ and $w = \rho_1^2/\bar{z}$ is $w = (\rho/\rho_1)^2 z$. This is a transformation of the form $w = kz$ $(k > 0)$, and the group can be identified with the 'multiplicative group of the positive reals'. □

1.13 The fundamental theorem of projective geometry

Theorem 1.43 *Any continuous transformation which maps the one dimensional projective space bijectively onto itself, and which transforms a harmonic sequence into a harmonic sequence, must be a general linear transformation.*

Proof. Let

$$w = f(z)$$

be such a transformation. Since any three points can be transformed to any other three points by a bilinear transformation, we may assume without loss that

$$0 = f(0), \quad 1 = f(1), \quad \infty = f(\infty).$$

Starting from the assumption

$$(\infty, z_2, z_3, z_4) = -1,$$

we find that $z_2 = \frac{1}{2}(z_3 + z_4)$. Thus, if $w_3 = f(z_3)$, $w_4 = f(z_4)$, then

$$\tfrac{1}{2}(w_3 + w_4) = f\big(\tfrac{1}{2}(z_3 + z_4)\big).$$

Consequently, for any two complex numbers a, b, we always have

$$f(a) + f(b) = 2f\big(\tfrac{1}{2}(a + b)\big).$$

Setting $b = 0$ we have $f(a) = 2f(\frac{1}{2}a)$, and the equation above becomes

$$f(a) + f(b) = f(a + b).$$

An inductive argument now gives $f(n) = nf(1) = n$ for every natural number n. Also, from $f(-n) + f(n) = f(0) = 0$, we have $f(-n) = -f(n)$. Again, from $pf(n/p) = f(n) = n$, we find that $f(n/p) = n/p$. Thus, for any rational r we have $f(r) = r$. By continuity, we have $f(x) = x$ for any real x.

Let $f(i) = j$. From $(1, -1, i, -i) = ((i - 1)/(1 + i))^2 = -1$ we deduce from the hypothesis that $(1, -1, j, -j) = -1$. Thus

$$\frac{j - 1}{-1 - j} \Big/ \frac{-j - 1}{-1 + j} = -1,$$

so that $h^2 = -1$. Therefore either $f(i) = i$ or $f(i) = -i$.

If $f(i) = i$ then the same method can be used to show that $f(ri) = ri$ for every rational r, and hence $f(yi) = yi$ for every real y. Thus

$$f(x + yi) = f(x) + f(yi) = x + yi,$$

that is

$$f(z) = z.$$

If $f(i) = -i$ then $f(yi) = -iy$ for every real y, and hence $f(z) = \bar{z}$. The theorem is proved. □

2
Non-Euclidean geometry

2.1 Euclidean geometry (parabolic geometry)

Consider a parabolic bundle of circles. Assume without loss that the circles in the bundle pass through ∞, so that the bundle is formed by the set of straight lines. We now consider the group of inversions associated with the circles (that is, straight lines) in the bundle.

Consider the inversion

$$w = e^{i\theta}\bar{z} + ie^{\frac{1}{2}i\theta}\lambda, \quad \lambda \text{ real} \tag{2.1}$$

(it is not difficult to show directly that this is indeed an inversion). The circle which is invariant with respect to the inversion is

$$\frac{1}{i}(ze^{-\frac{1}{2}i\theta} - \bar{z}e^{\frac{1}{2}i\theta}) = \lambda,$$

that is

$$y\cos\frac{1}{2}\theta - x\cos\frac{1}{2}\theta = \frac{\lambda}{2}.$$

This is a general straight line; thus, when $0 \le \theta \le 2\pi$, λ real, all possible inversions in the parabolic bundle are represented by (2.1). The product of two such inversions is given by

$$\begin{aligned} w_1 &= e^{i\psi}\bar{w} + ie^{\frac{1}{2}i\psi}\tau = e^{i\psi}(e^{-i\theta} - ie^{-\frac{1}{2}i\theta}\lambda) + ie^{\frac{1}{2}i\psi}\tau \\ &= e^{i(\psi-\theta)}z + ie^{\frac{1}{2}i\psi}\tau - ie^{i(\psi-\frac{1}{2}\theta)}\lambda, \end{aligned} \tag{2.2}$$

which has the form

$$w = e^{i\theta}z + q, \quad q \text{ complex}. \tag{2.3}$$

Conversely, we proved earlier that transformations of the form (2.3) can be represented as the product of two inversions. Therefore, in a group of transformations in a parabolic bundle, each of its members has the form (2.3), or $w = \bar{z}$, or a product of two such transformations.

We now examine the geometric meaning of the transformation (2.3). Since $w = e^{i\theta}z$ is a rotation, and $w = z + q$ is a translation, the effect of (2.3) is a rigid motion, and the inversion $w = \bar{z}$ is a reflection, the group so obtained is called the group of Euclidean transformations.

The geometry of the plane under this group is commonly called Euclidean geometry. Thus Euclidean geometry can be considered to be the geometry induced from transformations in a parabolic bundle.

The following are some of the important properties of Euclidean geometry:

(i) The whole space is reachable – any point can be transformed to any other point.
(ii) The distance between any two points is the only invariant; thus, if AB has the same distance as CD then there is a transformation which maps AB to CD.
(iii) Straight lines are mapped into straight lines (circles in a parabolic bundle are transformed into circles in the bundle).
(iv) The angle between two straight lines is the only invariant.
(v) There is a (unique) straight line through any two given points.
(vi) Through any point not on a given straight line, there is a (unique) straight line which is parallel to the given line.

Note. Let us reexamine (2.3). If $\theta \neq 0$ then

$$w - a = e^{i\theta}(z - a), \quad a = q/(1 - e^{i\theta}),$$

is a parabolic transformation with a as a fixed point – a rotation about a. If $\theta = 0$ then we have a translation – a parabolic transformation. Therefore, there can be no hyperbolic transformations in Euclidean geometry.

2.2 Spherical geometry (elliptic geometry)

Consider an elliptic bundle of circles. Assume without loss that the circles intersect the unit circle at antipodal points. Circles intersecting the unit circle at $\pm e^{i\theta}$ have equations of the form

$$\alpha(z\bar{z} - 1) - i\lambda e^{i\theta}\bar{z} + i\lambda e^{-i\theta}z = 0, \tag{2.4}$$

where α, λ, θ are real parameters. Equation (2.4) represents the whole elliptic bundle and, with respect to the circle, we have the inversion

$$w = \frac{i\lambda e^{i\theta}\bar{z} + \alpha}{\alpha\bar{z} + i\lambda e^{-i\theta}}, \tag{2.5}$$

the determinant for which is $-\alpha^2 - \lambda^2$. Divide the numerator and the denominator by $i\sqrt{\lambda^2 + \alpha^2}$ and let

$$\frac{\alpha}{\sqrt{\lambda^2 + \alpha^2}} = \sin\tau, \qquad \frac{\lambda}{\sqrt{\lambda^2 + \alpha^2}} = \cos\tau,$$

so that the transformation becomes

$$w = \frac{\bar{z}e^{i\theta}\cos\tau - i\sin\tau}{-i\bar{z}\sin\tau + e^{-i\theta}\cos\tau}, \tag{2.6}$$

which now has the determinant 1. Suppose that there is another inversion

$$z = e^{2i\psi}\bar{z}_1; \tag{2.7}$$

then the product of the two inversions (2.5) and (2.7) is

$$w = \frac{z_1 e^{i(\theta-\psi)}\cos\tau - ie^{i\psi}\sin\tau}{-iz_1 e^{-i\psi}\sin\tau + e^{-i(\theta-\psi)}\cos\tau}, \tag{2.8}$$

with the matrix of the transformation given by

$$M = \begin{pmatrix} e^{i(\theta-\psi)}\cos\tau & -ie^{i\psi}\sin\tau \\ -ie^{-i\psi}\sin\tau & e^{-i(\theta-\psi)}\cos\tau \end{pmatrix}. \tag{2.9}$$

This is a unitary matrix (that is, $M\bar{M}' = I$) with determinant 1, and is called a special unitary matrix. Conversely, any unitary matrix with determinant 1 can be so represented. Here is the proof: If

$$M = \begin{pmatrix} a & b \\ c & d \end{pmatrix},$$

where

$$|a|^2 + |b|^2 = 1, \quad a\bar{c} + b\bar{d} = 0, \quad |c|^2 + |d|^2 = 1, \quad ad - bc = 1,$$

then the general solution to the first equation is $a = e^{i(\theta-\psi)}\cos\tau$, $b = -ie^{i\psi}\sin\tau$. From the second equation we have $c = -\bar{b}t$, $d = \bar{a}t$, and from $ad - bc = (|a|^2 + |b|^2)t = 1$ we deduce that $t = 1$, so that

$$M = \begin{pmatrix} a & b \\ -\bar{b} & \bar{a} \end{pmatrix}, \quad |a|^2 + |b|^2 = 1, \tag{2.10}$$

and hence formula (2.9).

Theorem 2.1 *The group of inversions with respect to a circle from an elliptic bundle is a special unitary group (taking note of the fact that $\pm M$ represent only one transformation).*

Note. Equation (2.10) illustrates the relationship between unitary matrices and quaternions. Let $m = a + bj$ represent a quaternion. From $|a|^2 + |b|^2 = 1$ we see that $|m| = 1$. If $m_1 = a_1 + b_1 j$ then $mm_1 = aa_1 - bb_1 + (ab_1 + ba_1)j$, which also represents MM_1. Thus the special unitary group can be represented by the multiplicative group of unit quaternions.

Let us reexamine the meaning of the inversion (2.5) on the Riemann sphere. Corresponding to the circle (2.4) we have the plane

$$\alpha\zeta + \lambda\xi\sin\theta - \lambda\eta\cos\theta = 0. \tag{2.11}$$

This plane passes through the centre of the sphere, so that we have a great circle on the surface of the sphere.

Theorem 2.2 *An inversion (2.5) on the Riemann sphere represents a reflection with respect to the plane (2.11).*

Proof. If we can establish the theorem for $\theta = 0$ then the general case follows from a rotation on the plane. Now the plane (2.11) becomes

$$\zeta \sin\tau - \eta \cos\tau = 0, \tag{2.12}$$

and the transformation (2.6) becomes

$$w = \frac{\bar{z}\cos\tau - i\sin\tau}{-i\bar{z}\sin\tau + \cos\tau}, \tag{2.13}$$

which then gives

$$1 + |w|^2 = \frac{1+|z|^2}{|iz\sin\tau + \cos\tau|^2}.$$

With $z = x + iy$, $w = u + iv$, we then have

$$\begin{aligned} w &= \frac{(\bar{z}\cos\tau - i\sin\tau)(iz\sin\tau + \cos\tau)}{|-i\bar{z}\sin\tau + \cos\tau|^2} \\ &= \frac{i(z\bar{z}-1)\sin\tau\cos\tau + \bar{z}\cos^2\tau + z\sin^2\tau}{|-i\bar{z}\sin\tau + \cos\tau|^2}, \end{aligned}$$

so that

$$u = \frac{x}{|iz\sin\tau + \cos\tau|^2}, \qquad v = \frac{(z\bar{z}-1)\sin\tau\cos\tau - y(\cos^2\tau - \sin^2\tau)}{|iz\sin\tau + \cos\tau|^2}.$$

Therefore

$$\begin{aligned} \frac{2u}{1+|w|^2} &= \frac{2x}{1+|z|^2}, \qquad \frac{2v}{1+|w|^2} = \frac{|z|^2-1}{|z|^2+1}\sin 2\tau - \frac{2y}{1+|z|^2}\cos 2\tau, \\ \frac{|w|^2-1}{|w|^2+1} &= \frac{1}{1+|z|^2}(1+|z|^2 - 2|iz\sin\tau + \cos\tau|^2) \\ &= \frac{1}{1+|z|^2}(1+|z|^2 - 2|z|^2\sin^2\tau - 2\cos^2\tau + 4y\sin\tau\cos\tau) \\ &= \left(\frac{|z|^2-1}{|z|^2+1}\right)\cos 2\tau + \frac{2y}{1+|z|^2}\sin 2\tau. \end{aligned}$$

Let ξ, η, ζ be points on the sphere corresponding to z, ξ', η', and let ζ' correspond to w. Then

$$\begin{aligned} \xi' &= \xi, \\ \eta' &= -\eta\cos 2\tau + \zeta\sin 2\tau, \\ \zeta' &= \eta\sin 2\tau + \zeta\cos 2\tau. \end{aligned}$$

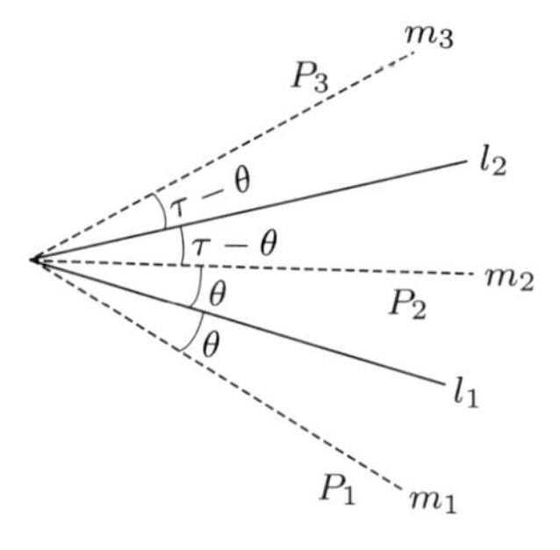

Figure 2.1.

The midpoint $\left(\frac{1}{2}(\xi+\xi_1), \frac{1}{2}(\eta+\eta_1), \frac{1}{2}(\zeta+\zeta_1)\right)$ of (ξ_1, η_1, ζ_1) and its image (ξ_2, η_2, ζ_2) must lie on the plane (2.12). This can be established directly from the calculation

$$\begin{aligned}&\sin\tau(\zeta_1+\eta_1\sin 2\tau+\zeta_1\cos 2\tau)-(\eta_1-\eta_1\cos 2\tau+\zeta_1\sin 2\tau)\cos\tau\\&\quad=\zeta_1(\sin\tau+\sin\tau\cos 2\tau-\cos\tau\sin 2\tau)+\eta_1(\sin\tau\sin 2\tau+\cos\tau\cos 2\tau-\cos\tau)=0.\end{aligned}$$

Therefore (ξ_1, η_1, ζ_1) and (ξ_2, η_2, ζ_2) are reflections of each other with respect to the plane (2.12). □

Theorem 2.3 *The result of two successive inversions on the Riemann sphere represents a rotation; the antipodal points of the axis of rotation correspond to the points of intersection of the circles with respect to which are the inversions.*

Proof. The two circles in the elliptic bundle intersect each other at two points, which are fixed points of the inversions. Moreover, the fixed points must be a pair of antipodal points for the sphere because the inversions are with respect to great circles. Therefore the inversions are reflections with respect to planes through a diameter. Viewed from a plane perpendicular to this diameter the picture (see Fig. 2.1) shows the following: There are two intersecting straight lines, with respect to each of which an inversion is applied once, and the result is a rotation. To see this we suppose that the angle between the two straight lines ℓ_1, ℓ_2 is τ. Let the original position of a point P_1 be on m_1, which makes the angle θ with ℓ_1. The inversion of P_1 with respect to ℓ_1 gives a point P_2 on m_2, which makes an angle $\tau-\theta$ with ℓ_2. The inversion of P_2 with respect to ℓ_2 gives a point P_3 on m_3, which makes an angle 2τ with m_1. Thus the result of the two inversions is a rotation by an angle 2τ. □

It is not difficult to prove that any rotation can be obtained in this way. Therefore elliptic geometry can be interpreted as the geometry of the rotation group on the surface of the sphere.

Note 1. The equation $M\bar{M}'=I$ can also be interpreted as the transformation which maps the imaginary circle $z\bar{z}+1=0$ onto itself.

Note 2. The transformations in an elliptic geometry are all elliptic. This is because the invariants for such a transformation form a pencil which is hyperbolic.

2.3 Some properties of elliptic geometry

It is not difficult to prove that elliptic geometry has the following properties:

(i) The whole space is reachable.
(ii) There is the following unique invariant associated with two points: There are two corresponding points on the Riemann sphere, and the angle subtended by the two corresponding vectors, with the centre of the sphere as origin, is invariant (noting that we may need to consider θ and $2\pi - \theta$ to be the same angle).
(iii) Circles in an elliptic bundle are transformed into circles within the bundle; such a circle is called a geodesic.
(iv) The angle subtended by two geodesics is a unique invariant.
(v) There is a unique geodesic through any two points (apart from points which correspond to antipodal points on the Riemann sphere).
(vi) Given a point not on a given geodesic, there is no geodesic through the given point which does not intersect the given geodesic.

2.4 Hyperbolic geometry (Lobachevskian geometry)

Consider a hyperbolic bundle, which we assume, without loss of generality, to be the set of circles which intersect the unit circle orthogonally. The condition for orthogonal intersection being $\alpha - \delta = 0$, the circles take the form

$$\alpha z\bar{z} + h\bar{z} + \bar{h}z + \alpha = 0, \quad |h|^2 - \alpha^2 > 0. \tag{2.14}$$

The corresponding inversion is then the transformation

$$w = \frac{-h\bar{z} - \alpha}{\alpha\bar{z} + \bar{h}}. \tag{2.15}$$

From

$$1 - w\bar{w} = \frac{(\alpha\bar{z} + \bar{h})(\alpha z + h) - (-h\bar{z} - \alpha)(-\bar{h}z - \alpha)}{|\alpha\bar{z} + \bar{h}|^2} = \frac{|h|^2 - \alpha^2}{|\alpha\bar{z} + \bar{h}|^2}(1 - z\bar{z})$$

we see that the inversion (2.15) maps the interior of the unit circle ($|z| < 1$) into the interior of the unit circle ($|w| < 1$); the circumference into the circumference; and the exterior into the exterior.

With $|h| > \alpha$, we now take an interior point $a = -\alpha/h$, and write $-h/\bar{h} = e^{i\theta}$, so that (2.15) becomes

$$w = e^{i\theta}\frac{\bar{z} - a}{-\bar{a}\bar{z} + 1}, \tag{2.16}$$

and a further transformation $z \mapsto \bar{z}$ then delivers the bilinear transformation

$$w = e^{i\theta} \frac{z - a}{1 - \bar{a}z}. \tag{2.17}$$

This transformation maps the unit circle onto itself, and the interior point a is mapped to $w = 0$. It is not difficult to prove that the product of any two inversions is a bilinear transformation leaving the unit circle invariant.

We now prove that, conversely, any bilinear transformation that leaves the unit circle invariant must have the form (2.17). Let $w = (az + b)/(cz + d)$ be such a transformation, which then maps the circle

$$|az + b|^2 + |cz + d|^2 = 0$$

into the unit circle $|w| = 1$. The necessary and sufficient conditions for the circle

$$(|a|^2 - |c|^2)z\bar{z} + (a\bar{b} - c\bar{d})z + (\bar{a}b - \bar{c}d)\bar{z} + |b|^2 - |d|^2 = 0$$

to be the unit circle are that $a\bar{b} - c\bar{d} = 0$ and $|a|^2 - |c|^2 - -(|b|^2 - |d|^2) \neq 0$. From the former we may write $a = \bar{d}t$, $c = \bar{b}t$, and substituting this into the latter we find that $(1 - |t|^2)(|b|^2 - |d|^2) = 0$, so that $|t| = 1$. The original transformation now becomes

$$w = \frac{\bar{d}tz + b}{\bar{b}tz + d} = \frac{\bar{d}t}{d}\left(z + \frac{b}{\bar{d}t}\right) \Big/ \left(\frac{\bar{b}t}{d}z + 1\right).$$

From $|\bar{d}t/d| = 1$ and $\overline{b/(\bar{d}t)} = \bar{b}/(d\bar{t}) = \bar{b}t/d$, we see that the transformation is indeed of the form (2.17).

Taking the interior of the unit circle to be the image space, the set of transformations of the form (2.17) now form a group. The resulting geometry is called Lobachevskian geometry, or hyperbolic geometry.

The space being considered is the interior of the unit circle. Had we started off from the upper half-plane instead then we would have a geometry with the same properties but with a different model. The group of transformation is now

$$w = \frac{az + b}{cz + d}, \quad ad - bc = 1, \tag{2.18}$$

and a, b, c, d are real numbers. The proof only requires us to check that any bilinear transformation of the upper half-plane onto itself must have the form (2.18). We need the image w to be real when $z = 0, 1, \infty$ in (2.18), so that a, b, c, d are real and that $ad - bc \neq 0$. Also, from

$$\frac{ai + b}{ci + d} = \frac{ac + bd + i(ad - bc)}{c^2 + d^2},$$

we find that if $ad - bc > 0$ then the image of i will be in the upper half-plane; otherwise the image of i will be in the lower half-plane. Therefore $ad - bc = \rho^2$, and on dividing the numerator and the denominator of the transformation formula by ρ we have the required result.

Henceforth we may consider either the interior of the unit circle, or the upper half-plane, depending on whichever one we find to be convenient; the result can be deduced easily by mapping from one to the other interior.

The model of geometry in the upper half-plane is called the Poincaré half-plane model of hyperbolic geometry.

2.5 Distance

With points in the unit circle (or on the upper half-plane) as objects in our geometry, we have the transformation

$$w = e^{i\theta}\frac{z-a}{1-\bar{a}z}, \qquad |a|<1, \quad 0 \le \theta < 2\pi, \tag{2.19}$$

which is also called a non-Euclidean motion. We construct the group Γ of such motions. First, any interior point a can be mapped to 0 by a member of Γ, so that hyperbolic space is reachable under (2.18). Not only that, but each point may also carry a direction, forming a set which is still reachable. The method is to map the point to the origin, and then rotate the given direction to that of the positive x-axis.

Consider next a geodesic; this is (the part of) a circle which intersects the unit circle orthogonally. There is one and only one such geodesic which passes through two given points. From the point of view of the upper half-plane, the proof is very simple: Let the perpendicular bisector of the two points intersect the x-axis at C; the circle through the two points with the centre at C is then the required unique circle through the two points.

The geodesics form a reachable set, and we can prove this via the upper half-plane again. Any geodesic crosses the x-axis at two points (intersecting points). Any two real numbers can be mapped into any two other real numbers by a real transformation $w = (az+b)/(cz+d)$, $ad-bc=1$, so we are done.

If two geodesics intersect then their angle of intersection is invariant.

It is imperative to examine now the problem of the invariant associated with two points. Let z_1, z_2 be two points, so that $\bar{z}_1^{-1} = z_1/|z_1|^2$, $\bar{z}_2^{-1} = z_2/|z_2|^2$ are their reflections with respect to the unit circle. The cross-ratio

$$g(z_1,z_2) = (z_1, \bar{z}_1^{-1}, z_2, \bar{z}_2^{-1}) = \frac{z_2 - z_1}{\bar{z}_1^{-1} - z_2}\Bigg/\frac{\bar{z}_2^{-1} - z_1}{\bar{z}_1^{-1} - \bar{z}_2^{-1}} = \left|\frac{z_1 - z_2}{1 - \bar{z}_1 z_2}\right|^2 \tag{2.20}$$

is the sought-after invariant. In other words, if z_1, z_2 are mapped into w_1, w_2 then

$$g(z_1,z_2) = (z_1, \bar{z}_1^{-1}, z_2, \bar{z}_2^{-1}) = (w_1, \bar{w}_1^{-1}, w_2, \bar{w}_2^{-1}) = g(w_1, w_2).$$

Conversely, if $g(z_1, z_2) = K$, a positive number, then there is a transformation which maps z_2 to 0 and z_2 to z, giving $g(z,0) = |z|^2 = K$. Through a rotation, the point 0 stays the same, while z becomes $K^{1/2}$. Therefore $g(z_1, z_2)$ is also the unique invariant for the two points. If we take the upper half-plane as our base then the invariance is given by

$$h(z_1, z_2) = \left|\frac{z_1 - z_2}{\bar{z}_1 - z_2}\right|^2.$$

If we take $g(z_1, z_2)$ directly as the distance between two points then we cannot have certain important properties associated with the notion of distance. Instead we use the following method to derive a distance function: Take $z_1 = z + dz$, $z_2 = z$, so that we have

$$\frac{|dz|^2}{(1 - |z|^2)^2} \tag{2.21}$$

as an invariant differential – it is invariant under (2.19). The direct proof is to differentiate (2.19), to yield

$$dw = e^{i\theta}\left(\frac{1}{1 - \bar{a}z} + \frac{\bar{a}(z - a)}{(1 - \bar{a}z)^2}\right)dz = e^{i\theta}\frac{1 - a\bar{a}}{1 - \bar{a}z}\,dz,$$

so that

$$|dw|^2 = \frac{(1 - |a|^2)^2}{|1 - \bar{a}z|^4}|dz|^2;$$

from

$$1 - |w|^2 = 1 - \frac{|z - a|^2}{|1 - \bar{a}z|^2} = \frac{(1 - |a|^2)(1 - |z|^2)}{|1 - \bar{a}z|^2},$$

we then have

$$\frac{|dz|^2}{(1 - |z|^2)^2} = \frac{|dw|^2}{(1 - |w|^2)^2}.$$

In hyperbolic geometry, an element of length is

$$ds = \frac{\sqrt{dx^2 + dy^2}}{1 - x^2 - y^2},$$

and an element of area is

$$d\sigma = \frac{dx\,dy}{(1 - x^2 - y^2)^2}.$$

In the upper half-plane, these elements become

$$ds = \frac{\sqrt{dx^2 + dy^2}}{y}, \quad d\sigma = \frac{dx\,dy}{y^2}.$$

Theorem 2.4 *Let C be any curve (assumed to be continuous and to have continuous tangents) joining two points z_1, z_2. Then the value for*

$$\int_C ds$$

is least when C is a geodesic.

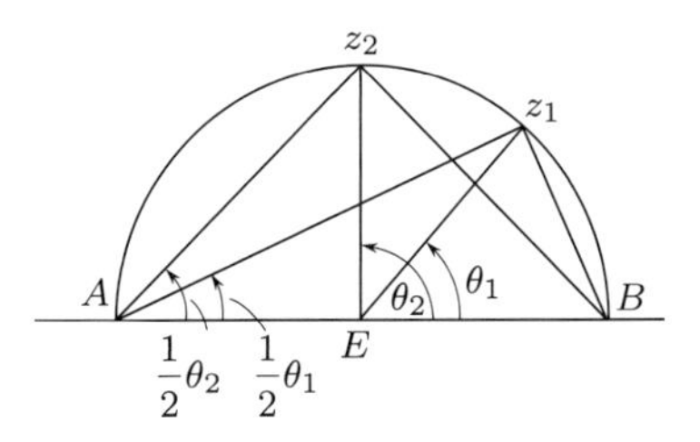

Figure 2.2.

Proof. Take the upper half-plane as the base. Construct a circle through z_1, z_2 with the centre on the x-axis. Let $E(t, 0)$ denote the centre of the circle, which can then be written in the form

$$x = t + \rho\cos\theta, \quad y = \rho\sin\theta,$$

and suppose that $\theta = \theta_1, \theta_2$ correspond to z_1, z_2. The curve C can then be written in the form

$$x = t + \rho(\theta)\cos\theta, \quad y = \rho(\theta)\sin\theta,$$
$$\rho(\theta_1) = \rho(\theta_2) = \rho, \quad 0 < \theta_1 < \theta_2 < \pi,$$

so that

$$\int_C \frac{\sqrt{dx^2 + dy^2}}{y} = \int_{\theta_1}^{\theta_2} \frac{\sqrt{(\rho'(\theta)\cos\theta - \rho(\theta)\sin\theta)^2 + (\rho'(\theta)\sin\theta + \rho(\theta)\cos\theta)^2}}{\rho(\theta)\sin\theta}\, d\theta$$
$$= \int_{\theta_1}^{\theta_2} \sqrt{1 + \left(\frac{\rho'(\theta)}{\rho(\theta)}\right)^2}\, \frac{d\theta}{\sin\theta} \geq \int_{\theta_1}^{\theta_2} \frac{d\theta}{\sin\theta} = \log \frac{\tan\frac{1}{2}\theta_2}{\tan\frac{1}{2}\theta_1}.$$

There is equality only when $\rho'(\theta) = 0$, that is when $\rho(\theta) = \rho$ is constant, and C becomes the circular arc, with centre on the x-axis, joining z_1, z_2. □

As a by-product of the proof, we find that the integral has the value

$$\log \frac{\tan\frac{1}{2}\theta_2}{\tan\frac{1}{2}\theta_1}$$

when C is part of the geodesic. The following is the geometric interpretation of this value: Let the circle through z_1, z_2, with centre E on the x-axis, cut the axis at A, B (see Fig. 2.2), so that

$$\tan\tfrac{1}{2}\theta_1 = \frac{Bz_1}{z_1A}, \quad \tan\tfrac{1}{2}\theta_2 = \frac{Bz_2}{z_2A}.$$

Thus

$$\log \frac{\tan\frac{1}{2}\theta_2}{\tan\frac{1}{2}\theta_1} = \log|(B, A, z_2, z_1)|.$$

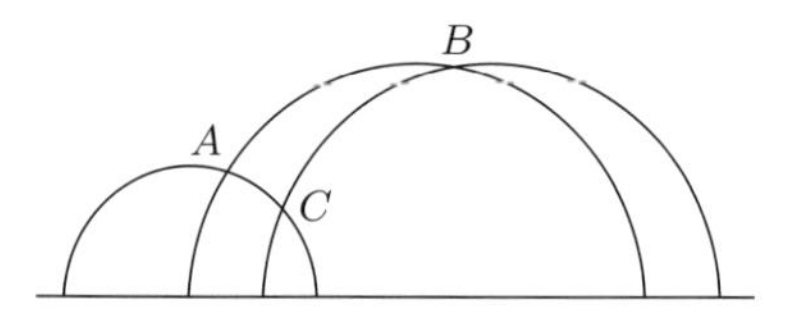

Figure 2.3.

It is not difficult to compute the following:

$$|(B, A, z_2, z_1)|^2 = \frac{1 + |z_1 - z_2|/|\bar{z}_1 - z_2|}{1 - |z_1 - z_2|/|\bar{z}_1 - z_2|} = \frac{1 + h^{1/2}(z_1, z_2)}{1 + h^{1/2}(z_1, z_2)}.$$

Therefore we may define

$$D(z_1, z_2) = \frac{1}{2} \log \frac{|\bar{z}_1 - z_2| + |z_1 - z_2|}{|\bar{z}_1 - z_2| - |z_1 - z_2|}, \quad \Im(z_1) > 0, \ \Im(z_2) > 0,$$

to be the non-Euclidean distance between z_1 and z_2.

The corresponding situation on the unit disc is

$$D(z_1, z_2) = \frac{1}{2} \log \frac{|1 - \bar{z}_1 z_2| + |z_1 - z_2|}{|1 - \bar{z}_1 z_2| - |z_1 - z_2|}, \quad |z_1| < 1, \ |z_2| < 1.$$

For such a definition we find that if

$$D(z_1, z_2) = 0$$

then

$$\frac{|1 - \bar{z}_1 z_2| + |z_1 - z_2|}{|1 - \bar{z}_1 z_2| - |z_1 - z_2|} = 1,$$

so that $|z_2 - z_1| = 0$, and hence $z_1 = z_2$. Consequently we have the following properties for the distance function so defined:

(i) A necessary and sufficient condition for $D(z_1, z_2) = 0$ is that $z_1 = z_2$.
(ii) $D(z_1, z_2) \geq 0$, which follows from $|1 - \bar{z}_1 z_2| + |z_1 - z_2| \geq |1 - \bar{z}_1 z_2| - |z_1 - z_2|$.
(iii) $D(z_1, z_2) = D(z_2, z_1)$.
(iv) $D(z_1, z_3) \leq D(z_1, z_2) + D(z_2, z_3)$, with equality if and only if z_1, z_2, z_3 lie on the same geodesic.

In (iv) we may say that the triangle inequality is valid for a non-Euclidean triangle – this can be deduced from the minimal property established in Theorem 2.4.

2.6 Triangles

By a triangle we mean the shape formed from the three geodesics shown in Fig. 2.3. It is not difficult to deduce various results such as 'side–angle–side implies congruence',

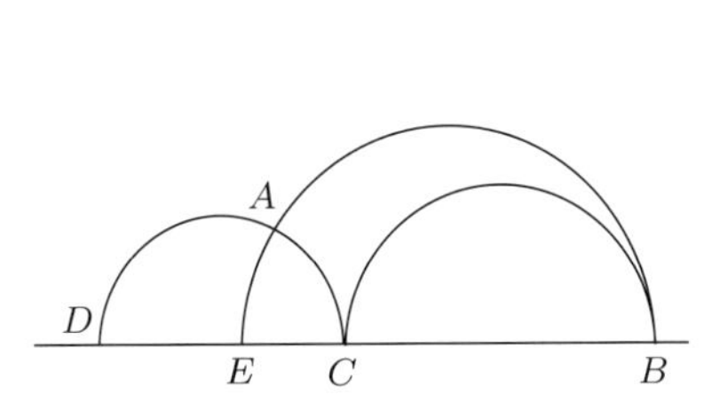

Figure 2.4.

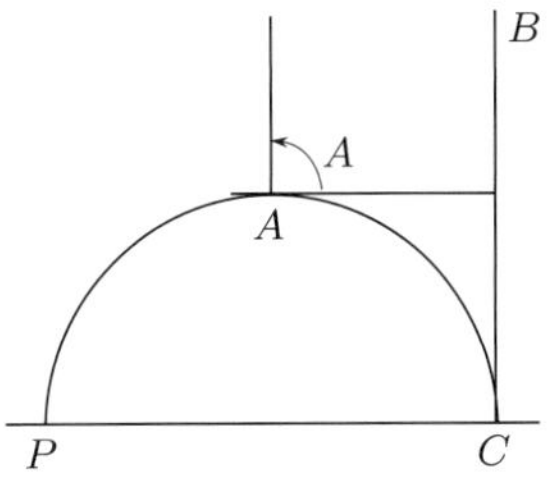

B
A
A
P
C

Figure 2.5.

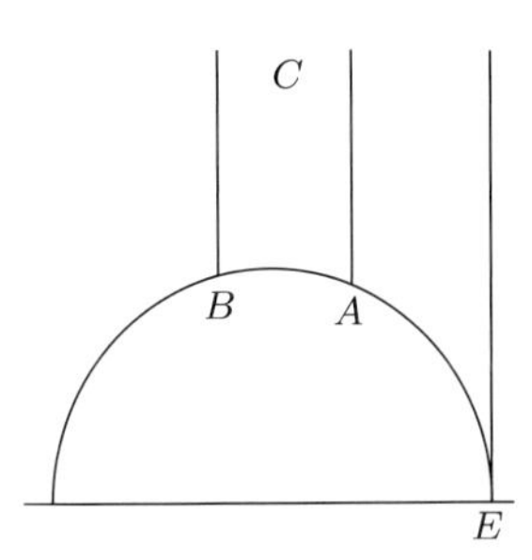

Figure 2.6.

'angle–side–angle implies congruence', 'three equal sides implies congruence', 'large angles correspond to long sides', etc. Let us now determine the area of a triangle.

Theorem 2.5 *The non-Euclidean area of a triangle ABC is $\pi - \angle A - \angle B - \angle C$.*

Proof. With the upper half-plane as the base, the arca is given by

$$\iint \frac{dx\,dy}{y^2}.$$

(1) Consider first the case $\angle B = \angle C = 0$ (see Fig. 2.4). It is not difficult to show that there is a real transformation mapping B, C, D to $\infty, 1, -1$ (or $\infty, -1, 1$), and that the corresponding determinant is positive. In fact,

$$z' = \pm\frac{(D-2B+C)z + (BC - 2DC + DB)}{(C-D)z + (D-C)B}$$

is such a transformation with the determinant having the value $\pm(D-C)(C-B)(B-D)$, and after such a transformation Fig. 2.4 becomes Fig. 2.5. If A has coordinates (x_0, y_0) then

$$\int_{x_0}^{1}\int_{\sqrt{1-x^2}}^{\infty} \frac{dx\,dy}{y^2} = \int_{x_0}^{1} \frac{dx}{\sqrt{1-x^2}} = \arcsin x\Big|_{x_0}^{1} = \frac{\pi}{2} - \arcsin x_0 = \pi - \angle A.$$

(2) Suppose only that $\angle C = 0$. Use a real transformation to map C to ∞ to give Fig. 2.6. From (1), we have

$$\triangle ABC = \triangle BEC - \triangle AEC = \pi - \angle B - (\pi - (\pi - \angle A)) = \pi - \angle A - \angle B.$$

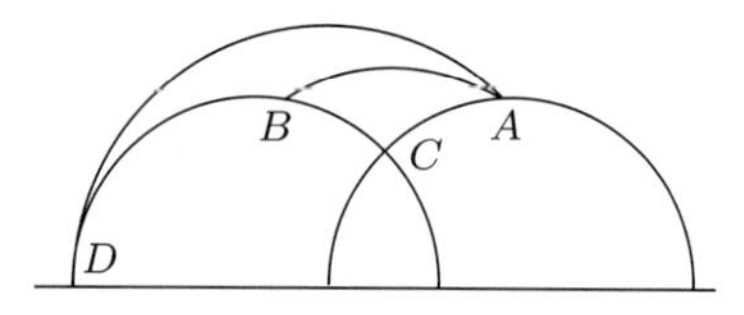

Figure 2.7.

(3) If none of $\angle A, \angle B, \angle C$ is 0, as in Fig. 2.7, say, then from (2) we have

$$\begin{aligned}\triangle ABC &= \triangle ADC - \triangle ABD \\ &= (\pi - \angle C - \angle A - \angle BAD) - (\pi - (\pi - \angle B) - \angle BAD) \\ &= \pi - \angle A - \angle B - \angle C.\end{aligned}$$

The theorem is proved. □

From this theorem we deduce at once the following.

Theorem 2.6 *The sum of the angles in a triangle does not exceed two right-angles; its value may take any value between* 0 *and* π.

2.7 Axiom of parallels

We summarise the fundamental differences between the three types of geometries.

Euclidean geometry (parabolic). Points on the plane (excluding the point at infinity) are our geometric objects. The group of motions is generated from $z \mapsto e^{i\theta}z + k$ and $z \mapsto \bar{z}$. Geodesics are straight lines. Given a point not on a given geodesic in the plane there is one (and only one) geodesic through the given point which does not intersect the given geodesic (Euclid's eleventh postulate[1]).

Riemannian geometry (elliptic). Points on the surface of a sphere are our geometric objects. The group of motions comprise rotation of the sphere and reflection with respect to planes through the centre of the sphere. Geodesics are great circles. Given a point not on a given geodesic, there is no geodesic through the given point which does not intersect the given geodesic.

Lobachevskian geometry (hyperbolic). Points on the unit disc are our geometric objects. The group of motions is generated from

$$z \mapsto e^{i\theta}\frac{z-a}{1-\bar{a}z}$$

and $z \mapsto \bar{z}$. Geodesics are (part of) circles intersecting orthogonally with the unit circle. There are infinitely many geodesics through a point not intersecting a given geodesic.

[1] Translator's note: Although Schopenhauer had written about Euclid's 'eleventh axiom', it is generally known as Euclid's 'fifth postulate'.

When the radius of the sphere, or the circle, is sufficiently large, say so large that our available instruments are not accurate enough for the purpose of detecting or determining the curvature of a large circle, there is no method for us to choose the appropriate geometry in order to study the relevant space properly. In ancient times we applied the notion that 'the Earth is just like a Go board[2]'. It was wrong to think that parabolic geometry could be applied to the surface of the spherical Earth. When considering space to be infinite or unbounded as a non-Euclidean entity in which the sun, the moon and the stars play only a small part, we suddenly come to realise as fact that geodesics may have curvatures.

2.8 Types of non-Euclidean motions

We now examine the types of non-Euclidean motions

$$w = e^{i\theta}\frac{z-a}{1-\bar{a}z}. \tag{2.22}$$

In elliptic geometry we have already seen that transformations can only be elliptic. In parabolic geometry, they can be parabolic or they can be elliptic. In hyperbolic geometry, the problem is even more complicated, because there can be hyperbolic, parabolic and also elliptic transformations.

(1) There is a fixed point in the circle. If (2.22) maps 0 to 0, then $w = e^{i\theta}z$, an elliptic transformation. Therefore

$$\frac{w-a}{1-\bar{a}w} = e^{i\theta}\frac{z-a}{1-\bar{a}z} \tag{2.23}$$

is a non-Euclidean transformation with the fixed point a and none other (inside the circle). Therefore a non-Euclidean motion with a fixed point inside the circle is elliptic, taking the form (2.23). Such a transformation is called a non-Euclidean rotation. If there is a fixed point b outside the circle, then $a = 1/\bar{b}$ is a fixed point inside the circle, so that the transformation is still elliptic. Therefore, henceforth we need only consider the situation when the fixed points are on the circumference of the circle.

(2) There are two fixed points on the circumference of the circle. We take the upper half-plane as our model instead, so that the fixed points are now on the x-axis. If they are 0 and ∞ then we have $w = kz$, $k > 0$, a hyperbolic transformation. If α, β are any two fixed points, then

$$\frac{w-\alpha}{w-\beta} = k\frac{z-\alpha}{z-\beta}$$

is the hyperbolic transformation leaving α, β fixed. To summarise, if there are two fixed points on the circle, then we have a hyperbolic transformation.

[2] Translator's note: The ancient game of Go is played on a board with a 19×19 grid.

(3) There is only one fixed point. Let the upper half-plane remain as our model. If the point concerned is ∞, then the transformation has to be $w = z + \lambda$ (λ real), and the general transformation has the form

$$\frac{1}{w-a} = \frac{1}{z-a} + \lambda.$$

This is a parabolic transformation.

3

Definitions and examples of analytic and harmonic functions

3.1 Complex functions

A complex function is a function of a complex variable z, with the value $w = f(z)$ also being complex. The domain for the function is a set M on the complex plane – there is a corresponding value w for each $z \in M$, and we say that $f(z)$ is a single-valued function. The set N of values w corresponding to $z \in M$ is called the range of the function.

Let $z = x + iy$ and $w = u + iv$. In fact, what we mean by a complex function $w = f(z)$ is really a pair of real functions of the real variables x, y:

$$u = u(x, y), \qquad v = v(x, y).$$

Letting z and w be placed on separate complex planes, we can consider a complex function to be the mapping of a point $(x, y) \in M$ to its image $(u, v) \in N$. If a single-valued function $w = f(z)$ has the additional property that different $z \in M$ produce different images $w \in N$, then we say that the function is injective, or that it is univalent. Such a function sets up a one-to-one correspondence, or a bijection, between the domain M and the range N. In this case we may define $z = \phi(w)$, which maps each $w \in N$ back to $z \in M$, where z satisfies $w = f(z)$; we call $\phi(w)$ the inverse function of $f(z)$.

If $w = f(z)$ maps M to N, and $w' = g(w)$ maps N to P, then

$$w' = h(z) = g\big(f(z)\big)$$

maps M to P, and we call h the composite function of f, g. In particular, when $w = f(z)$ is a bijection between M and N, its inverse function $z = \phi(w)$ satisfies

$$\phi\big(f(z)\big) = z.$$

Example 3.1 The bilinear transformations

$$w = \frac{az + b}{cz + d}, \qquad ad - bc \neq 0,$$

that we studied in Chapters 1 and 2 are complex functions (with M, N being the Gaussian plane).

Example 3.2 The function $w = z^n$ maps the interior of the unit disc onto itself, but it is not a bijection. This is because, for integers ℓ, all the points

$$r^{1/n} e^{(2\pi i\ell/n)+(i\theta/n)}$$

have the same image $re^{i\theta}$.

3.2 Conformal transformations

A complex function $w = f(z)$ can be considered as the mapping of a point (x, y) from its domain M to its image (u, v) in the range N, with

$$u = u(x, y), \quad v = v(x, y). \tag{3.1}$$

Suppose that u, v are differentiable functions of x, y in M. The relationship between the differential vectors (du, dv) and (dx, dy) is

$$\begin{cases} du = \dfrac{\partial u}{\partial x} dx + \dfrac{\partial u}{\partial y} dy, \\ dv = \dfrac{\partial v}{\partial x} dx + \dfrac{\partial v}{\partial y} dy. \end{cases} \tag{3.2}$$

Take another differential vector (d_1u, d_1v) corresponding to (d_1x, d_1y). The cosine of the angle between the vectors (du, dv) and (d_1u, d_1v) is given by

$$\begin{aligned} &\frac{du\, d_1u + dv\, d_1v}{\sqrt{(du^2 + dv^2)(d_1u^2 + d_1v^2)}} \\ &\quad = \frac{E\, dx\, d_1x + F(dx\, d_1y + d_1x\, dy) + G\, dy\, d_1y}{\sqrt{(E\, dx^2 + 2F dx\, dy + G\, dy^2)(E\, d_1x^2 + 2F d_1x\, d_1y + G\, d_1y^2)}}, \end{aligned} \tag{3.3}$$

where

$$E = \left(\frac{\partial u}{\partial x}\right)^2 + \left(\frac{\partial v}{\partial x}\right)^2, \quad F = \frac{\partial u}{\partial x}\frac{\partial u}{\partial y} + \frac{\partial v}{\partial x}\frac{\partial v}{\partial y}, \quad G = \left(\frac{\partial u}{\partial y}\right)^2 + \left(\frac{\partial v}{\partial y}\right)^2.$$

If, within a certain region, the magnitude of this angle is the same as that between (dx, dy) and (d_1x, d_1y), then taking (in particular) separately $dx = d_1y = 0$ and $dx = d_1x$, $dy = -d_1y$, we have the condition

$$F = 0, \quad E = G. \tag{3.4}$$

From $F = 0$ we have

$$\frac{\partial u}{\partial x}\frac{\partial u}{\partial y} = -\frac{\partial v}{\partial x}\frac{\partial v}{\partial y},$$

so that we may write

$$\frac{\partial u}{\partial x} = \phi\frac{\partial v}{\partial y}, \quad \frac{\partial v}{\partial x} = -\phi\frac{\partial u}{\partial y}.$$

Substituting these into $E = G$, we have

$$\left(\frac{\partial u}{\partial x}\right)^2 + \left(\frac{\partial v}{\partial x}\right)^2 = \phi^2\left(\left(\frac{\partial u}{\partial y}\right)^2 + \left(\frac{\partial v}{\partial y}\right)^2\right) = \left(\frac{\partial u}{\partial y}\right)^2 + \left(\frac{\partial v}{\partial y}\right)^2$$

so that $\phi^2 = 1$, and hence $\phi = \pm 1$. If we further require that the orientation of the angles is preserved then we need to have $\phi = 1$. We then arrive at the famous Cauchy–Riemann equations

$$\frac{\partial u}{\partial x} = \frac{\partial v}{\partial y}, \quad \frac{\partial v}{\partial x} = -\frac{\partial u}{\partial y}. \tag{3.5}$$

When two curves intersect, their angle of intersection is that between the two tangents at the point of intersection. Thus, if the mapping (3.1) is such that, for any two intersecting curves, the angle between them is unaltered by the transformation, then (3.5) must hold. Conversely, if a mapping (3.1) satisfies (3.5) in some region, then there is preservation of the magnitude and orientation of local angles (within the region).

A transformation which preserves the magnitude of local angles is called an isogonal mapping. A conformal transformation is an isogonal mapping that also preserves the orientation of local angles. However, we need to discard the points at which $E = F = G = 0$ in the above; in this case the right-hand side of (3.3) becomes $0/0$, and being conformal makes no sense there.

For example, let $u = x^2 - y^2$, $v = 2xy$. Then the transformation is not conformal at the origin, although it is so everywhere else.

If the transformation (3.1) satisfies (3.5) then

$$du^2 + dv^2 = \left[\left(\frac{\partial u}{\partial x}\right)^2 + \left(\frac{\partial v}{\partial x}\right)^2\right](dx^2 + dy^2). \tag{3.6}$$

This tells us that, in the neighbourhood of a point at which the transformation is conformal, an infinitesimal circle is mapped into another such circle. Naturally we also need to take note of the points at which

$$\frac{\partial u}{\partial x} = \frac{\partial v}{\partial x} = \frac{\partial u}{\partial y} = \frac{\partial v}{\partial y} = 0,$$

where the conclusion does not follow.

The transformations discussed in Chapters 1 and 2 are all isogonal, and the bilinear ones are conformal; indeed for these transformations the properties associated with ordinary circles are preserved.

From the Cauchy–Riemann equations we deduce (assuming that u, v have continuous second partial derivatives in a certain region D) that

$$\frac{\partial^2 u}{\partial x^2} + \frac{\partial^2 u}{\partial y^2} = \frac{\partial^2 v}{\partial x \partial y} - \frac{\partial^2 v}{\partial y \partial x} = 0.$$

Similarly we have

$$\frac{\partial^2 v}{\partial x^2} + \frac{\partial^2 v}{\partial y^2} = 0.$$

The operator $\Delta = \partial^2/\partial x^2 + \partial^2/\partial y^2$ is called the Laplacian operator, and the equation

$$\Delta u = \frac{\partial^2 u}{\partial x^2} + \frac{\partial^2 u}{\partial y^2} = 0$$

is called Laplace's equation. Functions u and v in (3.1), defined by a transformation which is conformal in a region D, are solutions to Laplace's equation.

Functions satisfying Laplace's equation in a region D are said to be harmonic in D, and the pair of functions u, v are called conjugate harmonic functions; clearly $-u$ is also a conjugate function to v.

We now consider the problem of the determination of a harmonic function which is a conjugate to a given harmonic function $u(x, y)$.

Let $u(x, y)$ be a harmonic function in a simply connected region D. Fix a point (x_0, y_0) in D and define, for (x, y) in D,

$$v(x, y) = \int_{(x_0, y_0)}^{(x,y)} \Big(-\frac{\partial u}{\partial y}\, dx + \frac{\partial u}{\partial x}\, dy \Big). \tag{3.7}$$

Then Laplace's equation, in the form

$$\frac{\partial}{\partial x}\Big(\frac{\partial u}{\partial x}\Big) = \frac{\partial}{\partial y}\Big(-\frac{\partial u}{\partial y}\Big),$$

is the condition for the integrand to be an exact differential, so that the integral in (3.7) is independent of the path leading from (x_0, y_0) to (x, y).

Now we have

$$\begin{aligned} \frac{\partial v}{\partial x} &= \lim_{h\to 0} \frac{v(x+h, y) - v(x, y)}{h} = \lim_{h\to 0} \int_{(x,y)}^{(x+h,y)} \Big(-\frac{\partial u}{\partial y}\, dx + \frac{\partial u}{\partial x}\, dy \Big) \\ &= \lim_{h\to 0} \frac{1}{h} \int_x^{x+h} -\frac{\partial u}{\partial y}\, dx = -\frac{\partial u}{\partial y}, \end{aligned}$$

and similarly $\partial v/\partial y = \partial u/\partial x$, so that $v(x, y)$ and $u(x, y)$ are conjugate functions. However, because $v(x, y)$ is determined from its partial derivatives, we need to add a constant for a general solution; that is, the general conjugate function to $u(x, y)$ is given by

$$v(x, y) = \int_{(x_0, y_0)}^{(x,y)} \Big(-\frac{\partial u}{\partial y}\, dx + \frac{\partial u}{\partial x}\, dy \Big) + C,$$

where C is a real number.

We next examine the situation when D is multiply connected. Here, the integral given above is not necessarily single-valued, because different paths may now give different values. If the two paths $L, \tilde{L}$ can be continuously deformed from one to the other within D,

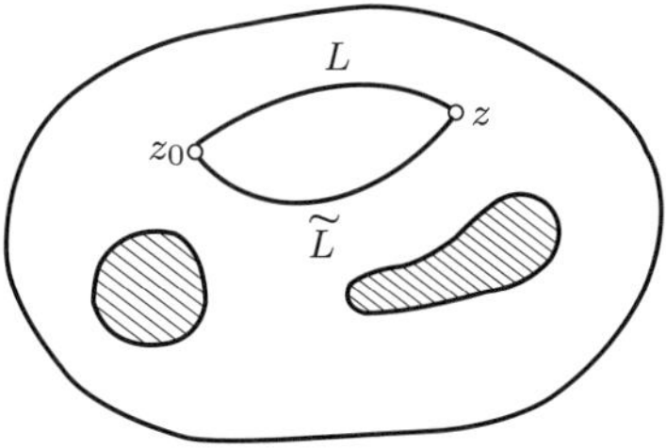

Figure 3.1.

Figure 3.2.

then the values for the integral are still the same (see Fig. 3.1). Otherwise the values may be different. Take, for example,

$$u = \log(x^2 + y^2),$$

which is harmonic in the annulus region $\rho_0^2 \leq x^2 + y^2 \leq R_0^2$. The annulus is connected, but not simply so. The conjugate function

$$v(x, y) = 2\int_{z_0}^{z} \frac{-y\,dx + x\,dy}{x^2 + y^2} + C = 2\arg z + C_2$$

has an unbounded number of values at each point inside the annulus. We have

$$\Gamma_1 = \int_{x^2+y^2=\rho^2} \frac{-y\,dx + x\,dy}{x^2 + y^2} = 2\pi,$$

and any curve which loops round $z = 0$ in the positive direction can be continuously deformed to the circle $x^2 + y^2 = \rho^2$ without moving outside the annulus. Thus, any path C from z_0 to z can be continuously deformed into another path $\tilde{C}$ from z_0 to z which loops round $z = 0$. We can construct $\tilde{C}$ so that it is made up of parts from (i) a curve $\tilde{C}_0$ which does not intersect itself, (ii) a curve looping along $|z| = \rho$ in a positive or negative direction N times (taking N to be negative when the direction is negative). For such a curve we find that

$$v(x, y) = 2\arg z + 2\pi N.$$

Returning to the general problem, suppose that there are m 'islands' $\gamma_1, \ldots, \gamma_m$ within the region D (see Fig. 3.2), and let

$$\Gamma_h = \int_{\bar{\gamma}_h} \frac{-y\,dx + x\,dy}{x^2 + y^2},$$

where $\bar{\gamma}_h$ is a closed loop round γ within D. Any path C leading from z_0 to z can be changed to $\tilde{C}$, which is made up of parts from (i) a simple (non-intersecting) curve $\tilde{C}_0$ leading from z_0 to z, (ii) loops round γ_h, say N_h times positively or negatively. We then have

$$v(x, y) = \int_{z_0:\tilde{C}}^{z} = \int_{z_0:\tilde{C}_0}^{z} + N_1\Gamma_1 + \cdots + N_m\Gamma_m,$$

where $N_1, \ldots, N_m$ are integers and the constants Γ_h are called the periods associated with γ_h. In this situation $v(x, y)$ is a multi-valued function. Nevertheless, its partial derivatives are single-valued, and they satisfy the Cauchy–Riemann equations.

3.3 Cauchy–Riemann equations

The Cauchy–Riemann equations derived in Section 3.2,

$$\frac{\partial u}{\partial x} = \frac{\partial v}{\partial y}, \quad \frac{\partial u}{\partial y} = -\frac{\partial v}{\partial x}, \tag{3.8}$$

are also known as the Euler–D'Alembert equations. Suppose that $u(x, y)$, $v(x, y)$ are partially differentiable functions in a region D in which (3.8) holds. Using polar coordinates, we have

$$\frac{\partial u}{\partial x} = \frac{\partial u}{\partial \theta}\frac{\partial \theta}{\partial x} + \frac{\partial u}{\partial \rho}\frac{\partial \rho}{\partial x} = -\frac{\sin\theta}{\rho}\frac{\partial u}{\partial \theta} + \cos\theta\frac{\partial u}{\partial \rho},$$
$$\frac{\partial u}{\partial y} = \frac{\partial u}{\partial \theta}\frac{\partial \theta}{\partial y} + \frac{\partial u}{\partial \rho}\frac{\partial \rho}{\partial y} = \frac{\cos\theta}{\rho}\frac{\partial u}{\partial \theta} + \sin\theta\frac{\partial u}{\partial \rho},$$

so that

$$-\frac{\sin\theta}{\rho}\frac{\partial u}{\partial \theta} + \cos\theta\frac{\partial u}{\partial \rho} = \frac{\cos\theta}{\rho}\frac{\partial v}{\partial \theta} + \sin\theta\frac{\partial v}{\partial \rho},$$
$$\frac{\cos\theta}{\rho}\frac{\partial u}{\partial \theta} + \sin\theta\frac{\partial u}{\partial \rho} = \frac{\sin\theta}{\rho}\frac{\partial v}{\partial \theta} - \cos\theta\frac{\partial v}{\partial \rho},$$

giving

$$\rho\frac{\partial u}{\partial \rho} = \frac{\partial v}{\partial \theta}, \quad \rho\frac{\partial v}{\partial \rho} = -\frac{\partial u}{\partial \theta}. \tag{3.9}$$

Also, Laplace's equation becomes

$$\frac{\partial^2 u}{\partial \theta^2} + \rho\frac{\partial}{\partial \rho}\left(\rho\frac{\partial u}{\partial \rho}\right) = 0, \tag{3.10}$$

that is

$$\rho^2\frac{\partial^2 u}{\partial \rho^2} + \rho\frac{\partial u}{\partial \rho} + \frac{\partial^2 u}{\partial \theta^2} = 0.$$

When applying (3.9) and (3.10) we need to note the special case $\rho = 0$.

Being harmonic is a linear property in the sense that if u, v are harmonic functions then so is $\alpha u + \beta v$ for any constants α, β.

Example 3.3 Separate the real and the imaginary parts of z^n to give

$$u = \rho^n \cos n\theta, \quad v = \rho^n \sin n\theta.$$

They satisfy (3.9), that is

$$\rho\frac{\partial u}{\partial \rho} = n\rho^n \cos n\theta = \frac{\partial v}{\partial \theta}, \quad \rho\frac{\partial v}{\partial \rho} = n\rho^n \sin n\theta = -\frac{\partial u}{\partial \theta}.$$

They are conjugate harmonic functions. We need to note that, as a transformation, it is not conformal at $\rho = 0$.

Example 3.4 From linearity, the real and imaginary parts of a polynomial also form a pair of conjugate harmonic functions.

Example 3.5 Supposc that the series

$$f(z) = \sum_{n=0}^{\infty} a_n z^n, \quad a_n = \alpha_n + i\beta_n,$$

is convergent in $|z| < R$. Then the real and the imaginary parts

$$u = \sum_{n=0}^{\infty} (\alpha_n \cos n\theta - \beta_n \sin n\theta)\rho^n,$$
$$v = \sum_{n=0}^{\infty} (\alpha_n \sin n\theta + \beta_n \cos n\theta)\rho^n$$

also form a pair of conjugate harmonic functions.

In particular, from $e^z = e^{\rho\cos\theta} \cdot e^{i\rho\sin\theta}$, we see that

$$e^{\rho\cos\theta}\cos(\rho\sin\theta) = \sum_{n=0}^{\infty} \frac{\rho^n \cos n\theta}{n!},$$
$$e^{\rho\cos\theta}\sin(\rho\sin\theta) = \sum_{n=0}^{\infty} \frac{\rho^n \sin n\theta}{n!}$$

form a pair of conjugate harmonic functions.

Again, for

$$\log z = \log|z| + i \arg z,$$

the function $u = \log|z|$ should be harmonic, and indeed we find that

$$\frac{\partial u}{\partial x} = \frac{x}{x^2+y^2}, \quad \frac{\partial^2 u}{\partial x^2} = \frac{y^2 - x^2}{(x^2+y^2)^2}, \quad \frac{\partial^2 u}{\partial x^2} + \frac{\partial^2 u}{\partial y^2} = 0.$$

The conjugate harmonic partner is the function $v = \arg z$.

Example 3.6 For positive integers n, the functions

$$u = \rho^{-n} \cos n\theta, \quad v = \rho^{-n} \sin n\theta$$

also form a conjugate harmonic pair (but the point $\rho = 0$ has to be excluded).

Example 3.7 Suppose that the series

$$\sum_{n=-\infty}^{\infty} a_n z^n, \quad a_n = \alpha_n + i\beta_n$$

converges in the annulus region $r < |z| < R$. Then

$$u = \sum_{n=-\infty}^{\infty} (\alpha_n \cos n\theta - \beta_n \sin n\theta)\rho^n,$$

$$v = \sum_{n=-\infty}^{\infty} (\alpha_n \sin n\theta + \beta_n \cos n\theta)\rho^n$$

form a pair of conjugate harmonic functions.

Suppose that

$$x = x(x_1, y_1), \quad y = y(x_1, y_1),$$

is a pair of conjugate functions. Then u, v, as functions of x_1, y_1, are also conjugate functions; thus

$$\frac{\partial u}{\partial x_1} = \frac{\partial u}{\partial x}\frac{\partial x}{\partial x_1} + \frac{\partial u}{\partial y}\frac{\partial y}{\partial x_1} = \frac{\partial v}{\partial y}\frac{\partial y}{\partial y_1} + \frac{\partial v}{\partial x}\frac{\partial x}{\partial y_1} = \frac{\partial v}{\partial y_1},$$

$$\frac{\partial u}{\partial y_1} = \frac{\partial u}{\partial x}\frac{\partial x}{\partial y_1} + \frac{\partial u}{\partial y}\frac{\partial y}{\partial y_1} = \frac{\partial v}{\partial y}\frac{\partial y}{\partial x_1} - \frac{\partial v}{\partial x}\frac{\partial x}{\partial x_1} = -\frac{\partial v}{\partial x_1}.$$

The relationship between conjugate pairs is preserved under a conformal transformation, or we may say that the product of two conformal transformations is still conformal. For Laplace's equation under a conformal transformation, we have

$$\frac{\partial^2 u}{\partial x_1^2} = \frac{\partial^2 u}{\partial x^2}\left(\frac{\partial x}{\partial x_1}\right)^2 + 2\frac{\partial^2 u}{\partial x \partial y}\left(\frac{\partial x}{\partial x_1}\right)\left(\frac{\partial y}{\partial x_1}\right) + \frac{\partial^2 u}{\partial y^2}\left(\frac{\partial y}{\partial x_1}\right)^2 + \frac{\partial u}{\partial x}\frac{\partial^2 x}{\partial x_1^2} + \frac{\partial u}{\partial y}\frac{\partial^2 y}{\partial x_1^2},$$

$$\frac{\partial^2 u}{\partial y_1^2} = \frac{\partial^2 u}{\partial x^2}\left(\frac{\partial x}{\partial y_1}\right)^2 + 2\frac{\partial^2 u}{\partial x \partial y}\left(\frac{\partial x}{\partial y_1}\right)\left(\frac{\partial y}{\partial y_1}\right) + \frac{\partial^2 u}{\partial y^2}\left(\frac{\partial y}{\partial y_1}\right)^2 + \frac{\partial u}{\partial x}\frac{\partial^2 x}{\partial y_1^2} + \frac{\partial u}{\partial y}\frac{\partial^2 y}{\partial y_1^2},$$

and on summing we obtain

$$\frac{\partial^2 u}{\partial x_1^2} + \frac{\partial^2 u}{\partial y_1^2} = \left(\left(\frac{\partial x}{\partial x_1}\right)^2 + \left(\frac{\partial x}{\partial y_1}\right)^2\right)\left(\frac{\partial^2 u}{\partial x^2} + \frac{\partial^2 u}{\partial y^2}\right).$$

Therefore being harmonic is also an invariant property under a conformal transformation.

In all such discussions we must take note of the region in which the functions are defined and operations are to be applied.

Conditions (3.8) and (3.9) are special cases of the following condition:

$$\frac{\partial u}{\partial s} = \frac{\partial v}{\partial n}, \quad \frac{\partial u}{\partial n} = -\frac{\partial v}{\partial s}.$$

In other words, for a curve C in the region of definition for a conformal transformation $u + iv$, the partial derivative of u along the tangent to the curve is equal to that of v along the normal, and that of u along the normal is the negative of that for v along the tangent. The proof is rather easy: We have

$$\frac{\partial u}{\partial n} = \frac{\partial u}{\partial x}\left(-\frac{dy}{ds}\right) + \frac{\partial u}{\partial y}\left(\frac{dx}{ds}\right) = -\frac{\partial v}{\partial y}\frac{dy}{ds} - \frac{\partial v}{\partial x}\frac{dx}{ds} = -\frac{\partial v}{\partial s}$$

and

$$\frac{\partial v}{\partial n} = \frac{\partial v}{\partial x}\left(-\frac{dy}{ds}\right) + \frac{\partial v}{\partial y}\left(\frac{dx}{ds}\right) = +\frac{\partial u}{\partial y}\frac{dy}{ds} + \frac{\partial u}{\partial x}\frac{dx}{ds} = \frac{\partial u}{\partial s}.$$

Note that, having chosen the direction of the tangent, the direction of the normal should be taken along a positive 90° turn. If C is the straight line $y = k$ parallel to the x-axis, then we have (3.8). If C is the circle $x^2 + y^2 = R^2$ then $\partial/\partial n = -\partial/\partial\rho$, $\partial/\partial s = (1/\rho)(\partial/\partial\theta)$, which is (3.9).

3.4 Analytic functions

Consider the complex function

$$f(z) = u(x, y) + iv(x, y),$$

where the pair of functions $u(x, y)$, $v(x, y)$ satisfy the Cauchy–Riemann equations in a region D. This function has the special property that the limit

$$\lim_{h\to 0} \frac{f(z+h) - f(z)}{h}$$

is always the same, regardless of the path h may take in approaching 0.

Let $h = s + it$. Then

$$u(x+s, y+t) - u(x, y) = \frac{\partial u}{\partial x}s + \frac{\partial u}{\partial y}t + o(|h|),$$
$$v(x+s, y+t) - v(x, y) = \frac{\partial v}{\partial x}s + \frac{\partial v}{\partial y}t + o(|h|),$$

so that

$$\begin{aligned} f(z+h) - f(z) &= \frac{\partial u}{\partial x}s + \frac{\partial u}{\partial y}t + i\left(\frac{\partial v}{\partial x}s + \frac{\partial v}{\partial y}t\right) + o(|h|), \\ &= \left(\frac{\partial u}{\partial x} + i\frac{\partial v}{\partial y}\right)(s + it) + o(|h|), \end{aligned}$$

and therefore

$$\lim_{h\to 0} \frac{f(z+h) - f(z)}{h} = \frac{\partial u}{\partial x} + i\frac{\partial v}{\partial y}$$

is a function which is independent of h.

Definition 3.8 If, at a point z in a region D, the limit

$$\lim_{h\to 0}\frac{f(z+h)-f(z)}{h}=f'(z)$$

exists as a unique number, then we say that $f(z)$ is differentiable. If $f(z)$ is differentiable throughout a neighbourhood of z then we say that $f(z)$ is analytic there.

We remind ourselves that, by the existence of the limit, we mean the following: For every $\epsilon > 0$ we can find $\delta > 0$ with the property that, for any h with $0 < |h| < \delta$, we have

$$\left|\frac{f(z+h)-f(z)}{h}-f'(z)\right|<\epsilon.$$

When we say that $f(z)$ is analytic in D we mean that $f(z)$ is analytic at every point of D. The sum, the difference and the product of two analytic functions in D are naturally still analytic in D; the quotient is also analytic provided there is no division by 0 within D.

It is also easy to see that if $g(z)$ is an analytic function in D, and $f(w)$ is an analytic function of w in a region Δ which is the range of $g(z)$, then the composite function $f\big(g(z)\big)$ is also analytic in D. If u, v are differentiable at z_0 and satisfy the Cauchy–Riemann equations, then $u+iv$ is analytic at z_0; conversely, if a function is analytic at z then its real and imaginary parts satisfy the Cauchy–Riemann condition and form a pair of conjugate harmonic functions. The proof is very easy: For real s and t, we have

$$\lim_{s\to 0}\frac{f(z+s)-f(z)}{s}=\lim_{s\to 0}\frac{f(z+it)-f(z)}{i}t,$$

so that

$$\frac{\partial u}{\partial x}+i\frac{\partial v}{\partial x}=-i\frac{\partial u}{\partial y}+\frac{\partial v}{\partial y},$$

and hence we obtain the required result.

If $f(z)$ is analytic then

$$\frac{df(z)}{dz}=\frac{\partial f(z)}{\partial x}=\frac{1}{i}\frac{\partial f(z)}{\partial y},$$

so that

$$\frac{df(z)}{dz}=\frac{1}{2}\Big(\frac{\partial}{\partial x}+\frac{1}{i}\frac{\partial}{\partial y}\Big)f(z).$$

Let us introduce the symbols

$$\frac{\partial}{\partial z}=\frac{1}{2}\Big(\frac{\partial}{\partial x}+\frac{1}{i}\frac{\partial}{\partial y}\Big),\quad \frac{\partial}{\partial \bar{z}}=\frac{1}{2}\Big(\frac{\partial}{\partial x}-\frac{1}{i}\frac{\partial}{\partial y}\Big),\tag{3.11}$$

so that we have

$$\frac{df(z)}{dz}=\frac{\partial f(z)}{\partial z},\quad \frac{\partial}{\partial \bar{z}}f(z)=\frac{1}{2}\big(f'(z)-f'(z)\big)=0.$$

Thus, if $f(z)$ is analytic then

$$\frac{\partial}{\partial \bar{z}} f(z) = 0, \tag{3.12}$$

and the converse also holds.

We may view such a result in the following way: Let $\zeta = x + iy$, $\eta = x - iy$, so that $f(z)$ is a function of ζ, and also a function of η, and if it is analytic as a function of z then it is independent of η.

Exercise 3.9 Show that Laplace's equation can be written as

$$\frac{\partial^2 u}{\partial z \partial \bar{z}} = 0.$$

Example 1.10 Since z is an analytic function, so is z^n, and hence any polynomial $p(z)$ of z is also analytic. If a polynomial $q(z)$ has no zero in D then $p(z)/q(z)$ is also analytic in D.

Example 1.11 If the function $f(z)$ has the power series expansion

$$f(z) = \sum_{n=0}^{\infty} a_n (z - z_0)^n,$$

which converges in the disc $|z - z_0| < R$, then it is an analytic function in the disc.

Therefore, the function

$$e^z = \sum_{n=0}^{\infty} \frac{z^n}{n!}$$

and $\sin z$, $\cos z$ are all analytic functions on the whole plane.

Again, from

$$\log(1 + z) = \sum_{n=1}^{\infty} (-1)^{n-1} \frac{z^n}{n}, \qquad |z| < 1,$$

we see that $\log(1 + z)$ is analytic inside the unit circle.

Example 3.12 If the function $f(z)$ has the expansion

$$f(z) = \sum_{-\infty}^{\infty} a_n (z - z_0)^n,$$

which converges in the annulus region $R_1 < |z - z_0| < R_2$, then it is an analytic function in the annulus region.

Definition 3.13 If the function $f(1/z)$ is analytic in the (punctured) neighbouhood of 0 then we say that $f(z)$ is analytic at ∞.

For example, $\sum_{n=0}^{\infty} 1/n!z^n$ is analytic everywhere on the whole plane, including ∞, but excluding 0.

Another example is that $\sum_{n=0}^{\infty} 1/z^n$ is an analytic function in the region exterior to the unit circle.

3.5 The power function

Let n be a natural number. Then

$$w = z^n \tag{3.13}$$

is an analytic function on the whole plane.

Using polar coordinates we may write

$$z = re^{i\phi}, \quad w = \rho e^{i\theta},$$

so that

$$\rho = r^n, \quad \theta = n\phi. \tag{3.14}$$

Corresponding to each z there is a w; however, corresponding to each $w \neq 0$, there are n lots of z with values given by

$$\rho^{1/n}, \quad \phi = \frac{\theta}{n} + \frac{2\pi}{n}k, \quad k = 0, 1, 2, \ldots, n-1.$$

Therefore (3.13) does not represent a bijection between the z-plane and the w-plane. However, if we consider the sector on the z-plane (with k being a fixed integer)

$$k\frac{2\pi}{n} < \phi < (k+1)\frac{2\pi}{n},$$

then (3.13) transforms the sector onto the w-plane with the positive real axis removed. A ray from the origin in the sector is transformed into a ray from $w = 0$. A circular arc with the centre at $z = 0$ is transformed into a circular arc with the centre at $w = 0$.

The curves corresponding to the straight lines $u = u_0$, $v = v_0$ are

$$u_0 = r^n \cos n\phi, \quad v_0 = r^n \sin n\phi.$$

Curves having the first equation and the second equation are shown in Fig. 3.3 with dashed and solid lines, respectively.

The inverse of the power function

$$w = z^{1/n}$$

is an n-valued function.

Take a curve C leading from $z = z_0$ to z_1, without crossing or enclosing 0. On the w-plane, select any one of the n values which correspond to z_0, and fix it as w_0. As z travels along C, there is now a curve D, leading from w_0 to w_1, which is uniquely determined. Obviously, there being n choices for w_0, there are n such curves $D_1, \ldots, D_n$ on the w-plane. Also, the curves $D_1, \ldots, D_n$ can be obtained from one of them with the remaining ones being rotated about the origin through an angle $2k\pi/n$ for $k = 1, 2, \ldots, n-1$.

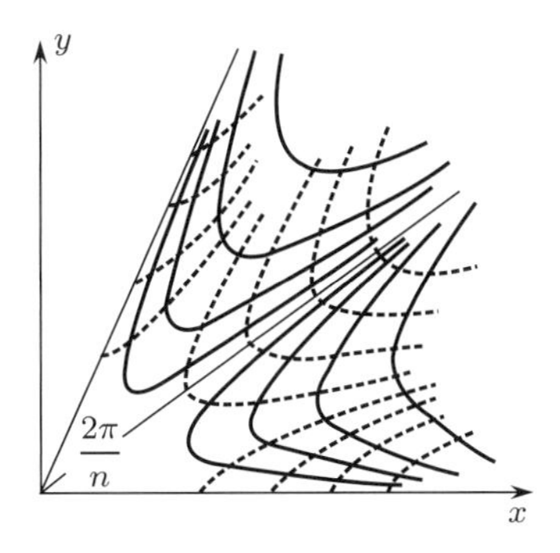

Figure 3.3.

If the domain of the function does not include $z = 0$, then $z^{1/n}$ can be separated into n single-valued continuous functions, with each of them taking one of the n values for $z^{1/n}$. These n functions are called the branches of the multi-valued function $z^{1/n}$. For each point, their values differ by the factor $e^{2\pi i/n}$. Each branch is a bijection, and also

$$(z^{1/n})' = \frac{1}{n} z^{1/n-1};$$

thus, each branch is an analytic function.

Suppose that the curve C loops round the origin once and returns to z_0. Then w_0 has been transformed to

$$w_0 e^{2\pi i/n}.$$

Consequently we cannot separate the n branches if the domain contains 0; after looping round once, the image curve travels from one branch to another.

3.6 The Joukowsky transform

The function

$$w = \frac{1}{2}\left(z + \frac{1}{z}\right) \tag{3.15}$$

is single-valued; we let $w = \infty$ correspond to $z = 0$. On differentiation we have

$$\frac{dw}{dz} = \frac{1}{2}\left(1 - \frac{1}{z^2}\right), \tag{3.16}$$

which takes the value 0 at $z = \pm 1$, and we may therefore expect the transformation to have some special properties at these points.

Let $z = \rho e^{i\theta}$, $w = u + iv$. Then

$$u = \frac{1}{2}\left(\rho + \frac{1}{\rho}\right)\cos\theta, \quad v = \frac{1}{2}\left(\rho - \frac{1}{\rho}\right)\sin\theta,$$

and, on eliminating θ, we obtain

$$\frac{u^2}{\frac{1}{4}(\rho + 1/\rho)^2} + \frac{v^2}{\frac{1}{4}(\rho - 1/\rho)^2} = 1. \tag{3.17}$$

For a fixed $\rho \neq 1$, this represents an ellipse on the w-plane. Thus, corresponding to the two circles $|z| = \rho$, $|z| = 1/\rho$, there is only one ellipse (3.17) on the w-plane. As $\rho \to 1$, the major axis of the ellipse tends to 1, while the minor axis tends to 0; as $\rho \to 0$ (or as $\rho \to \infty$), both axes tend to ∞. It follows that the interior and the exterior of the unit circle are mapped onto the whole w-plane, with the line segment from -1 to 1 along the real axis removed. The unit circle $|z| = 1$ itself corresponds to this segment, but taken twice over.

Solving for z we have the two-valued function

$$z_1, z_2 = w \pm \sqrt{w^2 - 1}.$$

If a point loops round $w = 1$ once, then the value $w + \sqrt{w^2 - 1}$ changes to $w - \sqrt{w^2 - 1}$. For the purpose of understanding this problem, we introduce the notion of a Riemann surface: Take two w-planes and make slits across the segments from -1 to 1. Surface I is used to represent $z = w + \sqrt{w^2 - 1}$, and surface II is used to represent $z = w - \sqrt{w^2 - 1}$. We now join the two sheets together in a crisscross manner across the slit: the upper part of the slit on surface I is joined to the lower part of the slit on surface II, and the lower part of the slit on surface I to the upper part of the slit on surface II (in practice this is an impossible task, but we are visualising it ideally). Such an ideal surface with more than one sheet is known as a Riemann surface, an excellent tool for the representation of a multi-valued function. On such a surface, the function

$$z = w + \sqrt{w^2 - 1}$$

becomes single-valued. As a point crosses the slit segment it moves from one surface to the other, and if the point crosses the slit again it returns to the original surface.

3.7 The logarithm function

Let

$$w = \log z.$$

As z varies in the sector $\alpha < \arg z < \beta$ its image w lies in the infinite strip $\alpha < v < \beta$. To see this we let $z = \rho e^{i\theta}$, so that $w = \log \rho + i\theta$, and hence

$$u = \log \rho, \quad v = \theta.$$

As ρ varies from 0 to ∞, the corresponding value for $\log \rho$ varies from $-\infty$ to ∞, so that the claim follows.

In general, we have

$$w = \log \rho + i(\theta + 2k\pi), \quad k = 0, \pm 1, \pm 2, \ldots .$$

Therefore we do not have just a single strip, but infinitely many corresponding strips.

On the other hand, the strip $\alpha < v < \beta$ corresponds to a sector on the z-plane. If $\beta - \alpha > 2\pi$ then (part of) the plane is covered more than once. How then do we resolve such a problem? The answer is to construct another Riemann surface. This time there are infinitely many sheets to the Riemann surface piling up on top of each other, and the cut is made along the negative real axis, from 0 to $-\infty$. The upper part of the cut is joined onto the lower part of the cut in the next sheet up in an infinite spiral staircase manner. Each strip with width 2π corresponds to a single sheet on the Riemann surface, and in this way each point on the Riemann surface corresponds to one point on the w-plane.

Exercise 3.14 The transformation $w = \tan^2 z/2$ maps the strip $0 < x < \pi/2$ on the z-plane into the interior of the unit circle, but with the line segment from $w = -1$ to $w = 0$ removed.

3.8 The trigonometric functions

With complex numbers at our disposal, the investigation of the trigonometric functions can be made to depend entirely on that of the exponential function. From Euler's formulae

$$\sin z = \frac{1}{2i}(e^{iz} - e^{-iz}), \qquad \cos z = \frac{1}{2}(e^{iz} + e^{-iz}), \tag{3.18}$$

we find that the mapping $w = \sin z$ can be considered as the product of the following four mappings:

$$iz = z_1, \qquad e^{z_1} = z_2, \qquad z_3 = -iz_2 \left(= \frac{e^{iz}}{i} \right) \tag{3.19}$$

and

$$w = \frac{1}{2}\left(z_3 + \frac{1}{z_3}\right) \ (= \sin z). \tag{3.20}$$

We first investigate the condition under which the mapping is injective. Let D be a region on the z-plane, which is transformed into D_1, D_2, D_3 by the mappings in (3.19). The first and the third mappings in (3.19) are injective, and the necessary and sufficient condition for the second mapping to be injective is that D_1 should not contain two points z_1', z_1'' satisfying

$$z_1' - z_1'' = 2k\pi i, \qquad k \text{ a non-zero integer.} \tag{3.21}$$

From Section 3.6 we already know that the necessary and sufficient condition for the mapping (3.20) to be injective is that D_3 should not contain two points z_3', z_3'' satisfying

$$z_3' z_3'' = 1. \tag{3.22}$$

Returning to the z-plane, we find that the necessary and sufficient condition for $w = \sin z$ to be injective is that the region D should not contain two points z', z'' satisfying

$$z' - z'' = 2k\pi, \qquad k \text{ a non-zero integer,}$$

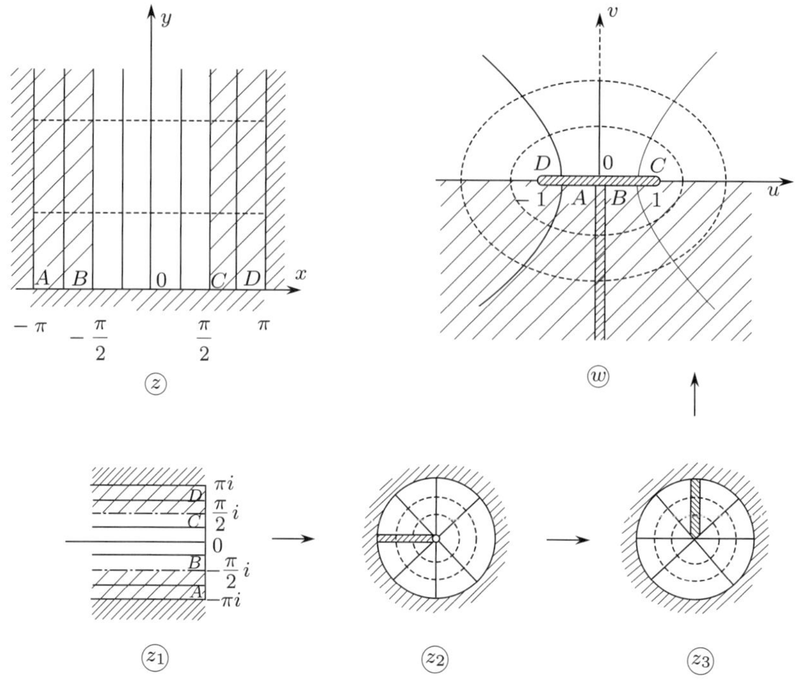

Figure 3.4.

and $e^{i(z'+z'')} = -1$, that is

$$z' + z'' = (2\ell + 1)\pi, \quad \ell \text{ an integer.}$$

For example, the half-strip $-\pi < x < \pi$, $y > 0$ satisfies such a condition imposed on D. The successive transformations of a point in such a region are given in Fig. 3.4. The image of the half-strip on the w-plane has two cuts: one along the u-axis from -1 to 1, and the other along the negative v-axis from 0 to $-\infty$. The region $-\pi/2 < x < \pi/2$, $y > 0$ is mapped into the upper half-plane.

The inverse function to sin is

$$w = \arcsin z = \frac{\pi}{2} - i \log(z + \sqrt{z^2 - 1}\,) \left(= \frac{\pi}{2} - \arccos z \right).$$

This too can be represented by a Riemann surface as follows: Reverse each of the mappings we have just been discussing; the domain of the mapping (3.20) is a Riemann surface with two sheets (see Section 3.6), and that of the second mapping in (3.19) is a Riemann surface with infinitely many sheets (see Section 3.7). Therefore there are infinitely many sheets piled up together with a certain labelling. Along the cut of the real axis from -1 to 1, we join a sheet with an odd label to the next sheet with an even label in a crisscross manner as described in Section 3.6, and along the cut from 0 to $-\infty$ on the imaginary axis we join a sheet with an even label to the next sheet also with an even label, and we do this sheet by sheet as described in Section 3.7. We do the same for the sheets with odd labels, so that there is now an infinite double-spiral staircase.

A point moving across the cut on the real axis will change the parity of the label on the sheet. Thus a circle round $z = 1$ will move from a sheet with an odd (or even) label to one with an even (or odd) label, and the value of the function changes from

$$\frac{\pi}{2} - i \log(z \pm \sqrt{z^2 - 1})$$

to $\pi/2 - i \log(z \mp \sqrt{z^2 - 1})$. If the loop is a large circle which includes both $z = 1$ and $z = -1$, then the point on the plane corresponding to sheet number $2n - 1$ will move to that with number $2n + 1$, or from number $2n$ to $2n + 2$.

The same discussion applies to the cos function.

The function

$$\arctan z = \frac{\pi}{2} - \operatorname{arccot} z = \frac{1}{2i} \log \frac{1 + iz}{1 - iz}$$

is also multi-valued.

Each of the hyperbolic functions also has an inverse:

$$w = \sinh z = \frac{e^z - e^{-z}}{2}, \qquad z = \operatorname{arcsinh} w = \log(w + \sqrt{w^2 + 1});$$

$$w = \cosh z = \frac{e^z + e^{-z}}{2}, \qquad z = \operatorname{arccosh} w = \log(w + \sqrt{w^2 - 1});$$

$$w = \tanh z = \frac{e^z - e^{-z}}{e^z + e^{-z}}, \qquad z = \operatorname{arctanh} w = \frac{1}{2} \log \frac{1 + w}{1 - w};$$

$$w = \coth z = \frac{e^z + e^{-z}}{e^z - e^{-z}}, \qquad z = \operatorname{arccoth} w = \frac{1}{2} \log \frac{w + 1}{w - 1}.$$

These inverse functions are all multi-valued, with log being allowed to take any branch. If a single-valued branch is taken then each branch is an analytic function.

3.9 The general power function

The function z^a may be defined by

$$z^a = e^{a \log z}.$$

Let $z = \rho e^{i\phi}$, $a = \alpha + i\beta$; then

$$z^a = e^{\alpha \log \rho - \beta(\phi + 2k\pi)} \cdot e^{i(\alpha(\phi + 2k\pi) + \beta \log \rho)}.$$

When $\beta \neq 0$, there are infinitely many values for z^a, and they all lie on the circle $|w| = \rho_k$, where the radius

$$\rho_k = |z^a| = e^{\alpha \log \rho - \beta\phi} \cdot e^{-2k\pi\beta}, \quad k = 0, \pm 1, \pm 2, \dots ;$$

thus ρ_k forms a geometric progression, with the common ratio $e^{-2\pi\beta}$. The argument for z^a is

$$\theta_k = \alpha\phi + \beta \log \rho + 2k\pi\alpha,$$

forming an arithmetic progression with the common difference $2\pi\alpha$.

When $\beta = 0$, that is when a is real, the value for z^a lies on the circle

$$|w| = e^{a \log \rho} = \rho^a.$$

If $a = \alpha = p/q$ is a reduced fraction then, for each z, the expression z^a may take q distinct values, given by

$$z^{p/q} = (z^p)^{1/q}$$

so that θ_k is essentially $2\pi k/q$ as in the extraction of the q-root of a complex number.

If $a = \alpha$ is an irrational number, then z^a may take infinitely many values. Moreover, we can prove that the points $e^{2k\pi a\iota}$ form a set which is dense in the unit circle. To prove this we only require the following theorem: If a is irrational then, for any $\epsilon > 0$, there are two integers h and q such that $|ah - q| < \epsilon$. The proof of the theorem is given in the Supplement to Chapter 1 of Volume I.

3.10 The fundamental theorem of conformal transformation

For a given analytic function $f(z)$ we have a conformal transformation. For any region D, once we know that $f(z)$ is injective, a conformal transformation with $u = u(x, y)$, $v = v(x, y)$ then maps the region D onto a region $D^\star$. The fundamental problem concerning conformal transformations is as follows: Given two regions D and $D^\star$, is there a conformal transformation which maps one region to the other? If there is one mapping from D to $D^\star$, we still wish to be able to determine the function $f(z)$.

It is clear that not every region D can be mapped to any other region $D^\star$. For example, it is not possible if D is multiply connected and $D^\star$ is simply connected. Consider a closed curve C in D which loops round points not belonging to D. If there were a conformal transformation mapping D onto $D^\star$ then there would be an image curve $C^\star$ in $D^\star$ which can be continuously deformed down to a single point $w_0 \in D^\star$, because the transformation is continuous. Thus the curve C itself would also need to be deformed down to a certain point in D, and this is clearly impossible because there are points inside C not belonging to D.

Concerning simply connected regions, there is the following fundamental theorem.

Theorem 3.15 (Riemann) *Let D be any simply connected region with more than one boundary point. Take a point z_0 interior to D, and a vector from z_0 pointing in any specified direction. Then we have a conformal transformation which maps D bijectively onto the interior of the unit circle, with z_0 being mapped to 0, and the specified direction of the* *vector being mapped to the direction of the positive real axis. Moreover, there is only one such transformation.*

There are two parts to this theorem: (i) The existence of the transformation, and (ii) the uniqueness of the transformation. The uniqueness part amounts to the following: If we take a conformal transformation of the unit disc onto itself, with the origin being a fixed point, and the image of a vector from the origin pointing in a certain direction, say toward the positive real axis, also pointing toward the positive real axis, then the whole transformation is simple, $w = z$. The reason for this equivalence is simple: If $w = f(z)$ and $w = f_1(z)$ are both conformal transformations that satisfy Theorem 3.15, then $w_1 = f\big(f_1^{-1}(w)\big)$ maps the unit disc onto itself, with the origin fixed, and the direction of the positive real axis preserved. The uniqueness problem is thus reduced to the one mentioned here.

It is not difficult to see that the substance of Theorem 3.15 is equivalent to the following, slightly more general, theorem.

Theorem 3.16 *Let D and $D^\star$ be two simply connected regions, each with more than one boundary point. Take any two points z_0, w_0 which are interior to D and $D^\star$, and also a real number α_0. Then there must be a conformal bijection*

$$w = f(z)$$

which maps D onto $D^\star$ with

$$w_0 f(z_0), \qquad \alpha_0 = \arg f'(z_0),$$

and there is only one such conformal transformation.

The special case when $D^\star$ is the unit disc, $w_0 = 0, \alpha_0 = 0$, corresponds to Theorem 3.15. Conversely, if Theorem 3.15 has been established, then we let $w = f(z)$, $w = f^\star(z)$, which map D, $D^\star$ onto the unit disc. Then $w = f^{\star-1}\big(f(z)\big)$ maps D onto $D^\star$, and it is not difficult to deduce Theorem 3.16 from the set up.

We do not give the proof of Theorem 3.15 at the moment. It is a very important result because it reduces the problem concerning a general simply connected domain to the unit disc. We first need to concentrate our efforts on investigating the theory concerning the unit disc, and we do this in the following chapter. It should tell us that if we have a clear understanding of the situation concerning the unit disc then it may be possible to investigate the more general theorem.

It will not be out of place to repeat the following: The whole (extended) complex plane (elliptic space) is a simply connected region with no boundary point. The Euclidean plane (parabolic space) is a simply connected region with one boundary point. The unit disc (hyperbolic space) is a simply connected region with more than one boundary point. Therefore complex function theory in these three types of spaces has its own important and fundamental features.

4

Harmonic functions

4.1 A mean-value theorem

Theorem 4.1 *Let $u(z)$ be a function which is continuous in the closed disc $|z| \leq \rho$ and harmonic in the interior of the disc. Then*

$$u(0) = \frac{1}{2\pi} \int_0^{2\pi} u(\rho e^{i\theta}) d\theta. \tag{4.1}$$

Proof. Since the closed disc $|z| \leq \rho$ is a simply connected region, $u(z)$ has a single-valued conjugate partner $v(z)$. For $\rho_1 < \rho$ we have

$$\begin{aligned}
\rho_1 \frac{\partial}{\partial \rho_1} \frac{1}{2\pi} \int_0^{2\pi} u(\rho_1 e^{i\theta}) d\theta &= \frac{1}{2\pi} \int_0^{2\pi} \rho_1 \frac{\partial}{\partial \rho_1} u \, d\theta \\
&= \frac{1}{2\pi} \int_0^{2\pi} \frac{\partial}{\partial \theta} z(z) d\theta = \frac{1}{2\pi} v(\rho_1 e^{i\theta}) \Big|_0^{2\pi} = 0.
\end{aligned}$$

Therefore $(1/2\pi) \int_0^{2\pi} u(\rho_1 e^{i\theta}) d\theta$ is a constant, and the required result follows from letting $\rho_1 \to 0$ and $\rho_1 \to \rho$. □

The following result is an important consequence of Theorem 4.1.

Theorem 4.2 *Let $f(z)$ be a function which is continuous in the closed disc $|z| \leq \rho$ and analytic in the interior of the disc. Then*

$$f(0) = \frac{1}{2\pi} \int_0^{2\pi} f(\rho e^{i\theta}) d\theta.$$

Proof. This follows from $f(z) = u(z) + iv(z)$, with $u(z)$ and $v(z)$ being harmonic, so that $f(z)$ can be considered as a (complex) harmonic function. The reader can also establish the theorem by considering $u(z)$ and $v(z)$ separately and then combining the result. □

Setting $z = \rho e^{i\theta}$, $dz = i\rho e^{i\theta}\, d\theta = iz\, d\theta$ in Theorem 4.2 we can deduce the following.

Theorem 4.3 *With the same hypothesis as in Theorem 4.2, we have*

$$f(0) = \frac{1}{2\pi i} \int_{|z|=\rho} \frac{f(z)}{z} dz.$$

Again, because $zf(z)$ is also analytic, we therefore have Theorem 4.4.

Theorem 4.4 *With the same hypothesis as in Theorem 4.2, we have*

$$\int_{|z|=\rho} f(z)dz = 0.$$

Replacing the disc with centre the origin to a general disc, we have Theorem 4.5.

Theorem 4.5 (Mean-value theorem) *Let $u(z)$ be continuous in a closed disc with radius ρ and centre z_0, and suppose that it is a harmonic function in the interior of the disc. Then the value of the function at the centre of the disc is the value of the function averaged over the circumference of the disc, that is*

$$u(z_0) = \frac{1}{2\pi} \int_0^{2\pi} u(z_0 + \rho e^{i\theta})d\theta.$$

Theorem 4.6 *Let $f(z)$ be continuous in a closed disc with radius ρ and centre z_0, and suppose that it is analytic in the interior of the disc. Then*

$$\int_C f(z)dz = 0;$$

here $\int_C$ represents the integral over the circumference of the circle.

4.2 Poisson's integral formula

The transformation

$$w = e^{i\psi} \frac{z-a}{1-\bar{a}z} \tag{4.2}$$

maps the unit disc onto itself, sending $z = a$ to $w = 0$. Also

$$dw = e^{i\psi} \frac{1-a\bar{a}}{(1-\bar{a}z)^2} dz, \tag{4.3}$$

and, on letting $w = e^{i\theta}$, $z = e^{i\tau}$,

$$d\theta = \frac{1-a\bar{a}}{|1-\bar{a}z|^2} d\tau. \tag{4.4}$$

Therefore

$$\frac{1}{2\pi}\int_0^{2\pi} u(w)d\theta = \frac{1}{2\pi}\int_0^{2\pi} \frac{1-a\bar{a}}{|1-\bar{a}z|^2} u_1(z)d\tau,$$

where $u_1(z) = u(w)$ with w being given by (4.2) and, by the mean-value theorem, the right-hand side here is just $u_1(a)$.

Letting $a = re^{i\theta}$ and $z = e^{i\psi}$, we then have the following.

Theorem 4.7 *Let $u(z)$ be a continuous function in the unit disc $|z| \le 1$ and harmonic in the interior of the circle. Then, for $0 < r < 1$,*

$$u(re^{i\theta}) = \frac{1}{2\pi}\int_0^{2\pi} \frac{1-r^2}{1-2r\cos(\theta-\psi)+r^2} u(e^{i\psi})d\psi. \tag{4.5}$$

In other words, for any z inside the unit circle, we have

$$u(z) = \frac{1}{2\pi}\int_0^{2\pi} \frac{1-|z|^2}{|1-ze^{-i\psi}|^2} u(e^{i\psi})d\psi. \tag{4.6}$$

More generally, on replacing the unit disc by $|z| \le R$, we have Poisson's integral formula

$$u(re^{i\theta}) = \frac{1}{2\pi}\int_0^{2\pi} \frac{R^2-r^2}{R-2rR\cos(\theta-\psi)+r^2} u(Re^{i\psi})d\psi. \tag{4.7}$$

The function

$$P(re^{i\theta}) = \frac{R^2-r^2}{R-2rR\cos(\theta-\psi)+r^2} \tag{4.8}$$

is called the Poisson kernel for $|z| \le R$.

Returning to the unit circle, we collect below some of the properties and results associated with the Poisson kernel:

(i) *Being positive inside the circle.* The Poisson kernel can be written as

$$P(re^{i\theta}) = \frac{1-|z|^2}{|w-z|^2}, \quad w = e^{i\psi},\ z = re^{i\theta}, \tag{4.9}$$

which is positive in $|z| < 1$.

(ii) *Being singular on the circle.* On the circle (that is, $r = 1$) we have $P(e^{i\theta}) = 0$ when $\psi \ne \theta$. When $\psi = \theta$ and $r < 1$ we have

$$P(re^{i\theta}) = \frac{1+r}{1-r} \to \infty \quad \text{as} \quad r \to 1_-.$$

More precisely: Let ψ_0 be a pre-assigned value and let $\epsilon > 0$. We then have an $r_0 < 1$ and a $\delta > 0$ with the property that, for $r_0 < r < 1$ and $|\psi_0 - \theta| < \delta$, any ψ satisfying $|\psi - \psi_0| > 2\delta$ also satisfies

$$0 \le P(re^{i\theta}) = \frac{1-r^2}{|1-re^{i(\psi-\theta)}|^2} < \epsilon.$$

To see this we observe that

$$\left|\frac{1-r^2}{1+r^2-2r\cos(\theta-\psi)}\right| \leq \frac{1-r^2}{1+r^2-r\cos\delta},$$

and the right-hand side is smaller than ϵ when r is sufficiently near 1.

(iii) *Having unit mean value*, that is

$$\frac{1}{2\pi}\int_0^{2\pi} P(re^{i\theta})d\theta = 1. \tag{4.10}$$

This follows from setting the substitution (4.4) into $(1/2\pi)\int_0^{2\pi} d\theta = 1$.

On combining (ii) and (iii) we find that $P(re^{i\theta})$ is a function with the following properties: It takes the value 0 everywhere on the circumference (with unit radius), apart from a certain point where it is infinite, and its value averaged over a circle with radius $r \to 1$ is equal to 1. Such behaviour is similar to that of the 'δ-function'.

(iv) *$P(re^{i\theta})$ is a harmonic function*. This follows at once from the Poisson kernel also having the representation

$$P(re^{i\theta}) = \frac{1}{2}\left(\frac{w+z}{w-z} + \frac{\bar{w}+\bar{z}}{\bar{w}-\bar{z}}\right) = \Re\frac{w+z}{w-z}. \tag{4.11}$$

(v) The conjugate function to $P(re^{i\theta})$ is

$$Q(re^{i\theta}) = \Im\frac{w+z}{w-z} = \frac{2r\sin(\theta-\psi)}{1-2r\cos(\theta-\psi)+r^2} + C. \tag{4.12}$$

Thus the conjugate function to u is

$$v(z) = \frac{1}{2\pi}\int_0^{2\pi} u(e^{i\psi})\frac{2r\sin(\theta-\psi)}{1-2r\cos(\theta-\psi)+r^2}\,d\psi + C,$$

and we deduce the Schwarz integral formula:

$$f(z) = u(z) + iv(z) = \frac{1}{2\pi}\int_0^{2\pi} u(e^{i\psi})\frac{e^{i\psi}+z}{e^{i\psi}-z}\,d\psi + iC. \tag{4.13}$$

Similarly, with $v(z)$ being a harmonic function having $-u(z)$ as its conjugate, we also have

$$v(z) - iu(z) = \frac{1}{2\pi}\int_0^{2\pi} v(e^{i\psi})\frac{e^{i\psi}+z}{e^{i\psi}-z}\,d\psi + iD. \tag{4.14}$$

Multiplying (4.14) by i and then adding it to (4.13), we find that

$$2f(z) = \frac{1}{2\pi}\int_0^{2\pi} f(e^{i\psi})\frac{e^{i\psi}+z}{e^{i\psi}-z}\,d\psi - D + iC. \tag{4.15}$$

Setting $z = 0$ we find from Theorem 4.1

$$f(0) = -D + iC = \frac{1}{2\pi}\int_0^{2\pi} f(e^{i\psi})d\psi,$$

and (4.15) now becomes

$$f(z) = \frac{1}{2\pi}\int_0^{2\pi} f(e^{i\psi})\frac{e^{i\psi}\,d\psi}{e^{i\psi} - z},$$

that is

$$f(z) = \frac{1}{2\pi i}\int_{|\zeta|=1} f(z)\frac{d\zeta}{\zeta - z}, \tag{4.16}$$

which is Cauchy's integral formula.

(vi) If $\phi(\zeta)$ is a continuous function on $|z| = 1$ then we call

$$F(z) = \frac{1}{2\pi i}\int_{|\zeta|=1} \phi(z)\frac{d\zeta}{\zeta - z}$$

a Cauchy type integral. It is clear that such an integral represents an analytic function inside, and also outside, the unit circle. If we set

$$z = re^{i\theta},\ 0 < r < 1, \quad z^\star = \frac{1}{\bar{z}} = \frac{1}{r}e^{i\theta},$$

then

$$F(z) - F(z^\star) = \frac{1}{2\pi i}\int_{|\zeta|=1} \phi(z)\Big(\frac{1}{\zeta - z} - \frac{1}{\zeta - z^\star}\Big)d\zeta.$$

If $\zeta = e^{i\psi}$ then

$$\frac{1}{2\pi i}\Big(\frac{1}{\zeta - z} - \frac{1}{\zeta - z^\star}\Big)d\zeta = \frac{1}{2\pi}P(re^{i\theta})d\psi,$$

giving

$$F(z) - F(z^\star) = \frac{1}{2\pi}\int_0^{2\pi} P(re^{i\theta})\phi(\zeta)d\psi.$$

Thus, as $r \to 1$ we have

$$\lim_{r\to 1}\big(F(z) - F(z^\star)\big) = \phi(e^{i\theta}).$$

This is the limit of the difference between the two Cauchy types of integrals, from inside and outside the circle, and its value is the original value of the function on the boundary point. The formula is known as Sokhotsky's formula.

(vii) *The Poisson kernel being the normal derivative of the logarithm of the bilinear transformation.* Consider the transformation

$$w_1 = \frac{w - z}{1 - \bar{z}w}, \quad w = \rho e^{i\psi},$$

which maps z to 0. The partial derivative with respect to ρ is

$$\frac{\partial}{\partial\rho}\log\Big(\frac{w - z}{1 - \bar{z}w}\Big) = \frac{e^{i\psi}}{w - z} + \frac{e^{i\psi}\bar{z}}{1 - \bar{z}w} = \frac{(1 - |z|^2)e^{i\psi}}{(w - z)(1 - \bar{z}w)},$$

and so, at $\rho = 1$,

$$\frac{\partial}{\partial \rho} \log \left(\frac{w - z}{1 - \bar{z}w} \right)_{\rho=1} = \frac{1 - |z|^2}{|w - z|^2} = P(re^{i\theta}).$$

Instead of introducing ρ, we can start directly from

$$\log \left(\frac{w - z}{1 - \bar{z}w} \right), \quad w = e^{i\psi},$$

which has $-P(re^{i\theta})$ as its derivative along its normal. Thus

$$P(re^{i\theta}) = -\frac{\partial}{\partial n} \log \left| \frac{w - z}{1 - \bar{z}w} \right| = \frac{\partial}{\partial n} \log \left| \frac{1 - \bar{z}w}{w - z} \right|.$$

We call $(\partial/\partial n) \log |(1 - \bar{z}w)/(w - z)|$ the Green's function. As a function of w it has the following special properties: It is harmonic inside the unit disc, except at the point $w = z$, and its value is 0 on the circumference $|w| = 1$.

(viii) *Test for being harmonic.* It is not difficult to show (see Section 4.5) that the integral

$$\mathbf{W}(z, \psi_1, \psi_2) = \frac{1}{2\pi} \int_{\psi_1}^{\psi_2} \frac{1 - r^2}{|1 - re^{i(\psi - \theta)}|^2} \, d\psi$$

represents the cross-ratio of the four points $e^{i\psi_1}, e^{i\psi_2}, z, \bar{z}^{-1}$.

4.3 Singular integrals

For the convenience of the reader we repeat some of the material already given in Volume I.

Theorem 4.8 *Let $Q(z, \zeta)$ be a real function, with*

$$z = re^{i\phi}, \quad \zeta = e^{i\psi}, \quad 0 \le r < 1, \quad 0 \le \phi, \quad \psi < 2\pi.$$

Suppose that it has the following properties.

(i) *Q is a non-negative continuous function.*
(ii) *For any z we have*

$$\frac{1}{2\pi} \int_0^{2\pi} Q(z, \zeta) d\psi = 1. \tag{4.17}$$

(iii) *As $z \to \zeta_0 = e^{i\psi_0}$ and $\zeta \neq \zeta_0$, the function $Q(z, \zeta)$ tends to 0 uniformly. (This means that, for any $\epsilon > 0$, there are $\rho < 1$ and $\delta > 0$ such that, for $r > 1 - \rho$ and $|\phi - \psi_0| < \delta$, we have $0 \le Q(z, \zeta) < \epsilon$ for all z and ζ with $|\psi - \psi_0| > 2\delta$.)*

Then we have the following conclusion: If $u(\zeta)$ is piecewise continuous, has discontinuities only of the first kind and is continuous at ζ_0 then

$$\lim_{z \to \zeta_0} \frac{1}{2\pi} \int_0^{2\pi} u(\zeta) Q(z, \zeta) d\psi = u(\zeta_0). \tag{4.18}$$

Proof. From property (ii), the proof amounts to proving

$$\Delta = \frac{1}{2\pi}\int_0^{2\pi} \big(u(\zeta) - u(\zeta_0)\big) Q(z,\zeta)d\psi \to 0 \quad \text{as} \quad z \to \zeta_0.$$

Let $\epsilon > 0$. From continuity we can choose $\delta > 0$ so that $|u(\zeta) - u(\zeta_0)| < \epsilon$ when $|\psi - \psi_0| < 2\delta$. Decompose Δ as the sum of Δ_1 and Δ_2 in which the ranges of integration are $|\psi - \psi_0| < 2\delta$ and $|\psi - \psi_0| > 2\delta$, respectively. First, we have

$$|\Delta_1| \le \frac{\epsilon}{2\pi}\int_{|\psi-\psi_0|<2\delta} Q(z,\zeta)d\psi \le \frac{\epsilon}{2\pi}\int_0^{2\pi} Q(z,\zeta)d\psi = \epsilon.$$

Next, suppose that $|\phi - \psi_0| < \delta$. Then $|\phi - \psi| > \delta$, because $|\psi - \psi_0| > 2\delta$ for Δ_2. From condition (iii) there is a positive $\rho < 1$ with the property that, for such ψ and $r > 1 - \rho$, we have $Q(z,\zeta) < \epsilon$. Thus

$$|\Delta_2| \le \frac{1}{2\pi}\int_{|\psi-\psi_0|>2\delta} \big|u(\zeta) - u(\zeta_0)\big| Q(z,\zeta)d\psi < \frac{\epsilon}{2\pi} 2M(2\pi - 2\delta) \le 2\epsilon M,$$

where M is the maximum of $|u(\zeta)|$ on the circle. The proof of the theorem is now clear. □

4.4 Dirichlet problem

Problem 4.9 Let $u(\zeta)$ be a function defined on the unit circle, on which it is continuous apart from a finite number of exceptional points $\zeta_1, \ldots, \zeta_n$, at which $u(\zeta)$ has discontinuities of the first type. Determine the bounded function $u(z)$ which is harmonic in the interior of the circle and assumes the values of $u(\zeta)$ on the circle, except at the exceptional points.

This is a particular case of the famous Dirichlet problem – in a more general case the unit circle is replaced by the boundary of a more complicated region. The method of solution is rather easy: From $u(\zeta)$ we apply Poisson's integral formula

$$u(z) = \frac{1}{2\pi}\int_0^{2\pi} P(re^{i\psi})u(\zeta)d\psi.$$

That this is indeed a solution to the problem follows from Theorem 4.8.

If we are to assume the fundamental theorem on conformal mapping then even the general Dirichlet problem will have been given a solution.

We first give some simple examples to explain the situation in the neighbourhood of an exceptional point. The function

$$f(z) = \arg(1 - z)$$

is harmonic inside the unit circle. Let us suppose that $f(z) = 0$ at $z = 0$. We now consider the situation concerning points inside the circle which are near 1. Let

$$z = 1 - \rho e^{i\theta}, \quad |\theta| \le \frac{\pi}{2}, \quad \rho > 0;$$

then

$$f(z) = \theta,$$

and $f(z)$ is continuous everywhere on the circle except at $z = 1$. As $\rho \to 0$ the expression for $f(z)$ may take any value between $-\pi/2$ and $\pm\pi/2$. There is a discontinuity of the first kind, with the jump $-\pi/2 - (\pi/2) = -\pi$.

We can now construct a function which is harmonic inside the circle, continuous on the circumference, but having jumps $\alpha_1, \ldots, \alpha_n$ at the discontinuities of the first kind at the exceptional points $\zeta_1, \ldots, \zeta_n$:

$$\sum_{k=1}^{n} \frac{\alpha_k}{\pi} \arg(z - \zeta_k).$$

If $u(z)$ is any harmonic function with the above stated properties, then

$$u(z) - \sum_{k=1}^{n} \frac{\alpha_k}{\pi} \arg(z - \zeta_k)$$

is a function which is continuous on the boundary. For this reason, at a point of discontinuity, we may let $u(z)$ take any value between its left-hand and right-hand limits.

4.5 Dirichlet problem on the upper half-plane

Let $u(t)$ be a bounded function on the real axis with a finite number of discontinuities. The problem is the determination of the harmonic function $u(z)$ on the upper half-plane, satisfying $u(z) = u(t)$ at the continuous points of the boundary.

We make the substitution

$$w = \frac{\zeta - z}{\zeta - \bar{z}}, \tag{4.19}$$

which maps the unit disc $|w| < 1$ onto the upper half-plane $\Im(\zeta) > 0$. If $w = e^{i\tau}$ then we may write $e^{i\tau} = (t - z)/(t - \bar{z})$, giving

$$\begin{aligned} e^{i\tau}\, d\tau &= \frac{2y}{(t - \bar{z})^2}\, dt \\ d\tau &= \frac{2y\, dt}{(t - z)(t - \bar{z})} = \frac{2y\, dt}{(t - x)^2 + y^2}. \end{aligned} \tag{4.20}$$

Let $U(w) = u(\zeta)$; the mean value formula

$$U(0) = \frac{1}{2\pi} \int_0^{2\pi} U(e^{i\tau}) d\tau,$$

after the transformation (4.19), becomes

$$u(z) = \frac{1}{\pi} \int_{-\infty}^{\infty} u(t) \frac{y\, dt}{(t - x)^2 + y^2}. \tag{4.21}$$

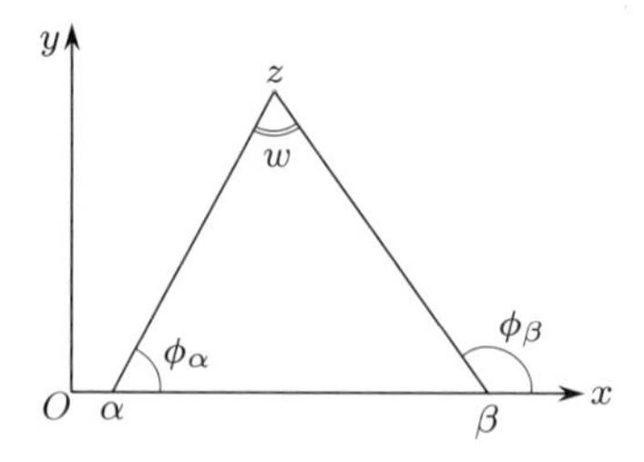

Figure 4.1.

This is the Poisson integral formula for the upper half-plane, and is precisely our desired solution.

From

$$\frac{y}{(t-x)^2+y^2} = \Re\frac{1}{i(t-z)},$$

the Schwarz integral formula becomes

$$f(z) = \frac{1}{\pi i}\int_{-\infty}^{\infty}\frac{u(t)dt}{t-z},$$

which is called a Hilbert integral. Do take note that the mere assumption of being bounded is not sufficient for the convergence of the integral. We need to assume that, as $|t| \to \infty$, the function needs to have certain other properties; for example, the rate of $u(t)$ tending to 0 should not be slower than $1/|t^\alpha|$, $\alpha > 0$.

Example 4.10 Suppose that $u(t) = 1$ for $\alpha \le t \le \beta$ and 0 elsewhere. The corresponding harmonic function is then given by

$$u(z) = \frac{1}{\pi}\int_\alpha^\beta \frac{y\,dt}{(t-x)^2+y^2} = \frac{1}{\pi}\left(\arctan\frac{\beta-x}{y} - \arctan\frac{\alpha-x}{y}\right).$$

The geometric interpretation here is the difference between the angles ϕ_α, ϕ_β that the vectors $z-\alpha$, $z-\beta$ make with the x-axis, divided by π (see Fig. 4.1); and this is just $\angle\alpha z\beta = \omega$ divided by π. In terms of the cross-ratio, we have the representation

$$(\alpha, \beta, z, \bar{z}) = \frac{z-\alpha}{\beta-z}\bigg/\frac{\bar{z}-\alpha}{\beta-\bar{z}} = e^{-2i\omega}.$$

At the same time, $u(z)$ is the real part of the function

$$f(z) = \frac{1}{\pi i}\int_\alpha^\beta \frac{dt}{t-z} = \frac{1}{\pi i}\log\frac{\beta-z}{\alpha-z}.$$

Example 4.11 Generalising the previous example, let $u(t)$ take the values $u_0, u_1, \dots, u_n$ in

$$(-\infty, a_1), (a_1, a_2), \dots, (a_n, \infty).$$

Then

$$\begin{aligned} u(z) &= \frac{\phi_1 u_0}{\pi} + \frac{u_1}{\pi}(\phi_2 - \phi_1) + \cdots + \frac{u_n}{\pi}(\pi - \phi_n) \\ &= \frac{\phi_1}{\pi}(u_0 - u_1) + \frac{\phi_2}{\pi}(u_1 - u_2) + \cdots + \frac{\phi_n}{\pi}(u_{n-1} - u_n) + u_n. \end{aligned}$$

Exercise 4.12 Use

$$w = \tanh \frac{\pi z}{4h}$$

to transform the disc $|w| < 1$ into the strip region $-h < \Im(z) < h$. Prove that the formula

$$f(w) = \frac{1}{2\pi} \int_{-\pi}^{\pi} u(e^{i\tau}) \frac{e^{i\tau} + w}{e^{i\tau} - w} \, d\tau + iC$$

becomes $F(z) = f(w)$ with $U(z) = u(w)$, where

$$\begin{aligned} F(z) = {} & \frac{1}{4h} \int_{-\infty}^{\infty} \frac{U_+(t) + U_-(t)}{\cosh\big(\pi(t - z)/4h\big)} \, dt \\ & - \frac{i}{4h} \sinh \frac{\pi z}{2h} \int_{-\infty}^{\infty} \frac{U_+(t) - U_-(t)}{\cosh\big(\pi t/2h\big) \cosh\big(\pi(t - z)/2h\big)} \, dt + iC, \end{aligned}$$

and $U_\pm(t)$ represent the values of $u(\zeta)$ at $\zeta = t \pm ih$. (This method for solving the Dirichlet problem is given by D. A. Grave in his Doctoral dissertation 'On the fundamental problems of the mathematical theory of constructing geographical maps' (St Petersburg University, 1896, in Russian).)

4.6 Expansions of harmonic functions

We continue with our discussion concerning the unit disc. The Poisson kernel has the expansion

$$\begin{aligned} \frac{1}{2\pi} \frac{1 - r^2}{1 - 2r\cos(\theta - \psi) + r^2} &= \frac{1}{4\pi} \Big(\frac{w + z}{w - z} + \frac{\bar{w} + \bar{z}}{\bar{w} - \bar{z}} \Big) \\ &= \frac{1}{2\pi} \big(1 + \bar{w}z(1 - \bar{w}z)^{-1} + w\bar{z}(1 - w\bar{z})^{-1} \big) \\ &= \frac{1}{2\pi} \Big(1 + \sum_{n=1}^{\infty} (\bar{w}z)^n + \sum_{n=1}^{\infty} (w\bar{z})^n \Big) \\ &= \frac{1}{\pi} \Big(\frac{1}{2} + \sum_{n=1}^{\infty} r^n \cos n(\theta - \psi) \Big), \end{aligned}$$

where $z = re^{i\theta}$, $w = e^{i\psi}$. The series is uniformly convergent in $0 \le r \le r_0 < 1$, so that

$$
\begin{aligned}
u(z) &= \frac{1}{2\pi}\int_0^{2\pi} \frac{1-r^2}{1-2r\cos(\theta-\psi)+r^2} u(e^{i\psi})d\psi \\
&= \frac{1}{\pi}\int_0^{2\pi} u(e^{i\psi})\Big(\frac{1}{2}+\sum_{n=1}^{\infty} r^n(\cos n\theta\cos n\psi+\sin n\theta\sin n\psi)\Big)d\psi \\
&= \frac{1}{2}a_0+\sum_{n=1}^{\infty}(a_n\cos n\theta+b_n\sin n\theta)r^n,
\end{aligned}
$$

where

$$
\begin{aligned}
a_0 &= \frac{1}{\pi}\int_0^{2\pi} u(e^{i\psi})d\psi, \quad a_n = \frac{1}{\pi}\int_0^{2\pi} u(e^{i\psi})\cos n\psi\, d\psi, \\
b_n &= \frac{1}{\pi}\int_0^{2\pi} u(e^{i\psi})\sin n\psi\, d\psi.
\end{aligned}
$$

Consequently, if $u(e^{i\theta})$ has the Fourier series expansion

$$
u(e^{i\theta}) \sim \frac{1}{2}a_0+\sum_{n=1}^{\infty}(a_n\cos n\theta+b_n\sin n\theta), \tag{4.22}
$$

then the harmonic function with boundary values $u(e^{i\theta})$ is

$$
u(re^{i\theta}) = \frac{1}{2}a_0+\sum_{n=1}^{\infty}(a_n\cos n\theta+b_n\sin n\theta)r^n. \tag{4.23}
$$

Note, however, the following: The assumption that $u(e^{i\theta})$ is continuous does not guarantee the convergence of its Fourier series in (4.22). Nevertheless, the convergence of the series (4.23) is guaranteed. On the other hand, if $\limsup |a_n|^{1/n} \le 1$ and $\limsup |b_n|^{1/n} \le 1$, then (4.23) converges, but (4.22) may diverge. Indeed, not only may the series diverge, but also sometimes we may not have any idea which summability method should be used to deliver a relevant function.

Let us examine again an analytic function $f(z)$ in the unit disc. If $|z| < 1$, $\zeta = e^{i\theta}$ then

$$
f(z) = \frac{1}{2\pi i}\int_{|\zeta|=1} \frac{f(\zeta)}{\zeta - z}d\zeta = \frac{1}{2\pi}\int_{|\zeta|=1} f(\zeta)\sum_{n=0}^{\infty}(\bar{\zeta}z)^n\, d\theta = \sum_{n=0}^{\infty} a_n z^n, \tag{4.24}
$$

where

$$
a_n = \frac{1}{2\pi}\int_0^{2\pi} f(e^{i\theta})e^{-ni\theta}\, d\theta.
$$

This shows that any function analytic in the interior of the unit circle has the power series representation (4.24). If $f(e^{i\theta})$ has the Fourier series expansion

$$
f(e^{i\theta}) \sim \sum_{n=0}^{\infty} a_n e^{in\theta},
$$

then, inside the unit circle, there is the analytic function

$$f(z) = \sum_{n=0}^{\infty} a_n z^n.$$

From this it follows that the derivative of an analytic function exists, and that it is also analytic.

4.7 Neumann problem

In certain applications we need to consider the so-called second boundary condition, or the Neumann problem. We only consider the situation of the unit disc. The problem is to find a harmonic function $u(z)$ satisfying the boundary condition

$$\left.\frac{\partial u}{\partial r}\right|_{r=1} = \phi(\theta). \tag{4.25}$$

Suppose that $\phi(\theta)$ has the Fourier series expansion

$$\phi(\theta) \sim \frac{1}{2}c_0 + \sum_{n=1}^{\infty}(c_n \cos n\theta + d_n \sin n\theta), \tag{4.26}$$

and that

$$u(re^{i\theta}) = \frac{1}{2}a_0 + \sum_{n=1}^{\infty}(a_n \cos n\theta + b_n \sin n\theta)r^n.$$

Substituting (4.25) into these relations we have, formally,

$$\sum_{n=1}^{\infty} n(a_n \cos n\theta + b_n \sin n\theta) = \frac{1}{2}c_0 + \sum_{n=1}^{\infty}(c_n \cos n\theta + d_n \sin n\theta).$$

Comparison of coefficients then gives $c_0 = 0$ and $na_n = c_n$, $nb_n = d_n$, and we deduce the following:

(i) A necessary condition to have a solution is that

$$\int_0^{2\pi} \phi(\theta)d\theta = 0;$$

(ii) a solution should have the form

$$u(re^{i\theta}) = \frac{1}{2}a_0 + \sum_{n=1}^{\infty} \frac{1}{n}(c_n \cos n\theta + d_n \sin n\theta)r^n. \tag{4.27}$$

We can establish the condition in (i) properly from

$$\int_0^{2\pi} \phi(\theta)d\theta = \left.\int_0^{2\pi} r\frac{\partial u}{\partial r}\, d\theta\right|_{r=1} = \int_0^{2\pi} \frac{\partial v}{\partial \theta}\, d\theta = 0,$$

so that it really is a necessary condition. With such a condition the solution (4.27) can be written as follows:

$$u(re^{i\theta}) = \frac{1}{2}a_0 + \frac{1}{\pi}\sum_{n=1}^{\infty}\frac{r^n}{n}\int_0^{2\pi}\phi(\psi)(\cos n\psi\cos n\theta + \sin n\psi\sin n\theta)d\psi$$
$$= \frac{1}{2}a_0 + \frac{1}{\pi}\int_0^{2\pi}\phi(\psi)\sum_{n=1}^{\infty}\frac{\cos n(\psi-\theta)}{n}r^n d\psi,$$

where a_0 is an arbitrary constant. From

$$\sum_{n=1}^{\infty}\frac{e^{in\tau}}{n}r^n = -\log(1-e^{i\tau}r),$$

we have

$$\sum_{n=1}^{\infty}\frac{\cos n\tau}{n}r^n = -\frac{1}{2}\log(1-e^{i\tau}r)(1-e^{-i\tau}r) = -\frac{1}{2}\log(1-2r\cos\tau + r^2).$$

Therefore

$$u(re^{i\theta}) = \frac{1}{2}a_0 - \frac{1}{2\pi}\int_0^{2\pi}\phi(\psi)\log(1-2r\cos(\theta-\psi)+r^2)d\psi. \qquad (4.28)$$

We leave it to the reader to find the condition under which this integral representation gives a solution, and that $\lim_{r\to 1}\partial u/\partial r = \phi(\theta)$.

Condition (4.25) can also be written as

$$r\frac{\partial u}{\partial r}\bigg|_{r=1} = \frac{\partial v}{\partial\theta}\bigg|_{r=1} = \phi(\theta).$$

This means that the boundary condition can be taken as

$$v\bigg|_{r=1} = \int_0^{\theta}\phi(\tau)d\tau = \psi(\theta),$$

and the determination of $v(re^{i\theta})$ can be dealt with in the same way as the Dirichlet problem.

4.8 Maximum and minimum value theorems

Theorem 4.13 *A non-constant harmonic function cannot take a maximum (or minimum) at an interior point in its region of definition.*

Proof. At an interior point z_0, we have, for sufficiently small r, the mean value formula

$$u(z_0) = \frac{1}{2\pi}\int_0^{2\pi}u(z_0+re^{i\theta})d\theta.$$

If the average value is the maximum or minimum then $u(z)$ has to be constant. Such a conclusion is the substance of the following theorem.

Theorem 4.14 *The mean value of a continuous function must lie between its maximum value and its minimum value. If the mean value takes on the value of the maximum, or the minimum, then the function is a constant.*

Proof. Let $f(x)$ be a continuous function in the interval $[a, b]$. Its mean value over the interval is

$$m = \frac{1}{b-a}\int_a^b f(x)dx.$$

If m is the maximum value and $f(x)$ is not constant, then there must be a point x_0 with $f(x_0) < m$. By continuity, there is an interval $|x - x_0| \leq \delta$ in which $f(x) < m$, and now

$$\begin{aligned} m = \frac{1}{b-a}\int_a^b f(x)dx &\leq \frac{1}{b-a}\int_{x_0-\delta}^{x_0+\delta} f(x)dx + \frac{m}{b-a}\left(\int_a^{x_0-\delta} + \int_{x_0+\delta}^{b}\right) \\ &< \frac{m}{b-a}\int_a^b dx = m, \end{aligned}$$

which is impossible. The theorem is proved. □

Returning to the proof of Theorem 4.13: If $u(z_0)$ is a maximum (or minimum) then $u(z_0 + re^{i\theta}) = u(z_0)$. This means that the function is constant round the circumference of a small circle, and so, by the Poisson integral formula, the function is constant throughout the interior of the circle. We proceed to show that the function is constant throughout its region of definition. Suppose that there exists z_1 such that $u(z_1) \neq u(z_0)$. Construct a curve leading from z_0 to z_1. There is now a point z_2 on the curve in the neighbourhood of which the points z are such that $u(z) \neq u(z_0)$. Take a small circle at z_2 and there will have to be a point at which the function takes the value $u(z_0)$. There must be a point inside the small disc which takes the maximum value. Therefore the function is also constant inside the small disc. This contradicts the hypothesis. □

Theorem 4.15 (Maximum modulus theorem) *A non-constant function analytic inside D and continuous on D cannot have its modulus taking the maximum value at an interior point of D.*

Proof. We already have

$$f(z_0) = \frac{1}{2\pi}\int_0^{2\pi} f(z_0 + re^{i\theta})d\theta,$$

so that

$$|f(z_0)| \leq \frac{1}{2\pi}\int_0^{2\pi} |f(z_0 + re^{i\theta})|d\theta.$$

The same method of proof for Theorem 4.14 delivers the required result. □

Note. Instead of an equation sign, there is only an inequality sign '$\le$' in the above. For this reason we can only deduce that there is no maximum for the modulus inside the region, and we cannot say that there is no minimum for the modulus inside the region.

4.9 Sequences of harmonic functions

Theorem 4.16 *Let*

$$u_0(z), u_1(z), \ldots, u_n(z), \ldots$$

be a sequence of functions which are harmonic in the interior of D, and continuous in $\bar{D}$. If $\sum_{k=0}^{\infty} u_k(z)$ converges uniformly on the boundary of D then it converges uniformly in $\bar{D}$, and the sum function is harmonic in the interior of D.

Proof. By hypothesis, for any $\epsilon > 0$, there exists N such that

$$|u_{n+1}(\zeta) + \cdots + u_{n+p}(\zeta)| < \epsilon$$

for all $n > N$ and all boundary points ζ. By the maximum and minimum theorem, the same inequality holds with ζ being replaced by any point $z \in \bar{D}$. Therefore $\sum_{k=0}^{\infty} u_k(z)$ converges uniformly in $\bar{D}$.

It remains to show that the sum function is harmonic. Let z_0 be any interior point of S, and take a circle centre at z_0, small enough so that it lies inside D. Each $u_k(z)$ is harmonic in the disc, and we let $U_k(z) = u_k(z + z_0)$, so that $U_k(z)$ is harmonic inside a circle centre 0. The Poisson integral formula now gives

$$\begin{aligned}\sum_{k=0}^{\infty} U_k(z) &= \sum_{k=0}^{\infty} \frac{1}{2\pi} \int_0^{2\pi} \frac{R^2 - r^2}{R^2 - 2rR\cos(\theta - \psi) + r^2} U_k(Re^{i\psi}) d\psi \\ &= \frac{1}{2\pi} \int_0^{2\pi} \frac{R^2 - r^2}{R^2 - 2rR\cos(\theta - \psi) + r^2} \sum_{k=0}^{\infty} U_k(Re^{i\psi}) d\psi.\end{aligned}$$

The series $\sum_{k=0}^{\infty} U_k(Re^{i\psi})$ is a continuous function, so that the right-hand side represents a harmonic function. This shows that the sum function is harmonic at each point inside D and is continuous in $\bar{D}$. □

4.10 Schwarz's lemma

Theorem 4.17 *Let the analytic function*

$$w = f(z)$$

map the unit disc onto itself, with the origin being a fixed point. Then it maps a disc with radius r (< 1) onto itself.

In other words: If $|f(z)| \leq 1$ for $|z| \leq 1$ and $f(0) = 0$, then $|f(re^{i\theta})| \leq r$ (that is, $|w| \leq |z|$). From the maximum modulus theorem, we may weaken the hypothesis to 'if $|f(z)| \leq 1$ for $|z| = 1$'.

Indeed we may add that there is equality only when $f(z) = e^{i\theta} z$.

Proof. The function $f(z)$ has a power series expansion in $|z| \leq 1$, and from $f(0) = 0$ we deduce that $f(z)/z$ also has such an expansion. Thus, with a suitable definition at $z = 0$, the function $f(z)/z$ is analytic. From the hypothesis that

$$\max_{|z|=1} \left| \frac{f(z)}{z} \right| \leq 1$$

we deduce from the maximum modulus theorem that

$$\left| \frac{f(re^{i\theta})}{re^{i\theta}} \right| \leq 1.$$

The theorem is proved. If there is equality then $f(z)/z$ is a constant with absolute value 1. □

After a simple substitution we can generalise Theorem 4.17 to the following.

Theorem 4.18 *Let $f(z)$ be analytic in the disc $|z| \leq R$, and suppose that*

$$\max_{|z|=R} |f(z)| \leq M, \qquad f(0) = 0.$$

Then, for $0 \leq r \leq R$, we have

$$|f(re^{i\theta})| \leq \frac{Mr}{R}.$$

Returning to the situation for the unit disc, if we use the non-Euclidean distance symbol, then the conclusion $|w| \leq |z|$ of Theorem 4.17 can be rewritten as

$$D(0, w) \leq D(0, z). \tag{4.29}$$

Since non-Euclidean distance is invariant and any point can be mapped to 0, we deduce the following.

Theorem 4.19 (Pick) *Let $f(z)$ be analytic inside the unit circle, and suppose that $|f(z)| \leq 1$ for $|z| \leq 1$. Let z_1, z_2 be any two points inside the unit circle. Then*

$$D(f(z_1), f(z_2)) \leq D(z_1, z_2), \tag{4.30}$$

with equality only when $w = f(z)$ is a non-Euclidean motion.

Proof. There is a non-Euclidean motion which maps (z_1, z_2) to $(0, z^\star)$, and another such motion which maps $w_1 = f(z_1)$, $w_2 = f(z_2)$ to $(0, w^\star)$. In this way $w = f(z)$ becoming $w^\star = f_1(z^\star)$ is also a transformation mapping the unit disc $|z^\star| < 1$ to $|w^\star| < 1$, with

$f_1(0) = 0$. Therefore $|w^\star| \le |z^\star|$, that is $D(0, w^\star) \le D(0, z^\star)$, and hence

$$D(w_1, w_2) = D(0, w^\star) \le D(0, z^\star) = D(z_1, z_2).$$

It is not difficult to deduce the final part concerning equality. □

Even more generally: The non-Euclidean length of a curve Γ lying inside the unit circle leading from a to b is given by

$$L(\Gamma) = \int_a^b \frac{|dz|}{1 - |z|^2}.$$

If the curve is divided into n parts, each of which is replaced by such a short length, then, as $n \to \infty$, the total length tends to $L(\Gamma)$. Consequently we deduce from Theorem 4.19 the following.

Theorem 4.20 (Pick) *Let $f(z)$ satisfy the hypothesis in Theorem 4.19. Suppose that $w = f(z)$ maps a curve Γ_z to a curve Γ_w. Then*

$$L(\Gamma_w) \le L(\Gamma_z), \tag{4.31}$$

with equality only when $w = f(z)$ is a non-Euclidean motion.

Writing out (4.30) explicitly we have

$$\frac{1}{2}\log\frac{|1 - \bar{w}_1 w_2| + |w_2 - w_1|}{|1 - \bar{w}_1 w_2| - |w_2 - w_1|} \le \frac{1}{2}\log\frac{|1 - \bar{z}_1 z_2| + |z_2 - z_1|}{|1 - \bar{z}_1 z_2| - |z_2 - z_1|},$$

which is equivalent to

$$\left|\frac{w_2 - w_1}{1 - \bar{w}_1 w_2}\right| \le \left|\frac{z_2 - z_1}{1 - \bar{z}_1 z_2}\right|. \tag{4.32}$$

Letting $z_2 \to z_1$ we find that

$$\lim_{z_2 \to z_1}\left|\frac{w_2 - w_1}{z_2 - z_1}\right| \le \frac{1 - |w_1|^2}{1 - |z_1|^2}.$$

Replacing the symbols, we then have Theorem 4.21.

Theorem 4.21 *Under the hypothesis of Theorem 4.19, for $|z| < 1$ we have*

$$|f'(z)| \le \frac{1 - |f(z)|^2}{1 - |z|^2}. \tag{4.33}$$

For the special case when $f(0) = 0$, we have $|f'(0)| \le 1$, which also follows directly from Theorem 4.17.

Note. If there is a point inside the circle at which equality holds in (4.32) then there is always equality in (4.32), and $w = f(z)$ is a non-Euclidean motion. The proof is left to the reader.

Again, from

$$\int_{z_1}^{z_2} f'(z)dz = f(z_2) - f(z_1), \qquad \frac{1-|f(z)|^2}{1-|z|^2} \leq \frac{1}{1-|z|^2},$$

we immediately deduce Theorem 4.22.

Theorem 4.22 *Under the hypothesis of Theorem 4.19, if* $|z_1| < r < 1$, $|z_2| < r < 1$ *then*

$$\left| \frac{f(z_1) - f(z_2)}{z_1 - z_2} \right| \leq \frac{1}{1-r^2}.$$

Exercise 4.23 Prove, under the hypothesis of Theorem 4.19, that if $|z| < r < 1$ then

$$|f(z)| \leq \frac{|z| + |f(0)|}{1 + |f(0)||z|}.$$

4.11 Liouville's theorem

As a first application of the Schwarz lemma, we have the following.

Theorem 4.24 *Let* $F(z)$ *be a bounded function which is analytic on the whole plane. Then* $F(z)$ *is a constant.*

Proof. Suppose that $|F(z)| < M$. Take any $R > 0$ and consider the function

$$f(u) = \frac{F(z) - F(0)}{2M}, \qquad z = Ru.$$

By the Schwarz lemma we have $|f(u)| \leq |u|$ for $|u| < 1$. This means that, for any z and any $R > |z|$, we have

$$|F(z) - F(0)| \leq \frac{2M}{R}|z|.$$

Letting $R \to \infty$, we deduce that $F(z) = F(0)$. The theorem is proved. □

More generally, we have Theorem 4.25.

Theorem 4.25 *Let* $F(z)$ *be an analytic function on the whole plane. Suppose that, for some positive integer* k,

$$F(z) = O(|z|^k) \quad as \quad |z| \to \infty.$$

Then $F(z)$ *is a polynomial with degree at most* k.

Proof. The analytic function $F(z)$ has an expansion about the origin

$$F(z) = a_0 + a_1 z + \cdots.$$

Let $P(z) = a_0 + a_1 z + \cdots + a_k z^k$, and consider the function

$$f(z) = \frac{F(z) - P(z)}{z^k}.$$

The required result follows from Theorem 4.24. □

Theorem 4.26 (Fundamental theorem of algebra) *Every polynomial with a positive degree must have a zero; or every algebraic equation must have at least one root.*

Proof. If $f(z)$ is a polynomial without a zero, that is $f(z) = 0$ has no solution, then we consider the function

$$F(z) = \frac{1}{f(z)},$$

which is analytic everywhere. Since it cannot be a constant, there is now a contradiction. □

4.12 Uniqueness of conformal transformations

Theorem 4.27 (Fundamental theorem of hyperbolic geometry) *A conformal bijection of the unit disc onto itself must be a non-Euclidean motion.*

Since the unit circle is reachable under the group of non-Euclidean motions, it suffices to prove that 'A conformal bijection of the unit disc onto itself, with the origin as a fixed point, must have the form $w = e^{i\theta} z$'. Let $w = f(z)$ be the transformation, so that $|f(z)| \le 1$ for $|z| \le 1$ and $f(0) = 0$. It follows from Section 4.10 that

$$|w| = |f(z)| \le |z|.$$

Since the conformal transformation $w = f(z)$ is a bijection, there is an inverse transformation $z = g(w)$, and we deduce that $|z| \le |w|$. Therefore $|w| = |z|$, and hence $w = e^{i\theta} z$.

From this we deduce Theorem 4.28.

Theorem 4.28 *A conformal bijection of the unit circle onto itself with the property that 'a point with a direction' is mapped to 'a point with a direction' is unique.*

Theorem 4.29 *Let D be any simply connected region. Suppose that*

$$w = f(z)$$

is a conformal transformation mapping D onto the unit disc $|w| < 1$, with an interior point z_0 of D carrying a direction being mapped to the origin and the direction pointing along the positive real axis. Then such a transformation is unique.

We have therefore established the uniqueness part of Riemann's fundamental theorem on conformal mapping.

4.13 Endomorphisms

Definition 4.30 Let D be a region. A conformal transformation

$$w = f(z)$$

mapping D onto D, or a part thereof, is called an endomorphism.

The theorems that we have been studying may be put into the following context: An endomorphism in a Lobachevsky space must reduce the non-Euclidean length, which is invariant only under non-Euclidean motions.

If D is the image region under the conformal transformation

$$w = w(z)$$

of the unit disc, then the measure in D space is

$$\frac{|dw|}{1-|w|^2} = \frac{|w'(z)|\,|dz|}{1-|w(z)|^2}.$$

The length of a curve Γ in D is then given by

$$L(\Gamma) = \int_\Gamma \frac{|w'(z)|\,|dz|}{1-|w(z)|^2},$$

and, by Theorem 4.21, any endomorphism into D does not make $L(\Gamma)$ any larger.

4.14 Dirichlet problem in a simply connected region

Let $w = w(z)$ be a conformal bijection from a region D to the unit disc $|w| < 1$, mapping z_0 to $w = 0$. Suppose that the boundary C of D is made up of a finite number of curves with continuous tangents and that, as z describes C, its image describes the circumference of the unit disc, that is the unit circle. Suppose further that the transformation of D has a continuous derivative with respect to the length parameter s for the boundary.

Let us now consider the mean value formula for a harmonic function:

$$u(0) = \frac{1}{2\pi}\int_0^{2\pi} u(\zeta)d\theta, \quad \zeta = e^{i\theta}. \tag{4.34}$$

What happens to this formula under the transformation $w = w(z)$? From

$$dw = w'(z)dz, \quad |dw| = |w'(z)|\,|dz|,$$

we have

$$d\theta = |w'(z)|ds.$$

Let $u(w) = U(z)$, $u(0) = U(z_0)$. Then

$$U(z_0) = \frac{1}{2\pi}\int_C U(z)|w'(z)|ds. \tag{4.35}$$

We now transform $|w'(z)|$.

Lemma 4.31 *On the curve* $|w(z)| = 1$ *we have*

$$-\frac{\partial}{\partial n}\log|w(z)| = |w'(z)|. \tag{4.36}$$

Proof. First, from an extension of the Cauchy–Riemann condition we can deduce that

$$\frac{\partial}{\partial n}w(z) = i\frac{\partial}{\partial s}w(z), \tag{4.37}$$

so that

$$\frac{\partial}{\partial n}\log|w(z)| = \Re\Big(\frac{\partial}{\partial n}\log w(z)\Big) = \Re\Big(i\frac{w'(z)}{w(z)}\frac{dz}{ds}\Big). \tag{4.38}$$

On the other hand, from

$$w(z)\overline{w(z)} = 1, \tag{4.39}$$

we deduce

$$w'(z)\overline{w(z)}\,\frac{dz}{ds} + w(z)\overline{w'(z)}\,\frac{d\bar{z}}{ds} = 0,$$

that is

$$\frac{w'(z)}{w(z)}\frac{dz}{ds} = -\frac{\overline{w'(z)}}{\overline{w(z)}}\frac{d\bar{z}}{ds}. \tag{4.40}$$

Applying the manipulation $a/c = b/d \Longrightarrow a/c = b/d = \sqrt{(ab)/(cd)}$ we have

$$\frac{w'(z)}{w(z)}\frac{dz}{ds} = i\frac{|w'(z)|}{|w(z)|}\left|\frac{dz}{ds}\right| = i|w'(z)|. \tag{4.41}$$

The required equation (4.36) in the lemma now follows. □

Substituting (4.36) into (4.35), we arrive at the formula

$$U(z_0) = \frac{1}{2\pi}\int_C U(z)\frac{\partial}{\partial n}\log\frac{1}{|w(\zeta)|}\,ds. \tag{4.42}$$

Therefore we have Theorem 4.32.

Theorem 4.32 *Let* D *have the properties stated at the beginning of this section. Let a function* $U(\zeta)$ *with only a finite number of discontinuities of the first kind be defined on* C*. Then the function defined by (4.42) is analytic in* D*. When* $U(\zeta)$ *is continuous on the boundary it takes the value* $u(\zeta)$*, and as* z_0 *approaches a point of discontinuity along various paths the behaviour of* $U(z_0)$ *can be dealt with in the same way as that in Section 4.4.*

Since $w(z)$ takes z_0 to 0, we write the function in the form

$$w = f(z, z_0),$$

and let

$$g(z, z_0) = \log \frac{1}{|f(z, z_0)|}.$$

Definition 4.33 We call $g(z, z_0)$ the Green's function of the region D.

From the property of the Poisson kernel it is not difficult to deduce Theorem 4.34.

Theorem 4.34 *The Green's function $g(z, z_0)$ takes the value 0 on C when $z_0 \in D$. It is harmonic in D, except at $z = z_0$, and $g(z, z_0) \to \infty$ as $z \to z_0$.*

4.15 Cauchy's integral formula in a simply connected region

Theorem 4.35 *Under the hypothesis stated in Section 4.14, if $f(z)$ is analytic in D and continuous in $\bar{D}$ then*

$$\int_C f(z)dz = 0.$$

Proof. Let $z = \tau(w)$ be the inverse of $w = w(z)$. Then

$$\int_C f(z)dz == \int_{|w|=1} f\big(\tau(w)\big)\tau'(w)dw.$$

Since the composition of analytic functions is analytic, and the derivative of an analytic function is analytic, it follows that the integrand on the right-hand side is an analytic function of w. The theorem is proved. □

Theorem 4.36 *With the same hypothesis as in Theorem 4.35, let $z_0 \in D$. Then*

$$\frac{1}{2\pi i} \int_C \frac{f(z)}{z - z_0} dz = f(z_0).$$

Proof. Let $g(w) = f\big(\tau(w)\big)$. Then

$$I = \frac{1}{2\pi i} \int_C \frac{f(z)}{z - z_0} dz = \frac{1}{2\pi i} \int_{|w|=1} \frac{g(w)\tau'(w)}{\tau(w) - \tau(w_0)} dw, \quad |w_0| < 1.$$

Since $\tau = \tau(w)$ is a conformal bijection, we have $\tau'(w_0) \neq 0$. Let

$$K(w) = \frac{1}{\tau(w) - \tau(w_0)} - \frac{1}{\tau'(w_0)(w - w_0)},$$

so that

$$I = \frac{1}{2\pi i} \int_{|w|=1} \frac{g(w)\tau'(w)}{(w - w_0)\tau'(w_0)} dw + \frac{1}{2\pi i} \int_{|w|=1} g(w)\tau'(w)K(w)dw.$$

The first term on the right-hand side is equal to

$$\frac{g(w_0)\tau'(w_0)}{\tau'(w_0)} = g(w_0),$$

and it remains to show that the second term has the value 0. This amounts to establishing $K(w)$ being analytic (with a suitable definition for $K(w_0)$), from which the required result follows from Theorem 4.35.

From

$$\tau(w) = \tau(w_0) + (w - w_0)\tau(w_0) + O(|w - w_0|^2),$$

we deduce that $K(w)$ is bounded as $w \to w_0$. Since $K(w)$ is clearly analytic at $w \neq w_0$, the theorem is proved. □

5

Point set theory and preparations for topology

5.1 Convergence

When considering convergence we immediately think of the notion of 'distance'. We have already come across several kinds of 'distances', the three more obvious ones being: (i) ordinary distance, that is the Euclidean distance $|z_1 - z_2|$ between two complex numbers z_1, z_2; (ii) the distance on the surface of a sphere in elliptic geometry, that is the chord distance between two points z_1, z_2, including ∞, given by

$$\frac{|z_1 - z_2|}{\sqrt{(1 + |z_1|^2)(1 + |z_1|^2)}};$$

(iii) in hyperbolic geometry there is the non-Euclidean distance, that is the pseudo-chord distance

$$\frac{|z_1 - z_2|}{|1 - \bar{z}_2 z_1|}$$

between the points z_1, z_2 with $|z_1| < 1$, $|z_2| < 1$. Let $d(z_1, z_2)$ represent any of these distances; they share the following properties:

(i) $d(z_1, z_2) \geq 0$, with equality if and only if $z_1 = z_2$;
(ii) $d(z_1, z_2) = d(z_2, z_1)$;
(iii) $d(z_1, z_2) \leq d(z_1, z_3) + d(z_3, z_2)$.

For (iii), there is equality when z_3 lies on the straight line between z_1 and z_2 in the first type of distance; in elliptic geometry and hyperbolic geometry, there is equality only when $z_3 = z_1$ or z_2. The distance function $d(z_1, z_2)$ having the three properties here is called a metric.

A sequence $z_1, z_2, \ldots$ is said to converge to a if

$$\lim_{n \to \infty} d(a, z_n) = 0.$$

We take note of the particular case in elliptic geometry concerning the point ∞, where

$$d(\infty, z_n) = \frac{1}{\sqrt{1 + |z_n|^2}};$$

in this situation, $z_n \to \infty$ means $1/|z_n| \to 0$. Also, we should not forget that, in the formula for distance in hyperbolic geometry, we need to have $|z_1| < 1$, $|z_2| < 1$. If the upper half-plane is taken as the model instead, then the pseudo-chord distance is given by

$$d(z_1, z_2) = \left|\frac{z_1 - z_2}{\bar{z}_1 - z_2}\right|, \quad \Im(z_1) > 0, \quad \Im(z_2) > 0.$$

We note that, apart form the chord distance to the point at infinity in elliptic geometry, there is really nothing new here! In elliptic geometry, if $|z_1| < R$, $|z_2| < R$, then

$$\frac{|z_1 - z_2|}{1 + R^2} \leq \frac{|z_1 - z_2|}{\sqrt{(1 + |z_1|^2)(1 + |z_1|^2)}} \leq |z_1 - z_2|.$$

Therefore, if z_n converges to a in the chord metric, then it also converges to a in the ordinary metric, and conversely. Also, if $|z_1| < \rho$, $|z_2| < \rho$, $\rho < 1$, then the pseudo-chord metric gives

$$\frac{|z_1 - z_2|}{1 - \rho^2} \geq \frac{|z_1 - z_2|}{1 - |z_1 \bar{z}_2|} \geq \frac{|z_1 - z_2|}{|1 - z_1 \bar{z}_2|} \geq \frac{|z_1 - z_2|}{1 + |z_1 \bar{z}_2|} \geq \frac{|z_1 - z_2|}{2}.$$

Thus, if a sequence converges in the pseudo-chord metric then it also converges in the ordinary metric, and conversely. What we have described here is an example of the notion of 'topological equivalence'.

From the above we may consider the chord distance to be the standard metric for convergence. Henceforth we use $d(a, b)$ to represent the distance. It is not difficult to establish the following: If

$$\lim_{n\to\infty} z_n = a, \quad \lim_{n\to\infty} w_n = b,$$

then

$$\lim_{n\to\infty} (z_n + w_n) = a + b, \quad \lim_{n\to\infty} z_n w_n = ab.$$

However, $a + b$ and ab have to be meaningful; for example, we do not allow $0 \cdot \infty$ or $\infty - \infty$, etc. (Since we already dealt with the situation when $a \neq \infty$, $b \neq \infty$, the reader only has to consider that (i) when $a = \infty$, $b \neq \infty$, the 'sum' formula holds and that (ii) when $a = \infty$, $b \neq 0$, the 'product' formula holds.)

5.2 Compact sets

We already know that any bounded infinite set must contain a convergent sequence (Weierstrass' theorem). If the set contains all such limits then we say that the set is compact. Besides all the complex numbers we also include the point ∞, and the purpose of choosing the chord metric is to ensure that the set of all such points is compact. That is, using the chord metric, the extended complex number system forms a compact set. In other words, any infinite set must contain a convergent sequence.

Let us be clear about this by considering a set M of complex points, which may, or may not, include ∞. If M is an infinite set then there are two possible situations:

(i) Corresponding to each natural number n, the set M contains points outside the circle $|z| = n$; (ii) there exists n such that all the points in M lie inside $|z| \leq n$. For the former, M contains a sequence which tends to ∞, and for the latter, by Weierstrass' theorem, there is a convergent sequence in the set M.

We therefore deduce the following. Let $\{z_n\}$ represent an infinite sequence. Then there must be a convergent subsequence $\{z'_n\}$ which converges to a. If the original sequence is not convergent, then there exists $\epsilon_0 > 0$ and infinitely many indices h_n such that

$$d(z_{h_n}, a) > \epsilon_0, \quad n = 1, 2, \ldots .$$

From the infinite set $\{z_{h_n}\}$ we can again select another subsequence with $z''_n \to b$, where $b \neq a$.

We therefore have the following theorem.

Theorem 5.1 *A necessary and sufficient condition for an infinite sequence of complex numbers to be convergent is that all the convergent infinite subsequences should have the same limit.*

From this theorem it is easy to deduce Cauchy's criterion, and in certain applications the new formulation may be more convenient.

5.3 The Cantor–Hilbert diagonal method

There are more than a few problems in which it is necessary for us to select, from an infinite array, a certain convergent sequence which is simultaneously a subsequence of other sequences. Take the infinite array

$$\begin{cases} a_{11}, \ a_{12}, \ldots, \ a_{1n}, \ldots \\ a_{21}, \ a_{22}, \ldots, \ a_{2n}, \ldots \\ \ldots\ldots\ldots\ldots\ldots\ldots\ldots \\ a_{m1}, a_{m2}, \ldots, a_{mn}, \ldots \\ \ldots\ldots\ldots\ldots\ldots\ldots\ldots \end{cases} \tag{5.1}$$

Our problem is to select an infinite array which is similar to (5.1) and in which each row now forms a convergent sequence.

We first select a convergent subsequence from the first row

$$a_{1j_{11}}, a_{1j_{12}}, \ldots, a_{1j_{1n}}, \ldots,$$

with the limit a_1; we use k_1 to represent the first coordinate j_{11}. We next consider the sequence

$$a_{2j_{11}}, a_{2j_{12}}, \ldots, a_{2j_{1n}}, \ldots,$$

from which we select a convergent subsequence

$$a_{2j_{22}}, a_{2j_{23}}, \ldots, a_{2j_{2n}}, \ldots ,$$

with the limit a_2; we use k_2 to represent j_{22}.

Continuing in this way, for each $i = 1, 2, \ldots$ we select

$$a_{ij_{ii}}, a_{ij_{ii+1}}, \ldots , \tag{5.2}$$

with the limit a_i, and each time using k_i for j_{ii}.

Let us now consider the array

$$\begin{array}{llll} a_{1k_1}, & a_{1k_2}, & \ldots, & a_{1k_n}, \ldots \\ a_{2k_1}, & a_{2k_2}, & \ldots, & a_{2k_n}, \ldots \\ \cdots & \cdots & \cdots & \cdots \\ a_{ik_1}, & a_{ik_2}, & \ldots, & a_{ik_n}, \ldots \\ \cdots & \cdots & \cdots & \cdots \end{array}$$

in which we have selected, from the array (5.1), k_1 from the first column, k_2 from the second column, etc. Here the sequence in the ith row differs from that in (5.2) by finitely many terms only, so that its limit is a_i.

Such a method of selection is known as the Cantor diagonal method.

5.4 Types of point sets

Let A be a set of complex numbers. If there are points z_n in A, distinct from z_0, which form a sequence with the limit z_0, then we say that z_0 is a *limit point*, or a point of accumulation, of A. If there is a positive ϵ such that all the points z satisfying

$$d(z, z_0) < \epsilon$$

belong to A, then we call z_0 an *interior point* of A (note that z_0 can be ∞).

Points not belonging to A form a set $C(A)$, called the *complement* of A. An interior point of the complement is called an *exterior point* of A. A point which is neither an interior point nor an exterior point of A is called a *boundary point* of A (note that ∞ is not excluded). The boundary points of A belonging to A form a set called the *boundary* of A, and those belonging to $C(A)$ form the boundary of $C(A)$.

A point of A which is not a limit point is called an *isolated point* of A.

A set in which all its points are interior points is called an *open set*. A set which contains all its limit points is called a *closed set*. A *dense set* is one in which every point is a limit point. A closed dense set is called a *perfect set*.

The complement of an open set is closed, and conversely the complement of a closed set is open. The only (non-empty) set which is both open and closed is the whole extended complex plane. This follows from the set having to include all its limit points, and that all its points have to be interior.

Let H_A denote the set of limit points of A. The set $A \cup H_A$ is called the *closure* of A, and is denoted by $\bar{A}$; it is the smallest closed set containing A. The complement of the closure of the complement, that is $C(\overline{C(A)})$, is the largest open set contained in A, and is called the *open kernel* of A.

It is easy to show that the union of open sets is open. A nested sequence

$$A_1 \supseteq A_2 \supseteq A_3 \supseteq \cdots$$

of non-empty closed sets has a non-empty closed intersection. The proof is as follows: Select a point z_i from A_i, and then select a convergent sequence

$$z_{n_1}, z_{n_2}, z_{n_3}, \ldots, \qquad n_1 < n_2 < n_3 < \cdots$$

which converges to ω. For any natural number m, we may take j large enough so that $n_j > m$, and the whole sequence

$$z_{n_j}, z_{n_{j+1}}, z_{n_{j+2}}, \ldots$$

belongs to A_m (because A_ν is nested). Since each A_m is closed, $\omega \in A_m$. This holds for $m = 1, 2, 3, \ldots$, so that ω belongs to the intersection D of all the A_ν; thus D is non-empty. Since the limit points of D belong to D, the set D is closed.

5.5 Mappings or transformations

Let A_z be a point set on the z-plane, and suppose that each member of the set has a corresponding point on the w-plane. Such a relationship is called a single-valued mapping, and the set A_z is transformed into a set A_w in the w-plane. Writing this down as a single-valued function, we have the relationship

$$w = F(z), \quad \text{or} \quad u = \phi(x, y), \quad v = \psi(x, y).$$

The point z is called the pre-image, and w is called the image of the mapping. The domain of the function is the point set A_z, and the range is A_w. Although each pre-image has a unique image, we do not rule out the possibility that an image may have more than one pre-image – there may be many, and even infinitely many of them. In the extreme case, the range consists of only a single point w_0 in the w-plane, so that the whole domain A_z is mapped onto w_0. In this case the function $F(z) = w_0$ is constant.

Let z_0 be a limit point of A_z, possibly not belonging to A_z. There is a sequence $\{z_n\}$ in A_z which converges to z_0. From the formula

$$w_n = F(z_n), \quad n = 1, 2, \ldots,$$

we have a sequence of images $\{w_n\}$, but it may not converge. We now suppose the situation in which, for any sequence $\{z_n\}$ in A_z which converges to z_0, its image sequence $\{w_n\}$ also converges. In that case, all such sequences must converge to the same number w_0. Here is the reason for this: Let $\{z'_n\}$, $\{z''_n\}$ both converge to z_0, and let their images $\{w'_n\}$, $\{w''_n\}$

converge to w_0', w_0''. The sequence

$$z_1', z_1'', z_2', z_2'', \ldots$$

also converges to z_0, and its image sequence

$$w_1', w_1'', w_2', w_2'', \ldots$$

has both w_0' and w_0'' as its limit, so that we need to have $w_0' = w_0''$ for uniqueness.

Suppose that $z_0 \in A_z$. Then we must have

$$w_0 = F(z_0),$$

which can be seen by taking $z_1 = z_0$, $z_2 = z_0, \ldots$. In this case we say that $F(z)$ is *continuous* at the point z_0. Even when $z_0 \notin A_z$, if the sequence of images $\{F(z_n)\}$ always has a unique limit for any sequence $\{z_n\}$ which converges to z_0, then we may define $F(z_0)$ to be $\lim_{n\to\infty} F(z_n)$. With the domain for F so extended, the function $F(z)$ is continuous at z_0.

Note that if the chord metric is to be taken, the definition of continuity in the above does not exclude the situation when either z_0 or w_0 equals ∞.

5.6 Uniform continuity

Let B be a closed subset of the point set A_z in which every point $\zeta \in B$ is a limit point of A_z, and suppose that $F(z)$ is continuous at each such ζ. For a positive δ, we consider all the complex numbers z, ζ satisfying

$$z \in A_z, \quad \zeta \in B, \quad d(z, \zeta) < \delta; \tag{5.3}$$

we may now specify

$$\epsilon(\delta) = \sup d\big(F(z), F(\zeta)\big). \tag{5.4}$$

First, as δ becomes smaller, the aggregate of z, ζ satisfying (5.3) is reduced, and so $\epsilon(\delta)$ also becomes smaller. Thus $\epsilon(\delta)$ is monotonically decreasing as δ decreases, so that

$$\lim_{\delta\to 0} \epsilon(\delta) = \epsilon_0 \tag{5.5}$$

exists. We proceed to determine ϵ_0. Let δ_n be a sequence with the following properties:

$$\delta_1 > \delta_2 > \delta_3 < \cdots, \quad \lim_{n\to\infty} \delta_n = 0; \tag{5.6}$$

corresponding to each δ_n there is a pair of points z_n, ζ_n such that

$$z_n \in A_z, \quad \zeta_n \in B, \quad d(z_n, \zeta_n) < \delta_n, \tag{5.7}$$

and

$$d\big(F(z_n), F(\zeta_n)\big) \geq \frac{\epsilon(\delta_n)}{2} \geq \frac{\epsilon_0}{2}.$$

By taking a subsequence if necessary, we may assume without loss that ζ_n converges to ζ_0. Since B is a closed set, we have $\zeta_0 \in B$, and thus ζ is a point of continuity of F. From (5.6) and (5.7) we have

$$\lim_{n\to\infty} z_n = \lim_{n\to\infty} \zeta_n = \zeta_0,$$

so that

$$\lim_{n\to\infty} F(z_n) = \lim_{n\to\infty} F(\zeta_n) = F(\zeta_0),$$

and hence

$$\frac{\epsilon_0}{2} \leq \lim_{n\to\infty} d\big(F(z_n), F(\zeta_n)\big) = 0;$$

thus

$$\epsilon_0 = \lim_{\delta\to 0} \epsilon(\delta) = 0.$$

This means that, given any small $\epsilon > 0$, we can find $\delta > 0$ with the property that for any $z \in A_z$ and $\zeta \in B$ satisfying

$$d(z, \zeta) < \delta,$$

we have

$$d\big(F(z), F(\zeta)\big) < \epsilon.$$

This property of being able to choose δ to depend only on ϵ, and not on (z, ζ), is called the uniform continuity of $F(z)$ in B.

If A_z itself is a closed set, in which $F(z)$ is continuous, then we have the following theorem.

Theorem 5.2 *A function which is continuous in a closed bounded set must be uniformly continuous there.*

Let A be a set on the complex plane, and let ζ be a boundary point, so that there is a sequence $z_1, z_2, \ldots, z_n \ldots$ in A which approaches ζ, and

$$\lim_{n\to\infty} F(z_n) = \alpha$$

exists. We call α the boundary value of $F(z)$ at the boundary point ζ. The set W of all boundary values is a closed set.

The proof is not difficult. Let α_0 be a limit point of W; this means that there is a sequence $\{\alpha_n\}$ of boundary values such that $d(\alpha_n, \alpha_0) < 1/n$. Thus there is at least one point z_n such that $d\big(F(z_n), \alpha_n\big) < 1/n$, and so $d\big(F(z_n), \alpha_0\big) < 2/n$. Therefore α_0 is a boundary value.

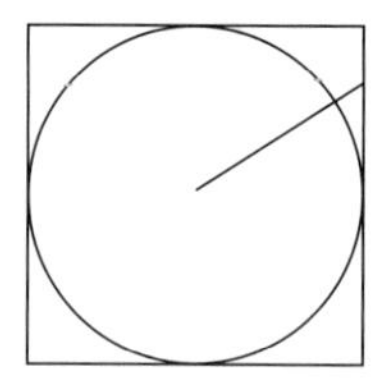

Figure 5.1.

5.7 Topological mappings

A function

$$w = F(z) \tag{5.8}$$

defines a mapping with domain A_z and range A_w. Suppose that we add the condition that

$$F(z_1) = F(z_2), \quad z_1, z_2 \in A_z \quad \text{implies} \quad z_1 = z_2,$$

so that there is only one $z \in A_z$ which corresponds to $w \in A_w$. Such a function is said to be *univalent*, or injective, and there is now a bijection between A_z and A_w. Thus we may also consider z to be a function of w, say a relationship

$$z = G(w), \tag{5.9}$$

which maps A_w back to A_z; we call this the inverse map to (5.8). If F and G are continuous then (5.8) is called a *topological mapping*, a bijection that is continuous both ways.

If $z = H(s)$ is a topological mapping from A_s to A_z, then

$$w = G\big(H(s)\big)$$

is also such a mapping from A_s to A_w.

The study of topology is the investigation of geometric properties under topological transformations. In the study of topology we need not distinguish between a disc and a square, because there is a topological transformation which maps a disc to a square. For example (see Fig. 5.1), we can map the circle in the $\theta\rho$-plane onto the square in the $\theta_1\rho_1$-plane by

$$\theta_1 = \theta, \quad \rho_1 = \frac{\tau(\theta)}{\rho(\theta)}\rho;$$

here $\rho(\theta)$ and $\tau(\theta)$ denote the distances from the circumference of the circle and the perimeter of the square, respectively, to the centre of the circle along a ray made with the angle θ. Under such mappings we can even make shears, similar to images in fun-fair mirrors, or even flip images over from left to right. However, we cannot split one piece into two pieces, fix discontinuous parts together or transform solid bodies into plane objects, such as treating a sphere as if it were a disc.

The Möbius transformation that we considered is a topological mapping of the Riemann sphere onto itself.

Note that the topology depends on the limiting process being applied. One definition of limit gives one kind of topology, and a change in the definition may result in a different kind of topology. Our definition of limit here makes use of the notion of distance, or the metric; given a certain metric there is an induced topology. However, sometimes two different definitions of limit may result in the same topology. For example, take two metrics $d(w, z)$, $d_1(w, z)$ satisfying the following relationship: there are positive ρ and ρ_1 such that

$$\rho_1 d_1(w, z) < d(w, z) < \rho d_1(w, z).$$

Then the topology induced by $d(w, z)$ is the same as that induced by $d_1(w, z)$. For example, various point sets formed inside the unit circle may depend on the ordinary metric, the chord metric and the pseudo-chord metric, but the induced topologies are equivalent. This is because if $z_n \to z_0$ under one metric, the same applies for any other equivalent metric.

5.8 Curves

By a continuous arc (or curve) on the complex plane (or the surface of a sphere) we mean the continuous image of the closed interval $0 \leq t \leq 1$. Thus a curve may be represented by

$$z = x + iy, \quad x = f(t), \quad y = g(t), \quad 0 \leq t \leq 1. \tag{5.10}$$

Here f, g are continuous functions of t, and 'continuity' may refer to the 'chord metric'. Thus, given $\epsilon > 0$, there exists $\delta > 0$ such that $d(z(s), z(t)) < \epsilon$ whenever $d(s, t) < \delta$. If $f(0) = f(1)$, $g(0) = g(1)$, then the curve is said to be closed.

It appears that we have a most suitable definition for our purposes. However, Peano had given an example of a curve that satisfies our condition for being one, and which passes through every point in a square – a space-filling curve. Thus, although intuition is a vital component when formulating ideas in geometry, it must be backed up with detailed analysis and proofs in order to put the concept on a firm and sound foundation.

Jordan was the first to state the important property associated with curves that do not intersect themselves.

Definition 5.3 By a Jordan arc we mean a curve defined by (5.10) satisfying the condition that

$$f(t) = f(t'), \quad g(t) = g(t'), \quad 0 \leq t \leq t' \leq 1 \quad \text{implies} \quad t = t'. \tag{5.11}$$

Such an arc does not intersect with itself; it is also called a *simple curve*.

If the hypothesis in (5.11) implies $t = 0$, $t' = 1$ instead, then we say that the point set forms a Jordan closed curve, a polygon being an example of such a curve. By abuse[1] of language, a Jordan closed curve is also called a *simple closed curve*.

Anyway, a simple curve is the topological image of the unit interval, and a simple closed curve is the topological image of the unit circle.

[1] A curve cannot be 'simple' and 'closed'.

Using the Heine–Borel theorem, we can prove the following: Let C be a simple curve in a region S. For any $\epsilon > 0$ there is a polygonal path (that is, a curve made up of sections of straight line segments) with the chord distance less than ϵ from C. (This means that for any point on the path, there is a point on C with chord distance less than ϵ, and for any point on C there is a point on the path with the same property.) In other words, a simple curve can be arbitrarily well approximated by a polygonal path, and a simple closed curve can be so approximated by a polygon.

5.9 Connectedness

A topological property is one which is invariant under topological transformations. One of the simplest such properties is connectedness.

Two points a and b in a set S are said to be (arcwise) connected in S if there is a curve C containing a, b and lying entirely in S. If every point in S is connected to a certain fixed point in S then we say that S is a connected set.

We can restrict ourselves to the use of simple curves in the definition of being connected. Let C be the curve

$$x = \phi(t), \quad y = \psi(t)$$

linking a, b in the connected set S. If it is not a simple curve then we have $0 \leq t_1 < t_2 \leq 1$ such that $\phi(t_1) = \phi(t_2)$, $\psi(t_1) = \psi(t_2)$, and we may discard the section of the curve corresponding to $t_1 < t < t_2$.

A connected open set is called a region.

A connected closed set is called a continuum.

The simplest example of a one dimensional region is the open line segment $0 < t < 1$, and that of a continuum is the closed line segment $0 \leq t \leq 1$. However, in our discussions on the complex plane the regions are two dimensional.

The interior of the circle $|z - a| < r$ is a region, and so is the exterior $|z - a| > r$. Similarly, if we use the chord metric, then $d(z, a) < \epsilon \leq 1$ also defines a region on the Riemann sphere.

If S is a region then connectedness can be defined using polygonal paths as follows: Let C be the curve linking a, b in S, so that all the points concerned are interior to S. For each point on C there is a disc with centre at the point which lies entirely inside S. The set of all such discs form a cover for C. Since C is a closed set, by the Heine–Borel theorem a finite number of discs with circumferences $C_1, C_2, \ldots, C_n$ already form a cover for C. We can now form a polygonal path joining adjacent centres of the discs for our desired result.

A connected open set containing a point is called a neighbourhood of the point.

We do not prove the following result here: Any topological transformation must map the neighbourhood of a point onto a neighbourhood of the image of the point. This property is more properly associated with complex function theory, and can be established directly from that theory.

Let S be a region, and assume that z_0 is not one of its interior points, but it is a limit point of the interior. Such a point is called a boundary point of S, and the set of boundary points is called the boundary of S.

The most important property associated with a simple closed curve is the following theorem.

Theorem 5.4 (Jordan curve theorem) *A simple closed curve divides the plane into two regions, with the curve as their common boundary.*

The theorem, which tells us that there is an inside and an outside associated with a simple closed curve, is intuitively clear, but the proof is tediously complicated, and will not be given here. However, we do give the proof for some special examples in the following section.

We shall also make use of the next theorem.

Theorem 5.5 *Let Q be a point inside a simple closed curve and let P be a point on it. Denote by $\alpha(P)$ the angle QP makes with the x-axis. As P describes the curve once round, the value of $\alpha(P)$ increases, or decreases, by 2π, depending on whether P describes the curve anti-clockwise or clockwise. If Q is on the outside of the curve then there is no change in the value of $\alpha(P)$.*

The proof of this theorem, similar to that of the Jordan curve theorem, is not at all simple, and will not be given in this book.

5.10 Special examples of Jordan's theorem

5.10.1 Convex regions

By a convex region we mean an open set with the property that if it contains the points A, B then it also contains the whole path (straight line segment) from A to B.

We already established in Volume I that a convex region has a continuous boundary. A convex region together with its boundary gives a continuum, which also has the defining convexity property.

It is obvious that a convex region is (connected and hence) a region. The set forming the exterior of a convex region is also an open set, and we proceed to show that it is connected. Let O be an interior point in the convex region, let P be an exterior point and consider the half-line from O through P. There are two parts to such a half-line. The first part leads from O to the boundary, and the second part leads from the boundary point through P to infinity. If there were a point R belonging to the convex set lying between P and ∞, then the path OR contains a point, namely P, not belonging to the convex set, which is impossible. Therefore any point P exterior to the convex set is connected to ∞, so that the set of exterior points forms a region.

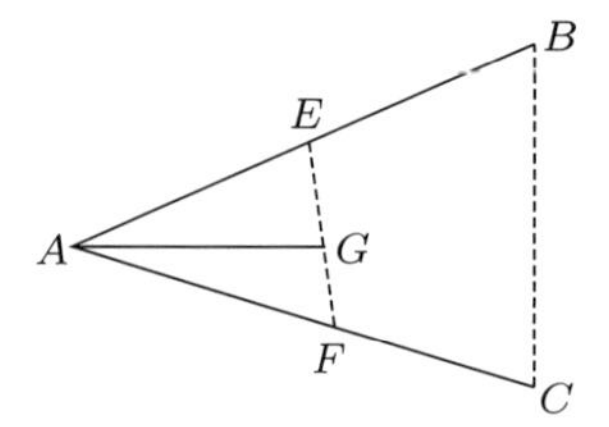

Figure 5.2.

5.10.2 Triangular regions

A triangular region is a convex region. Therefore the whole plane is divided into two regions, with the three sides of the triangle forming their common boundary. It is easy to see that if a line segment crosses only one side of the triangle, then one of the endpoints has to be inside, and the other outside, the triangle. By convexity, if both endpoints are inside then the path joining them does not cross a side of the triangle. If both points P, Q are outside and the path joining them contains an interior point R of the triangle, then both the paths $\overline{PR}$ and $\overline{RQ}$ have to cross the sides of the triangle, and these two points of intersection do not belong to the same side of the triangle.

5.10.3 Polygons

A simple (that is, not self-intersecting) closed polygon divides the plane into two regions, with the sides of the polygon forming the common boundary.

The result has already been established for triangles, and we now use induction on the number of sides of the polygon. Assume as the induction hypothesis that it holds for all polygons with fewer than n sides; the inductive step is to show that the result also holds for a polygon P with n sides.

By a diagonal we mean a straight path joining two vertices of the polygon, but not intersecting one of its sides. We proceed to show that there is a diagonal with the property that it divides P into two new polygons, with the diagonal as a common new side, and that both new polygons have fewer than n sides.

Let $\overline{AB}$ and $\overline{AC}$ be two adjacent sides, and join up $\overline{BC}$ to form a triangle (see Fig. 5.2). If the triangle and the path $\overline{BC}$ do not contain any other vertices of P, then $\overline{BC}$ is already such a diagonal. Otherwise we let G be a vertex in the interior of the triangle and the path $\overline{BC}$ which is furthest away from $\overline{BC}$, and if there is a choice then we just take any path $\overline{AG}$. This will then be the sought-after diagonal. Since the method of proof is the same, we shall only consider the latter situation.

Arrange the vertices of the polygon as a sequence, leading from A to G, and since we need to go back to A, there are now two polygonal paths leading from A to G, giving rise to two polygons P_1, P_2. Let r, s be the sides of these polygons, so that they exceed 2, and since AG is a common side we have $(r-1)+(s-1)=n$. Therefore $r<n$ and $s<n$.

By the induction hypothesis, these two polygons have 'inside' and 'outside' (the outside includes the point ∞) regions. We first prove that there are only two possibilities. The first is that P_1 lies inside P_2 (or P_2 lies inside P_1), and the other possibility is that they lie on the outside of each other.

If what we claim were not true, then P_1 and P_2 would have a common interior point, and yet the interior of P_1 does not include the whole of the interior of P_2, and the interior of P_2 does not include the whole of the interior of P_1. In other words, the interior of P_1 contains interior as well as exterior points of P_2, and the exterior of P_1 also contains interior as well as exterior points of P_2. Thus both the interior and the exterior of P_1 contain boundary points of P_2. Let x, y be boundary points of P_2 lying in the interior and the exterior of P_1. The part π of the boundary of P_2 linking x and y forms a simple curve (Jordan arc), which must intersect with the boundary of P_1. We already know that P_1, P_2 have $\overline{AG}$ as the common boundary, so that π must intersect P_1 on $\overline{AG}$, and not on any other part of its boundary. Consequently the boundary of P_2 has a branch point with at least three sides on $\overline{AG}$, and this contradicts the notion of a simple closed curve. Thus our claim is established.

Suppose now that the first situation holds, that is P_1 lies inside P_2. The interior of P_2, with the interior and the boundary of P_1 removed, is the interior of the original polygon P; the exterior of P_2, together with the interior of P_1, and the line segment $\overline{AG}$ (excluding the two points A, G), form the exterior of P. Under the second situation, the interior of P_1, the interior of P_2, together with the line segment $\overline{AG}$ (excluding the two points A, G), form the interior of P; the exterior of the union of P_1 and P_2 is the exterior of P. We only give the proof for the first situation, leaving the second for the reader.

We first prove that the interior, and the exterior, of P are open sets. Let I and II denote the interior and the exterior, respectively, and let Z represent the set formed by the compact complex plane, so that

$$I = Z \setminus \big((\text{exterior and boundary of } P_2) \cup (\text{interior and boundary of } P_1)\big).$$

The set in the larger parentheses, being the union of two closed sets, is closed. Thus I is the complement of a closed set, and is therefore open. Similarly, from

$$II = Z \setminus \big((\text{interior and boundary of } P_2) \cap (\text{interior and boundary of } P_1)\big),$$

we see that II is also the complement of a closed set, and is therefore open.

We now prove that the interior and the exterior of P are connected sets. That the exterior is connected is obvious, because $\overline{AG}$ is a common boundary for P_1, P_2, so that the exterior points of P_2 and interior points of P_1 can both be connected to $\overline{AG}$. The connectedness of the interior can be established as follows: Let x, y be any two interior points of P, so that they must be exterior points of P_1. There is a curve π lying entirely outside P_1, joining x to y. If π does not intersect the boundary of P then we are done. If π intersects the boundary of P then it must be on the boundary of P_2, and we may assume without loss that there are at least two such points of intersection. Let x_1, y_1 be the first and the last such points. Denote by $d(E, S)$ the distance of any vertex E of P_1 and P_2 to any non-adjacent side S, and let $\rho = \min d(E, S)$ be the least such values. Construct a simple closed polygon

Figure 5.3. Figure 5.4.

P_3 by running a point inside P_2 along and parallel to its boundary with a distance $\eta = \rho/k$, taking k to be sufficiently large. With x, y being in the interior of P_2, we may assume that their distances from the boundary exceed η, since otherwise we may take a smaller value for η. Thus, before π meets the boundary of P_2, it must intersect the boundary of P_3 at x_2, y_2. Letting k be large, the point x_2 approaches x_1, and y_2 approaches y_1. Since P_2 and P_1 have only $\overline{AG}$ as a common boundary, there are two polygonal paths for x_2 to go along P_3 to y_2. One of the paths does not need to traverse 'next to' $\overline{AG}$, and there is no need to go through P_1. Thus, the path from x through x_2 and y_2 to y lies entirely inside P_2 without meeting P_1 or its boundary, and is therefore entirely inside P. The Jordan curve theorem for the polygon is proved.

5.11 The connectivity index

In this section we follow up various intuitive notions arising from the Jordan curve theorem, and we do not give rigorous proofs for them.

The Jordan curve theorem has the following extension: Let γ be a simple closed curve inside a region G, dividing the plane into the two parts D', D''. Denote by G', G'' the intersection of G with D', D'', respectively. Then γ divides G into two parts G' and G'' which are regions, with γ as their common boundary.

Let $\gamma^{(1)}$, $\gamma^{(2)}$ be two non-intersecting simple closed curves (see Fig. 5.3). With $\gamma^{(1)}$ already dividing the plane into the two parts $G^{(1)}$ and $G^{(2)'}$, the curve $\gamma^{(2)}$ must lie in one of them, and $\gamma^{(2)}$ now also divides such a part into two parts. Therefore two non-intersecting simple closed curves divide the plane into three parts, which are all regions. By induction we find that n simple closed curves, no two of which intersect, divide the plane into $n + 1$ regions.

Note, however, that when $n \geq 3$, there are various situations which are not topologically equivalent; see, for example, Fig. 5.4.

In function theory the notion of being 'totally separated' is very important. We say that n simple closed curves are totally separated if, corresponding to any one of them, all the remaining $n - 1$ curves lie on either the interior or the exterior. In the first example in Fig. 5.4, the three curves are not totally separated, whereas the three in the second example are totally separated. The definition can also be restated as follows: Let the plane be divided into the $n + 1$ regions by the n simple closed curves $\gamma^{(\nu)}$ $(\nu = 1, 2, \ldots, n)$. One of the regions has these n curves $\gamma^{(\nu)}$ as a boundary, and the others have one of $\gamma^{(\nu)}$ as a boundary.

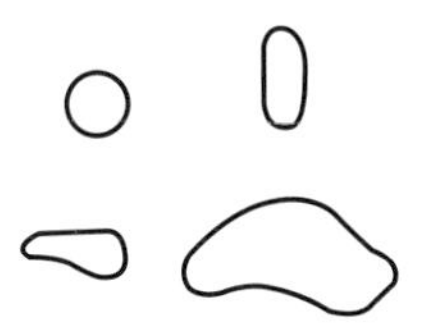
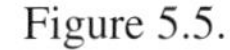

Figure 5.5.

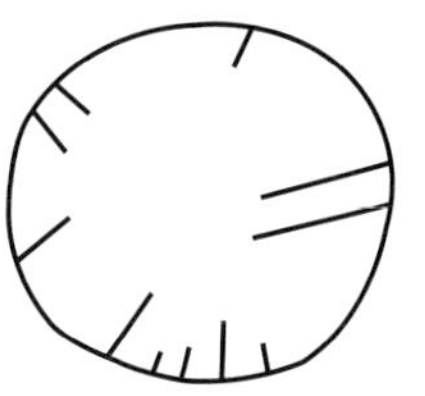

Figure 5.6.

Consider now a region G inside which there are three totally separated simple closed curves $\gamma^{(1)}, \gamma^{(2)}, \gamma^{(3)}$. We can construct a curve inside G linking $\gamma^{(2)}$ and $\gamma^{(3)}$ without intersecting $\gamma^{(1)}$. A (necessary and sufficient) condition for three simple closed curves to be totally separated in G is that any two of them can be so linked up without intersecting the remaining one.

In function theory it is often the case that we need only consider regions with boundaries forming finitely many continuums. Regions having a single continuum as the boundary are said to be simply connected. Those having two disjoint continuums as the boundary are said to be doubly connected. Generally speaking, regions having n disjoint continuums as the boundary are said to be n-connected (see Fig. 5.5).

The integer n is called the connectivity index, and is a topological invariant.

If the perimeter of a region is a simple closed curve, then it is a simply connected region. For example, the open disc $|z| < 1$ is a simply connected region, and so is the upper half-plane $\Im(z) > 0$. However, the unit disc with a radius removed is still a simply connected region, but its perimeter is no longer a simple closed curve! Indeed there are various ways to make cuts (see Fig. 5.6) so that the boundary may become very complicated but it remains a simply connected region.

As we shall see, however, any two simply connected regions (with more than one boundary point) can be topologically transformed into each other (and indeed, even by a conformal mapping). For this reason, topologically speaking, the complicated boundary situation is not such an important phenomenon because, as far as the (interior of a) region is concerned, it is not an invariant under topological mappings.

Let γ be a simple closed curve lying in a simply connected region G. Then γ divides the plane into two 'sides', with all the points inside one 'side' in G. If it were otherwise, part of the boundary of G would be inside γ, and part of it would be outside, and thus could not be a continuum. Conversely, if G is not simply connected, so that its boundary is formed by a certain number of continuums, then one can prove (for example, under a conformal transformation) that there is a simple closed curve inside G separating its boundary into the two sides. This leads us to another definition for being simply connected: A region G is simply connected if the whole of one of the sides of any simple closed curve in G lies inside G.

The connectivity index n of a multiply connected region G can also be defined as the maximum number of totally separated simple closed curves that can be constructed in G,

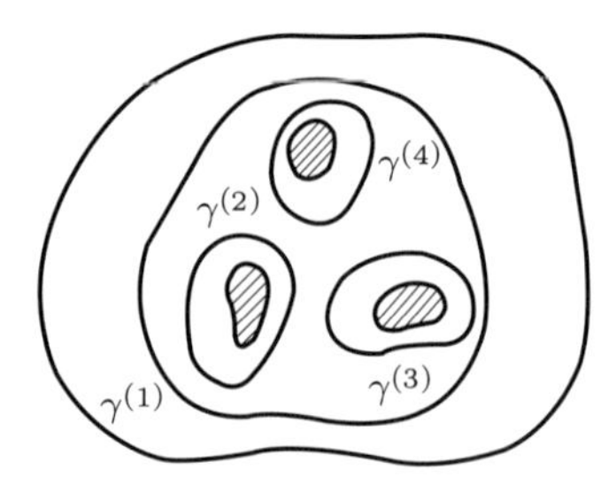

Figure 5.7.

with the curves having the property that there is a region with only these n curves as boundary.

There is yet another definition for n. Let G be a region. By a cut into G we mean a simple arc joining an interior point to the boundary (all other points on the arc being interior points). By a cut across G we mean a simple arc joining a boundary point to another boundary (all other points on the arc being interior points). A simply connected region is one in which a cut across the region divides the region into two regions.

If one can construct N cuts across G, which remains connected, but we have that any $N + 1$ cuts across G will make G no longer connected, then the connectivity index is $n = N + 1$ (see Fig. 5.7).

6
Analytic functions

6.1 Definition of an analytic function

Let $f(z)$ be a single-valued continuous function defined in a region D. If $f(z)$ satisfies any one of the following three (equivalent) conditions then it is called an analytic function.

(I) Cauchy–Goursat. Let z_0 be an interior point of D. We say that $f(z)$ is differentiable at $z = z_0$ if there is a number ℓ with the property that, for any $\epsilon > 0$, there exists $\delta > 0$ such that

$$\left|\ell - \frac{f(z) - f(z_0)}{z - z_0}\right| < \epsilon, \tag{6.1}$$

for all z satisfying $0 < |z - z_0| < \delta$. This is the definition given in Section 3.4. If $f(z)$ is differentiable at every interior point of D then we say that it is analytic in the interior of D. The number ℓ is defined to be the derivative of $f(z)$ at z_0, and is denoted by $f'(z_0)$.

(II) Riemann. Write

$$f(z) = u(x, y) + iv(x, y)$$

in the neighbourhood of z_0. Suppose that u, v are real functions with continuous first partial derivatives satisfying the Cauchy–Riemann equations

$$\frac{\partial u}{\partial x} = \frac{\partial v}{\partial y}, \quad \frac{\partial u}{\partial y} = -\frac{\partial v}{\partial x}. \tag{6.2}$$

Then we say that $f(z)$ is analytic at z_0.

(III) Weierstrass. If there is a neighbourhood of z_0 in which $f(z)$ can be expanded as a power series then $f(z)$ is analytic at z_0. In this case, there is a $\delta > 0$ such that

$$f(z) = a_0 + a_1(z - z_0) + a_2(z - z_0)^2 + \cdots$$

is convergent for $|z - z_0| < \delta$.

From Section 3.4 we know that definition (I) follows from (II) and (III), so that it remains to see why (I) implies (II) and (III). Henceforth, by 'analytic' we mean definition (I).

Note 1. The function

$$f(z) = \sqrt{|xy|}$$

takes the value 0 along the x- and the y-axes. At $z = 0$ we have

$$\frac{\partial u}{\partial x} = \frac{\partial u}{\partial y} = \frac{\partial v}{\partial x} = \frac{\partial v}{\partial y} = 0,$$

so that the Cauchy–Riemann equations hold. However, $f(z)$ does not have a derivative at $z = 0$; this is because

$$\frac{f(z)}{z} = \frac{\sqrt{|xy|}}{x + iy},$$

and with $x = \alpha r$, $y = \beta r$ and $r \to 0$,

$$\frac{\sqrt{|xy|}}{x + iy} \to \frac{\sqrt{|\alpha\beta|}}{\alpha + i\beta},$$

a value which depends on α, β.

To clarify, the Cauchy–Riemann equations do not deliver differentiability because they impose conditions on the limits in two directions only; more specifically, the equality of limits of the quotient

$$\frac{f(z) - f(z_0)}{z - z_0}$$

along two mutually orthogonal paths is not enough for differentiability. Indeed, even if the limit is the same along all straight lines, this approach is still not enough. For example, let

$$f(z) = \begin{cases} \dfrac{xy^2(x + iy)}{x^2 + y^4}, & \text{when } z \neq 0, \\ 0, & \text{when } z = 0. \end{cases}$$

It is not difficult to show that, as $z \to 0$ along any straight line,

$$\frac{f(z) - f(0)}{z} \to 0,$$

whereas, along the path $x = y^2$,

$$\frac{f(z) - f(0)}{z} = \frac{y^4}{y^4 + y^4} = \frac{1}{2} \neq 0,$$

so that $f(z)$ is not differentiable at $z = 0$.

Note 2. In definition (I), if we assume that $f'(z)$ is continuous then (II) follows immediately.

Note 3. The original first definition was given by Cauchy, who also made the assumption that $f'(z)$ has to be continuous – the assumption was shown to be unnecessary by Goursat. It was then asked whether there would be a similar improvement on definition (II). There is indeed such a result – the Looman–Menchoff theorem (the proof will not be given in this book).

Let $u(x, y)$ and $v(x, y)$ have partial derivatives everywhere in a region D, apart from an exceptional countable set. Suppose that

$$\frac{\partial u}{\partial x} = \frac{\partial v}{\partial y}, \quad \frac{\partial u}{\partial y} = -\frac{\partial v}{\partial x}$$

hold everywhere in D, apart from an exceptional set of measure 0. Then $f(z) = u(x, y) + iv(x, y)$ is analytic in D.

6.2 Certain geometric concepts

Let $x(t)$, $y(t)$ $(a \le t \le b)$ represent a simple curve C. For the purpose of the evaluation of integrals along a simple curve, we need to impose the condition that the curve is rectifiable. We shall assume that $x(t)$, $y(t)$ have continuous derivatives, and that $f(z)$ is a continuous function on C. Then

$$\begin{aligned}\int_C f(z)dz &= \int_a^b f\big(x(t) + iy(t)\big)\big(x'(t) + iy'(t)\big)dt \\ &= \int_a^b (ux' - vy')dt + i\int_a^b (uy' + vx')dt,\end{aligned}$$

and in Volume I we proved that

$$\left|\int_C f(z)dz\right| \le ML,$$

where M is the maximum of $|f(z)|$ on C, which has the length L.

A rectifiable simple closed curve is called a contour. For our own convenience we often consider only contours C of the following type: Suppose that there is an interval (a, b)[1] with the property that when $a < x' < b$ the line $x = x'$ intersects C at exactly two points, and when $x' < a$ and $x' > b$ the line $x = x'$ does not meet C. Similarly, there is an interval (α, β) such that when $\alpha < y' < \beta$ the line $y = y'$ intersects C at two points $x_1(y')$ and $x_2(y')$, $(x_1(y') < x_2(y'))$, and when $y' < \alpha$ and $y' > \beta$ the line $y = y'$ does not meet C. A point inside C satisfies $a < x < b$ and $y_1(x) < y < y_2(x)$. A point not inside C, nor on C, is outside C. We shall only deal with such contours.

Let C and C' be two such simple contours, with one or more common arcs between them; otherwise they lie outside of one another. On the removal of the common arcs we have a region C'', the interior of which is made up of the insides of C and C' and the common boundary points (the limit points of the common outside points of C and C' are not supposed to be included; see Fig. 6.1, with α, β, for example.).

Again, if the whole of the inside of C' is part of the inside of C, then the points on the inside of C which lie in the outside of C' form a new region; for example, the annulus

[1] Translator's note: Not to be confused with the parametric interval for the curve C.

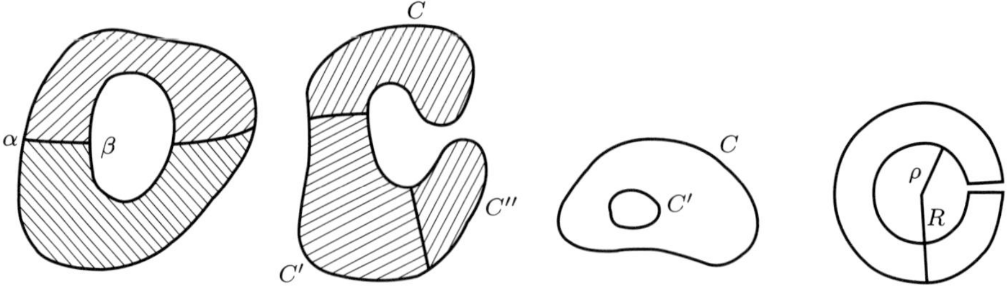

Figure 6.1. Figure 6.2.

$\rho \le |z| \le R$. Consider the part of the region with $y \ge 0$, reflect it along the real axis, and remove the segment $-R$ to $-\rho$ along the real axis (see Fig. 6.2). We then have a region in which the path from $z = \rho$ to $z = R$ is being run over twice in opposite directions.

6.3 Cauchy's theorem

Theorem 6.1 (Cauchy) *Let C represent a contour. Suppose that $f(z)$ is a single-valued analytic function on C and the two dimensional region inside C. Then*

$$\int_C f(z)dz = 0. \tag{6.3}$$

Here the integral is taken along the contour C.

The proof of the theorem is rather easy if we use definition (II) from Section 6.1. This is because

$$\int_C f(z)dz = \int_C (u + iv)(dx + i\,dy) = \int_C u\,dx - v\,dy + i\int_C u\,dy + v\,dy.$$

The required result then follows because, by the Cauchy–Riemann equations, $u\,dx - v\,dy$ and $u\,dy + v\,dy$ are both exact differentials. It can also be established by Green's theorem together with the Cauchy–Riemann equations.

In particular, the function $az + b$ has a continuous derivative so that, by Note 2 in Section 6.1, we know that the real and imaginary parts satisfy the Cauchy–Riemann equations. Therefore

$$\int_C (az + b)dz = 0.$$

The contribution from Goursat is the weakening of the hypothesis, namely the removal of the requirement of the continuity of $f'(z)$. He first showed that, from definition (I), the function $f(z)$ has the property of being 'uniformly differentiable', that is: Given $\epsilon > 0$ we can find $\delta > 0$ (independent of z) such that for all z, z' with $|z' - z| < \delta$ we have

$$|f(z') - f(z) - (z' - z)f'(z)| < \epsilon|z' - z|. \tag{6.4}$$

Here both z' and z lie on C or the region D surrounded by C.

We write $\bar{D} = D \cup C$, the set of points in D together with the points on the contour C.

For the proof of (6.4), we use the method of 'successive dissection'. We first cast a network of paths with lines parallel to the x- and y-axes, with distance ℓ between such parallel lines, so that the plane is partitioned into squares; we are only interested in those squares that contain a point of $\bar{D}$. Let $C_1, \ldots, C_M$ denote those squares which lie entirely inside D, and denote by $D_1, \ldots, D_N$ those parts of the squares that contain a point of C, so that

$$\bar{D} = \bigcup_{i=1}^{M} C_i \cup \bigcup_{j=1}^{N} D_j.$$

With such a division we now separate the sets C_i and D_j into two types. In one type, the inequality (6.4) holds for any z and z' belonging to a member with $|z - z'| < \delta_1$ (with $\delta_1 > 0$ being fixed). In the remaining type there are members for which (6.4) does not hold.

For each set of the second type we partition the region into four parts and set $\delta_2 = \delta_1/2$. If (6.4) still does not hold with δ_2, that is there are two points z, z' in one of the parts with $|z - z'| < \delta_2$ for which (6.4) does not hold, then we further subdivide it into four parts, and we continue in this way. There can only be two possibilities: After a finite number of divisions, the inequality becomes valid, and the required conclusion is established; otherwise, there is a nested sequence of small squares

$$R_1, R_2, R_3, \ldots, R_n, \ldots$$

in which (6.4) does not hold for $\delta_n = \delta_1/2^{n-1}$.

These squares shrink down to a single point z_0'. However, by hypothesis, there exists $\delta_0 > 0$ such that (6.4) holds when $|z - z_0'| < \delta_0$. Moreover, when n is sufficiently large, the whole square R_n lies inside the disc $|z - z_0| < \delta_0$, and $\delta_n < \delta_0$, so that we have a contradiction. Therefore $f(z)$ does have the property of being uniformly differentiable.

Remark Actually, in the above, we are repeating the argument set out in the proof of the Heine–Borel theorem (see Volume I, Section 4.17), and readers familiar with the theorem can apply it directly for the proof.

Proof of Theorem 6.1. We make use of the square grid introduced in the above, and evaluate the integral along the boundaries of $C_1, \ldots, C_M$ and $D_1, \ldots, D_N$, which are denoted by $\partial C_1, \ldots, \partial C_M$ and $\partial D_1, \ldots, \partial D_N$. Thus

$$\int_C f(z)dz = \sum_{m=1}^{M} \int_{\partial C_m} f(z)dz + \sum_{n=1}^{N} \int_{\partial D_n} f(z)dz.$$

Here, each integral is taken in the positive direction, and to see why this equation holds we take, for example, two adjacent squares $ABCD$ and $DCEF$, with the common side CD. In the first square, the integration is from D to C and in the second square the integration is

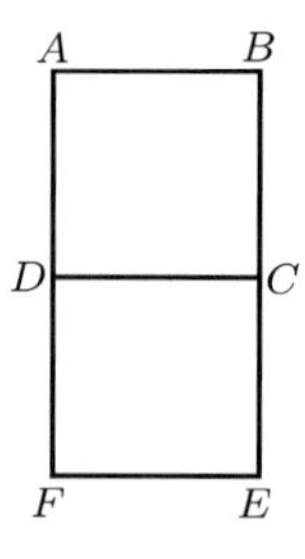

Figure 6.3.

from C to D, so that the two contributions to the sum cancel each other out (see Fig. 6.3). Consequently the total contribution of the two sums must be that from the integral along the curve C.

Now, by (6.4) we have

$$\int_{\partial C_m} f(z)dz = \int_{\partial C_m} [f(z_0') + (z - z_0')f'(z_0')]dz + \int_{\partial C_m} \phi(z)dz,$$

where $|\phi(z)| \leq \epsilon|z - z_0'|$ and z_0' is a point in C_m. The function

$$f(z_0') + (z - z_0')f'(z_0')$$

is linear in z, and thus has a continuous derivative. Therefore, by what we established at the beginning of the section,

$$\int_{\partial C_m} [f(z_0') + (z - z_0')f'(z_0')]dz = \int_{\partial D_n} [f(z_0') + (z - z_0')f'(z_0')]dz = 0.$$

Now, on C_m we have $|z - z_0'| \leq \sqrt{2}\ell$, where ℓ is the length of the side of the squares in the grid, so that

$$\left|\int_{\partial C_m} f(z)dz\right| = \left|\int_{\partial C_m} \phi(z)dz\right| \leq \epsilon\sqrt{2}\ell \cdot 4\ell.$$

Also, the length of D_n cannot exceed $4\ell + S_n$, where S_n is the length of the part of the curve C lying inside D_n, so that

$$\left|\int_{\partial D_n} \phi(z)dz\right| < \epsilon\sqrt{2}\ell(4\ell + S_n).$$

Summing over m, n we have

$$\left|\int_C f(z)dz\right| < 4\sqrt{2}\epsilon\ell^2(M + N) + \epsilon\sqrt{2}\ell \sum_n S_n. \tag{6.5}$$

Here $\ell^2(M + N) \leq (b - a)(\beta - \alpha)$, and $\sum_n S_n$ is the length of the curve C. Therefore the right-hand side of (6.5) is less than a fixed multiple of ϵ, and the left-hand side is independent of ϵ. The theorem is proved. □

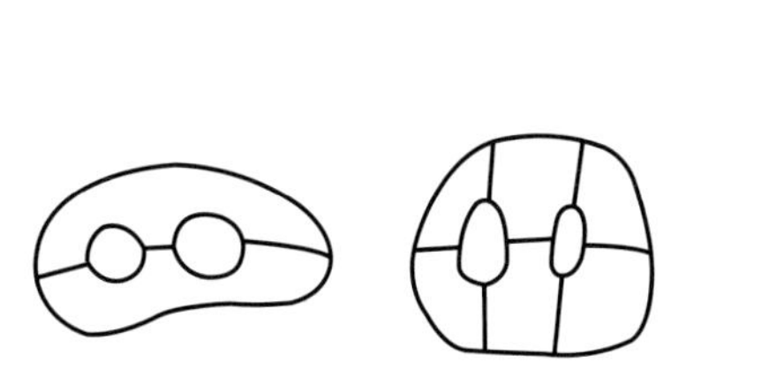

Figure 6.4.

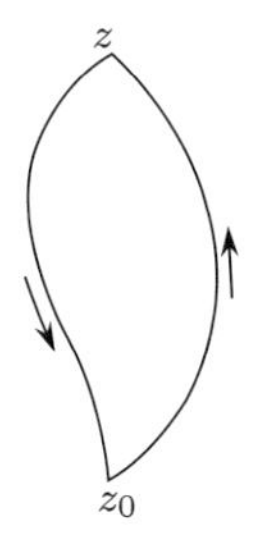

Figure 6.5.

Note 1. Cauchy's theorem still holds for any contour that can be partitioned into a finite number of simple closed curves (see Fig. 6.4). The proof depends on the cancellation of contributions from paths being described twice in opposite directions.

Note 2. The hypothesis for $f(z)$ on C can be somewhat weakened. We need only suppose that $f(z)$ is analytic in the interior of D and continuous in $\bar{D} = D \cup C$. This is because, if $f(z)$ is continuous in $\bar{D}$ then

$$\int_C f(z)dz = \lim_{C' \to C} \int_{C'} f(z)dz,$$

where C' is a contour inside C, but we shall neither clarify nor elaborate on what '$C' \to C$' means here.

A thought-provoking problem, starting from the unit circle, is whether $f(z)$ has to be continuous on C.

Theorem 6.2 *Under the hypothesis of Theorem 6.1, along any curve leading from z_0 to z in D, the function*

$$F(z) = \int_{z_0}^{z} f(w)dw$$

is single-valued and analytic.

Proof. First, if there are two curves leading from z_0 to z, they form a contour (see Fig. 6.5), so that $F(z)$ is single-valued. Next, we have

$$F(z + \delta z) - F(z) = \int_{z}^{z+\delta z} f(w)dw,$$

so that

$$\frac{F(z + \delta z) - F(z)}{\delta z} - f(z) = \frac{1}{\delta z} \int_{z}^{z+\delta z} \big(f(w) - f(z)\big)dw.$$

By continuity, for every $\epsilon > 0$, there exists $\delta > 0$ such that $|f(w) - f(z)| < \epsilon$ when $|w - z| < \delta$. Thus, for $|\delta z| < \delta$, we have

$$\left|\frac{F(z+\delta z) - F(z)}{\delta z} - f(z)\right| < \epsilon.$$

Therefore $F'(z) = f(z)$, and $F(z)$ is analytic in D. □

The function $F(z)$ is called the indefinite integral of $f(z)$.

It is not difficult to prove that if $G'(z) = f(z)$ then F and G differ by a constant, and that

$$\int_{z_0}^{z} f(w)dw = G(z) - G(z_0).$$

6.4 The derivative of an analytic function

Denote by D the inside of a simple contour C, and let $f(z)$ be analytic in D and on C. For a fixed z inside C, the function

$$\frac{f(w)}{w-z}$$

is analytic, except at the point $w = z$. Let $\epsilon > 0$. Take a circle r with centre z and radius ρ so small that $|z - w| = \rho$ lies in D and $|f(w) - f(z)| < \epsilon$ for such w. Then

$$\left(\int_C - \int_r\right)\frac{f(w)}{w-z}\,dw = 0,$$

and thus

$$\int_C \frac{f(w)}{w-z}\,dw = \int_r \frac{f(w)}{w-z}\,dw = f(z)\int_r \frac{f(w)}{w-z}\,dw + \int_r \frac{f(w)-f(z)}{w-z}\,dw.$$

Using polar coordinates

$$w = z + \rho e^{i\theta}, \quad 0 \le \theta \le 2\pi,$$

we have immediately

$$\int_r \frac{f(w)}{w-z}\,dw = \int_0^{2\pi} \frac{i\rho e^{i\theta}\,d\theta}{\rho e^{i\theta}} = 2\pi i$$

and

$$\left|\int_r \frac{f(w)-f(z)}{w-z}\,dw\right| < \frac{\epsilon}{\rho}2\pi\rho = 2\pi\epsilon$$

so that

$$\left|\int_C \frac{f(w)}{w-z}\,dw - 2\pi i f(z)\right| < 2\pi\epsilon.$$

Since the left-hand side does not depend on ϵ, we have

$$\frac{1}{2\pi i}\int_C \frac{f(w)}{w-z}\,dw = f(z). \tag{6.6}$$

This is the famous Cauchy's integral formula, giving the value of $f(z)$ at a point inside the contour in terms of values $f(w)$ on the contour.

In fact, it is enough to suppose that $f(z)$ is only continuous on C as a point approaches the contour from the inside.

Next, for a fixed $z \in D$, at a distance δ from the contour C, we have $z+h \in D$ for $0 < |h| < \delta$. Now

$$\frac{f(z+h)-f(z)}{h} = \frac{1}{2\pi i}\int_C \frac{f(w)}{(w-z)(w-z-h)}\,dw \tag{6.7}$$

and

$$\int_C \frac{f(w)dw}{(w-z)(w-z-h)} - \int_C \frac{f(w)dw}{(w-z)^2} = h\int_C \frac{f(w)dw}{(w-z)^2(w-z-h)}. \tag{6.8}$$

Let $|f(w)| \le M$ for w on C, and denote by ℓ the length of C. Then

$$\left|\int_C \frac{f(w)dw}{(w-z)^2(w-z-h)}\right| \le \frac{M\ell}{\delta^2(\delta-|h|)},$$

which is bounded as $|h| \to 0$. It now follows from (6.7) and (6.8) that

$$f'(z) = \frac{1}{2\pi i}\int_C \frac{f(w)}{(w-z)^2}\,dw.$$

This is Cauchy's formula for $f'(z)$.

The same method shows that

$$\frac{f'(z+h)-f'(z)}{h} = \frac{1}{2\pi i}\int_C \frac{2w-2z-h}{(w-z)^2(w-z-h)^2} f(w)dw$$

from which we deduce that

$$f''(z) = \frac{2}{2\pi i}\int_C \frac{f(w)}{(w-z)^3}\,dw,$$

which shows that $f''(z)$ exists. Thus, the special point about an analytic function is that, from its 'differentiability' we can deduce that it has higher differentiability. Therefore, we can deduce definition (II) from definition (I).

Proceeding inductively we find that, for $n = 1, 2, \dots,$

$$f^{(n)}(z) = \frac{n!}{2\pi i}\int_C \frac{f(w)}{(w-z)^{n+1}}\,dw.$$

Exercise 6.3 Suppose that $f(z)$ is analytic inside and on a circle with centre a and radius r. Show that

$$|f^{(n)}(a)| \le \frac{n!M}{r^n},$$

where M is a maximum value of $|f(z)|$ on the circle.

Exercise 6.4 Let $f(z)$ be an analytic function. Show that, for $f(z) \neq 0$,

$$\frac{\partial^2}{\partial z \partial \bar{z}} \log |f(z)| = 0,$$

and that, apart from those z satisfying $f(z) = 0$ or $f'(z) = 0$, we have

$$\frac{\partial^2}{\partial z \partial \bar{z}} |f(z)| > 0.$$

6.5 Taylor series

In the preceding section we saw that an analytic function has derivatives of all order. We now advance this further by establishing the following: Let $f(z)$ be analytic inside and on a simple closed curve, and let a be a point inside C at a distance δ from C. Then, for $|z - a| < \delta$, the function has the convergent series representation

$$f(z) = f(a) + (z - a) f'(a) + \cdots + \frac{(z - a)^n}{n!} f^{(n)}(a) + \cdots .$$

For this we start with Cauchy's integral formula:

$$f(z) = \frac{1}{2\pi i} \int_\Gamma \frac{f(w)}{w - z} \, dw,$$

where Γ is the circle centre a, radius $\rho < \delta$; the formula holds for $|z - a| < \rho$.

On the circle Γ we have

$$\frac{1}{w - z} = \frac{1}{w - a} + \frac{z - a}{(w - a)^2} + \cdots + \frac{(z - a)^n}{(w - a)^{n+1}} + \cdots ,$$

a uniformly convergent series. We may therefore multiply it by $f(w)/2\pi i$ and integrate along Γ term by term to give

$$f(z) = \frac{1}{2\pi i} \int_\Gamma \frac{f(w)}{w - a} \, dw + \frac{z - a}{2\pi i} \int_\Gamma \frac{f(w)}{(w - a)^2} \, dw + \cdots + \frac{(z - a)^n}{2\pi i} \int_\Gamma \frac{f(w)}{(w - a)^{n+1}} \, dw + \cdots .$$

Since this holds for any $\rho < \delta$, the required result follows by letting $\rho \to \delta$.

The result here shows that definition (I) can be used to deduce definition (III). Therefore the three definitions (I), (II), (III) are equivalent.

There is a very important difference between the Taylor series here and that for a real variable. When examining the Taylor series for a real function, we need to have an expansion up to n terms, and then investigate whether the remainder term will tend to 0 or not as $n \to \infty$. For complex functions, that the remainder tends to 0 is a consequence of the property of being analytic, and the determination of the radius of convergence for the series depends only on the region in which the function is analytic. For example, if ρ is the largest positive number for which $f(z)$ is analytic inside $|z - a| < \rho$, then the radius of convergence for its Taylor series at $z = a$ is precisely ρ. Take the example of the function

$$\frac{1}{1+z^2} \;(= 1 - z^2 + z^4 - z^6 + \cdots)$$

which is analytic except at the points $z = \pm i$. For this reason the expansion at the origin has a radius of convergence 1. However, viewing it only as a real function, we do not see the reason why such an expansion is valid only in the interval $-1 < x < 1$.

Remark In the above discussion there is no need to assume that $f(z)$ has to be analytic at points on C, as long as it is continuous on C.

An even deeper result is that, if ϕ is an integrable function on C, then we can define the Cauchy's type integral accordingly:

$$f(z) = \frac{1}{2\pi i}\int_C \frac{\phi(w)}{w - z}\,dw,$$

which will be analytic inside C, and there is also a Taylor expansion.

Exercise 6.5 Establish the following result, which is known as Morera's theorem – it is the converse of Cauchy's integral theorem.

Let D be a region in which $f(z)$ is a continuous function of z. Suppose that the integral

$$\int f(z)dz$$

is identically zero along any closed contour inside D. Then $f(z)$ is analytic in D.

Hint: Consider

$$F(z) = \int_{z_0}^{z} f(w)dw,$$

and deduce from

$$\frac{F(z+h) - F(z)}{h} - f(z) = \frac{1}{h}\int_z^{z+h} \big(f(w) - f(z)\big)dw$$

that $F(z)$ is analytic, and whence $f(z)$ is analytic.

6.6 Weierstrass' double series theorem

Theorem 6.6 *Let D be the region inside a simple closed curve C and let $\{f_n(z)\}$ be a sequence of analytic functions in $\bar{D} = D \cup C$. Suppose that*

$$\sum_{n=0}^{\infty} f_n(z) \tag{6.9}$$

converges uniformly in C. Then the series converges uniformly in $\bar{D}$, and the sum function is analytic in D.

Proof. Let

$$\Phi(z) = \sum_{n=0}^{\infty} f_n(z), \quad z \in C.$$

Then

$$\frac{1}{2\pi i} \int_C \frac{\Phi(z)}{z - w}\, dz$$

is an analytic function of w in D, which will also be denoted by $\Phi(w)$. Let a be any point in D. Then $1/(z-a)$ is a bounded function on C, so that the series $\sum_{n=0}^{\infty} f_n(z)/(z-a)$ also converges uniformly in C, and can therefore be integrated term by term. Thus

$$\Phi(a) = \frac{1}{2\pi i} \int_C \frac{\Phi(z)}{z - a}\, dz = \sum_{n=0}^{\infty} \frac{1}{2\pi i} \int_C \frac{f_n(z)}{z - a}\, dz = \sum_{n=0}^{\infty} f_n(a),$$

and since $a \in D$ is arbitrary we have

$$\sum_{n=0}^{\infty} f_n(w) = \Phi(w), \quad w \in D. \tag{6.10}$$

It remains to prove that the series converges uniformly in $\bar{D}$. Let $\epsilon > 0$. By hypothesis, for all large n, all positive integer m and all z on C,

$$\left| \sum_{k=n}^{n+m} f_k(z) \right| < \epsilon. \tag{6.11}$$

By the maximum modulus theorem we know that this inequality continues to be valid for $z \in \bar{D}$. The theorem is proved. □

Theorem 6.7 *Let*

$$f_1(z), f_2(z), \ldots, f_n(z), \ldots$$

be a sequence of analytic functions in a region D. Suppose that the series

$$\sum_{n=1}^{\infty} f_n(z)$$

converges uniformly in any closed region D' inside D. Then the sum function

$$f(z) = \sum_{n=1}^{\infty} f_n(z)$$

is analytic in D, and the derivative of any order can be obtained from term-by-term differentiation.

Proof. Let C be any simple closed curve inside D. We first prove that

$$f(z) = \frac{1}{2\pi i} \int_C \frac{f(w)}{w - z}\, dw. \tag{6.12}$$

From

$$f_n(z) = \frac{1}{2\pi i} \int_C \frac{f_n(w)}{w - z} dw$$

we have

$$f(z) = \sum_{n=1}^{\infty} f_n(z) = \sum_{n=1}^{\infty} \frac{1}{2\pi i} \int_C \frac{f_n(w)}{w - z} dw,$$

and, by uniform convergence, the sum can be taken inside the integral sign to give

$$f(z) = \frac{1}{2\pi i} \int_C \frac{1}{w - z} \sum_{n=1}^{\infty} f_n(w) dw = \frac{1}{2\pi i} \int_C \frac{f(w)}{w - z} dw,$$

which is the required equation (6.12).

It is not difficult to deduce from (6.12) that $f(z)$ has a derivative given by

$$f'(z) = \frac{1}{2\pi i} \int_C \frac{f(w)}{(w - z)^2} dw,$$

so that $f(z)$ is analytic.

Also, from uniform convergence, we have

$$\begin{aligned} \frac{1}{2\pi i} \int_C \frac{f(w)}{(w - z)^2} dw &= \frac{1}{2\pi i} \int_C \sum_{n=1}^{\infty} f_n(w) \frac{dw}{(w - z)^2} \\ &= \sum_{n=1}^{\infty} \frac{1}{2\pi i} \int_C \frac{f_n(w)}{(w - z)^2} dw = \sum_{n=1}^{\infty} f_n'(z), \end{aligned}$$

so that the derivative can indeed be obtained from term-by-term differentiation of the series, and the series of derivatives still converges uniformly inside any closed region of D. The reason for this is that if D' is such a closed region then we can include D' inside the closed curve C. Let δ denote the least distance between the points in D' and the points on C. Then, for any $z \in D'$,

$$\begin{aligned} \left| \sum_{n=N}^{N'} f_n'(z) \right| &= \left| \sum_{n=N}^{N'} \frac{1}{2\pi i} \int_C \frac{f_n(w)}{(w - z)^2} dw \right| \\ &= \left| \sum_{n=N}^{N'} \frac{1}{2\pi i} \int_C \sum_{n=N}^{N'} f_n(w) \frac{dw}{(w - z)^2} \right| \leq \frac{\epsilon \ell}{2\pi \delta^2}, \end{aligned}$$

where ℓ is the length of C and ϵ is the maximum of

$$\left| \sum_{n=N}^{N'} f_n(w) \right|, \quad w \text{ on } C.$$

Since these parameters do not depend on z, and $\epsilon \to 0$ as $N, N' \to \infty$, the required result follows.

Starting over again from $\sum_{n=1}^{\infty} f_n'(z)$, we find that

$$f''(z) = \sum_{n=1}^{\infty} f_n''(z)$$

is uniformly convergent. Continuing in this way, the general result is proved. □

Note 1. For term-by-term differentiation of a series of real functions we need to impose as a condition that the series of derivatives has to converge uniformly.

Note 2. We cannot change the conclusion of Theorem 6.7 to 'the sum function is analytic inside and on C'. Take, for example,

$$f(z) = \sum_{n=1} \frac{z^n}{n^2},$$

which converges uniformly in $|z| \leq 1$, but the function is not analytic at $z = 1$ because $f'(z) = -\log(1-z)/z$ and

$$f'(z) = \sum_{n=1} \frac{z^{n-1}}{n}$$

does not converge uniformly in $|z| < 1$.

Note 3. In Theorem 6.6, if C is not closed then we cannot deduce the result for term-by-term differentiation (see Exercise 6.9).

Note 4. The result for sequences is: Let $f_n(z)$ be a sequence of functions analytic inside and on C, and suppose that it converges uniformly on C. Then $f_n(z)$ converges uniformly to $f(z)$ which is analytic inside C, and each sequence of derivatives $f_n^{(\nu)}(z)$ also converges to $f^{(\nu)}(z)$.

Exercise 6.8 Show that the series

$$\zeta(s) = \sum_{n=1}^{\infty} n^{-s}$$

converges in $\Re(s) > 1$ and that, for $k = 1, 2, \ldots,$

$$\zeta^{(k)}(s) = (-1)^k \sum_{n=2}^{\infty} n^{-s} \log^k n.$$

Exercise 6.9 Determine the region in which the series

$$\sum_{n=1}^{\infty} (-1)^n \frac{\sin nz}{n^2}$$

represents an analytic function. The sum function converges uniformly on the real line, but the series for its derivative does not converge uniformly near $z = (2m+1)\pi$, and the series for the second derivatives does not even converge.

Exercise 6.10 Show that the series

$$\sum_{n=1}^{\infty} \frac{\sin nz}{n^2}$$

converges uniformly along the real axis, but that there is no region in the z-plane in which the series converges uniformly.

Exercise 6.11 Use Morera's theorem to deduce Theorem 6.7.

Hint: Use

$$\int_C f(z)dz = \sum_{n=1}^{\infty} \int_C f_n(z)dz = 0$$

to deduce that $f(z)$ is analytic.

Exercise 6.12 Assume the hypothesis in Theorem 6.6 or Theorem 6.7, and suppose that z_0 is a point inside. Expand $f_n(z)$ about z_0 into the power series

$$f_n(z) = \sum_{m=0}^{\infty} a_{mn}(z-z_0)^m$$

to show that

$$f(z) = \sum_{n=0}^{\infty} f_n(z) = \sum_{m=0}^{\infty} \Big(\sum_{n=0}^{\infty} a_{mn} \Big)(z-z_0)^m.$$

(This is the source of the nomenclature 'double series' for the theorem.)

6.7 Analytic functions defined by integrals

Theorem 6.13 *Let $f(z, w)$ represent a continuous function of the two complex variables z, w, with z varying in a region D, and w varying on a smooth curve C. Suppose that, for each w on C, $f(z, w)$ is an analytic function of $z \in D$. Then*

$$F(z) = \int_C f(z, w)dw$$

defines an analytic function in D, and

$$F'(z) = \int_C \frac{\partial f}{\partial z} dw;$$

there are similar results for the higher derivatives.

Proof. Let C be given by

$$w = u + iv, \quad u = u(t), \quad v = v(t), \quad t_0 \le t \le t_1,$$

where $u'(t)$, $v'(t)$ are continuous functions.

Take a smooth contour Γ in D, given by

$$z = x + iy,\ x = x(s),\ y = y(s),\ s_0 \le s \le s_1, \quad x(s_0) = x(s_1),\ y(s_0) = y(s_1),$$

where $x'(s)$, $y'(s)$ are continuous functions. For any point ζ inside Γ, we have

$$f(\zeta, w) = \frac{1}{2\pi i} \int_\Gamma \frac{f(z, w)}{z - \zeta} dz,$$

so that

$$F(\zeta) = \frac{1}{2\pi i} \int_C dw \int_\Gamma \frac{f(z, w)}{z - \zeta} dz.$$

We interchange the order of integration here, which is justified by the fact that the expression has the 'real' representation

$$\int_{t_0}^{t_1} dt \int_{s_0}^{t_1} (\phi(s, t) + i\psi(s, t))ds,$$

with ϕ, ψ being continuous functions. We then have

$$F(\zeta) = \frac{1}{2\pi i} \int_\Gamma \frac{dz}{z - \zeta} \int_C f(z, w)dw = \frac{1}{2\pi i} \int_\Gamma \frac{F(z)}{z - \zeta} dz.$$

Thus $F(z)$ satisfies Cauchy's integral formula, from which $F(z)$ can be shown to be analytic and can be differentiated under the integral sign. □

Example 6.14 Let $f(t)$ be continuous in (a, b). Then

$$F(z) = \int_a^b \cos zt\ f(t)dt, \quad G(z) = \int_a^b \sin zt\ f(t)dt,$$

are analytic functions of z in the whole plane.

The function

$$\int_a^b \frac{f(t)}{z - t} dt$$

is analytic, except for z lying in the real interval (a, b).

Exercise 6.15 Use Morera's theorem to prove Theorem 6.13.

Theorem 6.16 *If the curve C extends to infinity, but the integrals*

$$\int_C f(z, w)dw, \quad \int_C \frac{\partial^\nu f(z, w)}{\partial z^\nu} dw, \quad \nu = 1, 2, \ldots,$$

converge uniformly, then the results in Theorem 6.13 are still valid.

Proof. Let C_n be the part of C that lies in the circle $|z| \leq n$, and let

$$F_n(z) = \int_{C_n} f(z, w) dw.$$

Then, for any n, $F_n(z)$ is analytic and tends uniformly to $F(z)$ as $n \to \infty$, so that $F(z)$ is analytic. Finally,

$$F'(z) = \lim_{n\to\infty} F_n'(z) = \lim_{n\to\infty} \int_{C_n} \frac{\partial f}{\partial z} dw = \int_C \frac{\partial f}{\partial z} dw.$$

□

The same method can be used to deal with the case when D is an unbounded region, but uniform convergence has to be guaranteed.

Example 6.17 The integral

$$\Gamma(z) = \int_0^\infty e^{-w} w^{z-1} dw$$

represents an analytic function in $\Re(z) > 0$.

Exercise 6.18 In which regions do the integrals

$$\int_0^\infty e^{-zw^2} dw, \quad \int_0^\infty \frac{\sin w}{w^z} dw, \quad \int_0^\infty \frac{\cos w}{w^z} dw$$

represent analytic functions?

6.8 Laurent series

Theorem 6.19 *Let $f(z)$ be a single-valued analytic function in an annulus region $R' \leq |z - a| \leq R$. Denote by C' and C the positively oriented circles with radii R' and R, respectively, forming the boundary of the annulus. Then $f(z)$ can be expanded as a series of integer powers of $z - a$ which converges inside the annulus.*

Proof. Let z be any point inside the annulus, and consider the integral

$$\frac{1}{2\pi i} \int \frac{f(w)}{w - z} dw.$$

We take the following contour: First describe C and then move in along a radius to reach C', which is then described backward (that is, in the negative direction), and then move back out to C along that same part of the radius (see Fig. 6.6). Since $f(z)$ is single-valued, the contribution to the integral along the radius is cancelled, so that

$$f(z) = \frac{1}{2\pi i} \int_C \frac{f(w)}{w - z} dw - \frac{1}{2\pi i} \int_{C'} \frac{f(w)}{w - z} dw.$$

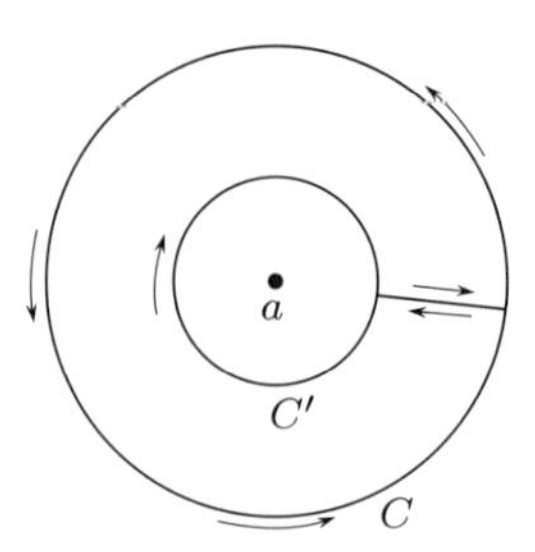

Figure 6.6.

It is easy to see that

$$\frac{1}{2\pi i}\int_C \frac{f(w)}{w-z}\,dw = \sum_{n=0}^{\infty} a_n(z-a)^n,$$

with

$$a_n = \frac{1}{2\pi i}\int_C \frac{f(w)}{(w-a)^{n+1}}\,dw.$$

Next, we have the series expansion

$$\frac{1}{z-w} = \frac{1}{z-a} + \frac{w-a}{(z-a)^2} + \cdots + \frac{(w-a)^{n-1}}{(z-a)^n} + \cdots,$$

which converges uniformly for $w \in C'$, so that

$$\begin{aligned}
-\frac{1}{2\pi i}\int_{C'} \frac{f(w)}{w-z}\,dw &= \frac{1}{z-a}\cdot\frac{1}{2\pi i}\int_{C'} f(w)dw + \cdots \\
&\quad + \frac{1}{(z-a)^n}\cdot\frac{1}{2\pi i}\int_{C'} (w-a)^{n-1} f(w)dw + \cdots \\
&= \sum_{n=1}^{\infty} \frac{a_{-n}}{(z-a)^n},
\end{aligned}$$

where

$$a_{-n} = \frac{1}{2\pi i}\int_{C'} (w-a)^{n-1} f(w)dw.$$

We have arrived at the Laurent expansion

$$f(z) = \sum_{n=-\infty}^{\infty} a_n(z-a)^n,$$

where

$$a_n = \frac{1}{2\pi i}\int \frac{f(w)}{(w-a)^{n+1}}\,dw, \quad n = 0, \pm 1, \pm 2, \ldots,$$

and the integral here is along any contour inside the annulus looping round the centre. It is easy to see that the part of the series with positive exponents for $z-a$ is convergent in $|z-a|<R$, while the part with negative exponents converges in $|z-a|>R'$. The theorem is proved. □

The problem becomes even clearer when we compare the expansion (taking $a=0$) with that for Fourier series: On the unit circle we have the general Fourier expansion

$$\psi(\theta)=\sum_{n=-\infty}^{\infty} a_n e^{ni\theta}.$$

In general, the power series

$$\sum_{n=1}^{\infty} a_n z^n$$

converges inside the disc $|z|<R$, where

$$R=\lim_{n\to\infty}|a_n|^{-1/n},$$

and the series

$$\sum_{n=1}^{\infty} a_{-n} z^{-n}$$

converges in $|z|>R'$, where

$$\frac{1}{R'}=\lim_{n\to\infty}|a_{-n}|^{-1/n}.$$

If $R'<R$ then, for $R'<|z|<R$, we have an analytic function

$$\sum_{n=-\infty}^{\infty} a_n z^n.$$

Example 6.20 We have

$$e^{\frac{1}{2}c(z-\frac{1}{z})}=\sum_{n=-\infty}^{\infty} a_n z^n, \quad \text{where} \quad a_n=\frac{1}{2\pi i}\int_0^{2\pi} e^{i(c\sin\theta-n\theta)}\,d\theta.$$

Example 6.21 When $c>0$, we have

$$e^{z+\frac{c^3}{2z^2}}=\sum_{n=-\infty}^{\infty} a_n z^n,$$

where

$$a_n = \frac{e^{-\frac{1}{2}c}}{2\pi c^n} \int_0^{2\pi} e^{c(\cos\theta+\cos^2\theta)+i(c\sin\theta(1-\cos\theta)-n\theta)}\, d\theta.$$

6.9 Zeros and poles

Definition 6.22 Let $f(z)$ be analytic at $z = a$, with the expansion

$$f(z) = \sum_{n=0}^{\infty} a_n (z-a)^n.$$

If $a_0 = a_1 = \cdots = a_{m-1} = 0$ and $a_m \neq 0$, then we call $z = a$ a zero of order m of the function $f(z)$; zeros of order 1 and 2 are usually called simple and double zeros, respectively.

Theorem 6.23 *Zeros are isolated. That is, if a non-constant analytic function $f(z)$ has a zero at $z = a$, then there exists $\rho_1 > 0$ such that $f(z)$ has no other zero in $|z-a| < \rho_1$.*

Proof. We may assume without loss that $a = 0$ and that

$$f(z) = \sum_{n=0}^{\infty} a_n z^n, \quad |z| < R.$$

For $f(z)$ to be non-constant, not all the coefficients a_n are 0, so there exists m such that $a_0 = a_1 = \cdots = a_{m-1} = 0$ and $a_m \neq 0$. Thus

$$f(z) = z^m(a_m + a_{m+1}z + \cdots), \quad |z| < R.$$

Let $0 < \rho < R$; then the series converges at $z = \rho$, so that $a_n\rho^n$ is bounded, say $|a_n|\rho^n < K$. Then, for $|z| < \rho$,

$$|f(z)| \geq |z|^m \left(|a_m| - \frac{K|z|}{\rho^{m+1}} - \frac{K|z|^2}{\rho^{m+2}} - \cdots \right) = |z|^m \left(|a_m| - \frac{K|z|}{\rho^m(\rho - |z|)} \right),$$

which is positive when

$$0 < |z| < \rho_1 = \frac{|a_m|\rho^{m+1}}{K + |a_m|\rho^m}.$$

□

As a corollary, we have the following:

(i) If an analytic function $f(z)$ is 0 either in a small region, or along a curve, then $f(z) \equiv 0$.

(ii) Let $f(z)$ be analytic in a region D, and suppose that $z_1, z_2, \ldots, z_n, \ldots$ is a set of points with a limit point z_0 inside D. If $f(z_k) = 0$ for $k = 1, 2, \ldots$ then $f(z) \equiv 0$.

(iii) Let $f(z)$ and $g(z)$ be analytic inside D, and suppose that

$$f(z_k) = g(z_k), \quad k = 1, 2, \ldots.$$

If z_k converges to a limit inside D, then $f(z) \equiv g(z)$.

Let a be a zero of order k of an analytic function $f(z)$, so that

$$f(z) = (z-a)^k g(z),$$

and assume there is a ρ such that $g(z)$ has no zero in $|z-a| < \rho$. Thus $1/g(z)$ is analytic in $|z-a| < \rho$, so that

$$\frac{1}{f(z)} = \frac{1}{(z-a)^k}\big(b_0 + b_1(z-a) + \cdots\big), \quad b_0 \neq 0, \quad 0 < |z-a| < \rho.$$

Definition 6.24 Let $h(z)$ be analytic in a punctured neighbourhood of a. Suppose that

$$\lim_{z\to a}(z-a)^m h(z) = C \neq 0.$$

Then we say that $h(z)$ has a pole of order m at $z = a$. Poles of order 1 and 2 are called simple and double poles, respectively.

It is clear that if $f(z)$ has a zero of order m at $z = a$, then $1/f(z)$ has a pole of order m at a, and conversely.

Example 6.25 The function $\sin z$ has zeros at $z = 0, \pm\pi, \pm 2\pi, \ldots$, and at no other place.

This follows from $|\sin(x+iy)| = \sqrt{\sin^2 x + \sinh^2 y}$, which is 0 if and only if $y = 0$ and $x = 0, \pm\pi, \pm 2\pi, \ldots$.

Also, $\cos z = \sin(\pi/2 - z)$ has zeros at $z = \pm\pi/2, \pm 3\pi/2, \ldots$, and at no other place.

Example 6.26 The functions $\cot z$ and $\operatorname{cosec} z$ have poles at $z = 0, \pm\pi, \pm 2\pi, \ldots$, and the functions $\tan z$ and $\sec z$ have poles at $z = \pm\pi/2, \pm 3\pi/2, \ldots$; there are no other poles for these functions.

Exercise 6.27 Find the poles for the functions

$$\frac{1}{\sin z \pm \sin a}, \quad \frac{1}{\cos z \pm \cos a}.$$

Exercise 6.28 Show that the function $\operatorname{cosec} z^2$ has a double pole and infinitely many simple poles.

Exercise 6.29 Let $f(z)$ and $g(z)$ be analytic in D, and suppose that $f(z)g(z) \equiv 0$. Show that either $f(z) \equiv 0$ or $g(z) \equiv 0$.

6.10 Isolated singularities

Let $f(z)$ be analytic in a punctured neighbourhood of a, say $0 < |z-a| < R$. The point a is called an isolated singularity of $f(z)$. Assuming that $f(z)$ is single-valued, it follows from Section 6.7 that there is the Laurent expansion

$$f(z) = \sum_{n=0}^{\infty} a_n (z-a)^n + \sum_{n=1}^{\infty} b_n (z-a)^{-n}, \quad 0 < |z-a| < R.$$

We call the second series the principal part of $f(z)$ at a. There are three possibilities:

(i) The coefficients b_n are all 0, so that there is no principal part. In this case the point a is not really a singularity, because if we define, or redefine, $f(a) = a_0$, then $f(z)$ becomes analytic at $z = a$.

(ii) There are only finitely many b_n not equal to 0, say $b_m \neq 0$ and $b_n = 0$ for $n > m$, so that the principal part is the finite sum

$$\sum_{n=1}^{m} b_n (z-a)^{-n}.$$

In this case $(z-a)^m f(z)$ is analytic in the neighbourhood of a, so that $f(z)$ has a pole of order m at a, with

$$\lim_{z \to a} f(z) = \infty, \qquad \lim_{z \to a} (z-a)^m f(z) = b_m.$$

(iii) There are infinitely many $b_n \neq 0$, so that the principal part is an infinite series

$$\sum_{n=1}^{\infty} b_n (z-a)^{-n}.$$

In this case we call a an isolated essential singularity of $f(z)$. There is the following famous theorem concerning such a singularity.

Theorem 6.30 (Weierstrass) *Let a be an isolated essential singularity of $f(z)$. Then, for any two positive numbers ρ, ϵ, and any complex number c, there is z such that $|f(z) - c| < \epsilon$ and $|z - a| < \rho$.*

In other words, given any complex number c, we can find a convergent sequence z_n with the limit a such that $f(z_n)$ has the limit c. The statement of the theorem does not cater for $c = \infty$, but we can also include it by having the conclusion $|f(z)| > 1/\epsilon$ with $|z - a| < \rho$ instead.

Proof. (1) We first deal with the case $c = \infty$. If the conclusion were false then there would be $\rho > 0$ and M such that $|f(z)| \leq M$ for $0 < |z - a| < \rho$. Let C be the circle centre at a with radius $r < \rho$; then

$$|b_n| = \left| \frac{1}{2\pi i} \int_C (w-a)^{n-1} f(w) dw \right| \leq M r^n.$$

For each fixed n, we may let $r \to 0$ to deliver $b_n = 0$, which contradicts the hypothesis that there is a principal part for $f(z)$.

(2) Now let c be any complex number. If $f(z) - c$ has a zero on a circle $|z - a| = \rho$ then there is nothing more to prove. We may therefore suppose that there is a ρ such that $f(z) - c$ has no zero in $|z - a| < \rho$, so that the function

$$\phi(z) = \frac{1}{f(z) - c}$$

is analytic in $0 < |z - a| < \rho$, and that $z = a$ is also an isolated essential singularity of $\phi(z)$. This is because, if $z = a$ were only a pole of $\phi(z)$, then

$$f(z) = \frac{1}{\phi(z)} + c$$

could not have an essential singularity at $z = a$.

From (1) we know that there is a z such that $|z - a| < \rho$ and $|\phi(z)| > 1/\epsilon$, and hence $|f(z) - c| < \epsilon$. The theorem is proved. □

The theorem of Weierstrass marks out clearly the distinction between an isolated essential singularity and a pole. The theorem led to a subsequent series of investigations, the discussions on which we shall return to in Chapters 13 and 14 of this volume.

Example 6.31 Each of the functions

$$e^{1/z}, \quad \sin\frac{1}{z}, \quad \cos\frac{1}{z},$$

has an isolated essential singularity at $z = 0$.

Example 6.32 The function cosec $1/z$ has poles at $z = 1/n\pi$ $(n = \pm1, \pm2, \ldots)$, and the point 0 is the limit point of $1/n\pi$, so that 0 is not an isolated singularity; we sometimes call it an essential singularity.

Example 6.33 Consider $w = e^{1/z}$. If $w \neq 0$ then

$$\frac{1}{z} = \log w + 2\pi mi, \quad m = 0, \pm1, \pm2, \ldots,$$

so that there are infinitely many values for z with $e^{1/z} = w$; on the other hand, we have

$$\lim_{x \to 0_+} e^{-1/x} = 0.$$

6.11 Being analytic at infinity

By the behaviour of $f(z)$ at $z = \infty$ we mean the behaviour of $g(w) = f(1/w)$ at $w = 0$.

For example, the functions

$$f(z) = \frac{1}{z^2}, \quad \frac{1}{z^2 + az + b}$$

have a double zero at infinity, and

$$z^3 + az + b$$

has a triple pole there; also e^z has an isolated essential singularity at infinity.

Theorem 6.34 *A function which is analytic everywhere, including the point at infinity, is a constant.*

Proof. Take the Laurent expansion about $z = 0$:

$$f(z) = \sum_{n=0}^{\infty} a_n z^n + \sum_{n=1}^{\infty} b_n z^{-n}.$$

Since $f(z)$ is analytic at $z = 0$, we need to have $b_n = 0$ for all n, so that there is no principal part; also, being analytic at $z = \infty$ implies $a_n = 0$ for all $n > 0$, so that $f(z) = a_0$, a constant. □

Theorem 6.35 *If all the singularities are poles then the function must be a rational function; the converse also holds.*

Proof. First, there can only be finitely many poles, since otherwise there would be a limit point, finite or infinite, for these poles, and the function cannot be analytic at such a limit point, which cannot be a pole either, contradicting the hypothesis.

Let the poles of $f(z)$ be at

$$a, b, \ldots, k,$$

with orders $\alpha, \beta, \ldots, \kappa$, respectively. Then

$$g(z) = f(z)(z-a)^{\alpha} \cdots (z-k)^{\kappa}$$

is a function which is analytic everywhere, except possibly infinity, which can only be a pole. Therefore we have the following expansions:

$$g(z) = \sum_{n=0}^{\infty} a_n z^n, \quad g\left(\frac{1}{w}\right) = \sum_{n=0}^{\infty} a_n \frac{1}{w^n}.$$

Since $g(1/w)$ can only have a pole at $w = 0$, it follows that $g(z)$ is a polynomial, and $f(z)$ is the ratio of two polynomials, a rational function.

The converse is trivial. □

We can say more about a rational function, say

$$f(z) = \frac{P(z)}{Q(z)},$$

where $P(z)$ and $Q(z)$ are polynomials with no common factors. The number of zeros (taking account of multiplicity) of $f(z)$ in the finite part of the plane is the degree n of $P(z)$, and the number of poles is the degree m of $Q(z)$; if $n > m$ then $f(z)$ has a pole of order $n - m$ at ∞, and if $m > n$ then it has a zero of order $m - n$ there.

Corresponding to each a on the complex plane, we may define

$$\ell(a) = \begin{cases} m, & \text{if } a \text{ is a zero of order } m \text{ for } f(z), \\ 0, & \text{if } f(a) \neq 0, \infty, \\ -m, & \text{if } a \text{ is a pole of order } m \text{ for } f(z). \end{cases}$$

Then a rational function $f(z)$ has the property that

$$\sum \ell(a) = 0,$$

where the sum is over all the complex numbers a and the point at infinity.

Theorem 6.36 (Liouville) *A bounded function which is analytic in the finite part of the plane is a constant.*

Proof. The point at infinity can be considered an isolated singularity, and since the function is bounded it is also analytic at infinity. The required result follows from Theorem 6.34. □

It is easy to deduce the following more general result.

Theorem 6.37 *Suppose that $f(z)$ is analytic in the finite part of the plane, and that*

$$f(z) = O(|z|^k), \quad |z| \to \infty.$$

Then $f(z)$ is a polynomial with degree at most k.

6.12 Cauchy's inequality

Theorem 6.38 *Let*

$$f(z) = \sum_{n=0}^{\infty} a_n z^n, \quad |z| < R,$$

and let $M(r)$ represent the maximum of $|f(z)|$ on $|z| = r$ $(r < R)$. Then

$$|a_n| r^n \le M(r), \quad n = 0, 1, 2 \dots .$$

Proof. This follows at once from the formula

$$a_n = \frac{1}{2\pi i} \int_{|z|=r} \frac{f(z)}{z^{n+1}}\, dz.$$

□

Theorem 6.39 *Let $A(r)$ denote the maximum of $\Re f(z)$ on the circle $|z| = r$ $(r < R)$. Then*

$$|a_n| r^n \le \max\big(4A(r), 0\big) - 2\Re f(0), \quad n = 0, 1, 2, \dots .$$

Proof. Let $z = re^{i\theta}$, $a_n = \alpha_n + i\beta_n$ and

$$f(z) = \sum_{n=0}^{\infty} a_n z^n = u(r, \theta) + iv(r, \theta);$$

then

$$u(r,\theta) = \sum_{n=0}^{\infty}(\alpha_n \cos n\theta - \beta_n \sin n\theta)r^n.$$

The power series is uniformly convergent, so that on multiplication by $\cos n\theta$ and $\sin n\theta$ and then integrating, we have

$$\frac{1}{\pi}\int_0^{2\pi} u(r,\theta)\cos n\theta\, d\theta = \alpha_n r^n, \quad \frac{1}{\pi}\int_0^{2\pi} u(r,\theta)\sin n\theta\, d\theta = -\beta_n r^n, \quad n > 0,$$

and

$$\frac{1}{2\pi}\int_0^{2\pi} u(r,\theta)d\theta = \alpha_0.$$

Therefore

$$a_n r^n = (\alpha_n + i\beta_n)r^n = \frac{1}{\pi}\int_0^{2\pi} u(r,\theta)e^{-ni\theta}\, d\theta, \quad n > 0,$$

so that

$$|a_n|r^n \le \frac{1}{\pi}\int_0^{2\pi} |u(r,\theta)|d\theta,$$

and hence

$$|a_n|r^n + 2\alpha_0 \le \frac{1}{\pi}\int_0^{2\pi} \big(|u(r,\theta)| + u(r,\theta)\big)d\theta.$$

If $u < 0$ then $u + |u| = 0$, so that when $A(r) < 0$, the right-hand side is 0. If $A(r) \ge 0$ then the right-hand side is at most

$$\frac{1}{\pi}\int_0^{2\pi} 2A(r)d\theta = 4A(r)).$$

The theorem is proved. □

There are similar results corresponding to the lower bound for $\Re f(z)$, and the upper and lower bounds for $\Im f(z)$.

Theorem 6.40 *Suppose that $f(z)$ is analytic in the finite part of the plane and that $A(r)$ is bounded. Then $f(z)$ is constant. If $A(r) = O(r^k)$ then $f(z)$ is a polynomial with degree at most k.*

Proof. We deduce from Theorem 6.39 that, for any fixed $n \ge 1$, the expression $|a_n|r^n$ is bounded for all large r, so that $a_n = 0$, giving $f(z) = a_0$. Next, if $|a_n|r^n = O(|z|^k)$ then we can deduce that $a_n = 0$ from $n > k$, and hence the required result. □

Theorem 6.41 *A bounded function which is harmonic in the finite part of the plane is a constant.*

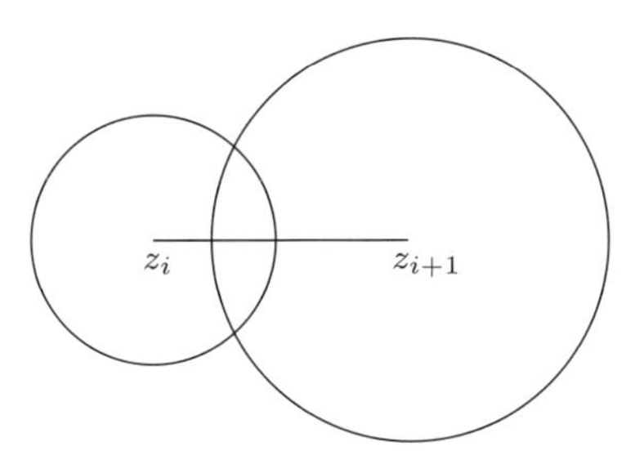

Figure 6.7.

6.13 Analytic continuations

We now consider the nature of an analytic function as a whole. Suppose that $f(z)$ is analytic at $z = z_0$, so that there is a power series in $z - z_0$ which represents $f(z)$ in the neighbourhood of z_0:

$$f(z) = \sum_{n=0}^{\infty} a_n (z - z_0)^n. \tag{6.13}$$

The radius of convergence ρ for the series is also the distance between z_0 and the nearest singularity of $f(z)$. Here is the reason: (i) It is clear that $f(z)$ is analytic in $|z - z_0| < \rho$; (ii) if there were no singularity on the boundary $|z - z_0| = \rho$ then there would be an $\epsilon > 0$ such that $f(z)$ is analytic in $|z - z_0| \leq \rho + \epsilon$, and the Taylor expansion (6.13) would then be convergent in $|z - z_0| < \rho + \epsilon$, which contradicts ρ being the radius of convergence.

Generally speaking, let us suppose that $f(z)$ is analytic at every point on a bounded curve Γ, leading from z_0 to z. Corresponding to each point on Γ, there is a disc with centre at the point in which $f(z)$ is analytic; these discs form a family of open cover for Γ, so that, by the Heine–Borel theorem, there is a finite sub-family which covers Γ already. Let the centres for these discs covering Γ be $z_0, z_1, z_2, \ldots, z_n = z$, with their corresponding circles

$$C_0, C_1, C_2, \ldots, C_n.$$

We now show that, if we know the power series expansion for $f(z)$ inside C_i, then we can use the following method to determine the power series expansion inside C_j. In effect the union of these discs forms a connected open set, so that we need only consider the situation when C_i and C_{i+1} are adjacent circles along Γ, with a non-empty intersection.

Let the distance between the two points of intersection of the circles be 2δ. Let η be the distance between their centres z_i and z_{i+1}, and divide this distance into $\ell = [2\eta/\delta] + 1$ equal parts. Each part has a length less than $\delta/2$, and we let the division points be (see Fig. 6.7)

$$z_i, z_i^{(1)}, z_i^{(2)}, \ldots, z_i^{(\ell)} = z_{n+1}.$$

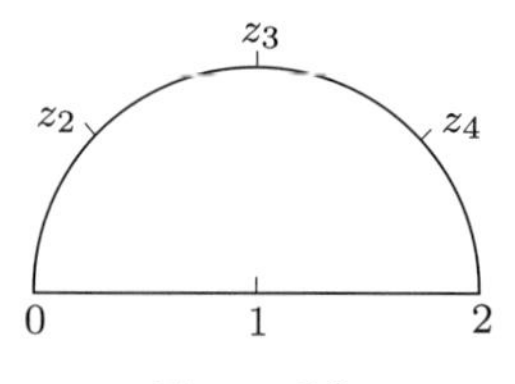

Figure 6.8.

At each $z_i^{(j)}$ the function has a power series expansion

$$f(z) = \sum_{n=0}^{\infty} a_n^{(j)}(z - z_i^{(j)})^n, \tag{6.14}$$

with a radius of convergence at least δ. We may therefore use (6.14) to compute $f^{(n)}(z_i^{(j+1)})$, and use them for the power series expansion about $z_i^{(j+1)}$; such a power series also has a radius of convergence which is at least δ, and we can use the expansion there to compute that at $z_i^{(j+2)}$. We continue in this way from z_0 along z_i to z_{i+1} until we reach z. Such a procedure is called an analytic continuation.

For example, the series

$$1 + z + z^2 + \cdots, \quad |z| < 1,$$

can be continued analytically along the upper semi-circle $|z - 1| = 1$ to the point $z = 2$ for a series expansion there (see Fig. 6.8). More specifically, in $|z| < 1$ we take $z_2 = 1 + e^{3\pi i/4}$, and compute a power series expansion in $z - z_2$ which converges in $|z - z_2| < 1$. In this disc we take $z_3 = 1 + i$, and compute the power series expansion there. We next take $z_4 = 1 + e^{\pi i/4}$, and then finally the power series expansion at the point $z = 2$.

In this way, we have the two series

$$1 + z + z^2 + \cdots, \quad |z| < 1,$$
$$-1 + (z - 2) - (z - 2)^2 + (z - 2)^3 - \cdots, \quad |z - 2| < 1,$$

with no common point z at which both series can converge; either series can be obtained from the other by means of analytic continuation. Actually, they both represent the function

$$\frac{1}{1 - z}$$

in their respective regions, and we can verify the expansions directly by finding the power series representations at the points $z = 0$ and $z = 2$.

More generally we have the following theorem.

Theorem 6.42 *Let D be the common region between D_1, D_2, in which the functions $f_1(z)$, $f_2(z)$, respectively, are single-valued and analytic, and take the same values in D. Then there is a unique single-valued analytic function $F(z)$ in $D_1 \cup D_2$ which takes the same values as $f_1(z)$ in D_1 and the same values as $f_2(z)$ in D_2.*

This can be deduced from Theorem 6.23, which states that the zeros of an analytic function are isolated.

6.14 Multi-valued functions

The following situation may occur from analytic continuation: Starting from z_0 and moving along a certain curve to z_1, we arrive at a certain value $f_1(z_1)$ and, moving along another curve to the same point z_1, we arrive at another value $f_2(z_1) \neq f(z_1)$. In this case we say that the function concerned is multi-valued, with multiple branches. For example, we take $\sqrt{z}$ on the plane with the cut along the negative real axis from 0 to ∞. Writing $z = re^{i\theta}$, the function $\sqrt{z}$ then has two branches:

$$\sqrt{r}e^{i\theta/2}, \quad -\sqrt{r}e^{i\theta/2}, \quad -\pi < \theta < \pi.$$

If we do not make the cut, then on looping round 0 once the first branch will change to the second branch; this can be shown explicitly using power series along the unit circles with centres at the points $1, e^{i\pi/4}, e^{i\pi/2}, e^{3i\pi/4}, -1, e^{i\pi/4}, e^{5i\pi/4}, e^{3i\pi/2}, e^{7i\pi/4}$, and the disc with centre at $e^{7i\pi/4}$ includes the point 1. We then find that, in this way, the value of the power series taken at $z = 1$ is different from that taken by the other branch.

Similarly, on the same cut plane, $\log z$ has infinitely many branches:

$$\log r + i(\theta + 2n\pi), \quad -\pi < \theta < \pi, \quad n = 0, \pm 1, \pm 2, \dots,$$

with each integer n corresponding to a branch. We still use $\log z$ to represent the function in the plane before the cut; obviously we need to specify the value to be taken by the function at an initial point, and we also need to know the chosen path to reach the point at which the functional value is to be determined.

If α is any real irrational number, then the definition

$$z^\alpha = e^{\alpha \log z}$$

also delivers a function with infinitely many branches (on the cut plane).

We call such a singularity a branch point. More precisely, z_0 is a branch point for $f(z)$ if its value at z is different after the point z has moved round z_0 through a sufficiently small loop.

Example 6.43 The function $z^{1/2}(1-z)^{1/3}$ has two branch points at $z = 0$ and $z = 1$. There are six branches:

$$z^{1/2}(1-z)^{1/3}, \quad e^{2i\pi/3}z^{1/2}(1-z)^{1/3}, \quad e^{4i\pi/3}z^{1/2}(1-z)^{1/3},$$
$$-z^{1/2}(1-z)^{1/3}, \quad -e^{2i\pi/3}z^{1/2}(1-z)^{1/3}, \quad -e^{4i\pi/3}z^{1/2}(1-z)^{1/3}.$$

After a loop round $z = 0$ the branches interchange 'up and down' along the same column in the display; and, after a loop round $z = 1$, they interchange 'side ways' along the same row.

Example 6.44 In the region $|z| < 1$ the function

$$\frac{1}{z}\log\frac{1}{1-z}$$

has a branch equal to

$$\frac{1}{z}\left(z+\frac{z^2}{2}+\frac{z^3}{3}+\cdots\right),$$

which is analytic at $z = 0$ (by setting its value to be 1 there), but the other branches have a pole there.

If the region being considered is simply connected, and the function is analytic at each of its points, then the analytic function in the whole region obtained by analytic continuation along any path from z_0 to z_1 from within is unique. The broad outline for the proof is that any path can be continuously deformed to another one.

Up to now the analytic functions have been so defined within a certain region. With analytic continuation we can now speak of the analytic function as a whole; that is, starting from its original region we can extend its domain again and again by analytic continuation. In this way we may eventually have the whole plane, or the plane apart from certain points. It is also possible that there is a certain region beyond which it is impossible to reach by analytic continuation; in this case, the region is called the region of existence for the function, and the boundary of the domain is called the natural boundary of the function. In the situation when the function is multi-valued we have multi-branches.

Therefore, for isolated singularities we have, besides poles and essential singularities, also branch points. We still have to consider the non-isolated singularities.

Example 6.45 The function

$$f(z) = 1 + z^2 + z^4 + z^8 + \cdots + z^{2^n} + \cdots$$

has the unit circle $|z| = 1$ as its natural boundary.

Since $f(z) \to \infty$ as $z \to 1_-$, the point $z = 1$ is a singularity. From $f(z) = z^2 + f(z^2)$ we see that if α is a singularity then so is $\alpha^{1/2}$, and hence

$$z = 1, -1, \pm i \pm \sqrt{\pm i}, \ldots$$

are also singularities. These singularities are everywhere dense on the unit circle, so that there is no way for the domain of the function to be extended beyond the unit disc.

We sometimes make use of the following definition.

Definition 6.46 The interior points of the circles used in the process of analytic continuation for a single-valued function are called the regular points, or we say that the function is regular at such points. Any limit point of regular points, itself not being a regular point, is called a singularity.

Being regular is more demanding than being analytic; take, for example, the function

$$f(z) = \begin{cases} e^{-1/z}, & \text{if } z \neq 0,\ |\arg z| \leq \frac{\pi}{4}, \\ 0, & \text{everywhere else.} \end{cases}$$

The function, according to our earlier consideration, is analytic at $z = 0$, and $f'(0) = 0$. However, when considering the triangle with vertices 0, $1 \pm i/2$, the function is analytic inside and on the triangle, although the point $z = 0$ is not a regular point. However, such a region is not important – the reason being that only certain specially constructed functions, such as that in the example, can give rise to such regions.

6.15 The location of singularities

We already know that the power series

$$f(z) = \sum_{n=0}^{\infty} a_n z^n \tag{6.15}$$

must have a singularity on its circle of convergence, which has the radius

$$R = \lim_{n\to\infty} |a_n|^{-1/n}.$$

Thus $f(z)$ has a singularity at z_0 with $|z_0| = R$.

Where exactly is the singularity? The answer is not that simple. More specifically, let $R = 1$, and we may ask if $z = 1$ is a singularity. Expanding $f(z)$ about the point $z = \frac{1}{2}$, we have

$$\sum_{b=0}^{\infty} b_n (z - \tfrac{1}{2})^n.$$

Now if

$$\lim_{n\to\infty} |b_n|^{-1/n} > \frac{1}{2},$$

then the disc with centre $\frac{1}{2}$ includes the point $z = 1$, so that it cannot be a singularity; if

$$\lim_{n\to\infty} |b_n|^{-1/n} = \frac{1}{2},$$

which means that there is no analytic continuation beyond $z = 1$, and yet every other point on the unit circle has a distance from $\frac{1}{2}$ exceeding $\frac{1}{2}$, we have that $z = 1$ is a singularity. We therefore have the following criterion: A necessary and sufficient condition for $z = 1$ to be a singularity of $f(z)$ is that

$$\lim_{n\to\infty} \left(\frac{1}{n!} \left| f^{(n)}(\tfrac{1}{2}) \right| \right)^{-1/n} = \frac{1}{2}.$$

Such a condition is rather awkward to manipulate in practice.

Let us consider

$$F(w) = \frac{1}{1-w} f\left(\frac{w}{1-w}\right).$$

If $\Re(w) < \frac{1}{2}$ then $|w| < |1-w|$, so that, in $|w| < \frac{1}{2}$, the function $F(w)$ can have the expansion

$$F(w) = \sum_{m=0}^{\infty} \frac{a_m w^m}{(1-w)^{m+1}} = \sum_{m=0}^{\infty} a_m w^m \sum_{r=0}^{\infty} \frac{(m+r)!}{m!\,r!} w^r$$
$$= \sum_{n=0}^{\infty} w^n \sum_{m=0}^{n} \frac{n!}{m!\,(n-m)!} a_m.$$

Let

$$b_n = \sum_{m=0}^{n} \frac{n!}{m!\,(n-m)!} a_m,$$

and we note that a necessary and sufficient condition for $f(z)$ to have a singularity at $z = 1$ is that $F(w)$ should have a singularity at $w = \frac{1}{2}$, that such a point is the only singularity on the circle $|w| = \frac{1}{2}$ and that all the other points are regular points, so that the necessary and sufficient condition becomes

$$\lim_{n\to\infty} |b_n|^{-1/n} = \frac{1}{2}.$$

It has to be said that this criterion is still far from easy to use.

It is not difficult to deduce the following theorem.

Theorem 6.47 *Let the power series*

$$\sum_{n=0}^{\infty} a_n z^n$$

have the radius of convergence 1, and let

$$b_n(e^{i\theta}) = \sum_{m=0}^{n} \frac{n!}{m!\,(n-m)!} a_m e^{im\theta}.$$

Then a necessary and sufficient condition for $z = e^{i\theta}$ to be a singularity of the sum function is that

$$\lim_{n\to\infty} |b_n(e^{i\theta})|^{-1/n} = \frac{1}{2}.$$

Theorem 6.48 *Suppose that $a_n \geq 0$ for $n = 0, 1, 2, \ldots,$ and that the unit circle is the circle of convergence for the power series $\sum_{n=0}^{\infty} a_n z^n$. Then $z = 1$ is a singularity of the sum function.*

Proof. Suppose, if possible, that $z = 1$ is not a singularity, so that

$$\lim_{n\to\infty} |b_n(1)|^{-1/n} = \tau > \frac{1}{2}.$$

From $|b_n(e^{i\theta})| \leq |b_n(1)|$ we deduce that

$$\lim_{n\to\infty} |b_n(e^{i\theta})|^{-1/n} \geq \lim_{n\to\infty} |b_n(1)|^{-1/n} = \tau > \frac{1}{2},$$

so that the sum function is analytic everywhere on its circle of convergence, which is impossible.

Obviously the same conclusion in Theorem 6.48 still holds if a_n is non-negative only for all sufficiently large n. □

Remark The mere convergence or divergence of a power series does not help us to clarify the situation concerning the regular points and the singularities.

Example 6.49 The function

$$f(z) = \sum_{n=0}^{\infty} \frac{(-1)^n z^n}{n} = \log \frac{1}{1+z}$$

has the regular point $z = 1$, at which the power series is convergent; the point $z = -1$ is a singularity, and the series is divergent.

Example 6.50 The function

$$f(z) = \sum_{n=0}^{\infty} (-1)^n z^n = \frac{1}{1+z}$$

has the regular point $z = 1$, at which the power series is divergent.

Example 6.51 The function

$$f(z) = \sum_{n=0}^{\infty} \frac{z^n}{n^2} = \int_0^z \frac{1}{w} \log \frac{1}{1-w}\, dw$$

has the singularity $z = 1$, at which the power series is convergent.

7

The residue and its application to the evaluation of definite integrals

7.1 The residue

Let $f(z)$ have an isolated singularity at $z = a$, and let it have the expansion

$$f(z) = \sum_{n=0}^{\infty} a_n(z-a)^n + \sum_{n=1}^{\infty} b_n(z-a)^{-n}.$$

The number b_1 has a particular significance: it is called the residue of $f(z)$ at $z = a$. From Laurent's formula we have

$$b_1 = \frac{1}{2\pi i}\int_{\gamma} f(z)dz,$$

where γ is a circle centre a, inside which $f(z)$ has no other singularity.

If a is a simple pole, then obviously

$$b_1 = \lim_{z\to a}(z-a)f(z).$$

Theorem 7.1 (Residue theorem) *Let $f(z)$ be analytic inside and on a contour C, apart from the singularities $z_1, z_2, \ldots, z_n$ inside C. Then*

$$\int_C f(z)dz = 2\pi i(R_1 + R_2 + \cdots + R_n),$$

where $R_1, R_2, \ldots, R_n$ are the residues of $f(z)$ at $z_1, z_2, \ldots, z_n$.

Proof. Draw small circles $\gamma_1, \gamma_2, \ldots, \gamma_n$ with centres at $z_1, z_2, \ldots, z_n$ so that they do not overlap and that they lie inside C. Then $f(z)$ is analytic inside C and outside these circles, so that

$$\int_C f(z)dz = \int_{\gamma_1} f(z)dz + \cdots + \int_{\gamma_n} f(z)dz = 2\pi i(R_1 + \cdots + R_n).$$

□

7.2 Integrals of rational functions along a circle

Let us evaluate the integral

$$I = \frac{1}{2\pi i}\int_{|z|=r} \frac{P(z)}{Q(z)}\,dz,$$

where P and Q are polynomials in z, with $P(z)$ and $Q(z)$ having no common factor, and $Q(z)$ does not have a zero on $|z| = r$. Write

$$Q(z) = \prod_{i=1}^{n}(z - a_i)^{m_i},$$

and set $P(z)/Q(z)$ in terms of its partial fraction decomposition

$$R(z) + \sum_{i=1}^{n}\Big(\frac{R_{i1}}{z - a_i} + \frac{R_{i2}}{(z - a_i)^2} + \cdots + \frac{R_{im}}{(z - a_i)^{m_i}}\Big).$$

If $a_1, \ldots, a_m$ are inside the circle, and $a_{m+1}, \ldots, a_n$ are outside it, then, by the residue theorem,

$$I = \sum_{i=1}^{m} R_{i1}.$$

Example 7.2 We have

$$\frac{1}{2\pi i}\int_{|z|=1} \frac{dz}{(z + \frac{1}{2})(z + 2)} = \lim_{z\to -\frac{1}{2}} \frac{1}{z + 2} = \frac{2}{3}.$$

An integral of the form

$$\int_0^{2\pi} R(\cos\theta, \sin\theta)d\theta,$$

where $R(x, y)$ is a rational function, can be transformed by the substitution $z = e^{i\theta}$,

$$\cos\theta = \frac{1}{2}(z + z^{-1}), \quad \sin\theta = \frac{1}{2i}(z - z^{-1}),$$

into

$$\frac{1}{i}\int_{|z|=1} \frac{1}{z} R\Big(\frac{1}{2}(z + z^{-1}), \frac{1}{2i}(z - z^{-1})\Big)dz,$$

which can then be evaluated as before.

Example 7.3 If $0 < p < 1$ then

$$\int_0^{2\pi} \frac{d\theta}{1 - 2p\cos\theta + p^2} = \int_{|z|=1} \frac{dz}{i(1 - pz)(z - p)}.$$

The only singularity inside the unit circle is at $z = p$, with the residue

$$\lim_{z\to p} \frac{z-p}{i(1-pz)(z-p)} = \frac{1}{i(1-p^2)},$$

so that

$$\int_0^{2\pi} \frac{d\theta}{1-2p\cos\theta+p^2} = \frac{2\pi}{1-p^2}.$$

Example 7.4 If $0 < p < 1$ then

$$\int_0^{2\pi} \frac{\cos^2 3\theta}{1-2p\cos\theta+p^2}\,d\theta = \int_{|z|=1} \frac{dz}{iz}\frac{(z^3+z^{-3})^2}{4}\frac{1}{(1-pz)(1-pz^{-1})},$$

which, according to the residue theorem, is given by $2\pi\sum R$, where R runs over the residues at the singularities of $(z^6+1)^2/[4z^6(1-pz)(z-p)]$ which lie inside the unit circle. There are two such singularities, namely the pole of order 6 at $z = 0$ and the simple pole at $z = p$. From the expansion of the expression about $z = 0$, that is

$$-\frac{1}{4}\left(z^6+2+\frac{1}{z^6}\right)(1+pz+p^2z^2+\cdots)\left(1+\frac{z}{p}+\frac{z^2}{p^2}+\cdots\right)\frac{1}{p},$$

we find that the residue at $z = 0$ is

$$\frac{1}{-4p}(p^{-5}+p^{-3}+p^{-1}+p+p^3+p^5).$$

The residue at the simple pole $z = p$ is $(p^6+1)^2/[4p^6(1-p^2)]$, so that the integral concerned is equal to

$$2\pi\left(\frac{(p^6+1)^2}{4p^6(1-p^2)} - \frac{1-p^{12}}{4p^6(1-p^2)}\right) = \pi\frac{1+p^6}{1-p^2}.$$

7.3 Evaluation of certain integrals over the reals

We now evaluate the integral

$$\int_{-\infty}^{\infty} Q(x)dx,$$

where $Q(x)$ satisfies the following conditions: (i) $Q(z)$ is single-valued and analytic in the upper half-plane and the x-axis, apart from a finite number of singularities in the upper half-plane; (ii) $zQ(z) \to 0$ uniformly as $z \to \infty$ in $0 \le \arg z \le \pi$; and (iii) the two improper integrals

$$\int_0^{\infty} Q(x)dx, \quad \int_{-\infty}^{0} Q(x)dx,$$

are convergent.

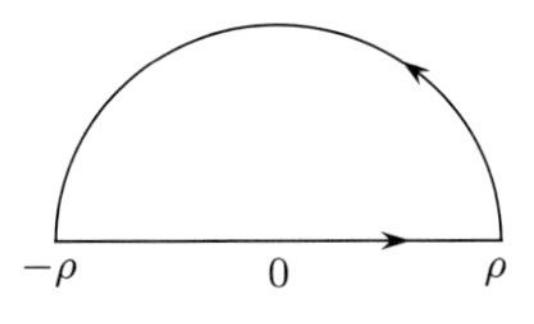

Figure 7.1.

Let C represent the contour from $-\rho$ to ρ along the x-axis, followed by the upper semi-circle γ with centre 0 and radius ρ (see Fig. 7.1). For sufficiently large ρ, we have, according to the residue theorem,

$$\int_C Q(z)dz = 2\pi i \sum R,$$

where the sum runs over the residues R of $Q(z)$ at the singularities in the upper half-plane. Therefore

$$\left|\int_{-\rho}^{\rho} Q(z)dz - 2\pi i \sum R\right| = \left|\int_r Q(z)dz\right|.$$

By condition (ii), for any $\epsilon > 0$, we may choose ρ so that on $|z| = \rho$

$$|zQ(z)| < \epsilon,$$

and hence

$$\left|\int_\gamma Q(z)dz\right| \leq \frac{\epsilon}{\rho}\int_r |dr| = \pi\epsilon.$$

That is

$$\lim_{\rho\to\infty}\int_{-\rho}^{\rho} Q(z)dz = 2\pi i \sum R,$$

and hence, by (iii),

$$\int_{-\infty}^{\infty} Q(z)dz = 2\pi i \sum R.$$

Note that if condition (iii) fails to hold then we still have the previous displayed equation.

Example 7.5 Show that

$$\int_0^\infty \frac{dx}{1+x^2} = \frac{\pi}{2}.$$

From the above we have

$$2\int_0^\infty \frac{dx}{1+x^2} = \int_{-\infty}^{\infty} \frac{dx}{1+x^2} = 2\pi i R,$$

where R is the residue of $1/(1+z^2)$ at the simple pole $z = i$, that is

$$R = \lim_{z \to i} \frac{z - i}{1 + z^2} = \frac{1}{2i},$$

and hence the required result.

Example 7.6 The function $(z^2+1)^{-3}$ has a triple pole at $z = i$ with the residue $-3i/16$, so that

$$\int_{-\infty}^{\infty} \frac{dx}{(1+x^2)^3} = \frac{3\pi}{8}.$$

Example 7.7 If $a > 0$, $b > 0$ then

$$\int_{-\infty}^{\infty} \frac{x^4\,dx}{(a+bx^2)^4} = \frac{\pi}{16a^{3/2}b^{3/2}}.$$

Exercise 7.8 Determine the values for

$$\int_0^{\infty} \frac{dx}{x^4+1}, \quad \int_0^{\infty} \frac{x^2\,dx}{x^4+1}, \quad \int_0^{\infty} \frac{dx}{x^6+1}.$$

Exercise 7.9 Show that

$$\int_0^{\infty} \frac{\log^2 x}{x^4+1}\,dx = \frac{\pi^3}{8}.$$

Exercise 7.10 Show that

$$\int_{-\infty}^{\infty} \frac{dt}{(1+t^2)^{n+1}} = \frac{1{\cdot}3{\cdot}5 \cdots (2n-1)}{2{\cdot}4{\cdot} \cdots 2n},$$

and determine the values for

$$\int_{-\infty}^{\infty} \frac{dt}{\left((t-\alpha)^2+\beta^2\right)^{n+1}}, \quad \int_{-\infty}^{\infty} \frac{dt}{(At^2+2Bt+C)^{n+1}}.$$

Exercise 7.11 Show that if $0 < \Re(a) < 1$ then

$$\int_{-\infty}^{\infty} \frac{e^{ax}\,dx}{1+e^x} = \frac{\pi}{\sin a\pi},$$

and if $0 < \Re(a) < 1$ and $0 < \Re(b) < 1$, then

$$\int_{-\infty}^{\infty} \frac{e^{ax} - e^{bx}}{1-e^x}\,dx = \pi(\coth a\pi - \coth b\pi).$$

7.4 Certain integrals involving sin and cos

If $Q(z)$ satisfies conditions (i) and (ii) in Section 7.3 then, for $m > 0$, so does $Q(z)e^{imz}$, and hence

$$\int_0^{\infty} \left(Q(x)e^{imx} + Q(-x)e^{-imx}\right)dx = 2\pi i \sum R', \tag{7.1}$$

where the sum runs over the residues R' of $Q(z)e^{imz}$ in the upper half-plane. Indeed, we can even weaken condition (ii) and replace it by (ii) $Q(z) \to 0$ uniformly as $z \to \infty$ in $0 \le \arg z \le \pi$.

To see this we need to modify the argument used in Section 7.3 by showing that we now have

$$\lim_{\rho\to\infty} \int_\gamma e^{imz} Q(z)dz = 0, \tag{7.2}$$

where γ is the upper semi-circle $z = \rho e^{i\theta}$, $0 \le \theta \le \pi$. For any $\epsilon > 0$, we choose ρ large enough so that $|Q(\rho e^{i\theta})| < \epsilon$, and hence

$$\left|\int_\gamma e^{imz} Q(z)dz\right| = \left|\int_0^\pi e^{im(\rho\cos\theta + i\rho\sin\theta)} Q(\rho e^{i\theta}) i\rho e^{i\theta}\, d\theta\right|$$
$$< \int_0^\pi \epsilon\rho e^{-m\rho\sin\theta}\, d\theta = 2\epsilon \int_0^{\pi/2} \rho e^{-m\rho\sin\theta}\, d\theta.$$

When $0 \le \theta \le \pi/2$ we have the lower estimate $\sin\theta \ge 2\theta/\pi$, so that

$$\left|\int_\gamma e^{imz} Q(z)dz\right| < 2\epsilon \int_0^{\frac{\pi}{2}} \rho e^{-2m\rho\theta/\pi}\, d\theta = (2\epsilon)\left(\frac{\pi}{2m}\right)\left[e^{-2m\rho\theta/\pi}\right]_{\frac{\pi}{2}}^0 < \frac{\epsilon\pi}{m}.$$

This gives (7.2), and hence (7.1).

In particular, if $Q(x)$ is an even function, that is $Q(x) = Q(-x)$, then

$$\int_0^\infty Q(x)\cos mx\, dx = \pi \sum R', \tag{7.3}$$

and if $Q(x)$ is odd then

$$\int_0^\infty Q(x)\sin mx\, dx = \pi \sum R'. \tag{7.4}$$

Example 7.12 For $a > 0$, we have

$$\int_0^\infty \frac{\cos x}{x^2 + a^2}\, dx = \frac{\pi}{2a} e^{-a};$$

this follows from (7.3) by taking $Q(z) = 1/(z^2 + a^2)$, so that $Q(z)e^{iz}$ has the residue $e^{-a}/(2ia)$ at the simple pole $z = ia$.

Example 7.13 We have

$$\int_0^\infty \frac{\sin x}{x}\, dx = \frac{\pi}{2}.$$

Here we take $Q(z) = 1/z$, so that

$$Q(z)e^{iz} = \frac{e^{iz}}{z}$$

has no singularity inside the upper half-plane. If we just blindly apply (7.4) then we arrive at the erroneous result 0. However, there is the matter of the simple pole at $z = 0$, which

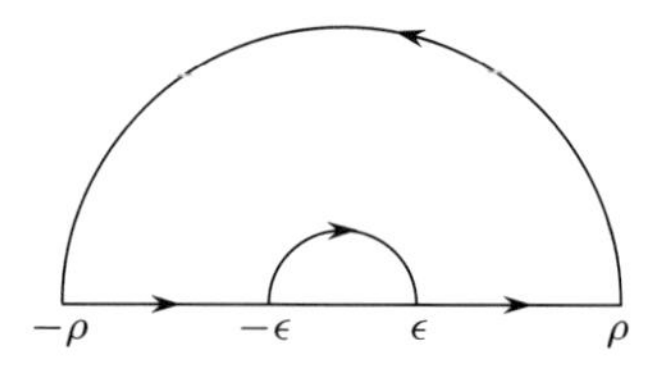

Figure 7.2.

is on the actual path of integration, and condition (i) is violated. In order to overcome this, we indent the contour at $z = 0$ (see Fig. 7.2). First, run along the x-axis from $-\rho$ to $-\epsilon$, where $0 < \epsilon < \rho$, then along the upper semi-circle $|z| = \epsilon$ to the point ϵ, then to ρ along the x-axis again, and then along the large upper semi-circle $|z| = \rho$ back to $-\rho$. Cauchy's theorem is now applicable, and on letting $\rho \to \infty$ and $\epsilon \to 0$ we find that the result follows from

$$\lim_{\epsilon \to 0} i \int_{\pi}^{0} \frac{e^{i\epsilon e^{i\theta}}}{\epsilon e^{i\theta}} \epsilon e^{i\theta}\, d\theta = -\pi i.$$

Example 7.14 We have, for $a \geq 0$, $b \geq 0$,

$$\int_0^{\infty} \frac{\cos 2ax - \cos 2bx}{x^2}\, dx = \pi(b - a),$$

where again we need to take some care with the singularity at $z = 0$.

Example 7.15 We have, for $\Re(z) > 0$,

$$\int_0^{\infty} (e^{-t} - e^{-tz}) \frac{dt}{t} = \log z.$$

From

$$\begin{aligned}
\int_0^{\infty} (e^{-t} - e^{-tz}) \frac{dt}{t} &= \lim_{\substack{\delta \to 0 \\ \rho \to \infty}} \left(\int_{\delta}^{\rho} \frac{e^{-t}}{t}\, dt - \int_{\delta}^{\rho} \frac{e^{-tz}}{t}\, dt \right) \\
&= \lim_{\substack{\delta \to 0 \\ \rho \to \infty}} \left(\int_{\delta}^{\rho} \frac{e^{-t}}{t}\, dt - \int_{\delta z}^{\rho z} \frac{e^{-u}}{u}\, du \right),
\end{aligned}$$

and the fact that e^{-t}/t is analytic over the quadrilateral with vertices $\delta, \delta z, \rho z, \rho$ we find that

$$\int_0^{\infty} (e^{-t} - e^{-tz}) \frac{dt}{t} = \lim_{\substack{\delta \to 0 \\ \rho \to \infty}} \left(\int_{\delta}^{\delta z} \frac{e^{-t}}{t}\, dt - \int_{\rho}^{\rho z} \frac{e^{-t}}{t}\, dt \right).$$

Since $\Re(z) > 0$, the required result follows from

$$\lim_{\rho \to \infty} \int_{\rho}^{\rho z} \frac{e^{-t}}{t}\, dt = 0$$

and

$$\lim_{\delta\to 0}\int_\delta^{\delta z}\frac{e^{-t}}{t}\,dt = \log z + \lim_{\delta\to 0}\int_\delta^{\delta z}\log t\, e^{-t}\,dt$$
$$= \log z + \lim_{\delta\to 0}\int_\delta^{\delta z}(t\log t - 1)e^{-t}\,dt = \log z.$$

Exercise 7.16 Determine the values of the integrals

$$\int_0^\infty \frac{\sin^2 x}{x^2}\,dx, \quad \int_0^\infty \frac{\sin^3 x}{x^3}\,dx.$$

Exercise 7.17 Show that if $c > 0$ then

$$\frac{1}{2\pi i}\int_{c-i\infty}^{c+i\infty}\frac{a^z}{z^2}\,dz = \begin{cases}\log a & (a>1),\\ 0 & (0<a<1).\end{cases}$$

Exercise 7.18 Show that

$$\int_0^\pi \frac{x\sin x\,dx}{1+a^2-2a\cos x} = \begin{cases}\dfrac{\pi}{a}\log(1+a) & (0<a<1),\\[2mm] \dfrac{\pi}{a}\log\dfrac{1+a}{a} & (a>1).\end{cases}$$

Exercise 7.19 Show that $f(x) = \operatorname{sech}(x\sqrt{\pi/2})$ satisfies the equation

$$f(t) = \sqrt{\frac{2}{\pi}}\int_0^\infty f(x)\cos xt\,dx.$$

Exercise 7.20 Show that if $0 < a < 1$, $0 < c < 1$ then

$$\frac{1}{2\pi i}\int_{c-i\infty}^{c+i\infty}\frac{dz}{a^z\sin\pi z} = \frac{1}{\pi(1+a)}.$$

7.5 The integral $\int_0^\infty x^{a-1}Q(x)dx$

Suppose that $Q(z)$ is analytic on the real axis, and that $x^aQ(x)\to 0$ as $x\to 0$ and as $x\to\infty$. We consider the contour integral

$$\int_C (-z)^{a-1}Q(z)dz,$$

where C is the contour shown in Fig. 7.3: The point describing the curve moves from ϵ to ρ along the x-axis, and then describes the circle around the origin (in the positive direction) coming back to ρ, then moves back to ϵ, and then describes the circle around the origin (in the negative direction) coming back to ϵ.

The important point to note now is that the contributions to the integral from ϵ to ρ, and then back from ρ to ϵ, do not cancel because $(-z)^{a-1}$ is not a single-valued function. The

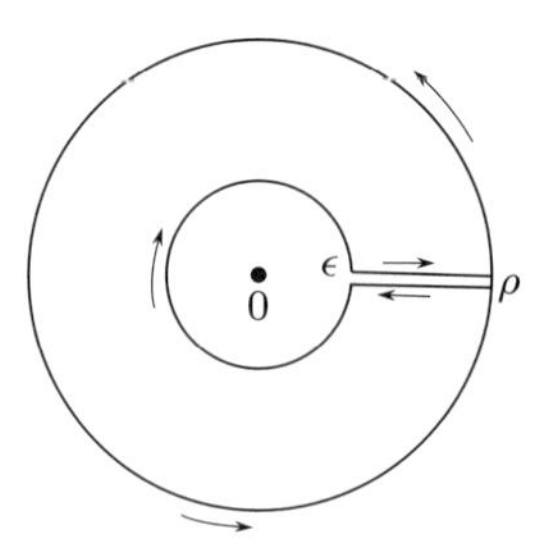

Figure 7.3.

meaning of the expression here is

$$\exp\big((a-1)\log(-z)\big),$$

and the complex plane is cut along the positive real axis from 0 to ∞, with the logarithm function taking the branch

$$\log(-z) = \log|z| + i\arg(-z), \quad -\pi \le (-z) \le \pi.$$

Under such a consideration, the function $(-z)^{a-1}$ is single-valued on C, and we suppose that $Q(z)$ has only finitely many singularities inside C in order to have

$$\int_C (-z)^{a-1} Q(z)dz = 2\pi i \sum R,$$

with R running over the residue of $(-z)^{a-1}Q(z)$ at the singularities inside C.

On the small circle we write $-z = \epsilon e^{i\theta}$, so that the contribution to the integral is

$$-\int_{\pi}^{-\pi} (-z)^{a-1} Q(z) i\, d\theta \to 0 \quad \text{as} \quad \epsilon \to 0;$$

similarly, on the large circle we write $-z = \rho e^{i\theta}$, so that the contribution to the integral is

$$-\int_{-\pi}^{\pi} (-z)^{a-1} Q(z) i\, d\theta \to 0 \quad \text{as} \quad \rho \to \infty.$$

As z moves from ϵ to ρ, we have $-z = xe^{-\pi i}$, and on its way back, after describing the large circle, we have $-z = xe^{\pi i}$; in other words, $(-z)^{a-1}$ takes the values $x^{a-1}e^{\pm(a-1)\pi i}$ when $\epsilon < x < \rho$, and the total contributions to the integral become

$$\lim_{\substack{\delta\to 0\\ \rho\to\infty}} \int_{\epsilon}^{\rho} \big(x^{a-1}e^{-(a-1)\pi i}Q(x) - x^{a-1}e^{(a-1)\pi i}Q(x)\big)dx = 2\pi i \sum R.$$

Thus

$$\int_0^{\infty} x^{a-1} Q(x)dx = \frac{\pi}{\sin a\pi} \sum R.$$

Example 7.21 We have, for $0 < a < 1$,

$$\int_0^\infty \frac{x^{a-1}}{1+x}\,dx = \frac{\pi}{\sin a\pi}.$$

To see this, we note that the function

$$\frac{(-z)^{a-1}}{1+z}$$

has the residue 1 at the simple pole $z = -1$.

Example 7.22 We have, for $0 < a < 1$,

$$\int_0^\infty \frac{x^{a-1}}{1-x}\,dx = \pi \cot a\pi.$$

Here we cannot just apply the formula blindly, because there is a singularity at $z = 1$, which is on the path of our contour; a small circular indentation about $z = 1$ is called for in order to arrive at the result.

Example 7.23 If $0 < z < 1$ and $-\pi < a < \pi$, then

$$\int_0^\infty \frac{t^{z-1}\,dt}{t + e^{ia}} = \frac{\pi\, e^{i(z-1)a}}{\sin \pi z}.$$

Example 7.24 If $-1 < z < 3$, then

$$\int_0^\infty \frac{x^z}{(1+x^2)^2}\,dx = \frac{\pi(1-z)}{4\cos\frac{1}{2}\pi z}.$$

Example 7.25 If $-1 < p < 1$ and $-\pi < \lambda < \pi$, then

$$\int_0^\infty \frac{x^{-p}}{1 + 2x\cos\lambda + x^2}\,dx = \frac{\pi}{\sin p\pi}\,\frac{\sin p\lambda}{\sin\lambda}.$$

7.6 The Γ function

Definition 7.26 The Γ function is defined by

$$\Gamma(z) = \int_0^\infty t^{z-1}e^{-t}\,dt; \tag{7.5}$$

it is an analytic function in $\Re(z) > 0$.

On integration by parts we find that, for $\Re(z) > 1$,

$$\Gamma(z) = [-t^{z-1}e^{-t}]_0^\infty + (z-1)\int_0^\infty t^{z-2}e^{-t}\,dt,$$

so that

$$\Gamma(z) = (z-1)\Gamma(z-1). \tag{7.6}$$

Since $\Gamma(1) = \int_0^\infty e^{-t}\,dt = 1$ it follows that, for any natural number n,

$$\Gamma(n) = (n-1)!.$$

Next, for $\Re(z) > 0$ and $\Re(w) > 0$, we have

$$\Gamma(z)\Gamma(w) = \int_0^\infty t^{z-1}e^{-t}\,dt \int_0^\infty s^{w-1}e^{-s}\,ds = \int_0^\infty t^{z-1}e^{-t}\,dt \int_0^\infty t^w v^{w-1}e^{-tv}\,dv,$$

and, on interchanging the order of integration, which is justified by uniform convergence,

$$\begin{aligned}\Gamma(z)\Gamma(w) &= \int_0^\infty v^{w-1}\,dv \int_0^\infty t^{z+w-1}e^{-t(1+v)}\,dt = \int_0^\infty v^{w-1}\,dv \int_0^\infty \frac{u^{z+w-1}e^{-u}\,du}{(1+v)^{z+w}}\\ &= \Gamma(z+w)\int_0^\infty \frac{v^{w-1}}{(1+v)^{z+w}}\,dv.\end{aligned}$$

We define the beta function by

$$B(z,w) = \frac{\Gamma(z)\Gamma(w)}{\Gamma(z+w)},$$

so that

$$B(z,w) = \int_0^\infty \frac{v^{w-1}}{(1+v)^{z+w}}\,dv, \quad (\Re(z) > 0,\ \Re(w) > 0). \tag{7.7}$$

Setting $v = \tan^2\theta$, we find that

$$B(z,w) = 2\int_0^{\frac{\pi}{2}} \sin^{2w-1}\theta\cos^{2z-1}\theta\,d\theta = \int_0^1 \lambda^{z-1}(1-\lambda)^{w-1}\,d\lambda. \tag{7.8}$$

Taking $w = 1-z$ in (7.7) we have

$$\Gamma(z)\Gamma(z-1) = \int_0^\infty \frac{v^{-z}}{1+v}\,dv,$$

and so, from Example 7.21,

$$\Gamma(z)\Gamma(z-1) = \frac{\pi}{\sin(1-z)\pi} = \frac{\pi}{\sin z\pi}.$$

At $z = \frac{1}{2}$ this gives $\Gamma(\frac{1}{2})^2 = \pi$, and since $\Gamma(\frac{1}{2}) > 0$ we have

$$\Gamma(\tfrac{1}{2}) = \sqrt{\pi}, \tag{7.9}$$

that is

$$\int_0^\infty t^{-1/2}e^{-t}\,dt = \sqrt{\pi},$$

and the substitution $t = x^2$ then gives

$$\int_{-\infty}^\infty e^{-x^2}\,dx = \sqrt{\pi}. \tag{7.10}$$

Next, by the definition in (7.5), we have

$$\int_0^\infty x^{\rho-1}e^{-ax}\,dx = \frac{\Gamma(\rho)}{a^\rho} \quad (\rho > 0,\ a > 0).$$

If we rashly make the 'substitution $x = it$' then

$$\int_0^\infty (it)^{\rho-1}e^{-ati}i\,dt = \frac{\Gamma(\rho)}{a^\rho},$$

and if we continue boldly to multiply by $(-i)^\rho = e^{\frac{1}{2}i\rho\pi}$ and then separate the real and the imaginary parts, we find that

$$\int_0^\infty t^{\rho-1}\begin{matrix}\cos\\ \sin\end{matrix}\,at\,dt = \frac{\Gamma(\rho)}{a^\rho}\begin{matrix}\cos\\ \sin\end{matrix}\,\frac{\rho\pi}{2}. \tag{7.11}$$

To justify our rash move we consider the integral

$$\int_C z^{\rho-1}e^{-az}\,dz \quad (a > 0,\ 0 < \rho < 1),$$

where C is the contour starting from ϵ along the real axis to ρ, and along the quarter circle $|z| = \rho$ in the first quadrant of the plane to reach $i\rho$ on the imaginary axis, then down to $i\epsilon$ along the axis, and then returning to ϵ along the circle $|z| = \epsilon$ in the first quadrant of the plane. The integral then has the value 0, and it is not difficult to show that the contributions from along $|z| = \rho$, and along $|z| = \epsilon$, tend to 0 as $\rho \to \infty$ and $\epsilon \to 0$. Such an argument shows that the substitution is allowed.

Taking $\rho = \frac{1}{2}$ and $a = 1$ in (7.11) we have

$$\int_0^\infty \cos x^2\,dx = \int_0^\infty \sin x^2\,dx = \frac{1}{2}\sqrt{\frac{\pi}{2}},$$

which are special values of the Fresnel integrals appearing in optics.

Exercise 7.27 Show that if $b > a > -1$ then

$$\int_0^{\frac{\pi}{2}} \cos a\theta \cos b\theta\,d\theta = \frac{\pi\Gamma(a+1)}{2^{a-1}\Gamma(\frac{a+b}{2}+1)\Gamma(\frac{a-b}{2}+1)}.$$

Exercise 7.28 Show that if $a > 0$ and $-\frac{\pi}{2} < a\lambda < \frac{\pi}{2}$ then

$$\int_0^\infty e^{-r^a\cos a\lambda}\begin{matrix}\cos\\ \sin\end{matrix}\,(r^a \sin a\lambda)dr = \begin{matrix}\cos\\ \sin\end{matrix}\,\frac{\lambda}{a}\,\Gamma\Big(\frac{1}{a}\Big).$$

Exercise 7.29 Show that if a is not an even number then, as $t \to \infty$ through the real numbers,

$$\int_0^\infty e^{-x^a}\cos xt\,dx \sim \frac{\Gamma(a+1)\sin\pi a/2}{t^{a+1}}.$$

7.7 Cauchy principal value

We define the improper integral

$$\int_{-\infty}^{\infty} f(t)dt$$

as

$$\lim_{\substack{\rho_1\to\infty \\ \rho_2\to\infty}} \int_{-\rho_2}^{\rho_1} f(t)dt$$

when the limit from the two independent limiting processes exists. If the limit does not exist, but

$$\lim_{\rho\to\infty} \int_{-\rho}^{\rho} f(t)dt$$

exists, then we call this limit the Cauchy principal value of the integral and, to clarify the situation, we sometimes write the integral as

$$(\mathrm{PV})\int_{-\infty}^{\infty} f(t)dt.$$

Again, the improper integral over a finite interval in which $f(t)$ is unbounded near $t=\xi$, we generally mean to have the form

$$\int_a^b f(t)dt = \lim_{\substack{\epsilon_1\to 0_+ \\ \epsilon_2\to 0_+}} \left(\int_a^{\xi-\epsilon_1} + \int_{\xi+\epsilon_2}^{b}\right);$$

but if this limit does not exist and

$$\lim_{\epsilon\to 0_+} \left(\int_a^{\xi-\epsilon} + \int_{\xi+\epsilon}^{b}\right)$$

does exist, then we also call it the Cauchy principal value, and write it as

$$(\mathrm{PV})\int_a^b f(t)dt.$$

Example 7.30 If $c > 1$ then

$$\frac{1}{2\pi i}(\mathrm{PV})\int_{c-i\infty}^{c+i\infty} \frac{a^z}{z}dz = \begin{cases} 1, & a > 1; \\ 0, & 0 < a < 1. \end{cases}$$

If $a > 1$, that is $\log a > 0$, then we take the contour along the line segment from $c - i\rho$ to $c + i\rho$, and then along the semi-circle on the left-hand side, with the segment as the diameter, to return to $c - i\rho$. The integrand has a pole at $z = 0$ with the residue 1 and, for large enough ρ, it lies inside the contour. It is not difficult to show that the contribution to the integral along the semi-circle tends to 0 as the radius $\rho \to \infty$.

If $a < 1$, then we take the semi-circle on the right-hand side instead, and there is no singularity inside the contour, so that the value for the integral has the Cauchy principal value 0.

Example 7.31 It can be shown directly that, for any real θ,

$$\frac{1}{2\pi i}(\text{PV})\int_{|\zeta|=1} \frac{d\zeta}{\zeta - e^{i\theta}} = \frac{1}{2}.$$

Example 7.32 If $f(\zeta)$ satisfies Lipschitz's condition on $|\zeta| = 1$ then

$$\frac{1}{2\pi i}(\text{PV})\int_{|\zeta|=1} \frac{f(\zeta)d\zeta}{\zeta - e^{i\theta}}$$

exists.

Example 7.33 If $f(\zeta)$ satisfies Lipschitz's condition on $|\zeta| = 1$ then the Cauchy form of integral

$$F(z) = \frac{1}{2\pi i}\int_{|\zeta|=1} \frac{f(\zeta)d\zeta}{\zeta - z}$$

takes the values

$$\frac{1}{2\pi i}(\text{PV})\int_{|\zeta|=1} \frac{f(\zeta)d\zeta}{\zeta - e^{i\theta}} \pm \frac{1}{2}f(e^{i\theta}),$$

as $z \to e^{i\theta}$, with the ambiguous signs being positive or negative depending on whether z approaches $e^{i\theta}$ from inside or outside, respectively, the unit circle $|\zeta| = 1$.

7.8 An integral related to a moment problem

Problem 7.34 Suppose that a continuous function $f(x)$ satisfies

$$\int_0^\infty x^n f(x)dx = 0, \qquad n = 0, 1, 2, \ldots .$$

Does it follow that $f(x) = 0$?

The answer is 'no', and a counter-example is

$$f(x) = e^{-x^{1/4}} \sin x^{1/4}.$$

With the substitution $x = t^4$, the moment integral becomes

$$4\int_0^\infty t^{4n+3}e^{-t} \sin t\, dt,$$

and we proceed to show that it has the value 0 for all n.

Consider the integral

$$\int_C z^{4n+3}e^{(i-1)z}\, dz,$$

where the contour C starts from the origin along the real axis to R, goes along the circle $|z| = R$ into the first quadrant to iR, and then returns to the origin along the imaginary axis. On the circular path, we have $z = R\cos\theta + iR\sin\theta$, with $0 \leq \theta \leq \pi/2$, so that

$$|e^{(i-1)z}| = e^{-R\cos\theta - R\sin\theta} \leq e^{-R},$$

and therefore, for each fixed n,

$$\left|\int z^{4n+3} e^{(i-1)z}\, dz\right| \leq \frac{\pi}{2} R^{4n+4} e^{-R} \to 0 \quad \text{as} \quad R \to \infty.$$

By Cauchy's theorem, we now have

$$\int_0^\infty x^{4n+3} e^{(i-1)x}\, dx - \int_0^\infty (iy)^{4n+3} e^{(i-1)iy} i\, dy = 0,$$

which combines to give

$$\int_0^\infty x^{4n+3} e^{-x}(e^{ix} - e^{-ix}) dx = 0,$$

and hence the sought-after result.

7.9 Counting zeros and poles

Suppose that $f(z)$ has a zero of order m at $z = a$. Then $f(z) = (z-a)^m g(z)$, with $g(a) \neq 0$, so that

$$\frac{f'(z)}{f(z)} = \frac{m}{z-a} + \frac{g'(z)}{g(z)},$$

which shows that $f'(z)/f(z)$ has a simple pole with the residue m at $z = a$. Similarly, if $f(z)$ has a pole of order m at $z = a$, then $-f'(z)/f(z)$ has a simple pole with the residue m at $z = a$. When counting zeros we take account of the multiplicity, so that a zero of order m is counted as m zeros; the same applies to the counting of poles.

The following theorem is an immediate consequence of our considerations in the above.

Theorem 7.35 *Let C be a contour, inside and on which $f(z)$ is analytic, except for a finite number of poles inside. If there is no zero on C, then*

$$\frac{1}{2\pi i}\int_C \frac{f'(z)}{f(z)}\, dz = N - P,$$

where N and P denote the number of zeros and poles, respectively, inside C.

The same argument also gives us Theorems 7.36 and 7.37.

Theorem 7.36 *Let $\phi(z)$ be analytic inside and on a contour C, and let $f(z)$ have the zeros $a_1, \ldots, a_m$ (with each zero repeated according to its order), and the poles $b_1, \ldots, b_n$ (with* *each pole repeated according to its order) inside C. Suppose that $f(z)$ is analytic elsewhere inside and on C, and has no zero on C. Then*

$$\frac{1}{2\pi i}\int_C \phi(z)\frac{f'(z)}{f(z)}\,dz = \sum_{\mu=1}^{m}\phi(a_\mu) - \sum_{\nu=1}^{n}\phi(b_\nu).$$

If $f(z)$ does not have poles then

$$\frac{1}{2\pi i}\int_C \frac{f'(z)}{f(z)}\,dz = N.$$

On the other hand, from

$$\frac{d}{dz}\log f(z) = \frac{f'(z)}{f(z)},$$

we also have

$$\int_C \frac{f'(z)}{f(z)}\,dz = \triangle_C \log f(z),$$

where $\triangle_C$ represents the change in the function $\log f(z)$ as z describes C. The logarithm function may take any initial value, but, having taken such a value, $\triangle_C$ records the change as z describes C. Also, from

$$\log f(z) = \log|f(z)| + i\arg f(z),$$

where $\log|f(z)|$ is single-valued, there is only a change in $\arg f(z)$. Consequently we have Theorem 7.37.

Theorem 7.37 *If $f(z)$ is analytic inside and on a contour C, with no zero on C, then the number of zeros inside C is given by*

$$N = \frac{1}{2\pi}\triangle_C \arg f(z).$$

Theorem 7.38 (Rouché) *Let $f(z)$ and $g(z)$ be analytic inside and on the contour C. Suppose that, for every z on C,*

$$|g(x)| < |f(z)|.$$

Then $f(z)$ and $f(z)+g(z)$ have the same number of zeros inside C.

Proof. Let $\phi(z) = g(z)/f(z)$, so that $|\phi(z)| < 1$ for z on C. We can then expand $\phi'(1+\phi)^{-1}$ as a power series in ϕ, and integrate term by term over C to give

$$\frac{1}{2\pi i}\int_C \frac{\phi'}{1+\phi}\,dz = 0.$$

This means that the number of zeros inside C for

$$1+\phi(z)=\frac{f(z)+g(z)}{f(z)}$$

is the same as the number of its poles. Since the poles come from the zeros of $f(z)$, the theorem is proved. □

7.10 The roots of an algebraic equation

Theorem 7.39 (Fundamental theorem of algebra) *There are n roots to a polynomial equation of degree n.*

Proof. Let

$$f(z)=a_0+a_1z+\cdots+a_nz^n, \quad a_n\neq 0,$$

and set

$$g(z)=-(a_0+a_1z+\cdots+a_{n-1}z^{n-1}).$$

Then, on a circle with centre the origin and a sufficiently large radius R, we have $|g(z)|<|f(z)|$. By Rouché's theorem the number of zeros for $f(z)$ is the same as the number of zeros for $f(z)+g(z)=a_nz^n$, which is n. □

The theorem can also be deduced from Liouville's theorem. For if $f(z)$ has no zero, then $1/f(z)$ is analytic, and since $1/f(z)\to 0$ as $z\to\infty$, it has to be constant. Therefore $f(z)=0$ has at least one root, and it can then be deduced further that it actually has n roots.

The result of Section 7.9 can be used to examine the positions of the complex roots for certain equations.

Example 7.40 How are the roots of the equation

$$z^4+z^3+4z^2+2z+3=0$$

distributed in the four quadrants of the plane?

Let $f(z)$ denote the given quartic, and we first show that $f(x)\neq 0$. There are various ways to do this; for example,

$$\begin{aligned}f(x)&=(x+\tfrac{1}{4})^4+\tfrac{29}{8}x^2+\tfrac{31}{16}x+3-\tfrac{1}{4}\\&=(x+\tfrac{1}{4})^4+\tfrac{29}{8}\left(x+\tfrac{31}{4\cdot 29}\right)^2+3-\tfrac{1}{4^4}-\tfrac{31^2}{8\cdot 4^2\cdot 29}>0.\end{aligned}$$

Next, it is clear that $f(iy)=y^4-iy^3-4y^2+2iy+3\neq 0$, and we proceed to show that $f(z)$ does not have a zero in the first quadrant of the plane. Let C be the contour starting from the origin along the real axis to the point R, and let it run along the circle with centre the origin into the first quadrant until it reaches the point iR, and then return to the origin along the imaginary axis. It suffices to show that $\Delta_C \arg f(z)=0$.

First, we have $\arg f(x) = 0$ throughout $0 \le x \le R$, so that there is no change in the argument. Next, as $z = Re^{i\theta}$ describes the circular arc part of C, so that $0 \le \theta \le \frac{\pi}{2}$, the change in $\arg f(z)$ is given by

$$\triangle \arg f(z) = \triangle \arg(R^4 e^{4i\theta}) + \triangle \arg\left(1 + O(R^{-1})\right) = 2\pi + O(R^{-1}).$$

Next, as $z = iy$ moves from iR back to 0, we need to trace the path of the image $f(iy) = u + iv$ in order to determine the variation in its argument. We have, for $R \ge y \ge 0$,

$$\arg f(iy) = \arg(u + iv), \quad u = y^4 - 4y^2 + 3, \quad v = -y^3 + 2y,$$

and we note that $u = 0$ when $y = \sqrt{3}$ and 1, and $v = 0$ when $y = \sqrt{2}$. Thus, starting from $f(iR)$ in the fourth quadrant, with $\arg f(iR) = O(R^{-1})$, the image curve moves into the third quadrant, cutting the negative v-axis when $y = \sqrt{3}$; it then moves into the second quadrant, cutting the negative u-axis when $y = \sqrt{2}$; it then moves into the first quadrant, cutting the positive v-axis when $y = 1$, and finally ending up at $f(0) = 3$. Thus, $\triangle \arg f(iy) = -2\pi + O(R^{-1})$. Therefore, for large R, the total change in $\arg f(z)$ as z describes C is

$$\triangle_C \arg f(z) = 2\pi - 2\pi + O(R^{-1}) = 0,$$

with no error term here because its value has to be a multiple of 2π. Therefore, by Theorem 7.37, there is no zero in the first quadrant of the plane. Since the coefficients for $f(z)$ are real, the zeros occur in conjugate pairs, so that there is no zero in the fourth quadrant either. Therefore there are two zeros in the second quadrant and two zeros in the third quadrant.

Exercise 7.41 Determine the number of roots for

$$z^6 + 6z + 10 = 0$$

in each of the four quadrants.

7.11 Evaluation of series

Contour integration can also be used to evaluate the sum of certain series

$$\sum f(n).$$

Let C be a contour which includes the points $m, m+1, \ldots, n$, and let $f(z)$ be analytic inside and on C, apart from a finite number of poles at $a_1, \ldots, a_k$ (distinct from $m, m+1, \ldots, n$), with corresponding residues $r_1, \ldots, r_k$. We consider

$$\int_C \pi \cdot f(z) \cdot \cot \pi z \, dz.$$

The function $\pi \cot \pi z$ has simple poles at the integers with residue 1, so that the integrand has simple poles at $z = m, m+1, \ldots, n$, with residues $f(m), f(m+1), \ldots, f(n)$, and therefore, by the residue theorem,

$$\int_C \pi \cdot f(z) \cdot \cot \pi z \, dz = 2\pi i \big(f(m) + f(m+1) + \cdots + f(n) + r_1 \pi \cot \pi a_1 + \cdots + r_k \pi \cot \pi a_k \big).$$

For example, let $f(z)$ be a rational function with no poles at the integers, and suppose that $f(z) = O(|z|^{-1})$ as $z \to \infty$. Take C to be the square with vertices at $(n + \frac{1}{2})(\pm 1 \pm i)$. Then it is not difficult to show that the integral along C tends to 0 as $n \to \infty$, so that

$$\lim_{n \to \infty} \sum_{m=-n}^{n} f(m) = -\pi (r_1 \cot \pi a_1 + \cdots + r_k \cot \pi a_k).$$

Take, in particular,

$$f(z) = \frac{1}{(a+z)^2}, \quad a \text{ not an integer},$$

which has a double pole at $z = -a$. From the Taylor expansion

$$\cot \pi z = \cot(-\pi a) + \pi (z + a)\{-\operatorname{cosec}^2(-\pi a)\} + \cdots$$

we see that the residue for $\cot \pi z / (z + a)^2$ at $z = -a$ is $-\pi \operatorname{cosec}^2 \pi a$, so that

$$\sum_{n=-\infty}^{\infty} \frac{1}{(a+n)^2} = \pi^2 \operatorname{cosec}^2 \pi a.$$

7.12 Linear differential equations with constant coefficients

Let

$$P(z) = a_0 + a_1 z + a_2 z^2 + \cdots + a_n z^n$$

be a polynomial with degree n, and consider

$$u(t) = \int_C \frac{e^{tz} f(z)}{P(z)} \, dz,$$

where t is real, $f(z)$ is a polynomial with degree at most n and C is a contour not passing through the zeros of $P(z)$.

By differentiation under the integral sign, we have

$$\frac{d^\ell u}{dt^\ell} = \int_C \frac{e^{tz} f(z) z^\ell}{P(z)} \, dz,$$

and so, by Cauchy's theorem,

$$\sum_{\ell=0}^{n} a_\ell \frac{d^\ell u}{dt^\ell} = \int_C e^{tz} f(z) dz = 0.$$

Thus $u(t)$ is a solution to the linear differential equation

$$\sum_{\ell=0}^{n} a_\ell \frac{d^\ell u}{dt^\ell} = 0. \tag{7.12}$$

Since we can take $f(z)$ to be any polynomial with degree $n-1$, so that there are n coefficients at our disposal, one may hope that the method will enable us to deliver the most general solution to the differential equation (7.12).

Take C to include all the zeros of $P(z)$, and we apply the method of residues to seek the solution. Suppose first that the roots for $P(z)$ are distinct, say $\alpha_1, \ldots, \alpha_n$. Then

$$\int_C \frac{e^{tz} f(z)}{P(z)} dz = 2\pi i \sum_{j=1}^{n} \lim_{z\to\alpha_j} \frac{e^{tz} f(z)}{P(z)}(z-\alpha_j) = 2\pi i \sum_{j=1}^{n} \frac{e^{\alpha_j t} f(\alpha_j)}{P'(\alpha_j)}.$$

We can then set $f(z)$ so that $2\pi i f(\alpha_j)/P'(\alpha_j)$ takes the values of the given arbitrary constants c_j, so that the solution has the form

$$u(t) = \sum_{j=1}^{n} c_j e^{\alpha_j t}.$$

If $z = \alpha_\nu$ is a repeated root of $P(z)$, with multiplicity r, then the corresponding integrand has a pole of order r, and the residue there can be obtained from the coefficient of $(z-\alpha_\nu)^{r-1}$ in the expansion of

$$t^{tz} f(z) = e^{t\alpha_\nu} \cdot e^{t(z-\alpha_\nu)} f(z)$$

about $z - \alpha_\nu$, that is

$$e^{t\alpha_\nu} \sum_{\ell=0}^{r-1} \frac{t^\ell}{\ell!} d_{r-1-\ell},$$

where

$$f(z) = \sum_{\ell=0}^{r} d_\ell (z-\alpha_\nu)^\ell,$$

giving the solutions

$$e^{t\alpha_\nu}, te^{t\alpha_\nu}, \ldots, e^{t^{r-1}\alpha_\nu}.$$

Exercise 7.42 Use this method to deal with systems of linear differential equations with constant coefficients.

7.13 The Bürmann and the Lagrange formulae

The purpose of this section is to discuss the expansion of functions in powers of a given function. In order to keep things simple, we shall not give proofs involving some aspects concerning convergence.

Let

$$w = \phi(z) \tag{7.13}$$

be a univalent transformation, giving a one-to-one mapping of a contour γ in the z-plane to a contour s in the w-plane, with matching interiors. Suppose that $\phi(z)$ is analytic inside and on γ, and let a be a point inside γ, with the image $b = \phi(a)$ being inside s. As $z = \zeta$ runs over γ, its image $w = \eta$ runs over s.

Since the transformation (7.13) is univalent, $\phi'(z) \neq 0$ inside γ, and we now assume that $\phi'(z)$ is also not 0 on γ. Let $f(z)$ be analytic inside and on γ, so that $f'(z)/\phi'(z)$ is also analytic. As a function inside and on s, the expression for w is also analytic. Therefore, by Cauchy's integral formula, we have

$$\frac{f'(z)}{\phi'(z)} = \frac{1}{2\pi i}\int_s \frac{f'(\zeta)}{\phi'(\zeta)}\frac{d\eta}{\eta - w} = \frac{1}{2\pi i}\int_\gamma \frac{f'(\zeta)}{\phi(\zeta) - \phi(z)}\, d\zeta, \tag{7.14}$$

and the expansion is given by

$$\frac{1}{\phi(\zeta) - \phi(z)} = \frac{1}{\phi(\zeta) - b}\left(1 - \frac{\phi(z) - b}{\phi(\zeta) - b}\right)^{-1} = \sum_{m=0}^{\infty} \frac{(\phi(z) - b)^m}{(\phi(\zeta) - b)^{m+1}}$$

(we shall not concern ourselves with the convergence aspect). Thus

$$\begin{aligned}
\frac{f'(z)}{\phi'(z)} &= \frac{1}{2\pi i}\int_\gamma f'(\zeta)\sum_{m=0}^{\infty}\frac{(\phi(z) - b)^m}{(\phi(\zeta) - b)^{m+1}}\, d\zeta \\
&= \sum_{m=0}^{\infty}(\phi(z) - b)^m \frac{1}{2\pi i}\int_\gamma f'(\zeta)\frac{d\zeta}{(\phi(\zeta) - b)^{m+1}} \\
&= \sum_{m=0}^{\infty}(\phi(z) - b)^m \frac{1}{2\pi i}\int_\gamma \frac{f'(\zeta)}{(\zeta - a)^{m+1}}\left(\frac{\zeta - a}{\phi(\zeta) - b}\right)^{m+1} d\zeta.
\end{aligned} \tag{7.15}$$

Since $\phi(z)$ is univalent, $\phi(\zeta) = b$ has no other roots besides $\zeta = a$, so $(\zeta - a)/(\phi(\zeta) - b)$ is analytic inside γ. Let

$$f'(z)\big(\psi(z)\big)^{m+1} \quad \left(\psi(z) = \frac{z - a}{\phi(z) - b}\right)$$

be expanded as a power series in $z - a$, with the coefficients for $(z - a)^m$ being

$$\frac{1}{m!}\frac{d^m}{dz^m}\left[f'(z)\big(\psi(z)\big)^{m+1}\right]_{z=a},$$

which will simply be denoted by

$$\frac{1}{m!}\frac{d^m}{da^m}\left[f'(a)\left(\psi(a)\right)^{m+1}\right],$$

so that, from (7.15),

$$f'(z)=\sum_{m=0}^{\infty}\frac{1}{m!}\frac{d^m}{da^m}\left[f'(a)\left(\psi(a)\right)^{m+1}\right]\left(\phi(z)-b\right)^m\phi'(z).$$

Integrating from a to z, we have Bürmann's formula

$$f(z)=f(a)+\sum_{m=0}^{\infty}\frac{1}{(m+1)!}\frac{d^m}{da^m}\left[f'(a)\left(\psi(a)\right)^{m+1}\right]\left(\phi(z)-b\right)^{m+1}. \tag{7.16}$$

If we set

$$w=\frac{z-a}{\tau(z)},\qquad b=0,$$

then

$$\psi(z)=\tau(z),$$

and (7.16) becomes Lagrange's formula

$$f(z)=f(a)+\sum_{n=1}^{\infty}\frac{w^n}{n!}\frac{d^{n-1}}{da^{n-1}}\left\{f'(a)\left[\tau(a)\right]^n\right\}. \tag{7.17}$$

(*Question:* Can (7.16) be derived from (7.17)?) We leave it to the reader to show that the condition for the validity for (7.17) is that $(z-a)/\tau(z)$ is analytic in the neighbourhood of a, and that the derivative at the point a is not 0; that is, for any function $f(z)$ which is analytic near a, the representation (7.17) is valid in a sufficiently small neighbourhood of a.

Example 7.43 The equation

$$z-a-\frac{w}{z}=0,\quad a\neq 0,$$

leads us to the expansion

$$z=a+\sum_{n=1}^{\infty}\frac{(-1)^n(2n-2)!}{n!(n-1)!a^{2n-1}}w^n,$$

which tells us that the solution z to the quadratic equation

$$z^2-az-w=0$$

can be expanded as a power series with constant coefficients. There is nothing new in this, of course: the quadratic equation has the solutions

$$z = \frac{a}{2}\left(1 \pm \left(1 + \frac{4w}{a^2}\right)^{1/2}\right),$$

and the binomial expansion applied here also gives the same result; such a direct method shows that the series is convergent for $|4w/a^2| < 1$.

Example 7.44 If z is a root of

$$z = 1 + wz^n,$$

then $z \to 1$ as $w \to 0$, and if $|w| < \frac{1}{4}$ then

$$z^n = 1 + nw + \frac{n(n+3)}{2!}w^2 + \frac{n(n+4)(n+5)}{3!}w^3 + \frac{n(n+5)(n+6)(n+7)}{4!}w^4 + \frac{n(n+6)(n+7)(n+8)(n+9)}{5!}w^5 + \cdots .$$

Example 7.45 If z is a root of

$$z = 1 + wz^\alpha,$$

then $z \to 1$ as $w \to 0$, giving

$$\log z = w + \frac{2\alpha - 1}{2}w^2 + \frac{(3\alpha - 1)(3\alpha - 2)}{2 \cdot 3}w^3 + \cdots ,$$

which is valid when

$$|w| < |(\alpha - 1)^{\alpha - 1} \cdot \alpha^{-\alpha}|.$$

Note. Kepler's equation

$$z - a = w \sin z$$

is a particular case of Lagrange's formula. Kepler's equation is useful in astronomy, and the method in this section enhances the method used in the solution to the equation (see Volume 1, Example 6.6).

7.14 The Poisson–Jensen formula

Let $f(z)$ be a function defined in the disc $|z| \le R$, with m zeros $a_1, \ldots, a_m$ and n poles $b_1, \ldots, b_n$ (repeated p times when the order is p), and analytic elsewhere in $|z| \le R$. We then have the following Poisson–Jensen formula:

$$\log|f(re^{i\theta})| = \frac{1}{2\pi}\int_0^{2\pi} \frac{R^2 - r^2}{R^2 - 2Rr\cos(\theta-\phi) + r^2} \log|f(Re^{i\phi})|d\phi - \sum_{\mu=1}^{m} \log\left|\frac{R^2 - \bar{a}_\mu re^{i\theta}}{R(re^{i\theta} - a_\mu)}\right| + \sum_{\nu=1}^{n} \log\left|\frac{R^2 - \bar{b}_\nu re^{i\theta}}{R(re^{i\theta} - b_\nu)}\right|. \tag{7.18}$$

Proof. (1) If $f(z)$ has neither zero nor pole, then $\log f(z)$ is analytic in $|z| \leq R$, and formula (7.18) is just Poisson's formula.

(2) Let

$$g(z) = f(z)\prod \frac{R(z - b_\nu)}{R^2 - \bar{b}_\nu z} \Big/ \prod \frac{R(z - a_\mu)}{R^2 - \bar{a}_\mu z},$$

so that $g(z)$ has neither zero nor pole in $|z| \leq R$. Also, apart from a finite number of points, $|g(z)| = |f(z)|$ when $|z| = R$. From (1) we then have

$$\log|g(re^{i\theta})| = \frac{1}{2\pi}\int_0^{2\pi} \frac{R^2 - r^2}{R^2 - 2Rr\cos(\theta-\phi) + r^2} \log|g(Re^{i\phi})|d\phi,$$

and so

$$\log|f(re^{i\theta})| = \frac{1}{2\pi}\int_0^{2\pi} \frac{R^2 - r^2}{R^2 - 2Rr\cos(\theta-\phi) + r^2} \log|f(Re^{i\phi})|d\phi - \sum_{\mu=1}^{m} \log\left|\frac{R^2 - \bar{a}_\mu re^{i\theta}}{R(re^{i\theta} - a_\mu)}\right| + \sum_{\nu=1}^{n} \log\left|\frac{R^2 - \bar{b}_\nu re^{i\theta}}{R(re^{i\theta} - b_\nu)}\right|.$$

When $a_\mu \neq 0$, $b_\nu \neq 0$, we may set $r = 0$ to give

$$\log|f(0)| = \frac{1}{2\pi}\int_0^{2\pi} \log|f(Re^{i\phi})|d\phi - \sum_{\mu=1}^{m} \log\left|\frac{R}{a_\mu}\right| + \sum_{\nu=1}^{n} \log\left|\frac{R}{b_\nu}\right|,$$

that is

$$\frac{1}{2\pi}\int_0^{2\pi} \log|f(Re^{i\phi})|d\phi = \log\frac{|f(0)|\prod_{\nu=1}^{n}|b_\nu|}{\prod_{\mu=1}^{m}|a_\nu|}R^{m-n}.$$

□

8

Maximum modulus theorem and families of functions

8.1 Maximum modulus theorem

The maximum principle has already been established in Chapter 4: If $f(z)$ is a single-valued regular function in a region, then its absolute value, its real part and its imaginary part cannot take their maximum values at an interior point of the region. The only exception is when $f(z)$ is constant.

The conditions can be relaxed a little: We let $|f(z)|$ be single-valued, and $f(z)$ may not be so; again, apart from $f(z)$ being constant, $|f(z)|$ does not assume a maximum at an interior point. There is no real need to prove this, because each branch of $f(z)$ is a single-valued function, to which the theorem applies, and the same result for $|f(z)|$, being single-valued, follows.

Let $M(r)$ denote the maximum value of $|f(z)|$ on the circle $|z| = r$. By the maximum principle, $M(r)$ is an increasing function and, apart from $f(z)$ being a constant, can even be considered as strictly increasing, in the sense that $r < r'$ implies $M(r) < M(r')$.

Let $A(r)$ denote the maximum value of $\Re f(z)$ on the circle $|z| = r$. Then $A(r)$ is also a strictly increasing function. There is the following inequality relating $M(r)$ and $A(r)$.

Theorem 8.1 (Borel, Carathéodory) *Let $f(z)$ be regular in $|z| \leq R$. Then, for $0 < r < R$, we have*

$$M(r) \leq \frac{2r}{R-r}A(r) + \frac{R+r}{R-r}|f(0)|.$$

Proof. The required result is trivial if $f(z)$ is constant. For non-constant $f(z)$, we first suppose that $f(0) = 0$, so that $A(R) > A(0) = 0$. Let

$$\phi(z) = \frac{f(z)}{2A(R) - f(z)},$$

which is regular because the denominator is never zero. Writing $f(z) = u + iv$, we deduce from $-2A(R) + u \leq u \leq 2A(R) - u$ that

$$|\phi(z)|^2 = \frac{u^2 + v^2}{(2A(R) - u)^2 + v^2} \leq 1,$$

so that, by Schwarz's lemma,

$$|\phi(z)| \le \frac{r}{R},$$

and therefore

$$|f(z)| = \left|\frac{2A(R)\phi(z)}{1+\phi(z)}\right| \le \frac{2A(R)r}{R-r},$$

as required.

If $f(0) \neq 0$, then we apply the established result to $f(z) - f(0)$ to give

$$|f(z) - f(0)| \le \frac{2r}{R-r} \max_{|z|=R} \Re(f(z) - f(0)) \le \frac{2r}{R-r}\big(A(R) + |f(0)|\big),$$

and the required result follows. □

We deduce immediately the following theorem.

Theorem 8.2 *If* $A(R) \ge 0$ *then*

$$M(R) \le \frac{R+r}{R-r}(A(r) + |f(0)|).$$

Theorem 8.3 *If* $A(R) \ge 0$ *then*

$$\max_{|z|=r} |f^{(n)}(z)| \le \frac{2^{n+2} n! R}{(R-r)^{n+1}}\big(A(R) + |f(0)|\big).$$

Proof. From

$$f^{(n)}(z) = \frac{n!}{2\pi i}\int_C \frac{f(w)}{(w-z)^{n+1}}\,dw,$$

where C is the circle centre $w = z$ and radius $\delta = \frac{1}{2}(R-r)$, on which $|w| \le r + \frac{1}{2}(R-r) = \frac{1}{2}(R+r)$, it follows from Borel–Carathéodory's inequality that

$$\max |f(w)| \le \frac{R + \frac{1}{2}(R+r)}{R - \frac{1}{2}(R+r)}\big(A(R) + |f(0)|\big) < \frac{4R}{R-r}\big(A(R) + |f(0)|\big),$$

so that

$$|f^{(n)}(z)| \le \frac{n!}{\delta^n}\frac{4R}{R-r}\big(A(R) + |f(0)|\big) = \frac{2^{n+2} n! R}{(R-r)^{n+1}}\big(A(R) + |f(0)|\big).$$

□

8.2 Phragmen–Lindelöf theorem

Theorem 8.4 *Suppose that* (i) $f(z)$ *is regular in a bounded region* D *and* $|f(z)|$ *is single-valued,* (ii) *for any* $\epsilon > 0$, *in the neighbourhood of a boundary point* ζ *and within* D,

$$|f(z)| < M + \epsilon. \tag{8.1}$$

Then $|f(z)| \leq M$ for every interior point of D; moreover, if there is equality at an interior point of D then $f(z)$ is constant.

Proof. Let G be the supremum of $|f(z)|$ in the interior of D (not excluding $G = \infty$). By the definition of a supremum there is a set of interior points z_n such that $|f(z_n)| \to G$ with z_0 being a limit point; we may then suppose that $z_k \to z_0$.

(1) Suppose that z_0 is an interior point of D. Then, by the continuity of $|f(z)|$, we have

$$\lim_{k \to \infty} |f(z_k)| = |f(z_0)| = G;$$

this means that $|f(z)|$ assumes its maximum value at an interior point, so that $f(z)$ is constant.

(2) Suppose that z_0 is a boundary point of D. There is a square q with centre z_0 such that (8.1) holds for $z \in q \cap D$. Now $z_k \in q$ for all large k, so that $G \leq M + \epsilon$; since $\epsilon > 0$ is arbitrary, we have $G \leq M$. This proves the first part of the theorem. The second part is a consequence of the maximum principle. □

8.3 Hardamard's three-circle theorem

Theorem 8.5 *Let $f(z)$ be analytic in the annulus region $r_1 \leq |z| \leq r_3$. Then $\log M(r)$ is a continuous convex function of $\log r$ in $r_1 < r < r_3$. More specifically, for $r_1 < r_2 < r_3$, we have*

$$\log M(r_2) \leq \frac{r_3 - \log r_2}{\log r_3 - \log r_1} \log M(r_1) + \frac{r_2 - \log r_1}{\log r_3 - \log r_1} \log M(r_3)$$

or, writing M_i for $M(r_i)$,

$$M_2^{\log\left(\frac{r_3}{r_1}\right)} \leq M_1^{\log\left(\frac{r_3}{r_2}\right)} M_3^{\log\left(\frac{r_2}{r_1}\right)}.$$

Proof. Let

$$\phi(z) = z^{\lambda} f(z),$$

where λ is a constant to be specified later. The function $\phi(z)$ is regular in the annulus, and $|\phi(z)|$ is single-valued. By the maximum modulus theorem,

$$|\phi(z)| \leq \max(r_1^{\lambda} M_1, r_3^{\lambda} M_3);$$

in particular, on $|z| = r_2$,

$$|f(z)| \leq \max\left[\left(\frac{r_1}{r_2}\right)^{\lambda} M_1, \left(\frac{r_3}{r_2}\right)^{\lambda} M_3\right].$$

The required result follows by choosing

$$\lambda = -\frac{\log M_3/M_1}{\log r_3/r_1}.$$

(*Question:* Why would we make this choice?) □

Note. There is equality only if $\phi(z)$ is constant, that is if $f(z) = cz^{\mu}$.

8.4 Hardy's theorem on the mean value of $|f(z)|$

Theorem 8.6 *Let $f(z)$ be regular in $|z| < R$, and let*

$$I_2(r) = \frac{1}{2\pi}\int_0^{2\pi} |f(re^{i\theta})|^2\, d\theta.$$

Then $I_2(r)$ is a strictly increasing function of r in $0 < r < R$, and $\log I_2(r)$ is a convex function of $\log r$.

Proof. Let

$$f(z) = \sum_{n=0}^{\infty} a_n z^n,$$

so that

$$I_2(r) = \sum_{n=0}^{\infty} |a_n|^2 r^{2n}.$$

Apart from $f(z) = a_0$, it is clear that $I_2(r)$ is a strictly increasing function. Let $u = \log r$; then

$$\frac{d^2}{du^2}\log I_2 = \frac{I_2 I_2'' - (I_2')^2}{I_2^2},$$

where I_2', I_2'' denote the derivatives of I_2 with respect to u. By the Cauchy–Schwarz inequality,

$$I_2'^2 = \Big(\sum_{n=0}^{\infty} |a_n|^2 (2n) e^{2nu}\Big)^2 \le \Big(\sum_{n=0}^{\infty} |a_n|^2 e^{2nu}\Big)\Big(\sum_{n=0}^{\infty} |a_n|^2 4n^2 e^{2nu}\Big) = I_2 I_2'',$$

so that

$$\frac{d^2}{du^2}\log I_2 \ge 0,$$

as required.
(*Question:* When does equality take place?) □

Theorem 8.7 *Let $f(z)$ be regular in $|z| < R$, and let*

$$I_1(r) = \frac{1}{2\pi} \int_0^{2\pi} |f(re^{i\theta})| d\theta.$$

Then $I_1(r)$ is a strictly increasing function of r in $0 < r < R$, and $\log I_1(r)$ is a convex function of $\log r$.

Proof. Let $0 < r_1 < r_2 < r_3$, and define

$$k(\theta) = \frac{|f(r_2 e^{i\theta})|}{f(r_2 e^{i\theta})}, \qquad F(z) = \frac{1}{2\pi} \int_0^{2\pi} f(ze^{i\theta}) k(\theta) d\theta.$$

Then $F(z)$ is regular in $|z| \leq r_3$, and $|F(z)|$ assumes a maximum value at a certain point $z = r_3 e^{i\lambda}$, so that

$$I_1(r_2) = F(r_2) \leq |F(r_3 e^{i\lambda})| \leq I_1(r_3).$$

Take α so that $r_1^\alpha I_1(r_1) = r_3^\alpha I_1(r_3)$; we then have

$$r_2^\alpha I_1(r_2) = r_2^\alpha F(r_2) \leq \max_{r_1 \leq |z| \leq r_3} |z^\alpha F(z)| \leq r_1^\alpha I_1(r_1) = r_3^\alpha I_1(r_3).$$

The rest of the argument is the same as that given in the proof of Theorem 8.5, the three-circle theorem. □

Note 1. Theorem 8.6 is an easy consequence of Theorem 8.7.

Note 2. There is equality when $f(z) = \lambda z^\alpha$.

Note 3. If $f(z)$ has no zero in $|z| < R$ then $(f(re^{i\theta}))^p$ is regular in $|z| < R$, and we deduce from Theorem 8.7 that

$$I_p(r) = \frac{1}{2\pi} \int_0^{2\pi} |f(re^{i\theta})|^p \, d\theta, \qquad p > 0,$$

is an increasing function of r in $0 < r < R$, and that $\log I_p(r)$ is a convex function of $\log r$. The purpose of the following sections is to show that this is still the case when there are zeros.

8.5 A lemma

Theorem 8.8 *The sum of several absolute values of analytic functions also assumes its maximum on the boundary. More precisely, let $f_1(z), \ldots, f_n(z)$ be single-valued functions regular inside and on D. Then the continuous function*

$$\phi(z) = |f_1(z)| + \cdots + |f_n(z)|$$

takes its maximum value only at the boundary of D; if it takes such a value inside D then each $f_\nu(z)$ is constant.

Proof. Let z_0 be an interior point of D. Then

$$|f_\nu(z_0)| \le \frac{1}{2\pi}\int_0^{2\pi} |f_\nu(z_0 + re^{i\theta})|d\theta, \quad \nu = 1, 2, \ldots, n,$$

so that

$$\phi(z_0) \le \frac{1}{2\pi}\int_0^{2\pi} \phi(z_0 + re^{i\theta})d\theta.$$

If there is inequality for one f_ν, there is inequality for ϕ. The rest of the argument is the same as that for the maximum principle. □

Theorem 8.9 *Let $f_1(z), \ldots, f_n(z)$ be single-valued functions regular inside and on D, and let $p_1, \ldots, p_n$ be positive numbers. Then the continuous function*

$$\phi(z) = |f_1(z)|^{p_1} + \cdots + |f_n(z)|^{p_n}$$

takes its maximum value only at the boundary of D; if it takes such a value inside D then $f_1, f_2, \ldots, f_n$ are all constants.

Proof. Suppose that z_0 is an interior point; choose r such that the disc $|z - z_0| \le r$ lies inside D and that $f_1, f_2, \ldots, f_n$ do not have zeros inside the punctured disc. If $f_\nu(z_0) = 0$ and $f_\mu(z_0) \ne 0$, then $(f_\mu(z))^{p_\mu}$ is regular in $|z - z_0| \le r$. By Theorem 8.7 there is a point z_1 on $|z - z_0| = r$ such that

$$\sum_\mu |f_\mu(z_1)|^{p_\mu} \ge \sum_\mu |f_\mu(z_1)|^{p_\mu},$$

and obviously

$$\sum_\nu |f_\nu(z_1)|^{p_\nu} \ge 0 = \sum_\nu |f_\nu(z_1)|^{p_\nu},$$

so that $\phi(z_1) \ge \phi(z_0)$. Moreover, if one of $f_\mu(z)$ or $f_\nu(z)$ is not constant, then $\phi(z_1) > \phi(z_0)$.

The set D together with its boundary form a closed set. By the compactness theorem the continuous function $\phi(z)$ assumes its maximum value at a certain point, which cannot be an interior point according to our argument here (unless each $f_\nu(z)$ is constant). □

Note. If we only assume that $f_m(z)$ is regular, but $|f_m(z)|$ is single-valued, then the theorem is still valid.

8.6 A general mean-value theorem

Theorem 8.10 (Pólya–Szegö) *Let $f(z)$ be regular in $|z| < R$, and let*

$$I_p(r) = \frac{1}{2\pi}\int_0^{2\pi} |f(re^{i\theta})|d\theta, \quad r < R, \ p > 0.$$

Then $I_p(r)$ is an increasing function of r, and $\log I_p(r)$ *is a convex function of* $\log r$.

Proof. (1) Being increasing. Let $w_\nu = e^{2\pi i\nu/n}$, $\nu = 1, 2, \ldots, n$, and $0 \le r_1 < r_2 < R$. By Theorem 8.9, there is a point $r_2e^{i\theta_2}$ on the circle $|z| = r_2$ such that

$$\frac{1}{n}\sum_{\nu=1}^{n}|f(r_1w_\nu)|^p \le \frac{1}{n}\sum_{\nu=1}^{n}|f(r_2w_\nu e^{i\theta_2})|^p.$$

Letting $n \to \infty$, we have $I_p(r_1) \le I_p(r_2)$.

(2) Convexity. Let $0 < r_1 < r_2 < r_3 < R$, and let α be a real number. Then the functions

$$z^{\alpha/p}f(w_1z),\ z^{\alpha/p}f(w_2z),\ \ldots,\ z^{\alpha/p}f(w_nz)$$

are regular in the annulus $r_1 \le |z| \le r_3$, and their absolute values are single-valued. From our results in Section 8.5, we have

$$r_2^\alpha\sum_{\nu=1}^{n}|f(r_2w_\nu)|^p \le \max_{\substack{0\le\theta_1\le 2\pi\\ 0\le\theta_3\le 2\pi}}\Big(r_1^\alpha\sum_{\nu=1}^{n}|f(r_1e^{i\theta_1}w_\nu)|^p,\ r_3^\alpha\sum_{\nu=1}^{n}|f(r_3e^{i\theta_3}w_\nu)|^p\Big).$$

On division by n and then letting $n \to \infty$ we have

$$r_2^\alpha I_p(r_2) \le \max\big(r_1^\alpha I_p(r_1),\ r_3^\alpha I_p(r_3)\big),$$

and the required result follows by setting α so that $r_1^\alpha I_p(r_1) = r_3^\alpha I_p(r_3)$. □

8.7 $(I_p(r))^{1/p}$

Obviously Theorems 8.6 and 8.7 follow from Theorem 8.10. Let

$$G(r) = \exp\Big(\frac{1}{2\pi}\int_0^{2\pi}\log|f(re^{i\theta})|d\theta\Big),$$

which may be called the geometric mean.

Theorem 8.11 *We have* $\lim_{p\to 0_+}(I_p(r))^{1/p} = G(r)$.

Proof. For small p, we have

$$\begin{aligned}
I_p(r) &= \frac{1}{2\pi}\int_0^{2\pi}|f(re^{i\theta})|\,d\theta = \frac{1}{2\pi}\int_0^{2\pi}e^{p\log|f|}\,d\theta\\
&= \frac{1}{2\pi}\int_0^{2\pi}\big(1 + p\log|f| + O(p^2(\log|f|)^2)\big)d\theta\\
&= 1 + \frac{p}{2\pi}\int_0^{2\pi}\log|f(re^{i\theta})|d\theta + O(p^2),
\end{aligned}$$

and so

$$\lim_{p\to 0_+} \frac{1}{p}\log I_p(r) = \lim_{p\to 0_+} \frac{1}{p}\log\left(1+\frac{p}{2\pi}\int_0^{2\pi}\log|f(re^{i\theta})|d\theta + O(p^2)\right)$$
$$= \frac{1}{2\pi}\int_0^{2\pi}\log|f(re^{i\theta})|d\theta = \log G(r),$$

as required. □

We deduce at once the following.

Theorem 8.12 *The function $G(r)$ is increasing, and* $\log G(r)$ *is a convex function of* $\log r$.

Theorem 8.13 *We have* $\lim_{p\to\infty}(I_p(r))^{1/p} = M(r)$.

Proof. Obviously, $(I_p(r))^{1/p} \le M(r)$. Let $\epsilon > 0$, and choose $re^{i\theta_1}$ so that $|f(re^{i\theta_1})| = M(r)$. Then there is a neighbouring region $|\theta - \theta_1| < \delta/2$ within which

$$|f(re^{i\theta})| > M(r) - \epsilon.$$

Therefore

$$(I_p(r))^{1/p} \ge (M(r)-\epsilon)\delta^{1/p},$$

and thus

$$\lim_{p\to\infty}(I_p(r))^{1/p} \ge M(r) - \epsilon.$$

This holds for every $\epsilon > 0$, so that the required result follows. □

We can now see that the Pólya–Szegö theorem includes Hadamard's three-circle theorem.

8.8 Vitali's theorem

Theorem 8.14 *Let $f_n(z)$ be a sequence of single-valued analytic functions in a region D. Suppose that there exists M, which does not depend on n and z, such that*

$$|f_n(z)| \le M, \quad n = 1, 2, \ldots, \; z \in D. \tag{8.2}$$

Suppose further that $\lim_{n\to\infty} f_n(z)$ *exists for every z belonging to some subset E of D, with a limit point being an interior of D. Then $f_n(z)$ converges uniformly in any compact subset of D to a function which is analytic in D.*

Proof. Suppose first that D is the disc with centre the origin and radius R, and suppose further that

$$z_1, z_2, \ldots \to 0$$

and that

$$\lim_{n\to\infty} f_n(z_\nu), \qquad \nu = 1, 2, \ldots,$$

all exist. We emphasise that $f_n(z)$ satisfies: (i) $f_n(z)$ is single-valued and regular in $|z| < R$, (ii) $|f_n(z)| < M$, (iii) $\lim_{n\to\infty} f_n(z_\nu)$ exists.

Let

$$f_1(z) = a_{01} + a_{11}z + a_{21}z^2 + \cdots + a_{k1}z^k + \cdots,$$

$$\ldots$$

$$f_\ell(z) = a_{0\ell} + a_{1\ell}z + a_{2\ell}z^2 + \cdots + a_{k\ell}z^k + \cdots,$$

$$\ldots$$

We first prove that each sequence of coefficients has a limit, that is

$$\lim_{\ell\to\infty} a_{k\ell} = a_k$$

exists, and then show that the Taylor series

$$\sum a_k z^k$$

converges in $|z| < R$, and so represents an analytic function $f(z)$ there. Finally we prove that, for any $\epsilon > 0$, the sequence of functions $f_n(z)$ converges to $f(z)$ uniformly in $|z| \le R - \epsilon$.

By hypothesis,

$$|f_n(z) - f_n(0)| \le |f_n(z)| + |f_n(0)| \le 2M,$$

so that, by Schwarz's lemma,

$$|f_n(z) - f_n(0)| \le \frac{2M|z|}{R}, \qquad |z| < R,$$

and therefore

$$\begin{aligned} |f_n(0) - f_{n+m}(0)| &\le |f_n(0) - f_n(z_\ell)| + |f_n(z_\ell) - f_{n+m}(z_\ell)| + |f_{n+m}(z_\ell) - f_{n+m}(0)| \\ &\le \frac{4M|z_\ell|}{R} + |f_n(z_\ell) - f_{n+m}(z_\ell)|. \end{aligned}$$

Let $\epsilon > 0$. Since $z_\ell \to 0$ and $\lim_{n\to\infty} f_n(z_\ell)$ exists, we can choose z_ℓ so that $4M|z_\ell|/R < \epsilon/2$ and N such that, for $n > N$,

$$|f_n(z_\ell) - f_{n+m}(z_\ell)| < \frac{\epsilon}{2}.$$

Therefore, for $n > N$,

$$|f_n(0) - f_{n+m}(0)| < \epsilon,$$

and so

$$\lim_{n\to\infty} f_n(0) = \lim_{n\to\infty} a_{0n}$$

exists; let us write

$$a_0 = \lim_{n\to\infty} a_{0n}.$$

We next examine

$$f_{\nu 1}(z) = \frac{f_\nu(z) - a_{0\nu}}{z} = a_{1\nu} + a_{2\nu}z + \cdots , \tag{8.3}$$

and we proceed to show that the functions also satisfy (i), (ii) and (iii), so that $\lim_{\nu\to\infty} a_{1\nu}$ exists.

Since the radius of convergence for the series (8.3) is R, condition (i) does hold. Also, by Cauchy's inequality,

$$|a_{k\nu}| \leq M/R^k, \tag{8.4}$$

so that, for $|z| \leq R - \epsilon$,

$$\begin{aligned} |f_{\nu 1}(z)| &\leq |a_{1\nu}| + |a_{2\nu}||z| + \cdots \\ &\leq M\left(1 + \frac{R-\epsilon}{R} + \left(\frac{R-\epsilon}{R}\right)^2 + \cdots\right)\Big/R = \frac{M}{\epsilon}. \end{aligned}$$

Thus condition (ii) can be considered to hold, albeit with $|f_{\nu 1}(z)| \leq M/\epsilon$ in $|z| \leq R - \epsilon$ instead. Finally,

$$\lim_{\nu\to\infty} f_{\nu 1}(z_\ell) = \lim_{\nu\to\infty} \frac{f_\nu(z_\ell) - a_{0\nu}}{z_\ell} = \frac{1}{z_\ell}\big(\lim_{\nu\to\infty} f_\nu(z_\ell) - a_{0\nu}\big), \quad z_\ell \neq 0,$$

exists. The first stage of the proof is accomplished.

Repeating the process, we deduce that each sequence of coefficients has a limit

$$\lim_{\nu\to\infty} a_{k\nu} = a_k, \quad k = 1, 2, \ldots ,$$

and, by (8.4), we have $|a_k| \leq M/R^k$ so that $\sum a_k z^k$ represents an analytic function in $|z| < R$.

Finally, for $|z| \leq R - \epsilon$, we have

$$\begin{aligned} |f(z) - f_\nu(z)| &\leq |(a_0 - a_{0\nu}) + \cdots + (a_0 - a_{0\nu})z^m| \\ &\quad + \left|\sum_{k=m+1}^{\infty} a_k z^k\right| + \left|\sum_{k=m+1}^{\infty} a_{k\nu} z^k\right|, \end{aligned}$$

so that

$$|f(z) - f_\nu(z)| \leq \sum_{\lambda=0}^{m} |a_\lambda - a_{\lambda\nu}||z^\lambda| + \frac{2M}{R^{m+1}}(R-\epsilon)^{m+1}\frac{R}{\epsilon}.$$

For a given $\delta > 0$, we can choose m so that

$$\frac{2RM}{\epsilon}\left(\frac{R-\epsilon}{R}\right)^{m+1} < \frac{\delta}{2}$$

and then take $\nu(\delta, m)$ so that, for $\nu > \nu(\delta, m)$, the $m+1$ numbers

$$|a_0 - a_{0\nu}|, \ldots, |a_m - a_{m\nu}|$$

are all smaller than

$$\frac{\delta}{2}\frac{1-(R-\epsilon)}{1-(R-\epsilon)^{m+1}},$$

giving

$$|f(z) - f_\nu(z)| \leq \frac{\delta}{2}\frac{1-(R-\epsilon)}{1-(R-\epsilon)^{m+1}}\sum_{\lambda=0}^{m}(R-\epsilon)^\lambda + \frac{\delta}{2} = \delta.$$

Therefore Vitali's theorem for the disc $|z| < R$ with $\lim_{\nu\to\infty} z_\nu = z_0 = 0$ is valid; it is also valid when $\lim_{\nu\to\infty} z_\nu = z_0 \neq 0$, because we can apply the Möbius transformation mapping the disc onto itself, and 0 to z_0.

For the general case of the theorem, let R be any closed region from the interior of D. Each $f_\nu(z)$ is regular everywhere in R; that is, corresponding to each point of R there is a circle with centre there inside which $f_\nu(z)$ is regular. Therefore, by the Heine–Borel theorem, there are finitely many open discs $C_1, \ldots, C_p$ forming a cover for R.

Let C_1 be such a disc, including z_0, the limit point of z_ν. From what has been proved, $f_\nu(z)$ converges uniformly to $f(z)$ in any closed region of C_1. Take the disc C_2 which intersects C_1 in a set K, and take a sequence of points $z'_1, \ldots, z'_n, \ldots$ with a limit point z'_0, which is still in K; then

$$\lim_{\nu\to\infty} f_\nu(z'_\ell) = f(z'_\ell).$$

Therefore, we can apply the theorem within C_2, that is $f_\nu(z)$ converges uniformly to a function as $\nu \to \infty$. Because the two limiting functions take the same value in K, the latter is an analytic continuation of the former. Using this method we can reach all p discs, so that the theorem holds for R. The proof of the theorem is now complete. □

8.9 Families of bounded functions

Definition 8.15 Let $\{f(z)\}$ be a set of single-valued functions, each of which is regular in a region D. If there is a constant M such that

$$|f(z)| < M$$

for every function $f(z)$ in the set, then we say that the set forms a bounded family.

We have the following important theorem on bounded families of functions.

Theorem 8.16 (Montel) *Let $\mathcal{F}$ be a bounded family of analytic functions in D. Then, corresponding to any closed subset of D, there is a uniformly convergent sequence of functions in $\mathcal{F}$ whose limit is an analytic function.*

Proof. Take any interior point z_0 as the limit point of a sequence

$$z_1, z_2, \ldots$$

and suppose that $f_\nu(z)$ form a bounded family. Since $|f_\nu(z_1)| < M$ we can take a sequence of functions

$$f_{\nu_{11}}(z), f_{\nu_{21}}(z), \ldots, f_{\nu_{n1}}(z), \ldots$$

(with $\nu_{n1} \geq n$) so that

$$f_{\nu_{11}}(z_1), f_{\nu_{21}}(z_1), \ldots, f_{\nu_{n1}}(z_1), \ldots$$

has a limit. We then consider, within the convergent sequence of functions,

$$f_{\nu_{11}}(z_2), f_{\nu_{21}}(z_2), \ldots,$$

which contains the convergent subsequence

$$f_{\nu_{12}}(z_2), f_{\nu_{22}}(z_2), \ldots .$$

Again, from the sequence we select $\nu_{13}, \nu_{23}, \ldots, \nu_{n3}, \ldots$ so that

$$f_{\nu_{13}}(z_3), f_{\nu_{23}}(z_3), \ldots$$

is convergent, etc. We then consider the sequence of functions (see the Hilbert–Cantor diagonal method in Section 3.5)

$$f_{\nu_{11}}(z), f_{\nu_{22}}(z), f_{\nu_{33}}(z), \ldots .$$

The sequence corresponding to $z = z_\ell$ is convergent for every ℓ. Since the limit of z_ℓ is an interior point of D, the theorem now follows from Vitali's theorem. □

The converse of the theorem is false; for example, the family of functions

$$\frac{1}{n(z-1)}$$

converges uniformly in any closed subset of the unit disc, but individual members of the family are not bounded in the unit disc.

Note. We can also use Montel's theorem to deduce Vitali's theorem. If $f_n(z)$ is not convergent, we can use Montel's theorem to select two sequences of functions tending to analytic functions $f(z)$ and $g(z)$, but with $f(z) = g(z)$ at a sequence of points having an interior point as a limit point, so that $f(z) \equiv g(z)$ by the 'identity theorem'.

8.10 Normal families

The result in Section 8.9 leads us to the important notion of a normal family.

Definition 8.17 Let $\{f_\nu(z)\}$ be a family of functions defined in a region D, with ν running over a certain index set. If, from any sub-family, we can always select a sequence $\{f_n(z)\}_{n=1,2,\ldots}$ such that $\{f_n(z)\}$ converges uniformly in any closed subset of D, with the limit function possibly $\equiv \infty$, then we say that $\{f_\nu(z)\}$ is a normal family in D.

By a sequence of functions $\{f_n(z)\}$ converging to ∞ uniformly in E, we mean that, for any $\epsilon > 0$, there exists N such that

$$\frac{1}{|f_n(z)|} < \epsilon$$

for all $n > N$ and all $z \in E$, with N being independent of z.

The limit functions associated with a normal family form a set called the derived set of the family; such a set is closed in the sense that the limit functions associated with the derived set also belong to the derived set, so that the derived set of a normal family is a normal family.

Theorem 8.15 may now be recast as follows: Any bounded family (of analytic functions) must be a normal family.

Theorem 8.18 *Let $\{f(z)\}$ be a normal family of regular functions in D. If it is bounded at a point z_0, then the family is a bounded family corresponding to any closed subregion D' of D.*

Proof. Suppose otherwise, so that $\{f(z)\}$ is not a bounded family in D'. For each positive integer n, there is a point z_n and a function $f_n(z)$ such that

$$|f_n(z_n)| > n.$$

We examine the sequence of functions

$$f_1(z), f_2(z), \ldots, f_n(z), \ldots.$$

By hypothesis we can select a subsequence which converges uniformly to $f(z)$ in D'. Since $f_n(z)$ has to converge at the point z_0, the limit function $f(z)$ cannot be ∞.

Thus $f(z)$ is bounded in the interior and the boundary of D', say

$$|f(z)| \leq K.$$

We have $|f_n(z) - f(z)| < 1$ for all sufficiently large n and $z \in D'$, so that

$$|f_n(z)| \leq K + 1, \quad z \in D'.$$

This contradicts $|f_n(z_n)| > n$. The theorem is proved. □

9

Integral functions and meromorphic functions

9.1 Definitions

By an *integral function*, or an *entire function*, we mean one which is analytic everywhere on the plane (excluding ∞). By a *meromorphic function*, we mean one which is analytic everywhere on the plane (excluding ∞), except for isolated poles.

An integral function can be considered as the generalisation of a polynomial, and a meromorphic function that of a rational function. It is easy to see that the sum, the difference and the product of two integral functions are also integral functions; the sum, the difference, the product and the quotient (the denominator not the zero function) of two meromorphic functions are meromorphic functions.

We first consider the properties of polynomials and rational functions, since they form the bases for our understanding of integral and meromorphic functions. First, polynomials must have zeros, and there is the unique factorisation theorem (which states that polynomials can be written as products of factors taking the value 0 at the roots). Can such results be generalised to integral functions?

First, we can give an example of an integral function without a zero, namely e^z. More generally, corresponding to any integral function $g(z)$, the integral function $e^{g(z)}$ does not have a zero. Conversely, if $f(z)$ is an integral function without a zero, then $\log f(z) = g(z)$ is also an integral function. Therefore, *a necessary and sufficient condition for an integral function $f(z)$ to have no zero is that it should be representable as $e^{g(z)}$, with $g(z)$ being an integral function.*

Secondly, more generally, corresponding to any complex number (excluding ∞) and a polynomial $p(z)$ with degree n, the equation $p(z) = a$ has n solutions. Now let $f(z)$ be an integral function, and we ask for what values a will the equation

$$f(z) = a \tag{9.1}$$

be soluble? The answer is given by the famous Picard's theorem: *The function $f(z)$ omits at most one value a_0, so that, for all $a \neq a_0, \infty$, equation (9.1) is soluble.* The original proof of Picard's theorem is very interesting, making use of the elliptic modular function $\omega(z)$. This function has the following properties: The non-constant function is regular everywhere apart from the points $0, 1, \infty$, and its imaginary part is never negative. If both $f(z) = a$

and $f(z) = b$ do not have a solution, then we set

$$g(z) = \frac{f(z) - a}{b - a},$$

so that both $g(z) = 0$ and $g(z) = 1$ do not have a solution. That is, $g(z) \neq 0, 1$, so that $\omega(g(z))$ is analytic everywhere except at $z = \infty$, and since the imaginary part of $\omega(z)$ is not negative we find that

$$\left|e^{\omega(g(z))}\right| \leq 1.$$

Thus $e^{\omega(g(z))}$ is a bounded integral function, and hence a constant according to Liouville's theorem. This means that $\omega(g(z))$ is constant, and since ω is not a constant function, we deduce that $g(z)$ has to be a constant.

Thirdly, we consider whether there is a generalisation of the factorisation theorem for polynomials to integral functions. If $f(z)$ has finitely many zeros $a_1, \ldots, a_n$, then it is not difficult to show that

$$f(z) = \prod_{\nu=0}^{n} (z - a_\nu) e^{g(z)},$$

where $g(z)$ is an integral function. If there are infinitely many zeros then we need to consider the convergence problem associated with the product

$$\prod_{\nu=0}^{\infty} \left(1 - \frac{z}{a_\nu}\right), \tag{9.2}$$

which amounts to the convergence of the series

$$\sum_{\nu=0}^{\infty} \frac{1}{|a_\nu|}.$$

In other words, given any n complex numbers, we can construct a polynomial with these numbers as roots, but given infinitely many $a_1, a_2, \ldots, a_\nu, \ldots$ we cannot simply use such zeros for the product in (9.2). Nevertheless, we may still be able to construct an integral function with these zeros. If the point set $\{a_\nu\}$ does not have a limit point in the finite plane, then the integral function so constructed is not identically zero, and this is the Weierstrass theorem that will be discussed in this chapter.

As for meromorphic functions, the first problem is as follows: Can it be represented as the ratio of two integral functions? This can be settled without too much fuss. Let $f(z)$ be a meromorphic function, with poles at

$$z_1, z_2, \ldots, z_\nu, \ldots,$$

with each pole being repeated according to its order. By Weierstrass' theorem we can construct an integral function $g(z)$ with zeros at $z_1, z_2, \ldots, z_\nu, \ldots$, so that $f(z)g(z)$ is an integral function $h(z)$, and hence $f(z) = h(z)/g(z)$.

Finally, rational functions have expansions in terms of partial fractions, and the generalisation of this is the Mittag–Leffler theorem that will be discussed in this chapter. If Weierstrass' theorem is considered to be the construction of an integral function with specified zeros, then Mittag–Leffler's theorem can be considered to be the construction of meromorphic functions with specified poles.

9.2 Weierstrass' factorisation theorem

Let $a_1, a_2, \ldots$ be a complex sequence with ∞ as its only limit point; each value in the sequence may occur only a finite number of times. *Then there is a non-identically zero integral function having these points as zeros.*

Suppose that ℓ of the numbers are 0; write $r_n = |a_n|$ and list the remaining numbers so that

$$0 < r_{\ell+1} \leq r_{\ell+2} \leq \cdots .$$

We first point out that there is a sequence of positive integers

$$p_{\ell+1}, p_{\ell+2}, \ldots$$

such that

$$\sum_{n=\ell+1}^{\infty} \left(\frac{r}{r_n}\right)^{p_n} \tag{9.3}$$

is convergent for all r. Indeed, $p_n = n$ will do because, for any given r, we have $r_n > 2r$ when n is large enough, so that $(r/r_n)^n < 1/2^n$, and therefore (9.3) converges. (Show as an exercise that taking $p_n > \log n$ will also satisfy the required condition.)

Definition 9.1 The functions

$$E(u, p) = \begin{cases} (1-u)e^{u+\frac{1}{2}u^2+\cdots+\frac{1}{p}u^p}, & p = 1, 2, \ldots, \\ (1-u), & p = 0, \end{cases}$$

are called elementary (or primary) factors.

If $|u| < 1$ then

$$\log E(u, p) = -\frac{u^{p+1}}{p+1} - \frac{u^{p+2}}{p+2} - \cdots ,$$

so that if $|u| < \frac{1}{2}$, then

$$\begin{aligned} |\log E(u, p)| &\leq |u|^{p+1} + |u|^{p+2} + \cdots \\ &\leq |u|^{p+1}\left(1 + \frac{1}{2} + \frac{1}{2^2} + \cdots\right) = 2|u|^{p+1}. \end{aligned} \tag{9.4}$$

Now take the infinite product

$$f(z) = z^{\ell} \prod_{n=\ell+1}^{\infty} E\Big(\frac{z}{a_n}, p_n - 1\Big).$$

From inequality (9.4) we know that, when $|a_n| > 2|z|$,

$$\Big|\log E\Big(\frac{z}{a_n}, p_n - 1\Big)\Big| \le 2\Big(\frac{r}{r_n}\Big)^{p_n}.$$

Therefore, in $|z| \le R$, the series

$$\sum_{|a_n|>2R} \Big|\log E\Big(\frac{z}{a_n}, p_n - 1\Big)\Big| \le 2\sum\Big(\frac{r}{r_n}\Big)^{p_n}$$

converges uniformly, so that the product

$$\prod_{|a_n|>2R} E\Big(\frac{z}{a_n}, p_n - 1\Big)$$

converges uniformly, delivering an analytic function in $|z| \le R$; it is clear that

$$z^{\ell} \prod_{|a_n|\le 2R, n>\ell} E\Big(\frac{z}{a_n}, p_n - 1\Big) \tag{9.5}$$

is an analytic function of z. Therefore $f(z)$ is analytic in $|z| \le R$, and has zeros coinciding with those in (9.5).

Since R can be arbitrarily large we have found an integral function with zeros at $a_1, a_2, \ldots$.

Note that there are no other zeros, because a convergent infinite product excludes its limit being zero. However, there is no uniqueness in the factorisation, because an integral function with no zero can always be inserted as an extra factor.

9.3 The order of an integral function

If there is a positive A such that

$$f(z) = O\big(e^{r^A}\big), \qquad r = |z| \to \infty,$$

then we say that the integral function $f(z)$ has finite order, and its order ρ is defined to be the infimum of the set of A; thus, for any $\epsilon > 0$, we have

$$f(z) = O\big(e^{r^{\rho+\epsilon}}\big) \tag{9.6}$$

and that this does not hold for any $\epsilon < 0$.

Let $M(r)$ denote the maximum of $|f(z)|$ on $|z| = r$. Then, equivalent to (9.6), we have

$$\rho = \limsup_{r\to\infty} \frac{\log\log M(r)}{\log r}.$$

Example 9.2 The functions e^z, $\sin z$ and $\cos z$ are all integral functions with order 1.

Example 9.3 The function $\cos\sqrt{z}$ is an integral function with order $\frac{1}{2}$.

Example 9.4 The integral function e^{e^z} does not have a finite order.

In order to avoid difficulties which are not of a fundamental nature, we shall assume henceforth that $f(z) \neq 0$; otherwise we can consider $f(z)/z^\ell$ instead.

Definition 9.5 Let a_n be the zeros of $f(z)$, with $|a_n| = r_n$, where $r_1 \leq r_2 \leq r_3 \leq \cdots$. The infimum ρ_1 of the set of numbers ρ for which the series

$$\sum_{n=1}^{\infty} \frac{1}{r_n^\rho}$$

is convergent is called the index of convergence of $f(z)$ for the zeros.

Theorem 9.6 *The index of convergence of an integral function is at most the order of the function; that is, $\rho_1 \leq \rho$, if the function has order ρ.*

Proof. By Jensen's formula we have, for $r_n \leq r \leq r_{n+1}$ and $\epsilon > 0$,

$$\log \frac{r^n}{r_1 \cdots r_n} = \frac{1}{2\pi} \int_0^{2\pi} \log|f(re^{i\theta})| d\theta - \log|f(0)| = O(r^{\rho+\epsilon}).$$

Denote by $n(r)$ the numbers of $r_j \leq r$, so that

$$\log \frac{r^n}{r_1 \cdots r_n} = \sum_{i=1}^{n} \log \frac{r}{r_i} \geq \sum_{r_i \leq \frac{r}{2}} \log \frac{r}{r_i} \geq n\left(\frac{r}{2}\right) \log 2,$$

and thus

$$n\left(\frac{r}{2}\right) = O(r^{\rho+\epsilon}), \quad \text{that is} \quad n(r) = O(r^{\rho+\epsilon}).$$

Take $r = r_n$, so that $n(r) = n$, giving $n = O(r_n^{\rho+\epsilon})$, that is

$$\frac{1}{r_n^{\rho+\epsilon}} = O\left(\frac{1}{n}\right).$$

Since $\sum 1/n^{1+\epsilon}$ converges, it follows that, for every $\epsilon > 0$,

$$\sum \frac{1}{r_n^{\rho+\epsilon}}$$

converges. The theorem is proved. □

We now have an important property of an integral function with a finite order with zeros at z_n: there is a positive integer p such that the product

$$\prod_{n=1}^{\infty} E\left(\frac{z}{z_n}, p\right) \tag{9.7}$$

converges for all z.

The smallest positive integer p for which (9.7) is convergent is called the *genus*. If ρ_1 is not an integer, then $p = [\rho_1]$; if ρ_1 is an integer, then

$$p = \begin{cases} \rho_1, & \text{if } \sum r_n^{-\rho_1} \text{ diverges,} \\ \rho_1 - 1, & \text{if } \sum r_n^{-\rho_1} \text{ converges.} \end{cases}$$

Therefore, in any case,

$$p \leq \rho_1 \leq \rho.$$

Theorem 9.7 *Let $z_n \neq 0$ be a sequence which tends to infinity, with ρ_1 as the index of convergence of the corresponding function having the genus p. Then the canonical product*

$$P(z) = \prod_{n=1}^{\infty} E\left(\frac{z}{z_n}, p\right)$$

is an integral function of order ρ_1.

Proof. Let ρ denote the order of the canonical product. By Theorem 9.6, we already have $\rho_1 \leq \rho$, so that it suffices to show that $\rho \leq \rho_1$. Write $r_n = |z_n|$, $r = |z|$, let $k > 1$ be an integer, and consider

$$\log|P(z)| = \sum_{r_n \leq kr} \log\left|E\left(\frac{z}{z_n}, p\right)\right| + \sum_{r_n > kr} \log\left|E\left(\frac{z}{z_n}, p\right)\right| = \Sigma_1 + \Sigma_2,$$

say. For Σ_2, we have

$$\Sigma_2 = O\left(\sum_{r_n > kr} \left(\frac{r}{r_n}\right)^{p+1}\right) = O\left(r^{p+1} \sum_{r_n > kr} \frac{1}{r_n^{p+1}}\right).$$

If $p = \rho_1 - 1$ then $\Sigma_2 = O(r^{p+1}) = O(r^{\rho_1})$; otherwise, for any sufficiently small positive ϵ, we have $\rho_1 + \epsilon < p + 1$, and hence

$$\begin{aligned} r^{p+1} \sum_{r_n > kr} \frac{1}{r_n^{p+1}} &= r^{p+1} \sum_{r_n > kr} \frac{1}{r_n^{\rho_1+\epsilon}} \frac{1}{r_n^{-\rho_1-\epsilon+p+1}} \\ &< r^{p+1}(kr)^{\rho_1+\epsilon-p-1} \sum \frac{1}{r_n^{\rho_1+\epsilon}} = O(r^{\rho_1+\epsilon}). \end{aligned}$$

Now let u denote z/z_n in Σ_1, so that $|u| \geq 1/k$. Thus

$$\log|E(u, p)| \leq \log(1 + |u|) + |u| + \cdots + \frac{|u|^p}{p} < K|u|^p,$$

where K depends on k and p; therefore

$$\Sigma_1 = O\Big(r^p \sum_{r_n \le kr} \frac{1}{r_n^p}\Big) = O\Big(r^p \sum_{r_n \le kr} r_n^{\rho_1+\epsilon-p}\, r_n^{-\rho_1-\epsilon}\Big)$$
$$= O\big(r^p (kr)^{\rho_1+\epsilon-p} \sum r_n^{-\rho_1-\epsilon}\big) = O(r^{\rho_1+\epsilon}),$$

and hence

$$\log|P(z)| = O(r^{\rho_1+\epsilon}).$$

Thus $\rho \le \rho_1 + \epsilon$ for all small ϵ, so that $\rho \le \rho_1$. The theorem is proved. □

9.4 Hadamard's factorisation theorem

Theorem 9.8 *Let $f(z)$ be an integral function of order ρ, with zeros at $z_1, z_2, \ldots$, ($f(0) \ne 0$). Then*

$$f(z) = e^{Q(z)} P(z), \tag{9.8}$$

where $P(z)$ is the canonical product formed with the zeros of $f(z)$, and $Q(z)$ is a polynomial with degree at most ρ.

Proof. We already know that $f(z)$ has the representation (9.8), where $Q(z)$ is an integral function, and it remains to show that $Q(z)$ is a polynomial with degree at most $\nu = [\rho]$.

Take the logarithm of (9.8), and differentiate it $\nu + 1$ times to give

$$\left(\frac{d}{dz}\right)^{\nu} \left\{\frac{f'(z)}{f(z)}\right\} = Q^{(\nu+1)}(z) - \nu! \sum_{n=1}^{\infty} \frac{1}{(z_n - z)^{\nu+1}}.$$

We proceed to show that $Q^{(\nu+1)}(z) = 0$, which will then deliver the required result that $Q(z)$ is a polynomial with degree at most ν.

Let

$$g_R(z) = \frac{f(z)}{f(0)} \prod_{|z_n| \le R} \left(1 - \frac{z}{z_n}\right)^{-1}.$$

On the circle $|z| = 2R$ we have $|1 - z/z_n| \ge 1$ when $|z_n| \le R$, so that

$$|g_R(z)| \le \left|\frac{f(z)}{f(0)}\right| = O\big(e^{(2R)^{\rho+\epsilon}}\big), \tag{9.9}$$

for any $\epsilon > 0$. Since $g_R(z)$ is an integral function, this holds also for $|z| < 2R$.

Let $h_R(z) = \log g_R(z)$, fixing the logarithm by taking the value $h_R(0) = 0$, so that $h_R(z)$ is regular in $|z| \le R$ and so, by (9.9),

$$\Re\{h_R(z)\} < K R^{\rho+\epsilon}. \tag{9.10}$$

By the Borel–Carathéodory inequality we now have, for $|z| = r < R$,

$$\left|h_R^{(\nu+1)}(z)\right| \le \frac{2^{\nu+3}(\nu+1)!R}{(R-r)^{\nu+2}} K R^{\rho+\epsilon},$$

and so, for $r = R/2$,

$$h_R^{(\nu+1)}(z) = O(R^{\rho+\epsilon-\nu-1}). \tag{9.11}$$

Therefore, for $|z| = r = R/2$, we have

$$\begin{aligned} Q^{(\nu+1)}(z) &= h_R^{(\nu+1)}(z) + \nu! \sum_{|z_n|>R} \frac{1}{(z_n - z)^{\nu+1}} \\ &= O(R^{\rho+\epsilon-\nu-1}) + O\Big(\sum_{|z_n|>R} |z_n|^{-\nu-1}\Big), \end{aligned}$$

which continues to hold even for $|z| < R/2$. For sufficiently small ϵ, the first term on the right-hand side tends to 0 as $R \to \infty$. Since the series $\sum |z_n|^{-\nu-1}$ is convergent, the second term also tends to 0 as $R \to \infty$. Therefore $Q^{(\nu+1)}(z) = 0$, and the theorem is proved. □

Exercise 9.9 Prove that if ρ is not an integer, then $\rho_1 = \rho$.

9.5 Mittag–Leffler's theorem

Let $a_1, a_2, \ldots, a_n, \ldots$ be a sequence with ∞ as the only limit point. We proceed to determine a meromorphic function with the following prescribed principal expansion at a_k:

$$g_k\Big(\frac{1}{z-a_k}\Big) = \frac{\alpha_{k1}}{z-a_k} + \cdots + \frac{\alpha_{k\ell_k}}{(z-a_k)^{\ell_k}}, \qquad k = 1, 2, \ldots. \tag{9.12}$$

We list the poles so that

$$|a_1| \le |a_2| \le |a_3| \le \cdots,$$

and we assume temporarily that 0 is not a pole. The expression (9.12) has the power series expansion

$$g_k\Big(\frac{1}{z-a_k}\Big) = \beta_0^{(k)} + \beta_1^{(k)} z + \beta_2^{(k)} z^2 + \cdots, \qquad |z| < |a_k|. \tag{9.13}$$

Let $\sum \epsilon_k$ be a convergent series of positive numbers ϵ_k. Since the power series (9.13) converges uniformly in $|z| < |a_k|/2$, we may take p_k so that

$$q_k(z) = \beta_0^{(k)} + \cdots + \beta_{p_k}^{(k)} z^{p_k}$$

satisfies

$$\left|g_k\Big(\frac{1}{z-a_k}\Big) - q_k(z)\right| < \epsilon_k; \tag{9.14}$$

we form the series

$$\phi(z) = \sum_{k=1}^{\infty} \left[g_k\left(\frac{1}{z - a_k}\right) - q_k(z) \right]. \tag{9.15}$$

Let C be the circle with centre the origin and radius R. Since $|a_k| \to \infty$, there exists N such that $R \leq |a_k|/2$ for all $k > N$. For such k the estimate (9.14) holds inside C so that the series (9.15) converges uniformly. For the initial part of the sum over $1 \leq k \leq N$, the poles $a_1, \ldots, a_N$ lie inside the circle, and the principal expansion for such points in (9.12), and the series (9.15) contains only these poles in C. Since R can be arbitrarily large the series (9.15) is the required meromorphic function.

If $z = 0$ is also a pole, then its principal expansion is $g_0(1/z)$, and all we need do is to adjoin this to (9.15).

Let $f(z)$ be any meromorphic function. The above method can be used to deliver $\phi(z)$, which then has the same principal expansions, so that $f(z) - \phi(z)$ is analytic in the whole plane. Therefore

$$f(z) = F(z) + g_0\left(\frac{1}{z}\right) + \sum_{k=1}^{\infty} \left[g_k\left(\frac{1}{z - a_k}\right) - q_k(z) \right],$$

where $F(z)$ is an integral function.

Readers should compare the proofs of the theorems of Weierstrass and Mittag–Leffler to see if they have the same fundamental elements, and to investigate whether we can set p_k explicitly, for example, $p_k = k$, as we did in Section 9.4.

Readers should also consider the factorisation theorem for functions analytic inside the unit circle; meromorphic functions are then ratios of analytic functions, and there should also be a similar theorem for such functions.

9.6 Representations for cot z and sin z

The function

$$F(z) = \cot z - \frac{1}{z} \tag{9.16}$$

has a simple pole at $z = k\pi$, with residue 1; here, k is a non-zero integer, and the function is analytic everywhere else.

Applying the method in Section 9.5, we take

$$F_1(z) = \sum_{k=-\infty}^{\infty}{}' \left(\frac{1}{z - k\pi} + \frac{1}{k\pi} \right),$$

where $\sum'$ denotes the omission of the term $k = 0$. It is clear that the series is convergent at every $z \neq k\pi$, and that there is a simple pole at $z = k\pi$, with the same residue as that in (9.16). From Section 9.5, we know that $F(z)$ and $F_1(z)$ differ by an integral function, and we now show that $F(z) = F_1(z)$.

Let C_n be the square with vertices at $z = x + iy$ with $x = \pm(n + \frac{1}{2})\pi$, $y = \pm(n + \frac{1}{2})\pi$. We use the method in Section 7.11 to consider the function

$$f(z) = \pi \cot \pi z\left(\frac{1}{x - z\pi} + \frac{1}{z\pi}\right)$$

on the boundary of C_n. Since $z = 0$ is a double pole, we cannot apply the original formula directly. The function $f(z)$ has no pole on C_n, but there are poles at $z = \pm 1, \pm 2, \ldots, \pm n$ inside C_n. The residue for $f(z)$ at $z = k$, a non-zero integer, is

$$\frac{1}{x - k\pi} + \frac{1}{k\pi}.$$

At the double pole $z = 0$, the function $f(z)$ has the expansion

$$\left(\frac{1}{z} - \frac{\pi^2 z}{3} + \cdots\right)\left(\frac{1}{\pi z} + \frac{1}{x} + \frac{\pi z}{x^2} + \cdots\right),$$

so that the residue there is

$$\frac{1}{x};$$

at the pole $z = x/\pi$ the residue is

$$-\cot x.$$

Therefore

$$\frac{1}{2\pi i}\int_{C_n} f(z)dz = \sum_{k=-n}^{n}{}^{\prime}\left(\frac{1}{x - k\pi} + \frac{1}{k\pi}\right) + \frac{1}{x} - \cot x,$$

and it remains to show that the left-hand side tends to 0 as $n \to \infty$.

From

$$|\cot(x + iy)|^2 = \frac{e^{2y} + e^{-2y} + 2\cos 2x}{e^{2y} + e^{-2y} - 2\cos 2x},$$

we find that, along the sides of C_n parallel to the y-axis, that is on $x = \pm(n + \frac{1}{2})\pi$, so that $\cos 2x = -1$, we have

$$|\cot z|^2 \le \frac{|e^y - e^{-y}|}{e^y + e^{-y}} < 1.$$

From $1/(x - z\pi) + 1/z\pi = O(1/|z|^2)$, it now follows that, for large n,

$$\int_{x=\pm(n+\frac{1}{2})\pi} f(z)dz = O\left(\frac{1}{n}\right).$$

Again, along the sides of C_n parallel to the x-axis, that is on $y = \pm(n + \frac{1}{2})\pi$, we have

$$|\cot(x + iy)|^2 \le \frac{e^{2y} + e^{-2y} + 2}{e^{2y} + e^{-2y} - 2} = \left(\frac{1 + e^{-2y}}{1 - e^{-2y}}\right)^2 = O(1),$$

so that the corresponding part of the integral tends to 0 as $n \to \infty$. We have therefore proved that

$$\cot x = \frac{1}{x} + \lim_{n\to\infty} \sum_{m=-n}^{n} \left(\frac{1}{x - m\pi} + \frac{1}{m\pi} \right) = \frac{1}{x} + 2x \sum_{n=1}^{\infty} \frac{1}{x^2 - n^2\pi^2}, \tag{9.17}$$

where the series is convergent.

Take any closed bounded region with a boundary not passing through any poles, including $z = 0$ inside. After the isolation of each pole inside the region with a small circle the series (9.17) is uniformly convergent in the remaining part of the region. Take z to be a point there, and integrate term by term from 0 to z, but not passing through any pole. We then have

$$\log \frac{\sin z}{z} = C + \sum_{n=1}^{\infty} \log \left(1 - \frac{z^2}{n^2\pi^2}\right),$$

that is

$$\frac{\sin z}{z} = c \prod_{n=1}^{\infty} \left(1 - \frac{z^2}{n^2\pi^2}\right).$$

Since $\lim_{z\to 0} \sin z/z = 1$, so that $C = 1$, we have the infinite product for $\sin z$:

$$\sin z = z \prod_{n=1}^{\infty} \left(1 - \frac{z^2}{n^2\pi^2}\right),$$

or

$$\frac{\sin \pi z}{\pi} = z \prod_{n=1}^{\infty} \left(1 - \frac{z^2}{n^2}\right).$$

Rewriting this as

$$\frac{\sin \pi z}{\pi} = z \lim_{m\to\infty} \prod_{n=-m}^{m} \left(1 - \frac{z}{n}\right) e^{z/n},$$

we have Weierstrass' form of the product.

Exercise 9.10 Prove that

$$\sec z = 2\pi \sum_{n=0}^{\infty} \frac{(-1)^n (n + \frac{1}{2})}{(n + \frac{1}{2})^2\pi^2 - z^2}, \qquad \operatorname{cosec} z = 2\pi \sum_{n=1}^{\infty} \frac{1}{(n - \frac{1}{2})^2\pi^2 - z^2}.$$

Exercise 9.11 Prove that

$$\cos z = \prod_{n=1}^{\infty} \left(1 - \frac{z^2}{(n - \frac{1}{2})^2\pi^2}\right).$$

Exercise 9.12 Prove that

$$\frac{1}{e^z - 1} = \frac{1}{z} - \frac{1}{2} + 2z \sum_{n=1}^{\infty} \frac{1}{z^2 + 4n^2\pi^2}.$$

Exercise 9.13 Prove that

$$\operatorname{cosec}^2 z = \sum_{n=-\infty}^{\infty} \frac{1}{(z - n\pi)^2}.$$

Exercise 9.14 Deduce from

$$\frac{\sin \pi z}{\pi} = z \prod_{n=1}^{\infty} \left(1 - \frac{z^2}{n^2}\right)$$

that $\sin z$ is periodic with period 2π.

9.7 The Γ function

Consider the Weierstrass product

$$z e^{\gamma z} \prod_{n=1}^{\infty} \left[\left(1 + \frac{z}{n}\right) e^{-\frac{z}{n}}\right], \tag{9.18}$$

where

$$\gamma = \lim_{m \to \infty} \left(1 + \frac{1}{2} + \frac{1}{3} + \cdots + \frac{1}{m} - \log m\right)$$

is Euler's constant. Since $\sum |z|^2/n^2$ converges, we may use the method of proof used in Section 9.2 by taking $p_n = 2$ so that the product

$$\prod E\left(-\frac{z}{n}, 1\right)$$

converges, and therefore (9.18) is an integral function.

We define

$$\frac{1}{\Gamma(z)} = z e^{\gamma z} \prod_{n=1}^{\infty} \left[\left(1 + \frac{z}{n}\right) e^{-z/n}\right], \tag{9.19}$$

so that it follows at once that $\Gamma(z)$ has simple poles at $0, -1, -2, \ldots$, and is analytic everywhere else.

Is such a definition for $\Gamma(z)$ consistent with our earlier definiiton

$$\int_0^{\infty} e^{-t} t^{z-1}\, dt \quad (\Re z > 0)?$$

To show that it is, we first deduce certain properties associated with (9.19).

Theorem 9.15 *Apart from $z = 0, -1, -2, \ldots$, we have*

$$\Gamma(z) = \frac{1}{z}\prod_{n=1}^{\infty}\left(1+\frac{1}{n}\right)^{z}\left(1+\frac{z}{n}\right)^{-1}.$$

Proof. Write (9.19) as

$$\begin{aligned}
\frac{1}{\Gamma(z)} &= z\lim_{m\to\infty} e^{(1+\frac{1}{2}+\frac{1}{3}+\cdots+\frac{1}{m}-\log m)z}\lim_{m\to\infty}\prod_{n=1}^{m}\left\{\left(1+\frac{z}{n}\right)e^{-z/n}\right\} \\
&= z\lim_{m\to\infty} m^{-z}\prod_{n=1}^{m}\left(1+\frac{z}{n}\right) = z\lim_{m\to\infty}\prod_{n=1}^{m-1}\left(1+\frac{1}{n}\right)^{-z}\prod_{n=1}^{m}\left(1+\frac{z}{n}\right) \\
&= z\lim_{m\to\infty}\prod_{n=1}^{m-1}\left\{\left(1+\frac{z}{n}\right)\left(1+\frac{1}{n}\right)^{-z}\right\}\left(1+\frac{1}{m}\right)^{z},
\end{aligned}$$

and the required result follows. □

Theorem 9.16 *For $\Re(z) > 0$, we have*

$$\lim_{n\to\infty}\int_0^n \left(1-\frac{t}{n}\right)^n t^{z-1}\,dt = \Gamma(z).$$

Proof. Let $t = n\tau$, so that

$$\int_0^n \left(1-\frac{t}{n}\right)^n t^{z-1}\,dt = n^z\int_0^1 (1-\tau)^n\tau^{z-1}\,d\tau.$$

Integrating by parts n times we find that

$$\begin{aligned}
\int_0^1 (1-\tau)^n\tau^{z-1}\,d\tau &= \frac{1}{z}\tau^z(1-\tau)^n\Big|_0^1 + \frac{n}{z}\int_0^1 (1-\tau)^{n-1}\tau^z\,d\tau \\
&= \cdots \\
&= \frac{n(n-1)\cdots 1}{z(z+1)\cdots(z+n)-1}\int_0^1 \tau^{z+n-1}\,d\tau \\
&= \frac{1\cdot 2\cdots n}{z(z+1)\cdots(z+n)},
\end{aligned}$$

and so, by Theorem 9.15,

$$\lim_{n\to\infty}\int_0^n\left(1-\frac{t}{n}\right)^n t^{z-1}\,dt = \lim_{n\to\infty}\frac{1\cdot 2\cdots n}{z(z+1)\cdots(z+n)}n^z = \Gamma(z).$$

□

Theorem 9.17 *For $\Re(z) > 0$, we have*

$$\Gamma(z) = \int_0^\infty e^{-t}t^{z-1}\,dt.$$

Proof. Let

$$\Gamma_1(z) = \int_0^\infty e^{-t} t^{z-1}\, dt.$$

Then, by Theorem 9.16, we have

$$\begin{aligned}\Gamma_1(z) - \Gamma(z) &= \lim_{n\to\infty}\left(\int_0^n e^{-t}t^{z-1}\,dt - \int_0^n \left(1-\frac{t}{n}\right)^n t^{z-1}\,dt\right)\\ &= \lim_{n\to\infty}\int_0^n \left[e^{-t} - \left(1-\frac{t}{n}\right)^n\right] t^{z-1}\,dt.\end{aligned}$$

As we shall see,

$$0 \le e^{-t} - \left(1-\frac{t}{n}\right)^n \le n^{-1}t^2e^{-t}, \tag{9.20}$$

so that, as $n\to\infty$,

$$\left|\int_0^n \left[e^{-t} - \left(1-\frac{t}{n}\right)^n\right] t^{z-1}\,dt\right| \le \int_0^n n^{-1}t^2e^{-t}t^{x-1}\,dt \le n^{-1}\int_0^\infty e^{-t}t^{x+1}\,dt \to 0.$$

To establish (9.20) we use $1+y \le e^y \le 1/(1-y)$, which holds for $0 \le y < 1$. With $y = t/n$ we then have

$$\left(1+\frac{t}{n}\right)^{-n} \ge e^{-t} \ge \left(1-\frac{t}{n}\right)^n,$$

and hence

$$0 \le e^{-t} - \left(1-\frac{t}{n}\right)^n = e^{-t}\left\{1 - e^t\left(1-\frac{t}{n}\right)^n\right\} \le e^{-t}\left\{1 - e^t\left(1-\frac{t^2}{n^2}\right)^n\right\}.$$

For $0 \le \alpha \le 1$ it can be proved by induction that $(1-\alpha)^n \ge 1 - n\alpha$. With $\alpha = t^2/n^2$ we then find that $1-(1-t^2/n^2)^n \le t^2/n$, and therefore (9.20) is established. The theorem is proved. □

Exercise 9.18 (Gauss–Legendre) Prove that, for any natural number n,

$$\Gamma(z)\Gamma\left(z+\tfrac{1}{n}\right)\cdots\Gamma\left(z+\tfrac{n-1}{n}\right) = n^{\frac{1}{2}-nz}(2\pi)^{\frac{1}{2}(n-1)}\Gamma(nz).$$

Suggestion. Use Theorem 9.15 to show that

$$\phi(z) = \frac{n^{nz}\Gamma(z)\Gamma\left(z+\frac{1}{n}\right)\cdots\Gamma\left(z+\frac{n-1}{n}\right)}{n\Gamma(nz)}$$

is independent of z, and then use $\Gamma(1-z)\Gamma(z) = \pi/\sin\pi z$ to evaluate $\Gamma(1/n)$.

The result here is sometimes called the multiplication formula of Gauss, and we often make use of the special case:

$$2^{2z-1}\Gamma(z)\Gamma\left(z+\tfrac{1}{2}\right) = \pi^{\frac{1}{2}}\Gamma(2z),$$

which is called the duplication formula of Legendre.

Exercise 9.19 Let $B(p,q) = \Gamma(p)\Gamma(q)/\Gamma(p+q)$. Show that

$$B(np, nq) = n^{-np} \frac{B(p,q)B(p+\frac{1}{n},q)\cdots B(p+\frac{n-1}{n},q)}{B(q,q)B(2q,q)\cdots B((n-1)q,q)}.$$

Exercise 9.20 Show that if $a_1 + \cdots + a_k = b_1 + \cdots + b_k$, then

$$\prod_{n=1}^{\infty} \frac{(n-a_1)\cdots(n-a_k)}{(n-b_1)\cdots(n-b_k)} = \prod_{m=1}^{k} \frac{\Gamma(1-b_m)}{\Gamma(1-a_m)};$$

for example,

$$\prod_{n=1}^{\infty} \frac{n(n+a+b)}{(n+a)(n+b)} = \frac{\Gamma(a+1)\Gamma(b+1)}{\Gamma(a+b+1)}.$$

Exercise 9.21 Show that

$$\prod_{n=1}^{\infty} \left(1 - \frac{x}{n^{\ell}}\right) = -1 \Big/ \prod_{n=0}^{\ell-1} \Gamma\left(-x^{1/m} w^m\right), \qquad w = e^{2\pi i/\ell}.$$

Exercise 9.22 Show that

$$\frac{\Gamma'(z)}{\Gamma(z)} = -\gamma - \frac{1}{z} + \lim_{n\to\infty} \sum_{m=1}^{n} \left(\frac{1}{m} - \frac{1}{z+m}\right).$$

9.8 The zeta function

Following tradition, in this section we change our notation by using s as the complex variable, and write $s = \sigma + it$, where σ and t are real. The Riemann zeta function is defined by

$$\zeta(s) = \sum_{n=1}^{\infty} \frac{1}{n^s}. \tag{9.21}$$

It is an analytic function in the half-plane $\sigma > 1$ because, for any $\delta > 0$, the series converges uniformly in the region $\sigma \geq 1 + \delta$.

The source of the function lies in number theory, particularly in the study of the distribution of prime numbers. However, its importance is not confined to number theory. The function is closely related to the Γ function and other advanced topics, for example in the study of integral equations.

The function is defined only in $\sigma > 1$ by (9.21), and our first task is to extend its domain to the whole complex plane to arrive at a meromorphic function. We start with

$$n^{-s}\Gamma(s) = \int_0^{\infty} x^{s-1} e^{-nx}\, dx, \qquad \sigma > 0.$$

For $\sigma \geq 1+\delta$, where $\delta > 0$, we have

$$\Gamma(s)\zeta(s) = \lim_{N\to\infty} \sum_{n=0}^{N} \int_0^\infty x^{s-1}e^{-nx}\,dx$$
$$= \lim_{N\to\infty} \left(\int_0^\infty \frac{x^{s-1}}{1-e^{-x}}\,dx - \int_0^\infty \frac{x^{s-1}}{1-e^{-x}} e^{-(N+1)x}\,dx \right).$$

Using the estimate $e^x \geq 1+x$, we find that, as $N \to \infty$,

$$\left| \int_0^\infty \frac{x^{s-1}}{1-e^{-x}} e^{-(N+1)x}\,dx \right| \leq \int_0^\infty x^{\sigma-2}e^{-Nx}\,dx = N^{1-\sigma}\Gamma(\sigma-1) \to 0,$$

so that

$$\zeta(s) = \frac{1}{\Gamma(s)} \int_0^\infty \frac{x^{s-1}}{1-e^{-x}}\,dx, \qquad \sigma > 1.$$

Consider the integral

$$I(s) = \int_C \frac{z^{s-1}}{e^z - 1}\,dz, \tag{9.22}$$

where C is the contour on the plane cut along the positive real axis: Starting from $+\infty$ the point z moves along the 'upper cut' until it reaches $z = \delta(> 0)$; the circle $|z| = \delta$ is then described in the positive direction to return to δ, and the contour is completed by running z along the 'lower cut' back to $+\infty$. We need to examine z^{s-1}, which is defined to be $e^{(s-1)\log z}$, with $\log z$ being real initially as z runs along the cut. After describing $|z| = \delta$, the imaginary part of $\log z$ increases from 0 to 2π; in other words, with $z^{s-1} = x^{s-1}$ before describing the circle, we need to have $z^{s-1} = (xe^{2\pi i})^{s-1}$ after describing the circle.

On the circle $|z| = \delta$ we have

$$|z^{s-1}| = e^{(\sigma-1)\log|z| - t\arg z} \leq |z|^{\sigma-1}e^{2\pi|t|}, \qquad |e^z - 1| > A|z|,$$

so that, for $\sigma > 1$, the integral along the circle tends to 0, as $\delta \to 0$. Therefore

$$I(s) = -\int_0^\infty \frac{x^{s-1}}{e^x - 1}\,dx + \int_0^\infty \frac{(xe^{2\pi i})^{s-1}}{e^x - 1}\,dx = (e^{2\pi i s} - 1)\Gamma(s)\zeta(s);$$

on making use of $\Gamma(s)\Gamma(1-s) = \pi/\sin\pi s$, we arrive at

$$\zeta(s) = \frac{e^{-i\pi s}\Gamma(1-s)}{2\pi i} \int_C \frac{z^{s-1}}{e^z - 1}\,dz. \tag{9.23}$$

This then is a representation of the previously defined ζ function in $\sigma > 1$. We note, however, that the integral $I(s)$ is convergent in every part of the finite s-plane. Therefore formula (9.23) delivers the analytic continuation of $\zeta(s)$ to the whole s-plane, and the only possible singularities are at the poles of $\Gamma(1-s)$, namely at $s = 1, 2, 3, \ldots$. Since we already know that $\zeta(s)$ is finite at $s = 2, 3, \ldots$, the only possible singularity is at $s = 1$, which is a simple pole. From

$$I(1) = \int_C \frac{dz}{e^z - 1} = 2\pi i \quad \text{and} \quad \Gamma(1-s) = -\frac{1}{s-1} + \cdots,$$

we see the residue for $\zeta(s)$ there is 1.

If s is an integer, then the integral $I(s)$ is single-valued, and the method of residues then delivers

$$\frac{z}{e^z - 1} = 1 - \frac{1}{2}z + B_1 \frac{z^2}{2!} - B_2 \frac{z^4}{4!} + \cdots,$$

where $B_1, B_2, \ldots$ are the Bernoulli numbers, and it can be shown that

$$\zeta(0) = -\frac{1}{2}, \quad \zeta(-2m) = 0, \quad \zeta(1-2m) = \frac{(-1)^m B_m}{2m}, \qquad m = 1, 2, \ldots .$$

9.9 The functional equation for $\zeta(s)$

Theorem 5.23 *The zeta function $\zeta(s)$ is finite everywhere on the plane, except at $s = 1$, where it has a simple pole with the residue 1. Moreover, it satisfies the functional equation*

$$\zeta(s) = 2^s \pi^{s-1} \sin \frac{\pi s}{2} \Gamma(1-s) \zeta(1-s).$$

Proof. Take the integrand in (9.22) but change the contour C on the cut plane to C_n: Starting from $+\infty$, the point z moves along the 'upper cut' until it reaches $z(2n+1)\pi$; the point z then describes, in the positive sense, the square with vertices at $(2n+1)\pi(\pm 1 \pm i)$, and then returns to $+\infty$ along the 'lower cut'. There are now poles at $\pm 2i\pi, \ldots, \pm 2i\pi n$ between the old contour C and C_n, and the sum of the residues at the poles $\pm 2mi\pi$ is

$$\begin{aligned}(2m\pi e^{i\pi/2})^{s-1} + (2m\pi e^{3i\pi/2})^{s-1} &= (2m\pi)^{s-1} e^{i\pi(s-1)} 2 \cos \frac{\pi(s-1)}{2} \\ &= -2(2m\pi)^{s-1} e^{i\pi s} \sin \frac{\pi s}{2}.\end{aligned}$$

Therefore, by the residue theorem,

$$I(s) = \int_C \frac{z^{s-1}}{e^z - 1} dz = \int_{C_n} \frac{z^{s-1}}{e^z - 1} dz + 4\pi i e^{i\pi s} \sin \frac{\pi s}{2} \sum_{m=1}^{n} (2m\pi)^{s-1}.$$

Now let $\sigma < 0$ and $n \to \infty$, so that, on C_n, the function $1/(e^z - 1)$ is bounded and $z^{s-1} = O(|z|^{\sigma-1})$. Thus the integral along C_n tends to 0 as $n \to \infty$, giving

$$I(s) = 4\pi i e^{i\pi s} \sin \frac{\pi s}{2} \sum_{m=1}^{\infty} (2m\pi)^{s-1} = 4\pi i e^{i\pi s} \sin \frac{\pi s}{2} (2\pi)^{s-1} \zeta(1-s).$$

The functional equation now follows from this and (9.23); that it holds also for $\sigma \geq 0$ is a consequence of the uniqueness theorem for analytic continuation. □

Several minor results can now be deduced immediately from the functional equation. From the values for $\zeta(1-2m)$ in Section 9.8, we now have

$$\zeta(2m) = 2^{2m-1}\pi^{2m}\frac{B_m}{(2m)!}, \quad m = 1, 2, \dots .$$

We next show that

$$\zeta'(0) = -\frac{1}{2}\log 2\pi.$$

Rewrite the functional equation in the form

$$\zeta(1-s) = 2^{1-s}\pi^{-s}\cos\frac{\pi s}{2}\Gamma(s)\zeta(s),$$

and then take logarithmic derivatives to give

$$-\frac{\zeta'(1-s)}{\zeta(1-s)} = -\log 2\pi - \frac{\pi}{2}\tan\frac{\pi s}{2} + \frac{\Gamma'(s)}{\Gamma(s)} + \frac{\zeta'(s)}{\zeta(s)}.$$

From expansions about the point $s = 1$, we have

$$\frac{\pi}{2}\tan\frac{\pi s}{2} = -\frac{1}{s-1} + O(|s-1|), \quad \frac{\Gamma'(s)}{\Gamma(s)} = \frac{\Gamma'(1)}{\Gamma(1)} + \cdots = -\gamma + \cdots$$

and

$$\frac{\zeta'(s)}{\zeta(s)} = \frac{-\frac{1}{(s-1)^2} + k + \cdots}{\frac{1}{s-1} + \gamma + k(s-1) + \cdots} = -\frac{1}{s-1} + \gamma + \cdots,$$

where k is a certain constant. Therefore, letting $s \to 1$ we have

$$-\frac{\zeta'(0)}{\zeta(0)} = -\log 2\pi,$$

and hence the required result for $\zeta'(0)$.

The functional equation can be written in a symmetrical form by introducing

$$\xi(s) = \frac{1}{2}s(s-1)\pi^{-s/2}\Gamma\left(\frac{s}{2}\right)\zeta(s)$$

to give

$$\xi(s) = \xi(1-s).$$

If we further write

$$\Xi(z) = \xi\left(\frac{1}{2} + iz\right),$$

then

$$\Xi(z) = \Xi(-z).$$

9.10 Convergence on the sphere

In order to extend the notions of uniform convergence and normal family to meromorphic functions, we now introduce the concept of uniform convergence on the sphere.

By the distance between two points (ξ, η, ζ) and (ξ', η', ζ') on the Riemann sphere we mean the least distance along a great circle. The spherical distance between two complex numbers z, z' (including $z = \infty$, $z' = \infty$) is the distance between the two corresponding points on the sphere, and this will be denoted by $[z, z']$.

A function $Z = f(z)$ is said to be spherically continuous at z_0 if, for any given $\epsilon > 0$, there exists $\delta > 0$ such that

$$[Z_0, Z] < \epsilon$$

for all z satisfying $[z_0, z] < \delta$. A similar definition is given for spherical uniform continuity. It is not difficult to show that, in a closed region, continuity implies uniform continuity.

9.10.1 Uniform convergence for meromorphic functions

Let

$$Z_1 = f_1(z), \ Z_2 = f_2(z), \ \ldots, \ Z_n = f_n(z), \ \ldots$$

be a convergent sequence of meromorphic functions in D, with the limit function $Z = f(z)$. If, for any given $\epsilon > 0$, we can find N (independent of z) such that

$$[Z_n, Z] < \epsilon$$

for all $n > N$ and all $z \in D$, then we say that $\{f_n(z)\}$ converges uniformly in D to $f(z)$.

The meaning of $[z_1, z_2]$ can be written out more explicitly: For z_i, the corresponding point on the surface of the sphere is given by

$$\xi_i = \frac{2x_i}{1+|z_i|^2}, \qquad \eta_i = \frac{2y_i}{1+|z_i|^2}, \qquad \zeta_i = \frac{|z_i|^2-1}{|z_i|^2+1}.$$

Considering the point (ξ_i, η_i, ζ_i) to be a unit vector on the surface of the sphere, we let θ denote the angle subtended between the vectors, so that

$$\begin{aligned}
\cos\theta &= \frac{4x_1x_2 + 4y_1y_2 + (1-|z_1|^2)(1-|z_2|^2)}{(1+|z_1|^2)(1+|z_2|^2)} \\
&= \frac{2(z_1\bar{z}_2 + z_2\bar{z}_1) + (1-|z_1|^2)(1-|z_2|^2)}{(1+|z_1|^2)(1+|z_2|^2)}, \\
\tan^2\frac{\theta}{2} &= \frac{1-\cos\theta}{1+\cos\theta} = \frac{|z_1|^2 + |z_2|^2 - z_1\bar{z}_2 - z_2\bar{z}_1}{1 + |z_1z_2|^2 + z_1\bar{z}_2 + z_2\bar{z}_1} = \left|\frac{z_1 - z_2}{1 + z_1\bar{z}_2}\right|^2,
\end{aligned}$$

giving

$$\theta = \pm 2\arctan\left|\frac{z_1 - z_2}{1 + z_1\bar{z}_2}\right|.$$

Previously we often used $1/|z| < \epsilon$ to denote a neighbourhood of the point ∞, and we can now see that there is some basis for this because

$$[z_1, \infty] = 2\arctan\frac{1}{|z_1|}.$$

Let

$$w = \frac{\alpha z + \beta}{\gamma z + \delta}$$

be a unitary transformation, that is one with

$$\alpha|^2 + |\gamma|^2 = |\beta|^2 + |\delta|^2 = 1, \quad \alpha\bar{\beta} + \gamma\bar{\delta} = 0. \tag{9.24}$$

For such a transformation, the spherical distance is invariant. To verify this we have

$$w_1 - w_2 = \frac{\alpha z_1 + \beta}{\gamma z_1 + \delta} - \frac{\alpha z_2 + \beta}{\gamma z_2 + \delta} = \frac{(\alpha\delta - \beta\gamma)(z_1 - z_2)}{(\gamma z_1 + \delta)(\gamma z_2 + \delta)},$$

$$1 + w_1\bar{w}_2 = 1 + \frac{\alpha z_1 + \beta}{\gamma z_1 + \delta}\frac{\bar{\alpha}\bar{z}_2 + \bar{\beta}}{\bar{\gamma}\bar{z}_2 + \bar{\delta}} = \frac{1 + z_1\bar{z}_2}{(\gamma z_1 + \delta)(\bar{\gamma}\bar{z}_2 + \bar{\delta})},$$

and, from $|\alpha\delta - \beta\gamma| = 1$, we find that $[w_1, w_2] = [z_1, z_2]$.

It follows that if

$$f_1, f_2, \ldots, f_n, \ldots$$

converges spherically and uniformly to f then, for $(\alpha, \beta, \gamma, \delta)$ satisfying condition (9.24), the sequence with

$$g_\nu = \frac{\alpha f_\nu + \beta}{\gamma f_\nu + \delta}$$

converges spherically and uniformly to

$$g = \frac{\alpha f + \beta}{\gamma f + \delta}.$$

We can also give such definitions without mentioning the spherical surface (or elliptic geometry) as follows.

(1) If z_0 is not a pole of $f(z)$, then the uniform convergence of $f_n(z)$ to $f(z)$ in a neighbourhood of z_0 has the usual meaning.
(2) If z_0 is a pole of $f(z)$, then the uniform convergence of $f_n(z)$ to $f(z)$ in a neighbourhood of z_0 means the uniform convergence of $1/f_n(z)$ to $1/f(z)$. (Note that the region should not contain zeros of $f_n(z)$.)

9.11 Normal families of meromorphic functions

Theorem 9.24 *Let $\{f_\nu(z)\}$ be a family of meromorphic functions in a region G. Suppose that there is a region D in the w-plane such that these functions do not take on values which are interior to D. Then $\{f_\nu(z)\}$ is a normal family.*

Proof. Let A be an interior point of D, and take a closed disc with centre A and radius δ so small that the disc lies inside D. Then

$$|f_\nu(z) - A| > \delta,$$

and we may set

$$g_\nu(z) = \frac{\delta}{f_\nu(z) - A}$$

so that $g_\nu(z)$ is regular and bounded ($|g_\nu(z)| < 1$) in G. Therefore $\{g_\nu(z)\}$ form a normal family.

Consider any infinite subset of $f_\nu(z)$. The corresponding $g_\nu(z)$ contain a uniformly convergent sequence $g_n(z)$, with limit $g_0(z)$, say. It then follows that $1/g_n(z)$ converges spherically to $1/g_0(z)$, and so $f_n(z)$ converges spherically to

$$f_0(z) = \frac{\delta}{g_0(z)} + A.$$

The theorem is proved. □

Note. We have the following deeper result, but the proof will be omitted: A family of meromorphic functions which omits values of points on a curve is a normal family. Still deeper results are as follows.

Theorem 9.25 *Let $\{f_\nu(z)\}$ be a family of analytic functions in D which omit two distinct values. Then $\{f_\nu(z)\}$ is a normal family.*

Theorem 9.26 *A family of meromorphic functions in D which omit three distinct values is a normal family.*

We now mention some easily deduced results.

Theorem 9.27 (Montel) *Let $f(z)$ be an analytic function of z which is regular in the half-strip $S = \{(x, y) : a < x < b,\ y > 0\}$. Suppose that $f(z)$ is bounded in S and that, for $a < \xi < b$,*

$$\lim_{y\to\infty} f(\xi + iy) = \ell.$$

Then $f(z) \to \ell$ uniformly in $a + \delta \le x \le b - \delta$ for any $\delta > 0$ as $y \to \infty$.

Proof. Consider the family of functions

$$f_n(z) = f(z + in)$$

in the rectangle $\{(x, y) : a < x < b,\ 0 < y < 2\}$. It is a bounded family, and for any λ with $0 < \lambda < 2$ we have $f_n(\xi + i\lambda) \to \ell$ as $n \to \infty$. By Vitali's theorem, $f_n(z) \to \ell$ in the rectangle $\{(x, y) : a + \delta \le x \le b - \delta,\ \frac{1}{2} \le y \le \frac{3}{2}\}$, so that the theorem follows. □

Under the conformal transformation $z = i \log w$, Theorem 9.27 becomes:

Theorem 9.28 *Let $\phi(w)$ be regular and bounded in the sector $\alpha < \arg w < \beta$ ($|\alpha - \beta| < \pi$). Suppose that, for $\alpha < \theta < \beta$,*

$$\lim_{r\to\infty} \phi(re^{i\theta}) = \ell.$$

Then $\phi(w) \to \ell$ uniformly in $\alpha + \delta \leq \arg w \leq \beta - \delta$ for any $\delta > 0$ as $r \to \infty$.

Theorem 9.29 (Picard) *A non-constant integral function $f(z)$ omits at most one finite value.*

Proof. Consider the family of functions

$$f_0(z) = f(z),\ f_1(z) = f(2z),\ \dots,\ f_n(z) = f(2^n z),\ \dots$$

in $|z| < 1$. The family is bounded at $z = 0$, and if such functions omit two values, then $\{f_n(z)\}$ is a normal family. By Theorem 8.17, there is a subsequence $f_{n_\nu}(z)$ which converges uniformly to an integral function $f(z)$ in any finite region of the z-plane. We may assume that $f_{n_\nu}(z)$ is just $f_n(z)$. Then $|f_n(z)| \leq M$ for $|z| \leq \frac{1}{2}$, so that $|f(z)| \leq M$ for $|z| \leq 2^{n-1}$. By Liouville's theorem, $f(z)$ has to be a constant, which contradicts the hypothesis. Therefore at most one value can be omitted. □

Theorem 9.30 (Schottky) *Suppose that $f(z)$ is regular in $|z| \leq 1$, and omits the values 0 and 1. Then, for $|z| \leq \frac{1}{2}$,*

$$|f(z)| < Q\big(f(0)\big),$$

where Q is a positive number depending only on $f(0)$.

Proof. We examine the family of functions $f(z)$ satisfying the hypothesis, and $f(0) = a_0$. By Theorem 9.25, we have a normal family, and the value taken at $z = 0$ is finite. The required result follows from Theorem 8.17. □

10
Conformal transformations

10.1 Important basic ingredients

The single most important result concerning conformal mapping is obviously the already alluded to Riemann mapping theorem.

Theorem 10.1 (Riemann mapping theorem) *Let G be any simply connected region with at least two distinct boundary points. There is a conformal transformation mapping the interior of G bijectively onto the interior of the unit circle. The mapping function is uniquely specified by the following condition: An interior point of G, together with a direction at the point, is mapped to a specific point together with a specific direction.*

More precisely, we may take $w = 0$ to be the specific point in the unit disc, and the direction there to be along the positive real axis. The condition then becomes that, given an interior point z_0 of G, there is a unique conformal transformation $f(z)$ mapping G bijectively onto the interior of the unit circle so that $f(z_0) = 0 < f'(z_0)$.

The inverse transformation

$$z = g(w)$$

which maps the unit disc back onto G has the following properties: (i) $g(w)$ is a single-valued analytic function in $|w| < 1$; (ii) if

$$g(w_1) = g(w_2), \quad |w_1| < 1, \quad |w_2| < 1,$$

then $w_1 = w_2$. Such a bijective analytic function is said to be univalent. Thus the study of univalent functions on the unit disc is, via the Riemann mapping theorem, equivalent to the study of such functions in all simply connected regions.

Let

$$g(w) = a_0 + a_1 w + a_2 w^2 + \cdots, \quad |w| < 1.$$

Then necessarily $a_1 \neq 0$, and so $h(w) = \big(g(w) - a_0\big)/a_1$ is also univalent. Therefore (abandoning the use, or notation, of $f(z)$ in the above) we need only study

$$f(z) = z + a_2 z^2 + a_3 z^3 + \cdots, \quad |z| < 1;$$

Figure 10.1.

such normalised univalent functions are sometimes called schlicht functions. These functions can be identified with a certain set of points $(a_2, a_3, \ldots)$ in an infinite dimensional space, and there are then various central problems which are fundamental to the set. For example, there is the famous Bieberbach conjecture[1] in mathematics, which states that the set concerned is contained in the infinite dimensional body

$$|a_n| \leq n.$$

The Riemann mapping theorem settles the problem of the image of the interior of a simply connected region, but the boundary of the region can be very complicated. Take, for example, the line segment from $\frac{1}{2}$ to 1, and join it to the unit circle to form the boundary, and the region concerned is still simply connected. Indeed, using the same idea the boundary can be as complicated as that shown in Fig. 10.1 and yet, according to the Riemann mapping theorem, as far as the interior is concerned, the complicated boundary makes no significant difference. The transformation of the interior, when applied to the boundary, may no longer be one-to-one, nor even be continuous. This is because the image of the unit circle under a continuous bijection has to be a Jordan contour. Associated with this, we have the following theorem.

Theorem 10.2 *Let G be the interior of a Jordan contour. There is a conformal transformation mapping the interior of G bijectively onto the interior of the unit circle; moreover, the mapping is continuous from the closure of G to the closure of the unit disc, so that there is a bijection between the Jordan contour and the unit circle with the same orientation. If any three points on the Jordan contour are to be taken, and any three corresponding points with the same direction are specified, then there is one and only one mapping satisfying the above conditions and having the three pairs of points in correspondence.*

This theorem not only clarifies the relationship between the boundaries, but also points out how a Jordan contour can be represented: The function $g(w)$ is analytic inside the unit circle, and $g(e^{i\theta})$, $0 \leq \theta \leq 2\pi$, describes the Jordan contour. Let

$$g(w) = \sum_{n=0}^{\infty} a_n w^n;$$

[1] Translator's footnote: As stated in the Introdution in Volume I by Professor Wang Yuan, the conjecture is now a theorem, proved by Louis de Branges in 1985.

then the Jordan contour must be representable as the Fourier series

$$\tau(\theta) = \sum_{n=0}^{\infty} a_n e^{in\theta}.$$

Thus, on letting $a_n = \alpha_n + i\beta_n$,

$$x = \phi(\theta) = \sum_{n=0}^{\infty} (\alpha_n \cos n\theta - \beta_n \sin n\theta),$$

$$y = \psi(\theta) = \sum_{n=0}^{\infty} (\alpha_n \cos n\theta + \beta_n \sin n\theta).$$

Note that the Fourier series for $\phi(\theta)$ is the conjugate to that for $\psi(\theta)$, and can therefore be uniquely specified by $\psi(\theta)$, apart from a constant. Given any continuous function $\psi(\theta)$, we can have a Fourier series; using the $(C, 1)$ summation method (see Section 11.1) the Fourier series represents $\psi(\theta)$. However, the problem is not that simple, because there is the one-to-one problem; that is, we need to deduce from

$$\phi(\theta_1) = \phi(\theta_2), \quad \psi(\theta_1) = \psi(\theta_2), \quad 0 \le \theta_1, \theta_2 < 2\pi,$$

that $\theta_1 = \theta_2$. For this reason the study of Jordan contours becomes that of continuous functions which are periodic with period 2π satisfying the above conditions.

There are extensive applications of conformal transformations in the study of fluid mechanics, electricity, ballistics, etc. In such applications we use explicit material in order to determine the required conformal transformations. Since any Jordan contour can be approximated by polygons, it is useful to have explicit transformations mapping polygons onto the unit disc (or the upper half-plane), and their methods of computation can also be very important, especially when the information on the boundary is given by only a certain number of points. For this reason, we introduce the Schwarz–Christoffel method, although the method does not deliver a simple procedure for numerical computations.

10.2 Univalent functions

Definition 10.3 An analytic function $f(z)$ in a region D is said to be univalent in D if it does not take on a value more than once, that is $f(z_1) = f(z_2)$ for $z_1, z_2 \in D$ implies $z_1 = z_2$.

The function $w = f(z)$ mapping the region D on the z-plane to the region D' in the w-plane is a one-to-one correspondence.

If $f(z)$ is univalent in D then $f'(z) \neq 0$ in D. We may state the following important point: If we suppose otherwise, say $f'(z_0) = 0$, so that the equation $f(z) - f(z_0) = 0$ has a repeated root of order n (≥ 2) at $z = z_0$. Since $f(z)$ is not a constant there is a punctured disc $0 < |z - z_0| \leq \delta$ inside which we have $f(z) - f(z_0) \neq 0$ and $f'(z) \neq 0$. Denote by m the minimum of $|f(z) - f(z_0)|$ on the circle. By Rouché's theorem (Theorem 7.38), if

$0 < |a| < m$ then $f(z) - f(z_0) - a$ has n zeros, which are distinct because $f'(z) \neq 0$ in the punctured disc. This contradicts $f(z)$ not taking any value more than once.

A univalent function of a univalent function is again univalent. More explicitly, if $f(z)$ is univalent in D, and its range lies inside D', in which $F(w)$ is univalent, then the composite function $F(f(z))$ is univalent in D.

10.3 The inverse from Taylor series

Theorem 10.4 *Let $f(z)$ be regular at $z = 0$, and let $f'(0) \neq 0$. Then there exists $\rho > 0$ such that $f(z)$ is univalent in $|z| \leq \rho$.*

We have the following even more precise result.

Theorem 10.5 (Landau) *Let*

$$w = f(z) = a_1 z + a_2 z^2 + \cdots, \qquad a_1 \neq 0,$$

be regular in $|z| \leq R$, and let $|w| < M$. Then there is an inverse function $z = g(w)$ which is regular and single-valued in $|w| < \phi(M, R|a_1|)$ with $|g(w)| < R$. Here ϕ is a function depending only on M and $R|a_1|$.

Proof. Let us first take $a_1 = 1$ and $R = 1$.

By the maximum modulus theorem,

$$\max_{|z|=r} |a_2 + a_3 z + a_4 z^2 + \cdots| \geq |a_2|, \qquad 0 < r < 1,$$

and the left-hand side is an increasing function of r. From

$$\max_{|z|=r} |f(z) - z| = r^2 \max_{|z|=r} |a_2 + a_3 z + a_4 z^2 + \cdots|,$$

we find that

$$\frac{1}{r} \max_{|z|=r} |f(z) - z| \to 0 \quad \text{as} \quad r \to 0.$$

Therefore there exists R' such that, for $0 < r < R'$,

$$\phi(r) = r - \max_{|z|=r} |f(z) - z| = r\left(1 - \frac{1}{r} \max_{|z|=r} |f(z) - z|\right) > 0,$$

and consequently, for $0 < |z| = r < R'$,

$$|f(z)| = |z - \{z - f(z)\}| \geq |z| - |z - f(z)| \geq \phi(r) > 0.$$

Thus the origin is the only zero for $f(z)$ in $|z| < R'$.

Let $|y| < \phi(r)$, where r is fixed, and consider the integral

$$I(y) = \frac{1}{2\pi i} \int_{|z|=r} \frac{f'(z)}{f(z) - y} \, dz.$$

We have $I(0) = 1$, because $f(z)$ has the single zero at the origin inside $|z| \le r < R'$, and there are no poles. On $|z| = r$, we have $|f(z)| \ge \phi(r) > |y|$, so that $I(y)$ is a continuous function, and therefore

$$\frac{1}{2\pi i}\int_{|z|=r} \frac{f'(z)}{f(z)-y}\,dz = 1$$

for $|y| < \phi(r)$. This means that, for such y, the equation $f(z) = y$ has a unique solution in $|z| < r$. This then defines an *inverse function $z(y)$ which is single-valued in $|y| < \phi(r)$.*

By the residue theorem we have

$$z(y) = \frac{1}{2\pi i}\int_{|z|=r} z\frac{f'(z)}{f(z)-y}\,dz,$$

so that $z(y)$ is a regular function of y in $|y| < \phi(r)$.

The problem reduces to one of finding r so that $\phi(r) > 0$. First, we prove that $M \ge 1$. Since $f(z)$ is regular in the disc $|z| \le 1$,

$$\max_{|z|=1}|f(z)| = \max_{|z|=1}\left|\frac{f(z)}{z}\right| = \max_{|z|=1}|1 + a_2 z + \cdots| \ge 1.$$

Next, we show that the objective is achieved by taking $r = 1/(4M)$. By Cauchy's estimate of the Taylor coefficients, we have

$$\phi\Big(\frac{1}{4M}\Big) \ge r - \sum_{2}^{\infty} Mr^n = r - \frac{Mr^2}{1-r},$$

so that, from

$$\frac{1}{1-r} = \frac{4M}{4M-1} \le \frac{4}{3},$$

we find that

$$\phi\Big(\frac{1}{4M}\Big) \ge r - \frac{4Mr^2}{3} = \frac{1}{6M}.$$

Thus, for $|w| < 1/(6M)$, the inverse function $z = g(w)$ is regular, and $|g(w)| \le 1$.

For the general case, we let

$$F(z) = \frac{f(Rz)}{Ra_1},$$

so that $F(0) = 0$ and $F'(0) = 1$, and when $|z| \le 1$

$$|F(z)| \le \frac{M}{R|a_1|},$$

so that $F(z)$ satisfies the special requirement if we replace M by $M/(R|a_1|)$. Consequently, when $|w| < R|a_1|/(6M)$, the function $w = F(z)$ has the inverse function $z = G(w)$ which is regular, and $|G(w)| \le 1$.

Let $z' = Rz$. Then

$$f(z') = Ra_1 F(z'/R),$$

so that

$$\frac{z'}{R} = G\left(\frac{w}{Ra_1}\right), \quad z' = RG\left(\frac{w}{Ra_1}\right).$$

This formula shows that, when

$$\left|\frac{w}{Ra_1}\right| < \frac{R|a_1|}{6M},$$

z' is a regular function and $|z'| < R$, that is

$$|w| < \frac{R^2|a_1|^2}{6M} \equiv \phi(M, R|a_1|),$$

which is the required result. □

Note. If $a_0 = f(0) \neq 0$, then we can still establish the theorem; all we need to do is to replace w by $w - a_0$. Similarly we can replace z by $z - z_0$ in order to have an inverse function theorem at any specific point.

10.4 The image of a region

Let $f(z)$ be analytic, with $f'(z) \neq 0$, in a region D, and let E be the image set for $f(z)$ in the w-plane. From Section 10.3 we know that, corresponding to any point of E, there is a sufficiently small neighbourhood in which there is an inverse function $z = g(w)$ which is regular and single-valued – that is, if $f(a) = b$ then there is a disc with centre b such that, corresponding to any point b' in the disc, there is a neighbouring point a' to a such that $f(a') = b'$. We now prove that *the set E corresponding to D is a region D' in the w-plane.*

Proof. From what has just been said, $D' = E$ is an open set, so that it remains to prove that it is connected. Let w_1 and w_2 be any two points in D', and let z_1 and z_2 be the points in D for which $w_1 = f(z_1)$ and $w_2 = f(z_2)$. Since D is connected, there is a curve C leading from z_1 to z_2 lying inside D, say $z(t_1) = z_1$ and $z(t_2) = z_2$, and we remark that each point on C is an interior point of D. Let C' be the image curve in the w-plane, which is represented by $f[z(t)]$. When $t = t_1, t_2$, we have $w_1 = f(z_1) = f[z(t_1)]$, $w_2 = f(z_2) = f[z(t_2)]$, and each point of C' is interior to D'. Therefore D' is connected. □

Note that, connected though D' is, there is every likelihood that it is a multi-valued region – this is because $f'(z) \neq 0$ does not guarantee $f(z)$ to be univalent. Let us examine again the situation $f'(a) = 0$, in which

$$w = f(z) = b + a_k(z - a)^k + \cdots, \quad a_k \neq 0, \ k > 1.$$

To simplify matters, let us write $S = w - b$ and $t = z - a$, so that

$$S = a_k t^k (1 + p_1 t + p_2 t^2 + \cdots);$$

in the neighbourhood of $t = 0$, we have

$$(1 + p_1 t + p_2 t^2 + \cdots)^{1/k} = 1 + q_1 t + q_2 t^2 + \cdots$$

so that

$$S = a_k t^k (1 + q_1 t + q_2 t^2 + \cdots)^k;$$

that is

$$S^{1/k} = a_k^{1/k} t (1 + q_1 t + q_2 t^2 + \cdots).$$

This function has the inverse function

$$t = g(S^{1/k}),$$

so that the inverse to the original function is

$$z = a + b_1 (w - b)^{1/k} + b_2 (w - b)^{2/k} + \cdots.$$

A loop round b will induce k separate branches with the k functions $(w - b)^{1/k}$, $(w - b)^{1/k} e^{2\pi i/k}, \ldots, (w - b)^{1/k} e^{(2\pi i/k)(k-1)}$. We call b a branch point of order k. Since the zeros of $f'(z)$ have to be isolated, any compact subset of D can have only finitely many such branch points, and if we consider the corresponding Riemann surface we can prove that $w = f(z)$ also transforms regions into regions.

Theorem 10.6 (Weierstrass uniqueness theorem) *If $f(z)$ is regular in a simply connected region, then $f(z)$ is single-valued.*

The theorem can be established by the method used in analytic continuation.

10.5 Sequences of univalent functions

Theorem 10.7 *The limit of a sequence of uniformly convergent sequence of univalent functions is univalent, or a constant.*

The sequence $f_n(z) = z/n$ tells us that it is possible that the limit is the constant function. More precisely: If $f_n(z)$ is univalent in D for any n, and converges uniformly to $f(z)$ in D, then $f(z)$ is also univalent in D, or else is constant.

Proof. We already know that $f(z)$ is analytic and single-valued in D. Suppose that it is not univalent. Then there are two points $z_1, z_2 \in D$ such that $f(z_1) = f(z_2) = w_0$. Take two non-overlapping circles with centres z_1, z_2 lying inside D on which $f(z) = w_0$ has no solution – this being possible unless $f(z)$ is a constant. Let m be the minimum of $|f(z) - w_0|$ on the two circles. For sufficiently large n, we have $|f(z) - f_n(z)| < m$ on these circles, so that, by Rouché's theorem, the function

$$f_n(z) - w_0 = \{f(z) - w_0\} + \{f_n(z) - w_0\}$$

has the same number of zeros as $f(z) - w_0$, which is at least two. This contradicts $f_n(z)$ being univalent, so that $f(z)$ has to be a constant. The theorem is proved. □

10.6 The boundary and the interior

Theorem 10.8 *Let $w = f(z)$ be an analytic function of z which is regular in $D \cup C$, where C is a Jordan contour with the interior region D. Suppose that $f(z)$ maps C bijectively to another Jordan contour C'. Then $f(z)$ is univalent in D.*

Proof. Let D' be the region inside C'. Take a point $z_0 \in D$ so that $f(z_0)$ is not equal to the values of $f(z)$ taken on C. Let Δ_C denote the change in value of an expression as a point loops round C, so that

$$\frac{1}{2\pi}\Delta_C \log(f(z) - f(z_0)) \tag{10.1}$$

is the number of zeros of $f(z) - f(z_0)$ inside C; it is a positive integer (because there is at least one root), but it is also the same as

$$\frac{1}{2\pi}\Delta_{C'} \arg(w - w_0), \qquad w_0 = f(z_0). \tag{10.2}$$

If w_0 lies outside C', then the value in (10.2) is 0, and if w_0 is inside C', the value is ± 1, with the sign depending on the orientation of C'. By (10.1) we know that w_0 has to be inside C', which has to be positively oriented, and it also follows that $f(z)$ takes on the value w_0 only once in D. Also, the image of $f(z)$ under D must be the whole of D', since otherwise there would exist boundary points besides those on C'. Therefore $f(z)$ is a univalent map from D to D'. □

Note 1. We may relax a little the hypothesis that $f(z)$ is analytic on the contour; the result still holds if there are certain points at which $f(z)$ is only continuous, but not analytic.

Suppose that z_1 is a singularity on the boundary. Replace the small part of the contour C near z_0 by a circular arc centre there, moving toward the inside of C, and label the resulting contour C_1. The number of zeros inside C_1 is equal to

$$\frac{1}{2\pi}\Delta_{C_1} \arg(f(z) - w_0) = \frac{1}{2\pi i}\int_{C_1} \frac{f'(z)}{f(z) - w_0}\, dz.$$

If $f'(z) = O(|z - z_1|^a)$ for $a > -1$ in the neighbourhood of z_1 then the integral along C_1 tends to that along C as '$C_1 \to C$'.

Note 2. It is also possible that $f(z)$ has a pole on the boundary, so that D' extends to ∞. Provided that the pole is simple and the function is just ordinary elsewhere on the boundary, the discussion in the above is still valid. We first transform the pole to $z = 0$, and map the region D and its boundary onto the upper half-plane $\Re(z) \geq 0$. We may assume without loss that the residue is 1, otherwise we need only divide the function by a constant, so that $f(z)$ has the representation

$$w = f(z) = \frac{1}{z} + g(z),$$

where $g(z)$ is regular and bounded in D; thus, for $z \in D$,

$$\Re(w) \geq \min \Re(g(z)).$$

Denote this by a, and let $b < a$. Then, for $z \in D$, we have $|w - b| \geq a - b$ so that

$$\zeta = \frac{1}{w - b} = \frac{z}{1 + zg(z) - bz}$$

is regular in D. What we have established can be applied directly to ζ, and since w is a univalent function of ζ, the theorem also applies to $w = f(z)$.

However, the result is no longer true for poles of a higher order.

Example 10.9 Let $w = -i(z + 1)/(z - 1)$. If $z = e^{i\theta}$ then

$$w = \frac{1}{i}\frac{e^{i\theta} + 1}{e^{i\theta} - 1} = -\cot\frac{\theta}{2}.$$

As z describes the unit circle, so that $0 \leq \theta \leq 2\pi$, the image w describes the real axis from $-\infty$ to $+\infty$. There is only a simple pole on the boundary $|z| = 1$, so that the unit disc on the z-plane is mapped by the univalent function onto the upper half-plane. We know all about this already, of course.

Example 10.10 Let

$$w = i\Big(\frac{z + 1}{z - 1}\Big)^3.$$

If $z = e^{i\theta}$ then $w = -\cot^3 \theta/2$, and this also establishes a bijection between the unit circle in the z-plane and the real axis in the w-plane. However, because the pole on the boundary has order 3, the regions corresponding to the two curves are not one-to-one. More specifically: Let $z = x + iy$ and $w = u + iv$, so that

$$w = i\Big(\frac{x + iy + 1}{x + iy - 1}\Big)^3 = u + i\frac{(x^2 + y^2 - 1)^3 - 12(x^2 + y^2 - 1)y^2}{\big((x - 1)^2 + y^2\big)^3},$$

and $v = 0$ when

$$(x^2 + y^2 - 1)(x^2 + y^2 \pm 2\sqrt{3}y - 1) = 0,$$

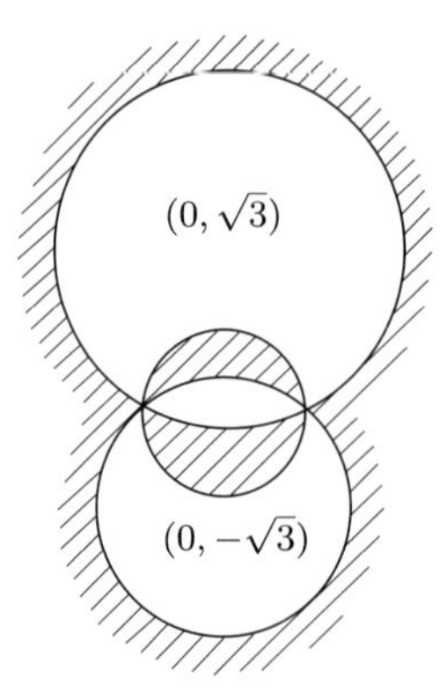

Figure 10.2.

an equation for three circles. If z lies outside all three circles then $v > 0$; if it lies outside one and inside the other two then $v > 0$ also, so that there are three regions all being mapped onto the upper half-plane (see the shaded regions in Fig. 10.2). Similarly there are three regions being mapped onto the lower half-plane (the blank regions in the figure).

10.7 The Riemann mapping theorem

Theorem 10.11 (Riemann) *Let D be any simply connected region in the w-plane with at least two boundary points. Then there is a univalent function $w = f(z)$ which maps the unit disc in the z-plane onto D.*

Proof. Suppose that $w = f(z)$ is such a function. Then its inverse $z = g(w)$ is regular and bounded ($|g(z)| < 1$) in D. Consider the family of regular and bounded functions in D satisfying $g(w_0) = 0$, $g'(w_0) = 1$, where w_0 is a fixed point in D. Denote by C the family of such functions which maps D to all sorts of other regions.

(1) The existence of such a family of functions. Let a, b be two boundary points of D, and consider the function

$$\zeta = \sqrt{\frac{w-a}{w-b}}.$$

As w runs over the simply connected region D, it cannot just describe a closed curve separating a and b; thus $\zeta(w)$ is single-valued in D, and, because the function under the square root sign is bilinear, it cannot assume the same value twice. Therefore, if we first take the square root sign at a point then $\zeta(w)$ also has the same property, so that $\zeta(w)$ is a univalent function mapping D onto a region D' in the ζ plane. If we take another square root (that is, another branch of $\sqrt{(z-a)/(z-b)}$), then D is mapped onto another region D'', so that D' and D'' are corresponding regions with respect to the original point. If c' is any interior point of D' then $c'' = -c'$ is an interior point of D''. The function $\tau = 1/(\zeta - c'')$

is a single-valued, regular and bounded function in D', and on subtraction of $\tau(w_0)$ and then division by $\tau'(w_0)$, we have a function in our family C.

(2) Let $g(w)$ be a function in the family C, denote by $M(g)$ the supremum of $|g(w)|$ in D, and let ρ denote the infimum of all $M(g)$ as g runs over C. We proceed to show that *there must be a function $h(w)$ in the family C with $M(h) = \rho$.*

Suppose, if possible, otherwise. There must be a sequence of functions $g_1, g_2, \ldots$ in C such that $M(g_n) \to \rho$ as $n \to \infty$; thus, for any $\epsilon > 0$, there exists N such that

$$M(g_n) < \rho + \epsilon$$

for all $n > N$. This shows that $g_1, g_2, \ldots$ have a common upper bound, which is denoted by B. We now show that $\{g_n\}$ has a subsequence which converges uniformly in any region contained in D. We can apply the previously described Cantor diagonal method to obtain a sequence which has a limit in a countable set which is everywhere dense in D, for example the set of rational points. By Vitali's theorem, the sequence of functions converges to a limit function uniformly in any compact subset of the region D. We may assume without loss that it is the original sequence of functions $\{g_n(z)\}$, and we let $g(w)$ be its limit. From Section 10.5 we know that $g(w)$ belongs to the family C, and $M(g) = \rho$. It also follows that $\rho > 0$.

(3) The function $g(w)$ maps D onto the disc $|z| < \rho$. Let D' be the image region of D under the mapping $z = g(w)$, which must include the disc $|z| < \rho$. We proceed to show that D' is just this disc. Suppose that D' has a boundary point ζ_0 satisfying $|\zeta_0| < \rho$. We deduce that there is a function $h(w)$ in the family C such that $M(h) < \rho$, which then contradicts the definition of ρ being the infimum of $M(g)$ as g runs over C.

Consider the function

$$\zeta_1(\zeta) = \rho\sqrt{\frac{\rho(\zeta - \zeta_0)}{\rho^2 + \bar{\zeta}_0\zeta}},$$

which is regular in D', and $M(\zeta_1) = \rho$. Take a fixed branch, and set

$$\zeta_2(w) = \frac{\rho^2(\zeta_1 - \zeta_1(0))}{\rho^2 - \bar{\zeta}_1(0)\zeta_1}.$$

Then $\zeta_2(w_0) = 0$, $M(\zeta_0) = \rho$ and $\zeta_2(w)$ is regular in D. At $w = w_0$, its derivative is $(\rho + |\zeta_0|)/2\sqrt{-\zeta_0\rho}$, and on division by $\zeta_2(w)$ we arrive at $\zeta_3(w)$, which then belongs to the family C, and

$$M(\zeta_3) = \rho\left|\frac{2\sqrt{-\zeta_0\rho}}{\rho + \sqrt{\zeta_0\zeta_0}}\right| < \rho.$$

This contradicts our assumption. The theorem is proved. □

Some discussion has already been given to Theorem 10.2; we shall not elaborate on it and will omit the proof of the theorem.

10.8 Estimating the second coefficient

Theorem 10.12 (Gronwall) *Let*

$$w = z + \frac{a_1}{z} + \frac{a_2}{z^2} + \cdots$$

be univalent in $|z| > 1$, *and be regular everywhere apart from the pole at infinity. Then*

$$\sum_{n=1}^{\infty} n|a_n|^2 \leq 1.$$

In particular, we have $|a_1| \leq 1$, with equality only when $w = z + e^{i\theta}/z$.

Proof. Fundamental area result. The function being univalent, it maps the circle $|z| = r > 1$ to a Jordan contour on the w-plane enclosing a region with a positive area. Let $w = u(\theta) + iv(\theta)$ describe the curve as $z = re^{i\theta}$ describes the circle. Then

$$u(\theta) + iv(\theta) = re^{i\theta} + \frac{a_1}{re^{i\theta}} + \frac{a_2}{r^2 e^{2i\theta}} + \cdots .$$

With $a_n = b_n + ic_n$, we then have

$$u(\theta) = r\cos\theta + \sum_{n=1}^{\infty} r^{-n}(b_n \cos n\theta + c_n \sin\theta),$$

$$v(\theta) = r\sin\theta - \sum_{n=1}^{\infty} r^{-n}(b_n \sin n\theta - c_n \cos\theta).$$

The area enclosed by the Jordan curve is equal to

$$\int_0^{2\pi} u(\theta)v'(\theta)d\theta = \int_0^{2\pi} \left\{r\cos\theta + \sum_{n=1}^{\infty} r^{-n}(b_n \cos n\theta + c_n \sin\theta)\right\}$$

$$\times \left\{r\cos\theta - \sum_{n=1}^{\infty} nr^{-n}(b_n \cos n\theta + c_n \sin\theta)\right\}d\theta$$

$$= \pi\left(r^2 - \sum_{n=1}^{\infty} nr^{-2n}(b_n^2 + c_n^2)\right) = \pi\left(r^2 - \sum_{n=1}^{\infty} n|a_n|^2 r^{-2n}\right).$$

As we remarked, the area has to be positive, so that

$$\sum_{n=1}^{\infty} n|a_n|^2 r^{-2n} < r^2,$$

and the required result follows by letting $r \to 1$.

If $|a_1| = 1$ then clearly $a_2 = a_3 = \cdots = 0$. □

Lemma 10.13 (Faber) *Let*

$$f(z) = z + a_2 z^2 + \cdots$$

be regular and univalent in the unit disc. Then $g(z) = \sqrt{f(z^2)}$ *is a regular univalent odd function in* $|z| < 1$.

Proof. (1) Regularity. We already know that

$$f(z^2) = z^2(1 + a_2 z^2 + \cdots)$$

represents a regular even function with no zero other than at $z = 0$, so that

$$h(z) = \sqrt{\frac{f(z^2)}{z^2}} = 1 + \frac{a_2}{2} z^2 \cdots$$

is a regular even function, and so

$$g(z) = zh(z) = z + \frac{a_2}{2} z^3 \cdots, \quad |z| < 1,$$

is a regular odd function.

(2) Being univalent. It is clear that 0 can be taken only once. If

$$g(z_1) = g(z_2), \quad 0 < |z_1| < 1,\ 0 < |z_2| < 1,$$

then

$$f(z_1^2) = (g(z_1))^2 = (g(z_2))^2 = f(z_2^2),$$

so that $z_1^2 = z_2^2$, giving $z_1 = \pm z_2$; but

$$g(-z_1) = -g(z_1), \quad g(z_1) \neq 0,$$

so that $z_1 = z_2$. □

Theorem 10.14 (Bieberbach) *If*

$$f(z) = z + a_2 z^2 + \cdots$$

is regular and univalent in $|z| < 1$, *then* $|a_2| \leq 2$, *and there is equality only when* $f(z) = z/(1 + e^{i\theta} z)^2$.

Proof. By Lemma 10.13, the function

$$\frac{1}{g(z)} = \frac{1}{z + \frac{a_2}{2} z^3 + \cdots} = \frac{1}{z} - \frac{a_2}{2} z + \cdots$$

is regular and univalent in $0 < |z| < 1$. On applying Theorem 10.12 to the function $1/g(1/z)$, we find that $|a_2| \leq 2$, with equality only when $1/g(z) = 1/z + e^{i\theta} z$, that is when

$$f(z) = \frac{z}{(1 + e^{i\theta} z)^2}.$$

□

Note. We leave it to the reader to show that this function is univalent.

10.9 Corollaries

Henceforth we assume that

$$f(z) = z + a_2 z^2 + \cdots$$

is regular and univalent in $|z| < 1$.

Theorem 10.15 (Faber) *The function $f(z)$ assumes every complex value with absolute value less than $\frac{1}{4}$; that is, the image of $|z| < 1$ under the mapping $w = f(z)$ includes the disc $|w| < \frac{1}{4}$.*

Proof. Suppose that $f(z) \neq \gamma$. Then the function

$$g(z) = \frac{f(z)}{1 - f(z)/\gamma} = \frac{z + a_2 z^2 + \cdots}{1 - z/\gamma + \cdots} = z + \left(a_2 + \frac{1}{\gamma}\right) z^2 + \cdots$$

is also regular and univalent in $|z| < 1$. By Theorem 10.14, we have

$$|a_2| \leq 2 \quad \text{and} \quad \left| a_2 + \frac{1}{\gamma} \right| \leq 2,$$

giving

$$\left| \frac{1}{\gamma} \right| \leq 2 + |a_2| \leq 4, \quad \text{that is} \quad |\gamma| \geq \frac{1}{4}.$$

□

Note 1. If

$$f(z) = \frac{z}{(1 + e^{i\alpha} z)^2}$$

then $f(e^{-i\alpha}) = e^{-i\alpha}/4$. It is easy to show that, conversely, if $f(z)$ assumes the value $e^{-i\alpha}/4$ on the unit circle, then it has to have this form.

Note 2. The last inequality in Theorem 10.15 can be replaced by

$$|w| < \frac{1}{2 + |a_2|}.$$

Theorem 10.16 (Bieberbach) *We have, for $|z| < 1$,*

$$\left| \frac{1 - |z|^2}{2} \frac{f''(z)}{f'(z)} - \bar{z} \right| \leq 2,$$

with equality only if

$$f(z) = \frac{(z - z_0)(1 - \bar{z}_0 z)}{[(1 - e^{i\alpha} z_0) + (\bar{z}_0 - e^{i\alpha} z)]^2},$$

and at $z = z_0$.

Proof. Let $|z_0| < 1$, and write

$$f\Big(\frac{z + z_0}{1 + \bar{z}_0 z}\Big) = A_0 + A_1 z + A_2 z^2 + \cdots, \qquad A_0 = f(z_0),$$

and

$$g(z) = \frac{1}{A_1}\Big(f\Big(\frac{z + z_0}{1 + \bar{z}_0 z}\Big) - A_0\Big) = z + \frac{A_2}{A_1} z^2 + \cdots,$$

so that $g(z)$ is also univalent, and therefore

$$\Big|\frac{A_2}{A_1}\Big| \le 2.$$

The required inequality then follows from the following computations:

$$A_1 = \frac{d}{dz} f\Big(\frac{z + z_0}{1 + \bar{z}_0 z}\Big)\Big|_{z=0} = f'\Big(\frac{z + z_0}{1 + \bar{z}_0 z}\Big)\frac{1 - |z_0|^2}{(1 - \bar{z}_0 z)^2}\Big|_{z=0} = f'(z_0)(1 - |z_0|^2),$$
$$A_2 = \frac{1}{2}\frac{d^2}{dz^2} f\Big(\frac{z + z_0}{1 + \bar{z}_0 z}\Big)\Big|_{z=0} = \frac{1}{2}\big(f''(z_0)(1 - |z_0|^2)^2 - 2f'(z_0)\bar{z}_0(1 - |z_0|^2)\big).$$

The situation under which there is equality is not difficult to establish, and the task is left to the reader. □

10.10 The Koebe quarter theorem

Theorem 10.17 (Koebe) *We have, for $|z| = r < 1$,*

$$\frac{1 - r}{(1 + r)^3} \le |f'(z)| \le \frac{1 + r}{(1 - r)^3},$$

with equality only if $f(z) = z/(1 + e^{i\alpha} z)^2$.

Proof. It is easy to see that

$$\log f'(z) + \log(1 - |z|^2) = \int_0^z \Big(\frac{f''(z)}{f'(z)} - \frac{2\bar{z}}{1 - |z|^2}\Big) dz,$$

where the path of integration is the straight line segment from 0 to z. By Theorem 10.16, we have

$$|\log f'(z) + \log(1 - |z|^2)| \le \int_0^r \frac{4}{1 - r^2}\, dr = 2\log\frac{1 + r}{1 - r}.$$

The real part then satisfies

$$-2\log\frac{1+r}{1-r} \leq \log|f'(z)| + \log(1-|z|^2) \leq 2\log\frac{1+r}{1-r},$$

from which the required inequality follows. The situation under which there is equality is left to the reader. □

Note. If we take the imaginary part instead, then we have

$$|\arg f'(z)| = |\Im \log f'(z)| \leq 2\log\frac{1+r}{1-r}.$$

This is not the best estimate, which was obtained by Golusin:

$$|\arg f'(z)| \leq \begin{cases} 4\arcsin|z|, & \text{when } |z| \leq 1/\sqrt{2}, \\ \pi + \log\dfrac{|z|^2}{1-|z|^2}, & \text{when } |z| \geq 1/\sqrt{2}. \end{cases}$$

By the maximum modulus theorem we deduce the following.

Theorem 10.18 *The inequalities in Theorem 10.17 hold for* $|z| \leq 1$.

Theorem 10.19 *Suppose that* $|z_1| \leq r$, $|z_2| \leq r$. *Then*

$$\left(\frac{1-r}{1+r}\right)^4 \leq \left|\frac{f'(z_1)}{f'(z_2)}\right| \leq \left(\frac{1+r}{1-r}\right)^4.$$

Theorem 10.20 (Bieberbach) *We have, for* $|z| \leq r$,

$$\frac{r}{(1+r)^2} \leq |f(z)| \leq \frac{r}{(1-r)^2},$$

with equality only if $f(z) = z/(1+e^{i\alpha}z)^2$.

Proof. (1) From Theorem 10.17, we have

$$\begin{aligned} |f(z)| &= \left|\int_0^z f'(z)dz\right| \leq \int_0^z |f'(\rho e^{i\theta})|d\rho \leq \int_0^r \frac{1+\rho}{(1-\rho)^3}\,d\rho \\ &= \int_0^r \left(\frac{2}{(1-\rho)^3} - \frac{1}{(1-\rho)^2}\right)d\rho = \frac{r}{(1-r)^2}. \end{aligned}$$

(2) Let the circle $C(|z| = r)$ be mapped to the contour Γ by $w = f(z)$, and let w_0 be the point on Γ with the least distance from the origin (see Figs 10.3 and 10.4). We then have

$$|w_0| = \int_0^{w_0} |dw| = \int_L |f'(z)||dz|,$$

where L is the curve in the z-plane leading from 0 to z_0 corresponding to the straight line segment leading from 0 to w_0 in the w-plane. With $\lambda = |z|$, so that $|dz| \geq d\lambda$, it now

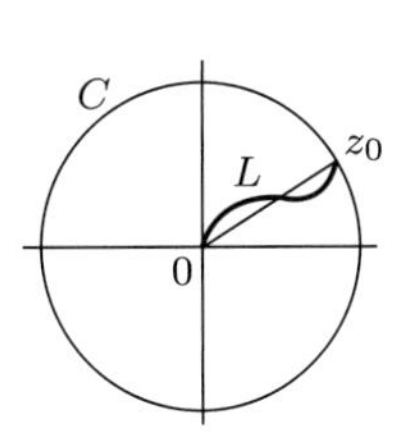

Figure 10.3.

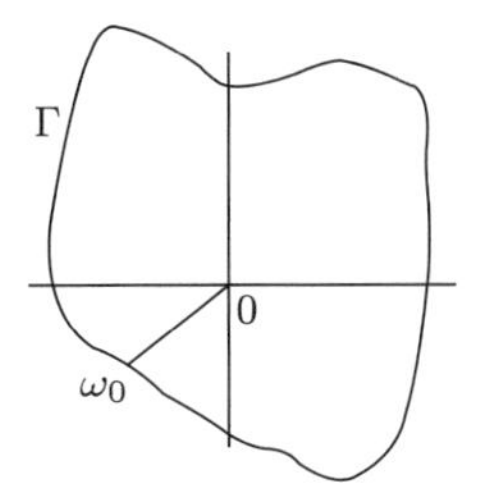

Figure 10.4.

follows from Theorem 10.17 that

$$|w_0| \geq \int_0^r |f'(\lambda e^{i\theta})| d\lambda \geq \int_0^r \frac{1-\lambda}{(1+\lambda)^3} d\lambda = \frac{r}{(1+r)^2}.$$

□

We deduce at once Theorem 10.21.

Theorem 10.21 *We have* $|f(z)| \geq r/4$ *on* $|z| = r$.

10.11 Littlewood's estimate

Theorem 10.22 (Littlewood) *We have, for* $n \geq 2$,

$$|a_n| \leq \left(1 + \frac{1}{n-1}\right)^{n-1} n < en.$$

Proof. By Lemma 10.13, we know that

$$g(z) = \sqrt{f(z^2)} = \sum_{n=1}^{\infty} b_n z^n, \quad b_1 = 1, \quad |z| < 1,$$

is a regular univalent function and, by Theorem 10.20,

$$|f(z)| \leq \frac{t}{(1-t)^2}, \quad |z| \leq t < 1,$$

so that

$$|g(z)| \leq \sqrt{\frac{t}{(1-t)^2}} = \frac{\sqrt{t}}{1-t}, \quad |z| \leq \sqrt{t} < 1.$$

This means that, under $w = g(z)$, the circle $|z| = \sqrt{t}$ is mapped to a Jordan contour Γ lying inside the circle $|w| = \sqrt{t}/(1-t)$, so that the area inside Γ does not exceed $\pi t/(1-t)^2$.

The area inside Γ is given by

$$\int_0^{2\pi} u(\theta)dv(\theta) = \int_0^{2\pi} \left\{\sum_{n=1}^{\infty}(b_n' \cos n\theta - b_n'' \sin\theta)t^{n/2}\right\}$$
$$\times \left\{\sum_{n=1}^{\infty}(b_n' \cos n\theta - b_n'' \sin\theta)nt^{n/2}\right\}d\theta$$
$$= \pi \sum_{n=1}^{\infty} n(b_n'^2 + b_n''^2)t^n = \pi \sum_{n=1}^{\infty} n|b_n|^2 t^n,$$

where we have written $b_n = b_n' + ib_n''$. Therefore

$$\sum_{n=1}^{\infty} n|b_n|^2 t^{n-1} \leq \frac{1}{(1-t)^2}.$$

Integrating over $0 \leq t \leq r$ we then have, for $0 < r < 1$,

$$\sum_{n=1}^{\infty} |b_n|^2 r^n \leq \int_0^r \frac{dt}{(1-t)^2} = \frac{r}{1-r}.$$

From

$$\sum_{n=1}^{\infty} a_n r^n = \left(\sum_{n=1}^{\infty} b_n r^{n/2}\right)^2,$$

we have that

$$|a_n| r^n = \left|\sum_{\nu=1}^{2n-1} b_\nu r^{n/2} b_{2n-\nu} r^{n-\nu/2}\right|,$$

and so, making use of $2|ab| \leq |a|^2 + |b|^2$,

$$|a_n| r^n \leq \frac{1}{2}\sum_{\nu=1}^{2n-1}(|b_\nu|^2 r^\nu + |b_{2n-\nu}|^2 r^{2n-\nu}) = \sum_{\nu=1}^{2n-1} |b_\nu|^2 r^\nu \leq \sum_{\nu=1}^{\infty} |b_\nu|^2 r^\nu,$$

and therefore

$$|a_n| \leq \frac{1}{r^{n-1}(1-r)}.$$

The required result follows by setting $r = 1 - 1/n$. □

10.12 Star regions

Definition 10.23 Suppose that there is a point z_0 inside a Jordan contour with the property that any half-line emitting from z_0 cuts the contour at precisely one point. Then the interior of the contour is called a star region, with respect to z_0.

Lemma 10.24 *Suppose that $w = g(z)$ transforms the circle $|z| = r$ into a contour with a star region with respect to $w = 0$. Then, on $|z| = r$, we have*

$$\Re\left(z\frac{g'(z)}{g(z)}\right) > 0;$$

the converse also holds.

Proof. From the definition, and the hypothesis of the lemma, we see that the angle associated with $g(re^{i\theta})$ is an increasing function of θ, so that

$$\frac{d}{d\theta}\Im \log g(re^{i\theta}) > 0,$$

that is

$$0 < \Im\frac{g'(z)}{g(z)}\frac{dz}{d\theta} = \Im\left(iz\frac{g'(z)}{g(z)}\right) = \Re\left(z\frac{g'(z)}{g(z)}\right).$$

□

Lemma 10.25 *If $g(z)$ is regular in $|z| < 1$ and*

$$g(z) = \frac{1}{2} + b_1 z + b_2 z^2 + \cdots, \qquad \Re g(z) > 0,$$

then $|b_n| \le 1$.

Proof. Let $g(z) = u + iv$ and $b_n = b_n' + ib_n''$. Then

$$u = \frac{1}{2} + \sum_{n=1}^{\infty} r^n(b_n' \cos n\theta - n_n'' \sin n\theta),$$

and also

$$b_n = b_n' + ib_n'' = \frac{1}{\pi r_n}\int_0^{2\pi} ue^{-in\theta}\, d\theta, \quad n \ge 1.$$

From $u > 0$, we deduce that

$$|b_n| \le \frac{1}{\pi r_n}\int_0^{2\pi} u\, d\theta = \frac{1}{r^n}.$$

The required result follows by letting $r \to 1$. □

Lemma 10.26 *Suppose that $f(z) = z + a_2 z^2 + \cdots$ is a univalent function mapping $|z| < 1$ to a star region with respect to $w = 0$. Then, for $|z| < 1$,*

$$\Re\left(z\frac{f'(z)}{f(z)}\right) > 0.$$

Proof. By Lemma 10.24 it is enough to prove that, for any $0 < r < 1$, the image B_r of $|z| < r$ under $w = f(z)$ is a star region with respect to $w = 0$. This means that, for any $w \in B_r$, we have $tw \in B_r$ for $0 < t < 1$.

The function $\psi(z) = f^{-1}[tf(z)]$ satisfies $\psi(0) = 0$, and is regular in $|z| < 1$ with $|\psi(z)| < 1$. Therefore $|\psi(z)| < |z|$ for $|z| < 1$ (see Section 4.10). Let $w_1 = f(z_1)$ be a point of B_r, that is $|z_1| < r$. By the above we find that $|f^{-1}(tw_1)| < |z_1| < r$. Since $tw_1 \in B_1, tw_1 = f(z_0)$ with $|z_0| < 1$. Feeding this into the inequality in the above, we find that $|z_0| < r$, that is $tw_1 \in B_r$. The lemma is proved. □

For the same reason concerning the extreme values for harmonic functions, the inequality sign in the lemma cannot be replaced by an equation sign anywhere inside $|z| < 1$.

Theorem 10.27 (R. Nevanlinna) *Let*

$$f(z) = z + a_2 z^2 + \cdots + a_n z^n + \cdots$$

be regular and univalent in $|z| < 1$. Suppose that $w = f(z)$ maps the unit disc to a star region with respect to $w = 0$. Then

$$|a_n| \le n, \quad n = 2, 3, 4, \ldots,$$

with equality only when $f(z) = z/(1 - e^{i\alpha}z)^2$.

In other words, the Bieberbach conjecture is valid in this special situation. In the proof below we use the non-standard notation of writing $A(z) \ll B(z)$ to mean that the functions have power series expansions $A(z) = \sum a_n z^n$ and $B(z) = \sum b_n z^n$ with $|a_n| \le |b_n|$. If $A(z) \ll B(z)$ then it can be shown, for example, that

$$\int_0^z A(w)dw \ll \int_0^z B(w)dw \quad \text{and} \quad e^{A(z)} \ll e^{B(z)}.$$

Proof. By Lemma 10.26, we have

$$\Re\left(z\frac{f'(z)}{f(z)}\right) > 0,$$

so that, by Lemma 10.25, we know that

$$z\frac{f'(z)}{f(z)} = z = 1 + b_1 z + \cdots + b_n z^n + \cdots, \quad |b_n| \le 2.$$

Therefore

$$z\frac{f'(z)}{f(z)} \ll 1 + 2z + 2z^2 + \cdots = \frac{1+z}{1-z},$$

and so

$$\frac{f'(z)}{f(z)} - \frac{1}{z} \ll \frac{2}{1-z},$$

giving

$$\log \frac{f(z)}{z} \ll \log \frac{1}{(1-z)^2}, \qquad \frac{f(z)}{z} \ll \frac{1}{(1-z)^2} = \sum_{n=1}^{\infty} nz^{n-1}.$$

Therefore $|a_n| \le n$, as required. □

10.13 Real coefficients

Theorem 10.28 (Dieudonné–Rogosinski) *Let*

$$f(z) = z + a_2 z^2 + \cdots + a_n z^n + \cdots$$

be regular and univalent in $|z| < 1$. Suppose that a_n are all real. Then

$$|a_n| \le n, \quad n = 2, 3, 4, \ldots .$$

Proof. Since $f(z)$ is univalent, we have

$$f(re^{i\theta}) - f(re^{-i\theta}) \ne 0$$

for $0 < r < 1$ and $\theta \ne 0, \pi$, that is

$$r(e^{i\theta} - e^{-i\theta}) + a_2 r^2 (e^{2i\theta} - e^{-2i\theta}) + \cdots + a_n r^n (e^{ni\theta} - e^{-ni\theta}) + \cdots \ne 0,$$
$$\sin\theta + a_2 r \sin 2\theta + \cdots + a_n r^{n-1} \sin n\theta + \cdots \ne 0.$$

On multiplication by $\sin\theta$ and making use of $\sin\theta \sin m\theta = \frac{1}{2}(\cos(m+1)\theta - \cos(m-1)\theta)$ we then have

$$\psi(r,\theta) = 1 + a_2 r \cos\theta + (a_3 r^2 - 1)\cos 2\theta + \cdots$$
$$+ (a_{n+1} r^n - a_{n-1} r^{n-2}) \cos n\theta + \cdots \ne 0.$$

Since $\psi(0,\theta) = 1 - \cos 2\theta > 0$, it follows that

$$\psi(r,\theta) > 0, \quad \theta \ne 0, \pi.$$

Therefore the function

$$F(z) = 1 + a_2 r z + (a_3 r^2 - 1) z^2 + \cdots + (a_{n+1} r^n - a_{n-1} r^{n-2}) z^n + \cdots$$

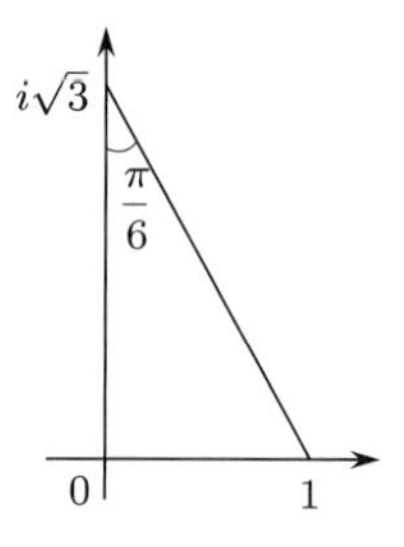

Figure 10.5.

is regular in $|z| < 1$, and $\Re[F(z)] \geq 0$. By Lemma 10.25,

$$|a_2 r| \leq 2, \quad |a_3 r^2 - 1| \leq 2, \quad \cdots, \quad |a_{n+1} r^n - a_{n-1} r^{n-2}| \leq 2,$$

so that

$$|a_2| \leq \frac{2}{r}, \quad |a_3| \leq \frac{3}{r^2}, \quad |a_4 r^3 - a_2 r| \leq 2, \quad \cdots,$$
$$|a_4| \leq \frac{2}{r^3} + \frac{|a_2|}{r^2}, \quad \cdots,$$
$$|a_{n+1}| \leq \frac{2}{r^n} + \frac{|a_{n-1}|}{r^2} \leq \frac{2 + |a_{n-1}| r^{n-2}}{r^n}.$$

Letting $r \to 1$ we have $|a_n| \leq n$. □

10.14 Mapping a triangle onto the upper half-plane

We seek a function $w = g(z)$ which maps the upper half of the z-plane onto the triangle with vertices $i\sqrt{3}, 0, 1$ in the w-plane (see Fig. 10.5).

By the Riemann mapping theorem there is a conformal transformation mapping the triangle onto the upper half-plane, sending the three points $(i\sqrt{3}, 0, 1)$ to three points (a, b, c) on the real axis, and there is a Möbius transformation which maps (a, b, c) onto $(-1, 0, 1)$. In other words, there is a function $w = g(z)$ such that $g(-1) = i\sqrt{3}$, $g(0) = 0$, $g(1) = 1$. The problem now is to find such a transformation in practice. We consider the integral

$$w = c \int_0^z (t+1)^{-5/6} t^{-1/2} (1-t)^{-2/3}\, dt,$$

and hope that it will do the job. We certainly have $w = 0$ when $z = 0$, and as z moves along the real axis from 0 to 1, the point w also moves along the real axis from 0 to

$$c \int_0^1 (t+1)^{-5/6} t^{-1/2} (1-t)^{-2/3}\, dt = 1,$$

if c is set by

$$\begin{aligned}\frac{1}{c} &= \int_0^1 t^{-1/2}(1-t)^{-2/3}\sum_{n=0}^{\infty}\frac{\Gamma(n+\frac{5}{6})}{\Gamma(\frac{5}{6})\Gamma(n+1)}(-t)^n\,dt\\ &= \sum_{n=0}^{\infty}\frac{(-1)^n\Gamma(n+\frac{5}{6})}{\Gamma(\frac{5}{6})\Gamma(n+1)}\cdot\frac{\Gamma(n+\frac{1}{2})\Gamma(\frac{1}{3})}{\Gamma(n+\frac{5}{6})}\\ &= \frac{\Gamma(\frac{1}{3})\Gamma(\frac{1}{2})}{\Gamma(\frac{5}{6})}\sum_{n=0}^{\infty}\frac{(-1)^n\Gamma(n+\frac{1}{2})}{\Gamma(\frac{1}{2})\Gamma(n+1)} = \frac{\Gamma(\frac{1}{3})\Gamma(\frac{1}{2})}{\Gamma(\frac{5}{6})}2^{-1/2}.\end{aligned}$$

In other words, the function

$$w = \frac{2^{1/2}\Gamma(\frac{5}{6})}{\Gamma(\frac{1}{2})\Gamma(\frac{1}{3})}\int_0^z (t+1)^{-5/6}t^{-1/2}(1-t)^{-2/3}\,dt$$

maps the line segment $0 \le z \le 1$ to $0 \le w \le 1$. Now let the path of z, instead of passing through 1, loop over it along a small upper semi-circle before moving toward ∞ along the real axis. In effect this means that $(z-1)$ becomes $(z-1)e^{-\pi i}$, so that

$$\begin{aligned}w &= \frac{2^{1/2}\Gamma(\frac{5}{6})}{\Gamma(\frac{1}{2})\Gamma(\frac{1}{3})}\left[\int_0^1 (t+1)^{-5/6}t^{-1/2}(1-t)^{-2/3}\,dt\right.\\ &\quad\left.+ e^{2\pi i/3}\int_1^x (t+1)^{-5/6}t^{-1/2}(1-t)^{-2/3}\,dt\right]\\ &= 1 + e^{2\pi i/3}\frac{2^{1/2}\Gamma(\frac{5}{6})}{\Gamma(\frac{1}{2})\Gamma(\frac{1}{3})}\int_1^x (t+1)^{-5/6}t^{-1/2}(1-t)^{-2/3}\,dt,\end{aligned}$$

which means that w moves from 1 along the straight line $1 + e^{2\pi i/3}t$ until it reaches

$$w_0 = 1 + e^{2\pi i/3}\frac{2^{1/2}\Gamma(\frac{5}{6})}{\Gamma(\frac{1}{2})\Gamma(\frac{1}{3})}\int_1^{\infty} (t+1)^{-5/6}t^{-1/2}(1-t)^{-2/3}\,dt,$$

where the integral is convergent because $\frac{5}{6}+\frac{1}{2}+\frac{2}{3}=2$.

Let us now examine the other end where z, instead of passing over 0, loops over it along a small upper semi-circle and then carries on along the real axis toward -1. Thus t becomes $te^{\pi i}$, and at $z=-1$ the point w arrives at

$$\begin{aligned}w &= c\int_0^{-1} (t+1)^{-5/6}t^{-1/2}(1-t)^{-2/3}\,dt\\ &= -c\int_0^1 (1-t')^{-5/6}(-t')^{-1/2}(1+t')^{-2/3}\,dt'\\ &= -e^{-i\pi/2}c\int_0^1 (1-t')^{-5/6}t'^{-1/2}(1+t')^{-2/3}\,dt' = i\sqrt{3}.\end{aligned}$$

(The last integral is computed in the same way as before, except that we need to use $\Gamma(z)\Gamma(1-z) = \pi/\sin \pi z$.) It is not difficult to show that, as z moves from -1 to $-\infty$, the point w moves from $i\sqrt{3}$ along the side of the triangle toward w_0.

Therefore, as z moves along the real axis from $-\infty$ to $+\infty$, the point w describes the triangle once in the positive direction. Meanwhile, the function $w = g(z)$ is analytic in the upper half-plane, and its inverse $z = f(w)$ maps the interior of the triangle onto the upper half-plane.

More generally, let the triangle ABC have the angles

$$\angle A = \alpha\pi, \quad \angle B = \beta\pi, \quad \angle C = \gamma\pi;$$

then

$$w = k\int_{z_0}^{z} (t-a)^{\alpha-1}(t-b)^{\beta-1}(t-c)^{\gamma-1}\,dt, \quad a, b < c,$$

maps the upper half-plane onto the triangle, with a, b, c corresponding to A, B, C. The constants k and z_0 can be determined from

$$A = k\int_{z_0}^{a} (t-a)^{\alpha-1}(t-b)^{\beta-1}(t-c)^{\gamma-1}\,dt,$$
$$B = k\int_{z_0}^{b} (t-a)^{\alpha-1}(t-b)^{\beta-1}(t-c)^{\gamma-1}\,dt,$$

so that k is given by $A - B = k\int_b^a (t-a)^{\alpha-1}(t-b)^{\beta-1}(t-c)^{\gamma-1}\,dt$, and z_0 can then be determined.

Exercise 10.29 Show that the function

$$w = \int_0^z \frac{dt}{(1-t^3)^{2/3}}$$

maps a half-plane onto an equilateral triangle.

Exercise 10.30 Show that the function

$$w = \int_0^z \frac{dt}{(1-t^2)^{5/6}t^{1/3}}$$

maps a half-plane onto a triangle with angles 30°, 30°, 120°.

10.15 The Schwarz reflection principle

Theorem 10.31 *Let G_1, G_2 be two disjoint regions with parts of their boundaries being a Jordan arc C, the interior points of which are not limit points of their boundaries apart from C. Suppose that $f_1(z)$ and $f_2(z)$ are analytic functions in G_1 and G_2, respectively, that they are continuous on $G_1 \cup C$ and $G_2 \cup C$ and that $f_1(z) = f_2(z)$ on C. Then there exists an analytic function $F(z)$ in $G_1 \cup G_2 \cup C$ which is $f_1(z)$ in G_1 and $f_2(z)$ in G_2.*

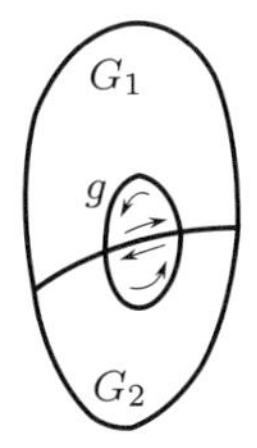

Figure 10.6.

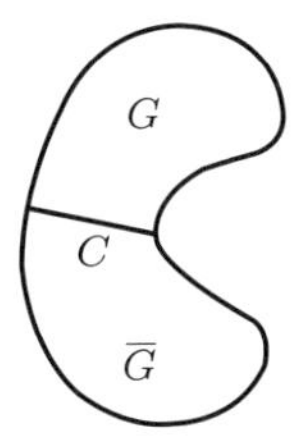

Figure 10.7.

Proof. Define in $G_1 \cup G_2 \cup C$ the single-valued continuous function

$$F(z) = \begin{cases} f_1(z), & \text{when } z \in G_1, \\ f_2(z), & \text{when } z \in G_2, \\ f_1(z) = f_2(z), & \text{when } z \in C. \end{cases}$$

We proceed to show that $F(z)$ is an analytic function. This is obviously so when $z \in G_1 \cup G_2$, so let $z \in C$. Take a rectifiable contour g straddling into G_1, G_2 (see Fig. 10.6), and consider the integral

$$\int_g F(z) dz. \tag{10.3}$$

Split the integral into two parts, with one part in G_1 and the remaining part in G_2, and then attach the relevant parts of C to them (see Fig. 10.6). Since $f_1(z)$ and $f_2(z)$ are continuous on $G_1 \cup C$ and $G_2 \cup C$, both integrals have the value zero, so that the integral (10.3) has the value zero. That $F(z)$ is analytic now follows from Morera's theorem. □

Theorem 10.32 (Schwarz reflection principle) *Let the region G in the upper half-plane have part of its boundary C on the real axis, and suppose that C is an interval with the same property as the Jordan arc in Theorem 10.31. Suppose that $f(z)$ is analytic in G, continuous in $G \cup C$, and takes real values on C. Then there is an analytic continuation for $f(z)$ into a region outside G through C (see Fig. 10.7).*

Proof. Take the mirror reflection of G along the real axis to obtain the region $\bar{G}$, and define there

$$g(z) = \overline{f(\bar{z})}, \quad z \in \bar{G} \cup C.$$

By definition, $g(z)$ is analytic in $\bar{G}$, continuous in $\bar{G} \cup C$, and $g(z) = f(z)$. The required result follows at once from Theorem 10.31. □

Exercise 10.33 Suppose that $f(z)$ in Theorem 10.32 does not take real values on C, but values lying on a certain straight line $\overline{AB}$ (so that $f(z)$ maps C continuously onto $\overline{AB}$). How should the proof be modified?

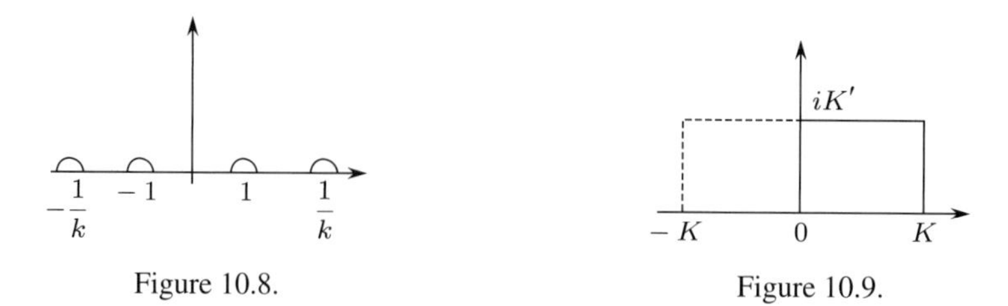

Figure 10.8.

Figure 10.9.

10.16 Mapping a quadrilateral onto the upper half-plane

Let $0 < k < 1$, and consider the integral

$$w = \int_0^z \frac{dt}{\sqrt{(1-t^2)(1-k^2t^2)}}.$$

As $z = x$ moves from 0 to 1 along the real axis, $w = u$ moves from 0 along the real axis to

$$K = \int_0^1 \frac{dt}{\sqrt{(1-t^2)(1-k^2t^2)}}.$$

As $z = x$ moves from 1 to $1/k$, we have

$$w = K + i\int_1^x \frac{dt}{\sqrt{(t^2-1)(1-k^2t^2)}},$$

so that the point w moves from $w = K$ to $K + iK'$, where

$$K' = \int_1^{1/k} \frac{dt}{\sqrt{(t^2-1)(1-k^2t^2)}}.$$

As $z = x$ moves from $1/k$ to $+\infty$, we have

$$w = K + iK' - \int_{1/k}^x \frac{dt}{\sqrt{(t^2-1)(k^2t^2-1)}},$$

so that the point $w = u + iv$ moves along the line $v = K'$ from $u = K$ to $u = 0$. To see this we need to have

$$\int_{1/k}^\infty \frac{dt}{\sqrt{(t^2-1)(k^2t^2-1)}} = K,$$

the proof of which is not difficult using the substitution $kt = 1/\tau$.

The same method shows that, as z moves along the real axis from 0 to -1, then to $-1/k$, and then to $-\infty$, the point w moves from 0 to $-K$, then to $-K + iK'$, and then to iK'.

Together, we see that, as z moves from $-\infty$ to ∞ (but taking care to make small indentations into the upper half-plane at ± 1, $\pm 1/k$; see Fig. 10.8) the point w runs over the perimeter of a rectangle with horizontal width $2K$ and vertical width K', so that the upper half-plane has been mapped into the rectangular region (see Fig. 10.9) with vertices

$$K,\ K + iK',\ -K + iK',\ -K.$$

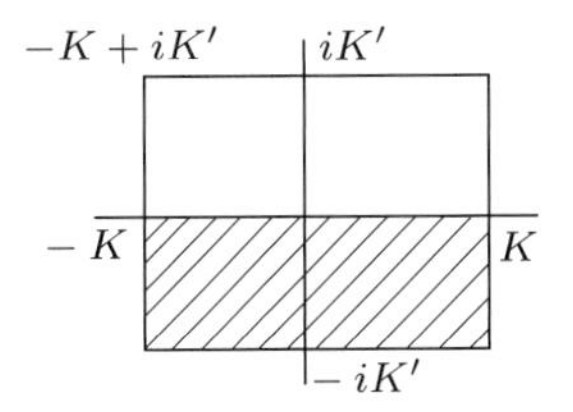

Figure 10.10.

The function

$$w = \int_0^z \frac{d\zeta}{\sqrt{(1-\zeta^2)(1-k^2\zeta^2)}}$$

is called an elliptic integral, which cannot be represented by elementary functions. Its inverse function is the Jacobi elliptic sine function

$$z = \operatorname{sn} w = \operatorname{sn}(w; k),$$

which maps the rectangle

$$[K,\ K+iK',\ -K+iK',\ -K]$$

onto the upper half-plane.

Suppose that we now wish to map a rectangle with (positive) side length L, M. Then we need to find C and k so that

$$L = CK, \quad M = CK'.$$

We therefore need to find k so that

$$\frac{M}{L} = \int_1^{1/k} \frac{dt}{\sqrt{(t^2-1)(1-k^2t^2)}} \Bigg/ \int_0^1 \frac{dt}{\sqrt{(1-t^2)(1-k^2t^2)}} = \phi(k);$$

here $\phi(k)$ is continuous in $0 < k < 1$ and

$$\lim_{k\to 1} \phi(k) = 0, \quad \lim_{k\to 0} \phi(k) = \infty,$$

so that there exists k such that $\phi(k) = M/L$ for any given ratio. The value for C can then be computed from either L or M.

Note. The function $z = \operatorname{sn} w$ is defined on the whole w-plane. When w lies in the interval $[-K, K]$ on the real axis, the image z lies in the real interval $[-1, 1]$, so that, by the Schwarz reflection principle, it has an analytic continuation into the rectangle (see Fig. 10.10)

$$[-K,\ -K+iK',\ K+iK',\ K],$$

with the values for z in the lower half-plane.

The line segment $[K, K+iK']$ corresponds to $[1, 1/k]$, so that the Schwarz reflection principle can again be applied to give an analytic continuation into the rectangle

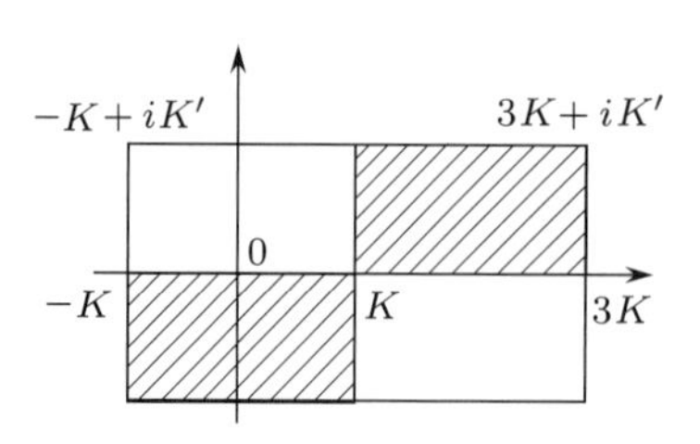

Figure 10.11.

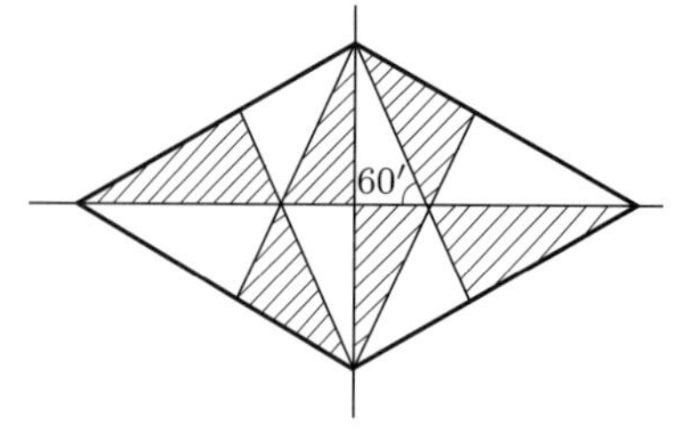

Figure 10.12.

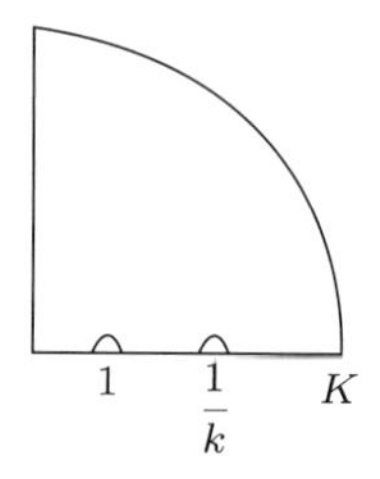

Figure 10.13.

$[K,\ 3K, 3K + iK',\ K + iK']$ (see Fig. 10.11). It is not difficult to prove that

$$\operatorname{sn}(w + 4K) = \operatorname{sn} w, \quad \operatorname{sn}(w + 2K'i) = \operatorname{sn} w,$$

so that sn w is a doubly periodic function with periods $2K'i$ and $4K$. Moreover, it is analytic everywhere except at $w = iK'$ and $iK' + 2K'im + 4Kn$, where m, n are integers.

Exercise 10.34 Use the Schwarz reflection principle to extend the region of existence of the inverse function in Section 10.14 across the triangle into the whole plane. Show that the function is also doubly periodic, with periods $4\omega + 2\omega'$, $2\omega + 4\omega'$, where (see Fig. 10.12)

$$\omega, \omega' = \frac{1}{2}(-1 \pm \sqrt{-3}).$$

Exercise 10.35 Show that the function

$$w = \int_0^z \frac{dt}{\sqrt{1 - t^4}}$$

maps the unit disc in the z-plane onto a square (let $z = (\zeta - i)/(\zeta + i)$).

Exercise 10.36 Show that

$$\int_0^{1/k} \frac{dt}{\sqrt{(t^2 - 1)(1 - k^2t^2)}} = \int_0^1 \frac{dt}{\sqrt{(1 + t^2)(1 + k^2t^2)}}.$$

(*Hint*: Consider the contour shown in Fig. 10.13.)

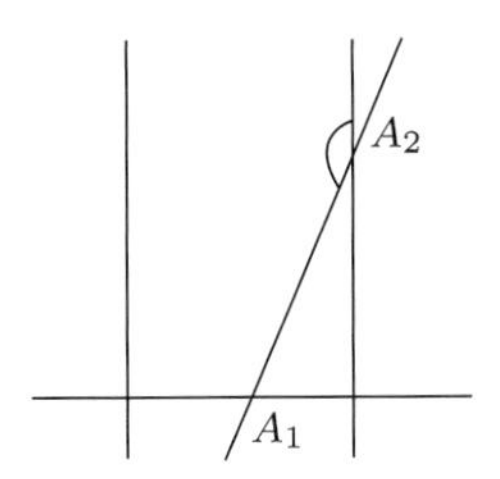

Figure 10.14.

10.17 Mapping polygons onto the upper half-plane – the Schwarz–Christoffel method

We consider, more generally, the integral

$$w = c \int_{-\infty}^{z} (t - a_1)^{\alpha_1 - 1}(t - a_2)^{\alpha_2 - 1} \cdots (t - a_n)^{\alpha_n - 1}\, dt + c_1, \tag{10.4}$$

where c, c_1 are constants, $-\infty < a_1 < a_2 < \cdots < a_n < \infty$ and

$$\alpha_1 + \alpha_2 + \cdots + \alpha_n = n - 2, \quad 0 < \alpha_\nu < 1, \quad \nu = 1, 2, \ldots, n.$$

We consider what happens as z moves along the real axis from $-\infty$ to $+\infty$, by-passing $a_1, a_2, \ldots, a_n$ with small upper semi-circular indentations in case they are singularities.

First, the integral (10.4) is convergent because, as $t \to \infty$, the integrand is $O(|t|^{\alpha_1+\alpha_2+\cdots+\alpha_n-n}) = O(|t|^{-2})$. Also, on a small upper semi-circle with centre a_ν, we have

$$t = a_\nu + \epsilon e^{i\theta}, \quad \epsilon > 0,\ 0 \le \theta \le \pi,$$

and the integral along there is $O(\epsilon^{\alpha_\nu}) = o(1)$, as $\epsilon \to 0$. As z varies in the upper half-plane, the integral (10.4) is analytic and single-valued.

Let us take $c = 1$, $c_1 = 0$ first, so that

$$w = \int_{-\infty}^{z} (t - a_1)^{\alpha_1 - 1}(t - a_2)^{\alpha_2 - 1} \cdots (t - a_n)^{\alpha_n - 1}\, dt. \tag{10.5}$$

As $z = x$ moves from $-\infty$ along the real axis toward a_1, we have

$$\begin{aligned} w &= e^{\pi i(\alpha_1+\alpha_2+\cdots+\alpha_n-n)} \int_{-\infty}^{x} (a_1 - t)^{\alpha_1 - 1} \cdots (a_n - t)^{\alpha_n - 1}\, dt \\ &= \int_{-\infty}^{x} (a_1 - t)^{\alpha_1 - 1} \cdots (a_n - t)^{\alpha_n - 1}\, dt, \end{aligned}$$

so that w moves from 0 along the real axis toward

$$A_1 = \int_{-\infty}^{a_1} (a_1 - t)^{\alpha_1 - 1} \cdots (a_n - t)^{\alpha_n - 1}\, dt.$$

After moving through the small upper semi-circle with centre a_1, and then moving toward a_2, the point w moves from A_1 to (see Fig. 10.14)

$$A_2 = A_1 + e^{-(\alpha_1 - 1)\pi i} B_1.$$

To be more explicit, as z moves from a_1 to a_2, the point w describes the line segment

$$w = A_1 + e^{-(\alpha_1 - 1)\pi i} t, \quad 0 \le t \le B_1,$$

where

$$B_1 = \int_{a_1}^{a_2} (t - a_1)^{\alpha_1 - 1}(a_2 - t)^{\alpha_2 - 1} \cdots (a_n - t)^{\alpha_n - 1}\, dt.$$

Similarly, as z moves from a_2 to a_3, the point w describes the line segment

$$w = A_2 + e^{-(\alpha_1 + \alpha_2 - 2)\pi i} t, \quad 0 \le t \le B_2,$$

where

$$B_2 = \int_{a_2}^{a_3} (t - a_1)^{\alpha_1 - 1}(t - a_2)^{\alpha_2 - 1}(a_3 - t)^{\alpha_3 - 1} \cdots (a_n - t)^{\alpha_n - 1}\, dt.$$

Let

$$A_3 = A_2 + e^{-(\alpha_1 + \alpha_2 - 2)\pi i} B_2,$$

and we find that the angle subtended by $\overline{A_1 A_2}$ and $\overline{A_2 A_3}$ is $\alpha_2 \pi$; this is because the two lines concerned make angles $-(\alpha_1 - 1)\pi$ and $-(\alpha_1 + \alpha_2 - 2)\pi$, with the x-axis.

Continuing in this way, the points $a_1, a_2, \ldots, a_n$ are mapped to $A_1, A_2, \ldots, A_n$, with $\overline{A_{\nu-1} A_\nu}$ and $\overline{A_\nu A_{\nu+1}}$ subtending an angle $\alpha_\nu \pi$. Finally, as z moves from a_n to ∞, the point w moves along the straight line

$$w = A_n + t, \quad 0 \le t \le B_n,$$

leading from A_n to $A_n + B_n$, with

$$B_n = \int_{a_n}^{\infty} (t - a_1)^{\alpha_1 - 1} \cdots (t - a_n)^{\alpha_n - 1}\, dt.$$

We proceed to show that $A_n + B_n = A_1$, which amounts to showing that

$$\int_{-\infty}^{\infty} (a_1 - 1)^{\alpha_1 - 1} \cdots (a_n - t)^{\alpha_n - 1}\, dt = 0,$$

where, obviously, there are the small semi-circular indentations at $a_1, \ldots, a_n$ on the path of integration.

Since the integrand is regular in the upper half-plane, we draw a circle with centre O and radius R, so that the integral along the upper part of the circle is $O(R^{\alpha_1+\alpha_2+\cdots+\alpha_n-n+1}) = O(R^{-1}) \to 0$ as $R \to \infty$. The required result follows from Cauchy's theorem.

Therefore, as z moves from $-\infty$ to ∞, the point w describes a polygon with the n vertices at

$$A_1, A_2, \ldots, A_n,$$

with the interior angle $\alpha_\nu \pi$ at A_ν.

By Theorem 10.8 we know that the transformation (10.5) maps the upper half-plane into the n-sided polygon. By the same token, the transformation (10.5) also maps the upper half-plane into the n-sided polygon with vertices at

$$A_\nu = c \int_{-\infty}^{a_\nu} (t-a_1)^{\alpha_1-1}(t-a_2)^{\alpha_2-1}\cdots(t-a_n)^{\alpha_n-1}\,dt + c_1,$$

and again with the interior angle $\alpha_\nu \pi$ at A_ν.

Exercise 10.37 For what values $\alpha_1, \alpha_2, \alpha_3$ will the inverse function $z = f(w)$ of

$$w = \int_{-\infty}^{z} (t-a_1)^{\alpha_1-1}(t-a_2)^{\alpha_2-1}(t-a_3)^{\alpha_3-1}\,dt$$

be a single-valued function on the whole w-plane? Try to show that there are only three possibilities. The angles for the triangle should be

$$\left(\frac{\pi}{6}, \frac{\pi}{3}, \frac{\pi}{2}\right), \left(\frac{\pi}{4}, \frac{\pi}{4}, \frac{\pi}{2}\right), \left(\frac{\pi}{3}, \frac{\pi}{3}, \frac{\pi}{3}\right).$$

Hint: Use the Schwarz reflection principle to loop round a point an even number of times to return to the same point; this should give $\alpha_i = 1/a_i$, where a_i is a natural number. Also, from $\alpha_1 + \alpha_2 + \alpha_3 = 1$ we need to have

$$\frac{1}{a_1} + \frac{1}{a_2} + \frac{1}{a_3} = 1, \quad a_1 \geq a_2 \geq a_3 \geq 2.$$

10.18 Continuation

We now prove that the one-to-one conformal mapping of the upper half-plane to a polygon must take the form given in Section 10.17.

First, by the Riemann mapping theorem, there must be a conformal transformation $w = f(z)$ mapping the upper half z-plane bijectively onto the interior of a polygon Δ. We agree to take three points (a_1, a_2, a_3, say) on the real axis which correspond to the three points (the vertices A_1, A_2, A_3, say) on the perimeter of Δ. By Theorem 10.2, the function $f(z)$ is uniquely specified, and there will be $n-3$ other points $a_4, \ldots, a_n$ on the x-axis which correspond to the remaining vertices $A_4, \ldots, A_n$. We proceed to find the analytic representation for such a function.

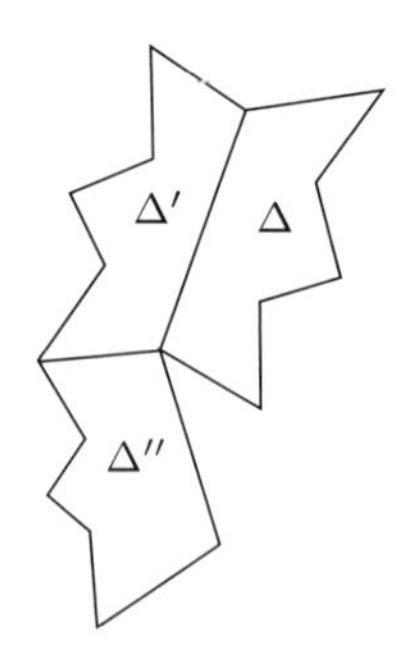

Figure 10.15.

As z moves along the real axis in the segment (a_k, a_{k+1}), the point $w = f(z)$ describes a line segment A_k, A_{k+1}. By the Schwarz reflection principle it is possible to make an analytic continuation across the segment into the lower half-plane. The resulting transformation is also conformal – it transforms the lower half-plane onto the reflection of $\triangle$, with respect to the line segment A_k, A_{k+1}. Its image $\triangle'$ is also a polygon, and, again by analytic continuation, the region can also be reflected through another segment $A_{k'}, A_{k'+1}$ to a new region $\triangle''$, with the corresponding segment $(a_{k'}, a_{k'+1})$ on the real axis, so that we are back onto the upper half-plane (see Fig. 10.15).

Suppose that we have completed the analytic continuation described in the above. Speaking generally, the result will be a multi-valued (possibly even infinitely many-valued) function $w = F(z)$. With respect to this function, the original $f(z)$ is only a single-valued branch on the upper half-plane.

Let $w = f^*(z)$ and $w = f^{**}(z)$ be any two branches of $F(z)$ on the upper half-plane. According to the construction of $F(z)$, there is an even number of corresponding (inverse) mappings from f^* to f^{**}, and the image of any two straight lines is the result of a Euclidean transformation, so that the result of the even number of mappings is

$$f^{**}(z) = e^{i\alpha} f^*(z) + a;$$

the same conclusion holds for the lower half-plane.

Consider the function

$$g(z) = \frac{f''(z)}{f'(z)} = \frac{d}{dz} \log f'(z).$$

Since $f(z)$ is a one-to-one conformal mapping on the upper half-plane, $f'(z) \neq 0$ there, so that $g(z)$ is analytic in the upper half-plane. Also, from

$$f^{**\prime}(z) = e^{i\alpha} f^{*\prime}(z), \quad f^{**\prime\prime}(z) = e^{i\alpha} f^{*\prime\prime}(z),$$

we find that $g(z)$ is single-valued, and the same conclusion can be drawn for the lower half-plane. Therefore, $g(z)$ is single-valued and analytic everywhere in the whole z-plane, apart from the points $z = a_k$. Since $z = \infty$ has been mapped onto a certain point on the perimeter of a polygon, and not a vertex, $g(z)$ is analytic even at the point at infinity.

We now examine the nature of $g(z)$ at $z = a_k$. Taking a branch $f(z)$, in the neighbourhood of $z = a_k$ we have

$$f(z) = A_k + (z - a_k)^{\alpha_k}\{c_0 + c_1(z - a_k) + \cdots\}. \tag{10.6}$$

To see this we consider

$$w(z) = (f(z) - A_k)^{1/\alpha_k};$$

as z describes a small upper semi-circle with centre a_k, $f(z)$ describes a loop $\alpha_k \pi$ round A_k, so that, as z passes through a_k along a straight line, $a(z)$ passes through A_k along a straight line segment. The corresponding $w(z)$ can thus be expanded as

$$w(z) = c_1(z - a_k) + c_2(z - a_k)^2 + \cdots ,$$

from which the expansion (10.6) follows.

We deduce from (10.6) that, in the neighbourhood of $z = a_k$,

$$\begin{aligned} g(z) = \frac{f''(z)}{f'(z)} &= \frac{(\alpha_k - 1)\alpha_k c_0 (z - \alpha_k)^{\alpha_k - 2} + \cdots}{\alpha_k c_0 (z - \alpha_k)^{\alpha_k - 1} + \cdots} \\ &= \frac{\alpha_k - 1}{z - \alpha_k} + c_0' + c_1'(z - a_k) + \cdots , \end{aligned}$$

so that $g(z)$ has a simple pole there, with the (non-zero) residue $\alpha_k - 1$.

The function $g(z)$ has n singularities in the whole plane, and

$$G(z) = g(z) - \frac{\alpha_1 - 1}{z - a_1} - \cdots - \frac{\alpha_n - 1}{z - a_n}$$

is analytic in the whole plane (including ∞), and is therefore a constant.

Again, since $f(z)$ is regular at $z = \infty$,

$$f(z) = d_0 + \frac{d_{-p}}{z^p} + \frac{d_{-p-1}}{z^{p+1}} + \cdots$$

and

$$g(z) = \frac{p(p+1)\frac{d_{-p}}{z^{p+2}} + \cdots}{-\frac{pd_{-p}}{z^{p+1}} + \cdots} = -\frac{p+1}{z} + \frac{d_2'}{z^2} + \cdots ,$$

so that $g(\infty) = 0$, and hence $G(\infty) = 0$, giving $G(z) \equiv 0$, and thus

$$g(z) = \frac{d}{dz} \log f'(z) = \frac{\alpha_1 - 1}{z - a_1} + \cdots + \frac{\alpha_n - 1}{z - a_n}.$$

Integrating along any path in the upper half-plane then gives

$$f'(z) = c(z - a_1)^{\alpha_1 - 1}(z - a_2)^{\alpha_2 - 1} \cdots (z - a_n)^{\alpha_n - 1}, \tag{10.7}$$

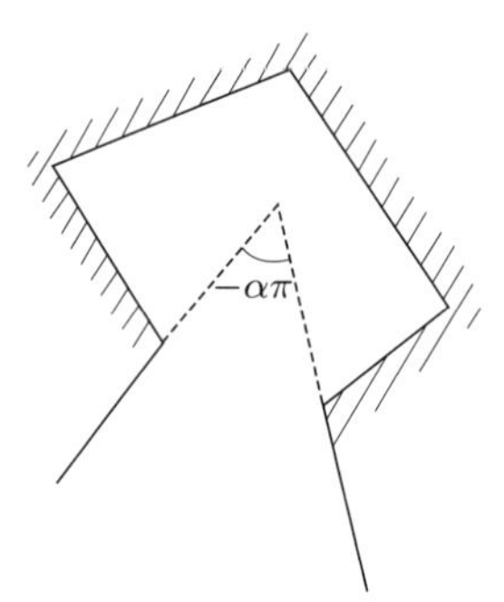

Figure 10.16.

and hence

$$f(z) = c \int_{-\infty}^{z} (t - a_1)^{\alpha_1 - 1} \cdots (t - a_n)^{\alpha_n - 1}\, dt + c_1, \tag{10.8}$$

which is the form for the transformation given in Section 10.17.

Note. From (10.7) we see that, for $z = x > a_n$, $\arg f'(z) = \arg c$; the line segment (a_n, a_1) is mapped under $w = f(z)$ to $A_n A_1$, so that $\arg c$ is the angle made by $A_n A_1$ with the x-axis.

Given the position of a vertex we can determine c_1. To determine a_k and $|c|$ we may make use of the lengths of the polygon:

$$\overline{A_k A_{k+1}} = |c| \int_{\alpha_k}^{\alpha_{k+1}} |f'(x)| dx, \quad k = 1, 2, \ldots, n-1.$$

Thus, in a practical application of the Schwarz–Christoffel formula, we need to solve a system of transcendental equations, and this is usually done by applying the multi-variable Newton's method.

Supplement

10.19 More on polygonal mappings

We do not give proofs for the following results.

(1) The transformation of the upper half-plane onto a polygon with a vertex corresponding to $a_n = \infty$ has the form

$$w = c \int_{z_0}^{z} (\zeta - a_1)^{\alpha_1 - 1} \cdots (\zeta - a_{n-1})^{\alpha_{n-1} - 1}\, d\zeta + c_1.$$

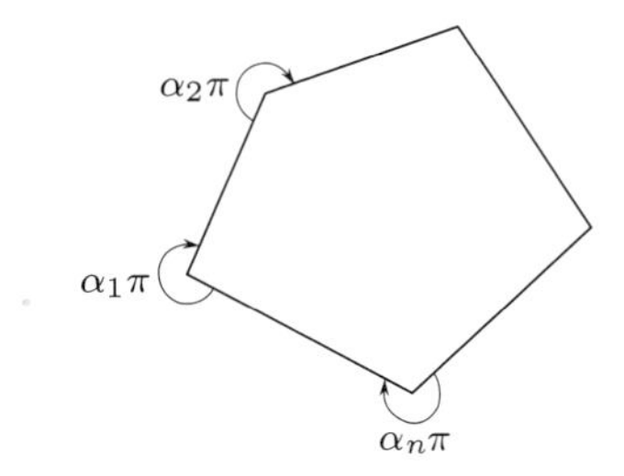

Figure 10.17.

(2) Polygons with one or more vertex at infinity. The Schwarz–Christoffel formula is still meaningful, but the angle at the vertex at infinity has to be interpreted externally, as in Fig. 10.16.

(3) Let the upper half-plane be mapped onto the exterior of a polygon (see Fig. 10.17), with the point ∞ corresponding to a. Then

$$w = c \int_{z_0}^{z} (\zeta - a_1)^{\alpha_1 - 1} \cdots (\zeta - a_{n-1})^{\alpha_{n-1} - 1} \times \frac{dt}{(\zeta - a)^2 (\zeta - \bar{a})^2} + c_1.$$

(4) Let the interior of the unit circle be mapped onto the polygon. Then

$$w = c \int_{z_0}^{z} (\zeta - a_1)^{\alpha_1 - 1} \cdots (\zeta - a_n)^{\alpha_n - 1} \, d\zeta + c_1,$$

where a_k are points on the unit circle corresponding to the vertices of the polygon.

11
Summability methods

11.1 Cesàro summability

Definitions 11.1 Let $a_0, a_1, \dots, a_n, \dots$ be a complex sequence, and let

$$\begin{aligned} s_n^{(0)} &= a_0 + \cdots + a_n, \\ s_n' &= s_0^{(0)} + \cdots + s_n^{(0)}, \\ s_n'' &= s_0' + \cdots + s_n', \\ &\dots \\ s_n^{(k)} &= s_0^{(k-1)} + \cdots + s_n^{(k-1)}, \\ &\dots . \end{aligned}$$

If

$$\lim_{n\to\infty} \frac{k! s_n^{(k)}}{n^k} = s, \tag{11.1}$$

then the series

$$a_0 + a_1 + \cdots + a_n + \cdots \tag{11.2}$$

is said to be kth order Cesàro summable to s, or simply that the series (11.2) is (C, k) summable, or (C, k) summable to s, and we write

$$a_0 + a_1 + \cdots + a_n + \cdots = s, \quad (C, k).$$

Let

$$c_n^{(k)} = \frac{k!\, n!}{(n+k)!} s_n^{(k)};$$

then (11.1) is equivalent to

$$\lim_{n\to\infty} c_n^{(k)} = s. \tag{11.3}$$

Theorem 11.2 *If a series is (C, k) summable to s, then it is also $(C, k+1)$ summable to s.*

Proof. From

$$s_n^{(k)} = s\frac{n^k}{k!} + o(n^k),$$

we find that

$$s_n^{(k+1)} = \sum_{\nu=0}^{n} s_\nu^{(k)} = \frac{s}{k!}\sum_{\nu=0}^{n} \nu^k + o\left(\sum_{\nu=0}^{n} \nu^k\right) = \frac{s}{k!}\left(\frac{n^{k+1}}{k+1} + o(n^{k+1})\right) + o(n^{k+1})$$
$$= \frac{sn^{k+1}}{(k+1)!} + o(n^{k+1}).$$

□

Note that being $(C, 0)$ summable is the same as ordinary convergence, so that if a series is convergent to a sum s, then it is Cesàro summable to s for any order.

Theorem 11.3 *If the series (11.2) is (C, k) summable, then*

$$a_n = O(n^k).$$

Proof. By the Definitions 11.1, we have

$$s_n^{(k)} = s\frac{n^k}{k!} + o\left(\frac{n^k}{k!}\right) = O(n^k),$$

from which we deduce that

$$s_n^{(k-1)} = s_n^{(k)} - s_{n-1}^{(k)} = O(n^k) + O(n^k) = O(n^k),$$
$$\cdots$$

and so on, until we have $s_n^{(0)} = O(n^k)$, and finally $a_n = O(n^k)$. □

Theorem 11.4 *Let $f(x) = \sum_{n=0}^{\infty} a_n x^n$. If the series (11.2) is (C, k) summable for some k to s, then*

$$\lim_{x \to 1_-} f(x) = s.$$

Proof. By Theorem 11.3, we know that

$$f(x) = \sum_{n=0}^{\infty} a_n x^n$$

is convergent for $|x| < 1$ and that

$$f(x) = (1-x)\sum_{n=0}^{\infty} s_n^{(0)} x^n = (1-x)^2 \sum_{n=0}^{\infty} s_n' x^n = \cdots = (1-x)^{k+1}\sum_{n=0}^{\infty} s_n^{(k)} x^n. \qquad (11.4)$$

Now

$$s_n^{(k)} = s\binom{n}{k} + o\left(\binom{n}{k}\right);$$

that is, corresponding to any $\delta > 0$, when $n \geq n_0(\delta)$ we have

$$\left| s_n^{(k)} - s\binom{n}{k} \right| < \delta \binom{n}{k}.$$

Thus, for $0 < x < 1$,

$$\begin{aligned}\left| \sum_{n=0}^{\infty} s_n^{(k)} x^n - s \sum_{n=0}^{\infty} \binom{n}{k} x^n \right| &\leq \sum_{n=0}^{\infty} \left| s_n^{(k)} - s\binom{n}{k} \right| x^n \\ &\leq \sum_{n=0}^{n_0 - 1} \left| s_n^{(k)} - s\binom{n}{k} \right| x^n + \delta \sum_{n=n_0}^{\infty} \binom{n}{k} x^n \\ &\leq \sum_{n=0}^{n_0 - 1} \left| s_n^{(k)} - s\binom{n}{k} \right| x^n + \delta \sum_{n=0}^{\infty} \binom{n}{k} x^n,\end{aligned}$$

giving

$$\limsup_{x \to 1} \frac{\sum_{n=0}^{\infty} s_n^{(k)} x^n - s \sum_{n=0}^{\infty} \binom{n}{k} x^n}{\sum_{n=0}^{\infty} \binom{n}{k} x^n} \leq \delta.$$

(Note that the denominator becomes ∞ as $x \to 1$.) Consequently, as $x \to 1$, we have

$$\frac{\sum_{n=0}^{\infty} s_n^{(k)} x^n - s \sum_{n=0}^{\infty} \binom{n}{k} x^n}{\sum_{n=0}^{\infty} \binom{n}{k} x^n} \to 0. \tag{11.5}$$

Now

$$\begin{aligned}\sum_{n=0}^{\infty} \binom{n}{k} x^n &= \frac{x^k}{k!} \sum_{n=k}^{\infty} n(n-1)\cdots(n-k+1)x^{n-k} = \frac{x^k}{k!} \frac{d^k}{dx^k} \sum_{n=0}^{\infty} x^n \\ &= \frac{x^k}{k!} \frac{d^k}{dx^k} \frac{1}{1-x} = \frac{x^k}{(1-x)^{k+1}},\end{aligned}$$

so that it follows from (11.5) that

$$\lim_{x \to 1} \frac{f(x) - sx^k}{x^k} = 0,$$

and hence

$$\lim_{x\to 1_-} f(x) = s.$$

□

Example 11.5 We have $1 - 1 + 1 - 1 + \cdots = \frac{1}{2}$, $(C, 1)$.

Example 11.6 We have $1 - 2 + 3 - 4 + \cdots = \frac{1}{4}$, $(C, 2)$.

These series are not $(C, 0)$ summable, however.

11.2 Hölder summability

Definition 11.7 Let $a_0, a_1, \ldots, a_n, \ldots$ be a complex sequence, and let

$$\begin{aligned} h_n^{(0)} &= a_0 + \cdots + a_n, \\ h_n' &= \frac{h_0^{(0)} + \cdots + h_n^{(0)}}{n+1}, \\ h_n'' &= \frac{h_0' + \cdots + h_n'}{n+1}, \\ &\ldots \\ h_n^{(k)} &= \frac{h_0^{(k-1)} + \cdots + h_n^{(k-1)}}{n+1}, \\ &\ldots \end{aligned}$$

be successive averages. If

$$\lim_{n\to\infty} h_n^{(k)} = s, \tag{11.6}$$

then the series

$$a_0 + a_1 + \cdots + a_n + \cdots \tag{11.7}$$

is said to be kth order Hölder summable to s, or simply that the series (11.7) is (H, k) summable, or (H, k) summable to s; we then write

$$a_0 + a_1 + \cdots + a_n + \cdots = s, \quad (H, k),$$

and s is called the (H, k) limit of $h_n^{(0)}$.

We now have the corresponding results to Theorems 11.2, 11.3 and 11.4, and in fact their proofs are easier.

Theorem 11.8 *If the series (11.7) is (H, k) summable to s, then it is also $(H, k+1)$ summable to s.*

This follows at once from the fact that if $h_n^{(k+1)} \to s$, then so do the averages.

Theorem 11.9 *If the series (11.7) is* (H,k) *summable, then*

$$a_n = O(n^k).$$

Proof. From $h_n^{(k+1)} \to s$, we have

$$\begin{aligned} h_n^{(k+1)} &= O(1), \\ h_n^{(k-1)} &= (n+1)h_n^{(k)} - nh_{n-1}^{(k)} = O(n) + O(n) = O(n), \\ h_n^{(k-2)} &= (n+1)h_n^{(k-2)} - nh_{n-1}^{(k-1)} = O(n^2) + O(n^2) = O(n^2), \\ &\dots \end{aligned}$$

and so on, until $h_n^{(0)} = O(n^k)$, and finally $a_n = h_n^{(0)} - h_{n-1}^{(0)} = O(n^k)$. □

Theorem 11.10 *Let* $f(x) = \sum_{n=0}^{\infty} a_n x^n$. *If the series (11.7) is* (H,k) *summable for some* k *to* s, *then*

$$\lim_{x \to 1_-} f(x) = s.$$

The theorem is the consequence of the corresponding Theorem 11.4 and the following.

Theorem 11.11 (Knopp–Schnee) *For any fixed natural number* k, *if* $c_n^{(k)} \to s$ *as* $n \to \infty$, *then* $h_n^{(k)} \to s$. *The converse also holds.*

In other words, Cesàro summability is equivalent to Hölder summability. We need a couple of lemmas in the following section before we prove Theorem 11.11.

Exercise 11.12 Examine the problem of summability for the series

$$1^k - 2^k + 3^k - 4^k + \cdots .$$

11.3 Two lemmas related to means

Lemma 11.13 *Let* $x_1, x_2, \dots, x_n, \dots$ *be a complex sequence, let* q *be a positive integer, and suppose that*

$$x_n + q\frac{x_1 + \cdots + x_n}{n} \to 0. \tag{11.8}$$

Then

$$x_n \to 0. \tag{11.9}$$

Proof. We write

$$y_n = q(x_1 + \cdots + x_n) + nx_n = q(x_1 + \cdots + x_{n-1}) + (n+q)x_n, \tag{11.10}$$

and proceed to establish, by induction, the identity

$$\sum_{\nu=1}^{n} y_\nu(\nu+1)\cdots(\nu+q-1) = (n+1)(n+2)\cdots(n+q)\sum_{\nu=1}^{n} x_\nu. \tag{11.11}$$

When $n = 1$, the identity gives $y_1 \cdot 2 \cdots q = 2 \cdots (q+1)x_1$, which is merely (11.10) with $n = 1$. Assume, as an induction hypothesis, that the identity holds for $n - 1$. Then, together with (11.10), we have

$$\begin{aligned}
&\sum_{\nu=1}^{n} y_\nu(\nu+1)\cdots(\nu+q-1) \\
&\quad = \sum_{\nu=1}^{n-1} y_\nu(\nu+1)\cdots(\nu+q-1) + y_n(n+1)\cdots(n+q-1) \\
&\quad = n(n+1)\cdots(n-1+q)\sum_{\nu=1}^{n-1} x_\nu + [q(x_1+\cdots+x_{n-1} \\
&\quad\quad + (n+q)x_n](n+1)\cdots(n+q-1), \\
&\quad = (n+1)\cdots(n+q-1)\left(\sum_{\nu=1}^{n-1}(n+q)x_\nu + (n+q)x_n\right),
\end{aligned}$$

which delivers the inductive step.

We have $y_n = o(n)$ by (11.8), so that

$$y_\nu(\nu+1)\cdots(\nu+q-1) = o(\nu^q).$$

Therefore

$$\sum_{\nu=1}^{n} y_\nu(\nu+1)\cdots(\nu+q-1) = o\left(\sum_{\nu=1}^{n}\nu^q\right) = o(n^{q+1}).$$

It now follows from the identity that

$$\sum_{\nu=1}^{n} x_\nu = \frac{1}{(n+1)\cdots(n+q)} o(n^{q+1}) = o(n),$$

and hence

$$nx_n = y_n - q\sum_{\nu=1}^{n} x_\nu = o(n) + o(n) = o(n).$$

The lemma is proved. □

Lemma 11.14 *Let k be a positive integer, and suppose that, as $n \to \infty$,*

$$\frac{1}{k}x_n + \frac{k-1}{k}\frac{x_1+\cdots+x_n}{n} \to \gamma.$$

Then $x_n \to \gamma$.

Proof. Write $z_n = x_n - \gamma$, so that the hypothesis becomes

$$\frac{1}{k} z_n + \frac{k-1}{k} \frac{z_1 + \cdots + z_n}{n} \to 0,$$

that is $z_n + (k-1)(z_1 + \cdots + z_n)/n \to 0$. By Lemma 11.13, we have $z_n \to 0$, so that $x_n \to \gamma$. □

11.4 The equivalence of (C, k) and (H, k) summabilities

Given a complex sequence

$$x_0, x_1, x_2, \ldots, x_n, \ldots,$$

we use $M(x_n)$ to denote the sequence of arithmetic means

$$x_0, \frac{x_0 + x_1}{2}, \frac{x_0 + x_1 + x_2}{3}, \ldots, \frac{x_0 + x_1 + \cdots + x_n}{n+1}, \ldots.$$

For any two constants a, b, we use $aM(x_n) + bx_n$ to denote the sequence

$$ax_0 + bx_0,\ a\frac{x_0 + x_1}{2} + bx_1,\ a\frac{x_0 + x_1 + x_2}{3} + bx_2,\ \ldots,\ a\frac{x_0 + x_1 + \cdots + x_n}{n+1} + bx_n.$$

The formal setting of $y_n = aM(x_n) + bx_n$ transforms the sequence $\{x\}$ to a sequence $\{y\}$, with

$$y = \frac{a}{n+1} x_0 + \cdots + \frac{a}{n+1} x_{n-1} + \left(\frac{a}{n+1} + b\right) x_n.$$

This is a special example of the following more general setting:

$$\begin{aligned} y_0 &= c_{00} x_0, \\ y_1 &= c_{10} x_0 + c_{11} x, \\ &\cdots \\ y_n &= c_{n0} x_0 + c_{n1} x_1 + \cdots + c_{nn} x_n, \\ &\cdots. \end{aligned}$$

If we obtain a sequence $\{y\}$ from $\{x\}$ under such a transformation, and then obtain a sequence $\{z\}$ from $\{y\}$, we then have a sequence $\{z\}$ from $\{x\}$, which can be considered to be the product of the two transformations; this then allows us to combine transformations, although we cannot expect such a product to commute. However, the two transformations

$$aM(x_n) + bx_n, \qquad a'M(x_n) + b'x_n$$

do commute in the formation of the product:

$$\begin{aligned} &a'M(aM(x_n)+bx_n)+b'(aM(x_n)+bx_n)\\ &\qquad = aa'MM(x_n)+(a'b+b'a)M(x_n)+b'bx_n,\\ &aM(a'M(x_n)+b'x_n)+b(a'M(x_n)+b'x_n)\\ &\qquad = a'aMM(x_n)+(ab'+ba')M(x_n)+bb'x_n, \end{aligned}$$

are the same.

Corresponding to a positive integer k, we define

$$T_k(x_n) = \frac{k-1}{k}M(x_n) + \frac{1}{k}x_n,$$

so that T_k and T commute, and hence T_k and T_ℓ also commute.

From $x_n \to s$, we deduce that

$$T_k(x_n) \to \frac{k-1}{k}s + \frac{1}{k}s = s,$$

and, by Lemma 11.14, we see that the converse also holds.

We first prove the identity

$$M(c_n^{(k-1)}) = T_k(c_n^{(k)}), \quad k \geq 1, \tag{11.12}$$

that is

$$\frac{c_0^{(k-1)}+\cdots+c_n^{(k-1)}}{n+1} = \frac{k-1}{k}\frac{c_0^{(k)}+\cdots+c_n^{(k)}}{n+1} + \frac{1}{k}c_n^{(k)}. \tag{11.13}$$

From

$$s_n^{(k)} = s_{n-1}^{(k)} + s_n^{(k-1)},$$

we deduce that

$$\begin{aligned} \frac{(n+k)!}{n!\,k!}c_n^{(k)} &= \frac{n(n+k-1)!}{n!\,k!}c_{n-1}^{(k)} + \frac{(n+k-1)!}{n!\,(k-1)!}c_n^{(k-1)},\\ kc_n^{(k-1)} &= (n+k)c_n^{(k)} - nc_{n-1}^{(k)} = (k-1)c_n^{(k)} + ((n+1)c_n^{(k)} - nc_{n-1}^{(k)}), \end{aligned}$$

that is

$$k\sum_{n=0}^{m} c_n^{(k-1)} = (k-1)\sum_{n=0}^{m} c_n^{(k)} + (m+1)c_m^{(k)},$$

which is the identity (11.13).

We now have

$$\begin{aligned} h_n^{(0)} &= s_n^{(0)} = c_n^{(0)},\\ h_n' &= \frac{s_n'}{n+1} = c_n', \end{aligned}$$

which will simply be written as

$$h' = c'.$$

Then

$$\begin{aligned}
h'' &= M(h') = M(c') = T_2(c''),\\
h''' &= M(h'') = MT_2(c'') = T_2M(c'') = T_2T_3(c'''),\\
&\cdots\\
h^{(k)} &= M(h^{(k-1)}) = MT_2T_3\cdots T_{k-1}(c^{(k-1)})\\
&= T_2T_3\cdots T_{k-1}M(c^{(k-1)}) = T_2\cdots T_{k-1}T_k(c^{(k)}).
\end{aligned}$$

If

$$\lim_{n\to\infty} c_n^{(k)} = s,$$

then we have successively

$$\begin{aligned}
&T_k(c^{(k)}) \to s,\\
&T_{k-1}T_k(c^{(k)}) \to s,\\
&\cdots\\
&h^{(k)} = T_2T_3\cdots T_k(c^{(k)}) \to s.
\end{aligned}$$

We have thus established the following (Schnee's theorem).

Theorem 11.15 (Schnee) *From* $\lim_{n\to\infty} c_n^{(k)} = s$, *we have* $\lim_{n\to\infty} h_n^{(k)} = s$.

Conversely, if $h_n^{(k)} \to s$ then, by Lemma 11.14, we successively deduce that

$$\begin{aligned}
&T_3T_4\cdots T_k(c^{(k)}) \to s,\\
&T_4\cdots T_k(c^{(k)}) \to s,\\
&\cdots\\
&T_k(c^{(k)}) \to s\\
&c^{(k)} \to s,
\end{aligned}$$

which is Knopp's theorem, as follows.

Theorem 11.16 (Knopp) *From* $\lim_{n\to\infty} h_n^{(k)} = s$, *we have* $\lim_{n\to\infty} c_n^{(k)} = s$.

11.5 (C, α) summability

Recall the definition of $s_n^{(k)}$ in Section 11.1. Let

$$f(x) = \sum_{n=0}^{\infty} a_n x^n,$$

so that

$$f(x) = (1-x)^{k+1}\sum_{n=0}^{\infty} s_n^{(k)} a_n x^n.$$

Since

$$(1-x)^{-(k+1)} = \sum_{n=0}^{\infty}\binom{n+k}{k}x^n,$$

we see that from

$$\sum_{n=0}^{\infty} s_n^{(k)} a_n x^n = (1-x)^{-(k+1)} f(x) = \sum_{m=0}^{\infty}\binom{m+k}{k}x^m \sum_{\ell=0}^{\infty} a_\ell x^\ell \tag{11.14}$$

we have

$$s_n^{(k)} = \sum_{m+\ell=n}\binom{m+k}{k}a_\ell = \sum_{\ell=0}^{n}\binom{n-\ell+k}{k}a_\ell.$$

Therefore

$$c_n^{(k)} = \frac{\sum_{\ell=0}^{n}\binom{n-\ell+k}{k}a_\ell}{\binom{n+k}{k}}. \tag{11.15}$$

It is not difficult to show that

$$\sum_{\ell=0}^{n}\binom{n-\ell+k}{k} = \binom{n+k+1}{k+1}$$

(setting all $a_\ell = 1$ in (11.14)).

Equation (11.15) can be rewritten as

$$\begin{aligned} c_n^{(k)} &= \sum_{\ell=0}^{n}\frac{(n-\ell+k)\cdots(n-\ell+1)}{(n+k)\cdots(n+1)}a_\ell \\ &= \sum_{\ell=0}^{n}\frac{\Gamma(n-\ell+k+1)\Gamma(n+1)}{\Gamma(n+k+1)\Gamma(n-\ell+1)}a_\ell, \end{aligned}$$

so that we may define (C,k) summability even when k is not an integer.

When considering (C,α) summability, we assume that $\alpha > -1$. If

$$c_n^{(k)} = \sum_{\ell=0}^{n}\frac{\Gamma(n-\ell+k+1)\Gamma(n+1)}{\Gamma(n+k+1)\Gamma(n-\ell+1)}a_\ell \to s,$$

then we write

$$a_0 + a_1 + \cdots + a_n + \cdots = s,\ (C,\alpha).$$

We do not delve into problems concerning (C, α) summability, and leave it to the reader to prove the following: If $\alpha' > \alpha$ then from

$$a_0 + a_1 + \cdots + a_n + \cdots = s,\ (C, \alpha),$$

we can deduce that

$$a_0 + a_1 + \cdots + a_n + \cdots = s,\ (C, \alpha').$$

11.6 Abel summability

Definition 11.17 Let

$$f(x) = a_0 + a_1 x + \cdots + a_n x^n + \cdots .$$

If

$$\lim_{x \to 1_-} f(x)$$

exists as a number s, then we say that the series

$$a_0 + a_1 + \cdots + a_n + \cdots$$

is Abel summable, the Able sum is then s, and we then write

$$a_0 + a_1 + \cdots + a_n + \cdots = s,\ (A).$$

From the preceding sections we know that being (C, k) summable (or (H, k) summable) implies being Abel summable; but not conversely.

Take, for example,

$$f(x) = e^{1/(1+x)} = \sum_{n=0}^{\infty} a_n x^n,$$

which converges in $|x| < 1$, and

$$\lim_{x \to 1_-} f(x) = e^{\frac{1}{2}}.$$

However, if

$$a_0 + a_1 + a_2 + \cdots$$

were (C, k) summable then $a_n = O(n^k)$, so that

$$|a_n| < P\binom{n+k}{k}$$

for some constant P. Consequently, for $0 \le r < 1$,

$$e^{1/(1-r)} = f(-r) \le \sum_{n=0}^{\infty} |a_n| r^n < P \sum_{n=0}^{\infty} \binom{n+k}{k} r^n = P(1-r)^{-(k+1)},$$

which cannot be true when $r \to 1$.

11.7 Introduction to general summability

Advancing one step further we let

$$t_m = \sum_{n=0}^{\infty} C_{mn} s_n, \quad m = 0, 1, 2, \dots . \tag{11.16}$$

If, from

$$\lim_{n\to\infty} s_n = s,$$

it can be deduced that

$$\lim_{m\to\infty} t_m = s,$$

then (11.16) can be used to define a summability method; (C, α) summability described in Section 11.5 is of this type.

Again, let

$$t(x) = \sum_{n=0}^{\infty} C_n(x) s_n; \tag{11.17}$$

if, from

$$\lim_{n\to\infty} s_n = s,$$

it can be deduced that

$$\lim_{x\to 1} t(x) = s,$$

then (11.17) can also be used to define a summability method; Abel summability, in which

$$t(x) = \sum_{n=0}^{\infty} a_n x^n = \sum_{n=0}^{\infty} x^n (1-x) s_n,$$

is an example.

The summability methods in (11.16), and (11.17), are represented by the matrix (C_{mn}), and $C_n(x)$, respectively.

The process then involves the search, under (11.16), for necessary and sufficient conditions for the convergence of a sequence to be transformed to that of another, and whether convergence to s is also being preserved.

If, for two summability methods P and Q, summability under P always implies summability under Q, then we say that Q is a stronger method than P, and we write

$$P < Q.$$

It is established in the above that if $k \leq k'$ then $(C, k) < (C, k')$, and that $(C, k) < A$. The relation $<$ is obviously transitive, that is

$$\text{if} \quad P < Q \text{ and } Q < R, \quad \text{then} \quad P < R.$$

We then have the following problems:

(1) We need to find some meaningful summability methods.
(2) We must investigate the relative strengths of the various summability methods.
(3) We must ask what type of conditions can be added to a summability method for it to be weakened.
(4) What happens if we replace convergence with being asymptotic, such as $s_n \sim an^\beta$, say?

11.8 Borel summability

Let

$$J(x) = \sum_{n=0}^{\infty} p_n x^n$$

represent an integral function, which is not a polynomial, with $p_n \geq 0$. We introduce a summability method: if

$$\frac{s(x)}{J(x)} = \frac{\sum_{n=0}^{\infty} p_n s_n x^n}{\sum_{n=0}^{\infty} p_n x^n} \to s \quad \text{as} \quad x \to \infty,$$

then we write $s_n \to s, (J)$, or

$$\sum_{n=0}^{\infty} a_n = s, \ (J);$$

that is, the series $\sum a_n$ is summable J.

Taking

$$p_n = \frac{1}{n!},$$

so that the condition becomes

$$e^{-x} \sum_{n=0}^{\infty} s_n \frac{x^n}{n!} \to s,$$

we have the well-known Borel summability method, which is denoted by $s_n \to s, \ (B)$.

Theorem 11.18 *If*

$$\sum_{n=0}^{\infty} a_n = s$$

then

$$\sum_{n=0}^{\infty} a_n = s, \ (J);$$

that is, $s_n \to s$ *implies* $s_n \to s, \ (J)$.

Proof. We may assume, without loss of generality, that $s = 0$. Let $\epsilon > 0$, and choose N so that $|s_n| < \epsilon$ for all $n > N$. Then

$$\left|\sum_{n=0}^{\infty} p_n s_n x^n\right| \le \left|\sum_{n=0}^{N} p_n s_n x^n\right| + \epsilon \sum_{n=N+1}^{\infty} p_n x^n \le \left|\sum_{n=0}^{N} p_n s_n x^n\right| + \epsilon \sum_{n=0}^{\infty} p_n x^n.$$

Since $J(x)$ is not a polynomial and $p_n \ge 0$, for any fixed k, we have

$$\lim_{x\to\infty} \frac{x^k}{J(x)} = 0,$$

so that

$$\sum_{n=0}^{\infty} p_n s_n x^n = o\big(J(x)\big),$$

and the theorem is proved. □

We now introduce Borel's second summability method. If

$$\int_0^{\infty} e^{-x} \sum_{n=0}^{\infty} a_n \frac{x^n}{n!}\, dx = \lim_{X\to\infty} \int_0^{X} e^{-x} \sum_{n=0}^{\infty} a_n \frac{x^n}{n!}\, dx = s,$$

then we define

$$\sum_{n=0}^{\infty} a_n = s,\ (B').$$

Theorem 11.19 *Let*

$$a(x) = \sum_{n=0}^{\infty} a_n \frac{x^n}{n!}.$$

If $e^{-x}a(x) \to 0$ as $x \to \infty$, then the two summability methods B and B' are equaivalent.

Proof. Let

$$s(x) = \sum_{n=0}^{\infty} s_n \frac{x^n}{n!}.$$

It is not difficult to show that if $a(x)$ is convergent for all x then so is $s(x)$, and the converse is also true. Moreover, we have

$$a'(x) = \sum a_{n+1} \frac{x^n}{n!}, \quad s'(x) = \sum s_{n+1} \frac{x^n}{n!},$$

$$\int_0^x e^{-t} a'(t) dt = e^{-x} a(x) - a_0 + \int_0^x e^{-t} a(t) dt, \tag{11.18}$$

and

$$\begin{aligned} e^{-x}s(x) - a_0 &= \int_0^x \frac{d}{dt}\big(e^{-t}s(t)\big)dt = \int_0^x e^{-t}\big(s'(t) - s(t)\big)dt \\ &= \int_0^x e^{-t}\sum_{n=0}^{\infty}(s_{n+1} - s_n)\frac{t^n}{n!}\,dt \\ &= \int_0^x e^{-t}\sum a_{n+1}\frac{t^n}{n!}\,dt = \int_0^x e^{-t}a'(t)dt. \end{aligned} \tag{11.19}$$

Comparing (11.18) and (11.19), we have

$$e^{-x}s(x) = e^{-x}a(x) + \int_0^x e^{-t}a(t)dt, \tag{11.20}$$

which proves the theorem. □

A deeper result can also be deduced from (11.20).

Theorem 11.20 *If a series is summable (B), then it must be summable (B'), that is*

$$B < B'.$$

Writing

$$\phi(x) = \int_0^x e^{-t}a(t)dt,$$

if the series is summable (B) then, by (11.20), we have

$$\phi(x) + \phi'(x) \to s \quad as \quad x \to \infty,$$

so that the proof of Theorem 11.20 follows from the following lemma.

Lemma 11.21 *Suppose that $\phi(x)$ has a continuous derivative. If*

$$\phi(x) + \phi'(x) \to 0 \quad \text{as} \quad x \to \infty,$$

then $\phi(x) \to 0$ as $x \to \infty$.

Proof. (1) If $\phi'(x) \geq 0$ for all large x, then $\phi(x)$ is an increasing function for large x, and so has a limit ℓ (which may be ∞). By hypothesis, $\phi'(x) \to -\ell$, so that from $\phi'(x) \geq 0$ we find that $\ell \leq 0$. Since

$$\phi(x+1) - \phi(x) = \int_x^{x+1} \phi'(t)dt \to -\ell,$$

the only possibility is $\ell = 0$.

(2) The same argument applies if $\phi'(x) \leq 0$ for all large x.

(3) We are left with the case when $\phi'(x)$ changes signs infinitely often, and thus with infinitely many zeros, as $x \to \infty$, and so $\phi(x)$ has infinitely many maxima and infima.

Suppose that $\phi(x)$ has maxima at $x = x_1, x_2, \ldots$. Then from

$$\phi(x_i) + \phi'(x_i) = \phi(x_i),$$

and the hypothesis of the lemma, we deduce that $\lim_{i\to\infty} \phi(x_i) = 0$. The same argument shows that $\lim_{i\to\infty} \phi(y_i) = 0$, where y_i are the minima of $\phi(x)$. Therefore $\lim_{x\to\infty} \phi(x) = 0$, as required. □

There are series which are summable (B') but not summable (B). For example, if

$$a_n = \sum_{p=0}^{\infty} \frac{(-1)^p(2p+2)^n}{(2p+1)!},$$

then

$$a(x) = \sum_{n=0}^{\infty} \frac{x^n}{n!} \sum_{p=0}^{\infty} \frac{(-1)^p(2p+2)^n}{(2p+1)!} = \sum_{p=0}^{\infty} \frac{(-1)^p}{(2p+1)!} e^{(2p+2)x} = e^x \sin e^x,$$

$$\int_0^\infty e^{-x} a(x)dx = \int_0^\infty \sin e^x \, dx = \int_1^\infty \frac{\sin u}{u} \, du,$$

so that the series $\sum a_n$ is summable (B'). However, $e^{-x}a(x)$ does not have a limit, so that the series fails to be summable (B).

Nevertheless, we can use (11.19) to deduce the following theorem.

Theorem 11.22 *The following two formulae are equivalent:*

$$a_0 + a_1 + a_2 + \cdots = s, \quad (B),$$
$$a_1 + a_2 + \cdots = s - a_0, \quad (B').$$

Note, however, that

$$a_1 + a_2 + \cdots = s - a_0, \quad (B),$$

and

$$a_0 + a_1 + \cdots = s, \quad (B'),$$

are not equivalent.

11.9 Hardy–Littlewood theorems

Theorem 11.23 *Suppose that $\sum a_n = s$, (C, k), for an integer $k > 1$ and that*

$$a_n = O(1/n).$$

Then $\sum a_n$ converges to s.

Theorem 11.24 *Suppose that $\sum a_n = s$, (C, k), for an integer $k > 1$, where a_n are real, and that there is a positive constant H such that*

$$na_n > -H.$$

Then $\sum a_n$ converges to s.

By arguing separately for the real and the imaginary parts for a_n, we see that Theorem 11.23 is a corollary of Theorem 11.24.

Setting $b_n = na_n$, the definition for T_n^k applied to b_n is the same as that for s_n^k applied to a_n. The proof of Theorem 11.24 depends on the following two theorems.

Theorem 11.25 *Let $\sum a_n$ be summable (C, k) for some $k > 0$. Then a necessary and sufficient condition for $\sum a_n$ to be summable $(C, k-1)$ is that $T_n^{k-1} = O(n^k)$.*

Theorem 11.26 *A necessary and sufficient condition for $\sum a_n$ to be summable (C, k) is the convergence of the series*

$$\sum_{n=0}^{\infty} \frac{T_n^{(k-1)}}{\binom{n+k}{n} n} = \sum_{n=0}^{\infty} \frac{(n-1)!}{(k+1)(K+2)\cdots(n+k)} T_n^{(k-1)},$$

or, equivalently, the convergence of

$$\sum n^{-k-1} T_n(k-1).$$

Proof. From

$$\sum_{n=0}^{\infty} a_n x^n = (1-x)^k \sum_{n=0}^{\infty} s_n^{(k-1)} x^n$$

and

$$\sum_{n=0}^{\infty} na_n x^n = (1-x)^k \sum_{n=0}^{\infty} T_n^{(k-1)} x^n,$$

we have

$$\begin{aligned}
(1-x)^k \sum_{n=0}^{\infty} T_n^{(k-1)} x^n &= -xk(1-x)^{k-1} \sum_{n=0}^{\infty} s_n^{(k-1)} x^n + (1-x)^k \sum_{n=0}^{\infty} n s_n^{(k-1)} x^n \\
&= (1-x)^k \sum_{n=0}^{\infty} (k+n) s_n^{(k-1)} x^n - k(1-x)^{k-1} \sum_{n=0}^{\infty} s_n^{(k-1)} x^n \\
&= (1-x)^k \left(\sum_{n=0}^{\infty} (k+n) s_n^{(k-1)} x^n - k \sum_{n=0}^{\infty} s_n^{(k)} x^n \right),
\end{aligned}$$

so that

$$T_n^{(k-1)} = (k+n) s_n^{(k-1)} - k s_n^{(k)}. \tag{11.21}$$

Again, from

$$s_n^{(k-1)} = s_n^{(k)} - s_{n-1}^{(k)},$$

we have

$$T_n^{(k-1)} = ns_n^{(k)} - (n+k)s_{n-1}^{(k)}. \tag{11.22}$$

From (11.21) and (11.22) we deduce that

$$\frac{s_n^{(k-1)}}{\binom{n+k-1}{k-1}} - \frac{s_n^{(k)}}{\binom{n+k}{k}} = \frac{T_n^{(k-1)}}{k\binom{n+k}{k}} \tag{11.23}$$

and

$$\frac{s_n^{(k)}}{\binom{n+k}{k}} - \frac{s_{n-1}^{(k)}}{\binom{n+k-1}{k}} = \frac{T_n^{(k-1)}}{n\binom{n+k}{k}}.$$

Summing over $n = 1, 2, \ldots, N$ we arrive at

$$\frac{s_N^{(k)}}{\binom{N+k}{k}} = a_0 + \sum_{n=1}^{N} \frac{T_n^{(k-1)}}{n\binom{n+k}{k}}. \tag{11.24}$$

Theorems 11.25 and 11.26 now follow from (11.23) and (11.24), respectively, and the equivalence of the two conditions in Theorem 11.26 is a consequence of Stirling's formula. □

Proof of Theorem 11.24. We may assume that $H = 1$. If $T_n^{(k-1)} \neq O(n^k)$ then there is a number $c > 0$ such that either

$$T_n^{(k-1)} > cn^k \tag{11.25}$$

or

$$T_n^{(k-1)} < -cn^k \tag{11.26}$$

holds for infinitely many n. Suppose first that there are infinitely many N for which (11.25) holds with $n = N$.

If $\eta > 1$ and $N \le n \le \eta N$, then

$$\begin{aligned} T_n^{(k-1)} - T_N^{(k-1)} &= \sum_{\nu=0}^{N} \left\{ \binom{n-\nu+k-1}{k-1} - \binom{N-\nu+k-1}{k-1} \right\} b_\nu \\ &\quad + \sum_{\nu=N+1}^{n} \binom{n-\nu+k-1}{k-1} b_\nu. \end{aligned} \tag{11.27}$$

Since $b_\nu > -1$, and the coefficients for b_ν are positive, we have

$$T_n^{(k-1)} - T_N^{(k-1)} > -\sum_{\nu=0}^{N} \left\{ \binom{n-\nu+k-1}{k-1} - \binom{N-\nu+k-1}{k-1} \right\} - \sum_{\nu=N+1}^{n} \binom{n-\nu+k-1}{k-1},$$

so that

$$T_n^{(k-1)} - T_N^{(k-1)} > -\binom{n+k}{k} + \binom{N+k}{k}.$$

From

$$\binom{n+k-1}{k} \sim \frac{n^k}{k!}, \quad \binom{N+k-1}{k} \sim \frac{N^k}{k!},$$

we find that, for any $\epsilon > 0$ and sufficiently large N, those n in the interval $N \le n \le \eta N$ will satisfy

$$T_n^{(k-1)} - T_N^{(k-1)} > -\frac{1}{k!}\big((1+\epsilon)\eta^k - (1-\epsilon)\big)N^k.$$

We take ϵ and η so that

$$T_n^{(k-1)} - T_N^{(k-1)} > -\frac{1}{2}cN^k.$$

From (11.25) (with '$n = N$') we now have

$$T_n^{(k-1)} > \frac{1}{2}cN^k, \quad N \le n \le \eta N.$$

Therefore, for sufficiently large N we have

$$\sum_{N \le n \le \eta N} \frac{T_n^{(k-1)}}{n^{k+1}} > \frac{1}{2}cN^k \sum_{N \le n \le \eta N} \frac{1}{n^{k+1}} > \frac{1}{2}cN^k \frac{(\eta-1)N}{(\eta N)^{k+1}} = \frac{c(\eta-1)}{2\eta^{k+1}}.$$

Since this holds for infinitely many N, the series in Theorem 11.26 is divergent, so that $\sum a_n$ is not summable (C, k). Therefore, (11.25) cannot hold for infinitely many n.

The same method (but taking the interval $(\zeta N, N)$ with $\zeta < 1$) shows that (11.26) cannot hold for infinitely many n.

Therefore $T_n^{(k-1)} = O(n^k)$, and $\sum a_n$ is summable $(C, k-1)$ according to Theorem 11.25. On successive reductions of k we arrive at the convergence of the series $\sum a_n$. □

11.10 Tauber's theorem

Theorem 11.27 (Tauber) *If*

$$a_0 + a_1 + \cdots + a_n + \cdots = s, \ (A),$$

and $a_n = o(1/n)$, *then*

$$a_0 + a_1 + \cdots + a_n + \cdots = s.$$

We may assume without loss of generality that $s = 0$. The theorem then states that if

$$f(x) = \sum_{n=0}^{\infty} a_n x^n$$

is convergent in $|x| < 1$, and $f(x) \to 0$ as $x \to 1_-$, then the condition $na_n \to 0$ as $n \to \infty$ guarantees

$$\sum_{n=0}^{\infty} a_n = 0.$$

Proof. Let

$$s_m = \sum_{n=0}^{m} a_n, \quad m = 1, 2, \ldots,$$

so that, for $0 \le x < 1$,

$$s_m - f(x) = \sum_{n=0}^{m} a_n(1 - x^n) - \sum_{n=m+1}^{\infty} a_n x^n.$$

From $1 - x^n = (1 - x)(1 + x + \cdots + x^{n-1}) \le n(1 - x)$, we have

$$|s_m - f(x)| \le (1 - x) \sum_{n=0}^{m} n|a_n| + \sum_{n=m+1}^{\infty} |a_n| x^n.$$

Let

$$\epsilon_m = \max_{n>m} n|a_n|,$$

so that $\epsilon_m \to 0$ as $m \to \infty$, and that

$$\sum_{n=m+1}^{\infty} |a_n| x^n = \sum_{n=m+1}^{\infty} n|a_n| \frac{1}{n} x^n \le \sum_{n=m+1}^{\infty} \frac{\epsilon_m}{m} x^n \le \frac{\epsilon_m}{m(1 - x)}.$$

On the other hand, from $n|a_n| \to 0$ as $n \to \infty$, we deduce that its arithmetic mean also tends to 0, that is

$$\frac{1}{m} \sum_{n=1}^{m} n|a_n| \to 0 \quad \text{as} \quad m \to \infty.$$

Setting $x = 1 - 1/m$ and letting $m \to \infty$, we find that

$$|s_m - f(1 - 1/m)| \le \frac{1}{m}\sum_{n=1}^{m} n|a_n| + \epsilon_m \to 0.$$

The required result $s_m \to 0$ follows from the hypothesis that $f(1 - 1/m) \to 0$. □

Note 1. It appears that we have not made full use of the hypothesis that '$f(x) \to 0$', only that '$f(1 - 1/m) \to 0$'. However, with the condition $na_n \to 0$, the latter implies the former. To see this, we let $|na_n| < c$ so that

$$|f'(x)| = \left|\sum_{n=1}^{\infty} na_n x^{n-1}\right| < c\sum_{n=1}^{\infty} x^{n-1} = \frac{c}{1-x}, \qquad 0 < x < 1.$$

Thus, for $1 - 1/m < x < 1 - 1(m + 1)$,

$$|f(x) - f(1-1/m)| = \left|\int_{1-1/m}^{x} f(y)dy\right| < \int_{1-1/m}^{1-1/(m+1)} \frac{c}{1-(1-1/(m+1))}\,dy$$

$$= c(m+1)\left(\frac{1}{m} - \frac{1}{m+1}\right) = \frac{c}{m}.$$

The bound does not depend on x, and tends to 0 as $m \to \infty$.

Exercise 11.28 Under what hypothesis can

$$a_0 + a_1 + \cdots + a_n + \cdots = s,\ (A)$$

be used to deduce that

$$a_0 + a_1 + \cdots + a_n + \cdots = s,\ (C, k)?$$

Note 2. Littlewood strengthened Tauber's theorem by weakening the hypothesis from $o(1/n)$ to $O(1/n)$; see Theorem 11.37. Hardy and Littlewood[1] further showed that if a_n are real numbers then the condition $a_n \ge H/n$ $(H < 0)$ is enough. Such a deep result may appear perhaps even more spectacular if considered in the absence of the result in Section 11.9.

11.11 Asymptotic properties on the circle of convergence

Theorem 11.29 *Let $a_n \ge 0$ and $b_n \ge 0$, and suppose that the series*

$$f(x) = \sum a_n x^n, \quad g(x) = \sum b_n x^n$$

are convergent in $0 < x < 1$ and divergent at $x = 1$. If

$$a_n \sim cb_n \quad as \quad n \to \infty,$$

[1] Translator's note. In 1897, A. Tauber proved his version (Theorem 11.27) of the converse of Abel's theorem. Such types of converse theorems concerning the convergence of series are now called *Tauberian theorems*, a termed coined by G. H. Hardy and J. E. Littlewood when they made their deep generalisations of Tauber's theorem; see Sections 11.12 and 11.13.

then

$$f(x) \sim cg(x) \quad as \quad x \to 1.$$

Proof. Given $\epsilon > 0$, there exists N such that $|a_n - cb_n| \le \epsilon b_n$ for all $n > N$, so that

$$|f(x) - cg(x)| = \left|\sum_{n=0}^{\infty}(a_n - cb_n)x^n\right| \le \sum_{n=0}^{N}|a_n - cb_n| + \epsilon \sum_{n=N+1}^{\infty} bx^n$$
$$\le \sum_{n=0}^{N}|a_n - cb_n| + \epsilon g(x).$$

Since $g(x) \to \infty$ as $x \to 1$, we may take $\delta > 0$ so that, for $1 - \delta < x < 1$,

$$\sum_{n=0}^{N}|a_n - cb_n| < \epsilon g(x)$$

and hence $|f(x) - cg(x)| < 2\epsilon g(x)$. The theorem is proved. □

Theorem 11.30 *Let the two series*

$$f(x) = \sum a_n x^n, \quad g(x) = \sum b_n x^n$$

be convergent in $0 < x < 1$, *and let*

$$s_n = a_0 + a_1 + \cdots + a_n, \quad t_n = b_0 + b_1 + \cdots + b_n.$$

Suppose that $s_n \ge 0$, $t_n \ge 0$, *and that* $\sum s_n$, $\sum t_n$ *are divergent. If*

$$s_n \sim ct_n \quad as \quad n \to \infty,$$

then

$$f(x) \sim cg(x) \quad as \quad x \to 1.$$

Proof. For $0 < x < 1$, we have

$$f(x) = (1 - x)\sum_{n=0}^{\infty} s_n x^n, \quad g(x) = (1 - x)\sum_{n=0}^{\infty} t_n x^n,$$

and, by Theorem 11.29,

$$\sum_{n=0}^{\infty} s_n x^n \sim c \sum_{n=0}^{\infty} t_n x^n,$$

so that the required result follows. □

The reader should try to deduce the corresponding theorem for summability with a higher order.

A special example. If $s_n \sim cn$ then $f(x) \sim \dfrac{c}{1 - x}$.

Example 11.31 If $p < 1$ then we have

$$\sum_{n=1}^{\infty} \frac{x^n}{n^p} \sim \frac{\Gamma(1-p)}{(1-x)^{1-p}}, \quad \text{as} \quad x \to 1.$$

This follows from

$$(1-x)^{1-p} = \frac{1}{\Gamma(1-p)} \sum_{n=0}^{\infty} \frac{\Gamma(n-p+1)}{\Gamma(n+1)} x^n$$

together with

$$\frac{1}{n^p} \sim \frac{\Gamma(n-p+1)}{\Gamma(n+1)} \quad \text{as} \quad x \to 1.$$

Example 11.32 Let

$$F(\alpha, \beta, \gamma, x) = 1 + \frac{\alpha}{1}\frac{\beta}{\gamma} x + \frac{\alpha(\alpha+1)}{1{\cdot}2}\frac{\beta(\beta+1)}{\gamma(\gamma+1)} x^2 + \cdots .$$

If $\alpha + \beta > \gamma$ then, as $x \to 1$,

$$F(\alpha, \beta, \gamma, x) \sim \frac{\Gamma(\gamma)\Gamma(\alpha+\beta-\gamma)}{\Gamma(\alpha)\Gamma(\beta)} \frac{1}{(1-x)^{\alpha+\beta-\gamma}}$$

and

$$F(\alpha, \beta, \alpha+\beta, x) \sim \frac{\Gamma(\alpha+\beta)}{\Gamma(\alpha)\Gamma(\beta)} \log \frac{1}{(1-x)}.$$

Can there be a converse to Theorem 11.30? Perhaps not at first sight; for example,

$$f(x) = \frac{1}{(1+x)^2(1-x)} = (1-x)\sum_{n=0}^{\infty}(n+1)x^{2n}$$

$$= \sum_{n=0}^{\infty}(n+1)(x^{2n} - x^{2n+1}).$$

Here $s_{2m+1} = -(m+1)$ and $s_{2m} = m+1$, so that s_n oscillates unboundedly, whereas

$$f(x) \sim \frac{1}{4(1-x)} \quad \text{as} \quad x \to 1.$$

In this example, the coefficients take both positive and negative values. Can there be a converse to the theorem when the coefficients are all positive? The answer is yes.

11.12 A Hardy–Littlewood theorem

Theorem 11.33 (Hardy–Littlewood). *If* $a_n \geq 0$ *and*

$$f(x) = \sum_{n=0}^{\infty} a_n x^n \sim \frac{1}{1-x} \quad \textit{as} \quad x \to 1,$$

then

$$s_n = \sum_{\nu=0}^{n} a_\nu \sim n \quad \textit{as} \quad n \to \infty.$$

For comparison, we first prove an easier theorem.

Theorem 11.34 *Let*

$$f(x) = \sum_{n=0}^{\infty} a_n x^n, \qquad 0 < x < 1,$$

where $a_n \geq 0$. *If*

$$\frac{1}{1-x} \ll f(x) \ll \frac{1}{1-x} \quad \textit{as} \quad x \to 1,$$

then

$$n \ll s_n \ll n \quad \textit{as} \quad n \to \infty.$$

The converse is obviously true.

Proof. (1) We have

$$s_n x^n \leq \sum_{m=0}^{n} a_m x^m \leq f(x) \ll \frac{1}{1-x},$$

and, on setting $x = e^{-1/n}$, we find that

$$\frac{s_n}{e} \ll \frac{1}{1-e^{-1/n}} \ll n,$$

that is

$$s_n \ll n. \tag{11.28}$$

(2) From (11.28) we have

$$\begin{aligned} f(x) &= (1-x)\sum_{m=0}^{\infty} s_m x^m \\ &\ll (1-x)s_n \sum_{m=0}^{n-1} x^m + (1-x)\sum_{m=n}^{\infty} m x^m \\ &\ll s_n + n x^n + \frac{x^{n+1}}{1-x}, \end{aligned}$$

so that

$$\frac{1}{1-x} \ll s_n + n x^n + \frac{x^{n+1}}{1-x}.$$

Taking $x = e^{-\lambda/n}$, we have

$$1 - x = 1 - e^{-\lambda/n} = \int_0^{\lambda/n} e^{-t}\, dt \le \frac{\lambda}{n},$$

and hence

$$\frac{n}{\lambda} \ll s_n + ne^{-\lambda} + \frac{ne^{-\lambda}}{\lambda}.$$

When λ is sufficiently large this gives $s_n \gg n$, so that the theorem is proved. □

The proof of the Hardy–Littlewood theorem, however, is not so easy. The following proof is by J. Karamata.

Proof of Theorem 11.33. (1) We first prove that, for any polynomial $p(x)$, we have

$$\lim_{x\to 1}(1-x)\sum_{n=0}^{\infty} a_n x^n p(x^n) = \int_0^1 p(t) dt. \tag{11.29}$$

Since the result is 'additive', we need only consider the case $p(x) = x^\ell$, which is fine because

$$\begin{aligned}\lim_{x\to 1}(1-x)\sum_{n=0}^{\infty} a_n x^{n+\ell n} &= \lim_{x\to 1}\frac{(1-x)}{1-x^{\ell+1}}(1-x^{\ell+1})\sum_{n=0}^{\infty} a_n x^{(\ell+1)n}\\ &= \frac{1}{\ell+1} = \int_0^1 t^\ell\, dt.\end{aligned}$$

(2) We next show that $p(x)$ in (11.29) can be replaced by a function $g(x)$ which is continuous in the interval $(0, 1)$. We apply the approximation theorem of Weierstrass: Given $\epsilon > 0$, there are two polynomials $p_1(t)$, $p_2(t)$ such that

$$p_1(t) < g(t) < p_2(t)$$

and

$$\int_0^1 [g(t) - p_1(t)] dt < \epsilon, \quad \int_0^1 [p_2(t) - g(t)] dt < \epsilon.$$

Therefore

$$(1-x)\sum_{n=0}^{\infty} a_n x^n p_1(x^n) < (1-x)\sum_{n=0}^{\infty} a_n x^n g(x^n) < (1-x)\sum_{n=0}^{\infty} a_n x^n p_2(x^n),$$

and as $x \to 1$ the left-hand side and the right-hand side become

$$\int_0^1 p_1(t) dt, \quad \int_0^1 p_2(t) dt.$$

We deduce that

$$\int_0^1 g(t)dt - \epsilon \leq \liminf_{x\to 1}(1-x)\sum_{n=0}^{\infty} a_n x^n g(x^n)$$
$$\leq \limsup_{x\to 1}(1-x)\sum_{n=0}^{\infty} a_n x^n g(x^n) \leq \int_0^1 g(t)dt + \epsilon.$$

The required result follows.

(3) Let

$$h(t) = \begin{cases} 0, & \text{if } 0 \leq t < e^{-1}, \\ \dfrac{1}{t}, & \text{if } 0 \leq e^{-1} \leq t \leq 1, \end{cases}$$

a function with a discontinuity at $t = e^{-1}$. Given $\epsilon > 0$, we can construct two continuous functions $g_1(t)$ and $g_2(t)$ so that

$$g_1(t) \leq h(t) \leq g_2(t)$$

and

$$\int_0^1 [h(t) - g_1(t)]dt < \epsilon, \qquad \int_0^1 [g_2(t) - h(t)]dt < \epsilon,$$

and in this way we prove that

$$\lim_{x\to 1}(1-x)\sum_{n=0}^{\infty} a_n x^n h(x^n) = \int_0^1 h(t)dt = \int_{e^{-1}}^1 \frac{dt}{t} = 1,$$

that is

$$(1-x)\sum_{n\leq 1/\log(1/x)} a_n \sim 1.$$

Taking $x = e^{-1/N}$, we then have

$$\frac{1}{N}\sum_{n\leq N} a_n \sim 1.$$

The theorem is proved. □

Theorem 11.35 *Suppose that* $a_n \geq 0$, $\alpha > 1$ *and*

$$f(x) \sim (1-x)^{-\alpha} \quad \textit{as} \quad x \to 1.$$

Then

$$s_n \sim \frac{n^\alpha}{\Gamma(1+\alpha)} \quad \textit{as} \quad n \to \infty.$$

Proof. Let

$$f_{\alpha-1}(x) = \frac{1}{\Gamma(\alpha-1)} \int_0^x (x-t)^{\alpha-2} f(t)dt$$

(this is called the fractional integral of order $\alpha - 1$ for $f(x)$). By hypothesis,

$$\begin{aligned} f_{\alpha-1}(x) &= \frac{1}{\Gamma(\alpha-1)} \sum_{n=0}^{\infty} a_n \int_0^x (x-t)^{\alpha-2} t^n \, dt \\ &= \sum_{n=0}^{\infty} a_n \frac{\Gamma(n+1)}{\Gamma(n+\alpha)} x^{\alpha+n-1} \sim \sum_{n=1}^{\infty} \frac{a_n x^n}{n^{\alpha-1}}, \end{aligned}$$

and, on the other hand,

$$\begin{aligned} f_{\alpha-1}(x) &\sim \frac{1}{\Gamma(\alpha-1)} \int_0^x (x-t)^{\alpha-2}(1-t)^{-\alpha} \, dt \\ &= \frac{x^{\alpha-1}}{\Gamma(\alpha-1)} \int_0^1 (1-u)^{\alpha-2}(1-xu)^{-\alpha} \, du \\ &= \frac{x^{\alpha-1}}{\Gamma(\alpha-1)} \sum_{\ell=0}^{\infty} \frac{\alpha(\alpha+1)\cdots(\alpha+\ell-1)}{\ell!} x^\ell \int_0^1 (1-u)^{\alpha-2} u^\ell \, du \\ &= \frac{x^{\alpha-1}}{\Gamma(\alpha-1)} \sum_{\ell=0}^{\infty} \frac{\alpha(\alpha+1)\cdots(\alpha+\ell-1)}{\ell!} \frac{\Gamma(\alpha-1)\Gamma(\ell+1)}{\Gamma(\alpha+\ell)} x^\ell \\ &= \frac{x^{\alpha-1}}{\Gamma(\alpha)(1-x)} \sim \frac{1}{\Gamma(\alpha)(1-x)}, \end{aligned}$$

so that

$$\sum_{\nu=1}^{n} \frac{a_\nu}{\nu^{\alpha-1}} \sim \frac{n}{\Gamma(\alpha)}.$$

The proof of the theorem follows from this and the following lemma. □

Lemma 11.36 *Let $\alpha > 0$. Then*

$$\sum_{n=1}^{N} \frac{a_n}{n^{\alpha-1}} \sim \frac{N}{\Gamma(\alpha)}$$

implies

$$\sum_{n=1}^{N} a_n \sim \frac{N^\alpha}{\Gamma(\alpha+1)}.$$

Proof. Let

$$t_0 = 0, \quad t_N = \sum_{n=1}^{N} \frac{a_n}{n^{\alpha-1}}, \quad N = 1, 2, \ldots .$$

Then, by partial summation,

$$\sum_{n=1}^{N} a_n = \sum_{n=1}^{N} (t_n - t_{n-1}) n^{\alpha-1} = \sum_{n=1}^{N-1} t_n (n^{\alpha-1} - (n+1)^{\alpha-1}) + t_N N^{\alpha-1}.$$

By hypothesis, given $\epsilon > 0$, there exists X such that, for $N > X$,

$$\left| t_N - \frac{N}{\Gamma(\alpha)} \right| < \epsilon N,$$

so that there exists $c > 0$ such that

$$\begin{aligned}
&\left| \sum_{n=1}^{N-1} \left(t_n - \frac{n}{\Gamma(\alpha)} \right) (n^{\alpha-1} - (n+1)^{\alpha-1}) + \left(t_n - \frac{n}{\Gamma(\alpha)} \right) N^{\alpha-1} \right| \\
&\leq c \sum_{n=1}^{X} n |n^{\alpha-1} - (n+1)^{\alpha-1}| + \epsilon \sum_{n=X+1}^{N-1} n |n^{\alpha-1} - (n+1)^{\alpha-1}| + \epsilon N^{\alpha} \\
&\leq c \sum_{n=1}^{X} n |n^{\alpha-1} - (n+1)^{\alpha-1}| + \epsilon \left| \sum_{n=X+1}^{N-1} n^{\alpha} - \sum_{n=X+2}^{N} (n-1) n^{\alpha-1} \right| + \epsilon N^{\alpha}.
\end{aligned}$$

Therefore, by the partial summation equation,

$$\begin{aligned}
\sum_{n=1}^{N} a_n &= \frac{1}{\Gamma(\alpha)} \left\{ \sum_{n=1}^{N-1} n [n^{\alpha-1} - (n+1)^{\alpha-1}] + N^{\alpha} \right\} + o(N^{\alpha}) \\
&= \frac{1}{\Gamma(\alpha)} \left\{ \sum_{n=1}^{N} n^{\alpha} - \sum_{n=2}^{N} (n-1) n^{\alpha-1} \right\} + o(N^{\alpha}) \\
&= \frac{1}{\Gamma(\alpha)} \sum_{n=1}^{N} n^{\alpha-1} + o(N^{\alpha}).
\end{aligned}$$

Since

$$\sum_{n=1}^{N} n^{\alpha-1} \sim \frac{N^{\alpha}}{\alpha},$$

the lemma (and Theorem 11.35) are proved. □

11.13 Littlewood's Tauberian theorem

Theorem 11.37 (Littlewood) *Suppose that*

$$\lim_{x \to 1} \sum_{n=0}^{\infty} a_n x^n = s \tag{11.30}$$

and that $a_n = O(1/n)$. *Then*

$$\sum_{n=0}^{\infty} a_n = s. \tag{11.31}$$

In other words, from

$$\sum_{n=0}^{\infty} a_n = s,\ (A),$$

and $a_n = O(1/n)$, we can deduce that the series converges to s.

We first establish a lemma.

Lemma 11.38 *Suppose that $f(x)$ is twice differentiable in $0 \le x \le 1$, and that, as $x \to 1$,*

$$f(x) = o(1), \quad f''(x) = O\Big(\frac{1}{(1-x)^2}\Big).$$

Then

$$f'(x) = o\Big(\frac{1}{1-x}\Big).$$

Proof. Let $x' = x + \delta(1-x)$, where $0 < \delta < \frac{1}{2}$. Then

$$f(x') = f(x) + \delta(1-\delta)f'(x) + \frac{1}{2}\delta^2(1-x)^2 f''(\xi),$$

where $x < \xi < x'$. Therefore,

$$\begin{aligned}(1-x)f'(x) &= \frac{f(x') - f(x)}{\delta} - \frac{1}{2}\delta(1-x)^2 f''(\xi)\\ &= \frac{f(x') - f(x)}{\delta} + O(\delta);\end{aligned} \tag{11.32}$$

we have made use of

$$f''(\xi) = O\Big(\frac{1}{(1-\xi)^2}\Big) = O\Big(\frac{1}{(1-x')^2}\Big) = O\Big(\frac{1}{(1-x)^2}\Big).$$

Taking δ to be sufficiently small, and then letting $x \to 1$, we find that the left-hand side of (11.32) can be make arbitrarily small. The lemma is proved. □

Proof of Theorem 11.37. (The reader should appreciate that the main difficulty for the proof has been dealt with in Theorem 11.33.) We may assume, without loss, that $s = 0$, so that

$$f(x) = \sum_{n=0}^{\infty} a_n x^n = o(1), \quad x \to 1.$$

From $a_n = O(1/n)$, we have

$$f''(x) = \sum_{n=2}^{\infty} n(n-1)a_n x^{n-2} = O\Big(\sum_{n=2}^{\infty}(n-1)a_n x^{n-2}\Big) = O\Big(\frac{1}{(1-x)^2}\Big),$$

and so, by Lemma 11.38,

$$f'(x) = o\Big(\frac{1}{1-x}\Big).$$

Suppose that $|na_n| \le c$. Then

$$\sum_{n=1}^{\infty}\Big(1 - \frac{na_n}{c}\Big)x^{n-1} = \frac{1}{1-x} - \frac{f'(x)}{c} \sim \frac{1}{1-x}.$$

The coefficients for the power series are positive (we may assume that a_n are real) so that, by Theorem 11.33, we have

$$\sum_{\nu=1}^{n}\Big(1 - \frac{\nu a_\nu}{c}\Big) \sim n, \quad n \to \infty,$$

giving

$$\sum_{\nu=1}^{n} \nu a_\nu = o(n). \tag{11.33}$$

Let

$$\begin{aligned}
w_n &= \sum_{\nu=1}^{n} \nu a_\nu, \quad n > 0,\ w_0 = 0,\\
f(x) - a_0 &= \sum_{n=1}^{\infty} \frac{w_n - w_{n-1}}{n} x^n = \sum_{n=1}^{\infty} w_n\Big(\frac{x^n}{n} - \frac{x^{n+1}}{n+1}\Big)\\
&= \sum_{n=1}^{\infty} w_n\Big(\frac{x^n - x^{n+1}}{n+1} + \frac{x^n}{n(n+1)}\Big)\\
&= (1-x)\sum_{n=1}^{\infty}\frac{w_n}{n+1}x^n + \sum_{n=1}^{\infty}\frac{w_n}{n(n+1)}x^n.
\end{aligned}$$

From $w_n = o(n)$ we find that the first term on the right-hand side tends to 0 as $x \to 1$. Since we know that $f(x) \to 0$ as $x \to 1$, it follows that

$$\sum_{n=1}^{\infty}\frac{w_n}{n(n+1)}x^n \to -a_0,$$

and because the coefficients are $o(1/n)$, the original theorem by Tauber yields

$$\sum_{n=1}^{\infty} \frac{w_n}{n(n+1)} = -a_0.$$

The required result now follows, because the left-hand side here is given by

$$\lim_{N\to\infty} \sum_{n=1}^{N} \frac{w_n}{n(n+1)} = \lim_{N\to\infty} \sum_{n=1}^{N} w_n\left(\frac{1}{n} - \frac{1}{n+1}\right)$$

$$= \lim_{N\to\infty} \left(\sum_{n=1}^{N} \frac{w_n - w_{n-1}}{n} - \frac{w_N}{N+1} \right) = \lim_{N\to\infty} \sum_{n=1}^{N} a_n.$$

□

Remark The result here obviously includes the result in Section 11.9, with the corresponding $O(1/n)$, but the thought process would involve knowledge of what is in Section 11.9 in order to arrive at such a deep result.

11.14 Being analytic and being convergent

The function defined by the series

$$f(z) = \sum_{n=0}^{\infty} (-1)^n z^n = \frac{1}{1+z}$$

is regular at $z = 1$, although at this point the series

$$\sum_{n=0}^{\infty} (-1)^n$$

is divergent. Nevertheless, we do have

$$\sum_{n=0}^{\infty} (-1)^n = \frac{1}{2}, \ (C, 1).$$

Take the function

$$\frac{1}{(1+z)^n}$$

and consider the following situation: There is a function $f(z)$ which is regular at $z = 1$, but the power series representation is not summable (C, k) (for $k < n$). Abel summation may best reflect the analytic properties in that if $f(z)$ is regular at $z = 1$ then $\lim_{z\to 1} f(z)$ exists (conversely, there are examples in which $\lim_{z\to 1} f(z)$ exists, but $f(z)$ is not analytic at $z = 1$).

Our present problem is concerned with the condition that has to be imposed so that there is convergence (or summability) at a regular point.

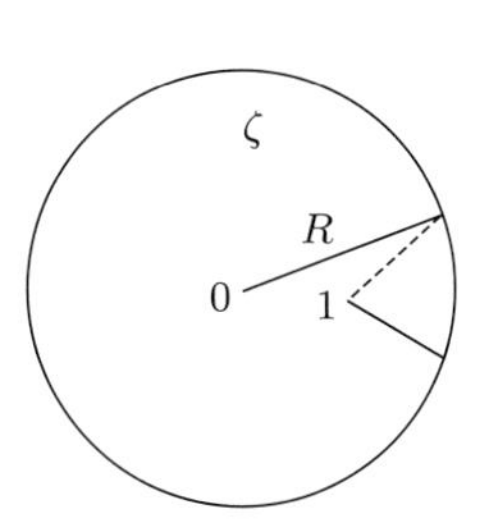

Figure 11.1.

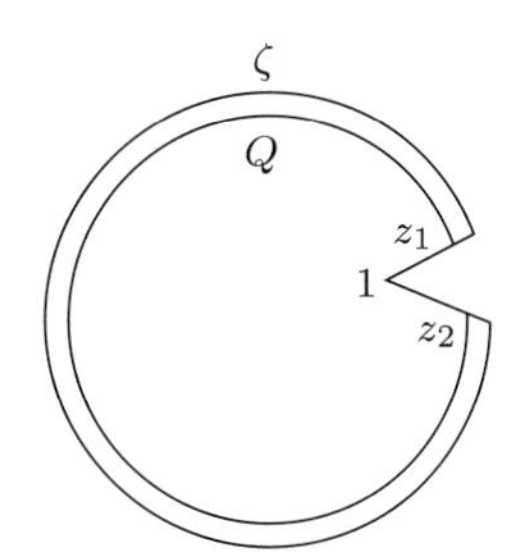

Figure 11.2.

Theorem 11.39 (M. Riesz) *Let ζ represent the fan-shaped region (see Fig. 11.1):*

$$|z| \leq R, \quad \theta \leq \arg(z-1) \leq 2\pi - \theta \quad (R > 1,\ 0 < \theta < \pi/2).$$

Suppose that $f(z)$ is continuous in ζ, and is regular everywhere except $z = 1$. Then the series

$$f(z) = \sum_{n=0} a_n z^n$$

converges uniformly on $|z| = 1$.

Proof. We need the following lemma.

Lemma 11.40 *If $a_n = o(1/n)$, and $\lim_{r \to 1} f(re^{i\phi}) = g(\phi)$ (exists) with the limit being uniform in ϕ, then $\sum_n a_n e^{in\phi}$ converges uniformly to $g(\phi)$.*

The proof of the lemma is the same as that for Tauber's theorem (but note the uniformity aspect). By this lemma, we need only establish that $a_n = o(1/n)$.

Assume, without loss of generality, that $f(1) = 0$. Let M be the maximum of $|f(z)|$ in ζ. Given $\delta > 0$ we can take $r(= r(\delta))$ so that on the straight line boundary of (see Fig. 11.2)

$$Q : |z| \leq r, \quad \theta \leq \arg(z-1) \leq 2\pi - \theta,$$

we have

$$|f(z)| < \delta.$$

Since $f(z)$ is continuous in Q, by Cauchy's theorem

$$a_n = \frac{1}{2\pi i}\left(\int_1^{z_1} \frac{f(z)}{z^{n+1}}\, dz + \int_{z_1}^{z_2} \frac{f(z)}{z^{n+1}}\, dz + \int_{z_2}^{1} \frac{f(z)}{z^{n+1}}\, dz \right).$$

When $n > 0$ we have

$$\begin{aligned}
\left| \int_1^{z_1} \frac{f(z)}{z^{n+1}} \, dz \right| &\le \delta \int_0^{|z_1 \;\; 1|} \frac{dy}{|1 + e^{i\theta} y|^{n+1}} \\
&\le \delta \int_0^{|z_1 - 1|} \frac{dy}{(1 + y \cos\theta)^{n+1}} \\
&< \delta \int_0^{\infty} \frac{dy}{(1 + y \cos\theta)^{n+1}} = \frac{\delta}{n \cos\theta},
\end{aligned}$$

and similarly

$$\left| \int_{z_2}^{1} \frac{f(z)}{z^{n+1}} \, dz \right| < \frac{\delta}{n \cos\theta}.$$

On the circular arc from z_1 to z_2, we have

$$\left| \int_{z_1}^{z_2} \frac{f(z)}{z^{n+1}} \, dz \right| \le 2\pi r \frac{M}{r^{n+1}} = \frac{2\pi M}{r^n},$$

so that

$$|a_n| < \frac{\delta}{\pi \cos\theta} \frac{1}{n} + \frac{M}{r^n}.$$

Thus, for any $\delta > 0$, we have

$$\limsup_{n\to\infty} n|a_n| \le \frac{\delta}{\pi \cos\theta}.$$

Therefore $a_n = o(1/n)$, and the theorem is proved. □

Theorem 11.41 (Fejér) *If*

$$\sum_{n=0}^{\infty} n|a_n|^2$$

converges, then it can be deduced from

$$\sum_{n=0}^{\infty} a_n e^{in\theta} = s(\theta), \ (A), \tag{11.34}$$

that

$$\sum_{n=0}^{\infty} a_n e^{in\theta} = s(\theta). \tag{11.35}$$

Moreover, if there is uniform convergence in a certain region for (11.34) then there is also uniform convergence in the same region for (11.35).

Proof. The theorem is obviously true for a polynomial. Let

$$\sum_{n=\nu}^{\infty} n|a_n|^2 = \epsilon_\nu,$$

so that $\epsilon_\nu > 0$ and $\epsilon_\nu \to 0$ as $\nu \to \infty$. Let ν be so large that

$$r_\nu = 1 - \frac{\sqrt{\epsilon_\nu}}{\nu} > 0,$$

and so, for all real ϕ,

$$\begin{aligned}\left|\sum_{n=0}^{\nu} a_n e^{n\phi i} - f(r_\nu e^{\phi i})\right| &= \left|\sum_{n=0}^{\nu} a_n e^{n\phi i}(1 - r_\nu^n) - \sum_{n=\nu+1}^{\infty} a_n r_\nu^n e^{n\phi i}\right| \\ &\le (1 - r_\nu)\sum_{n=0}^{\nu} n|a_n| + \sum_{n=\nu+1}^{\infty} |a_n| r_\nu^n.\end{aligned}$$

The bound on the right-hand side is independent of ϕ, so that the theorem will follow if we can show that it tends to 0 as $\nu \to \infty$.

By the Cauchy–Schwarz inequality the bound is

$$\begin{aligned}&\le (1 - r_\nu)\sum_{n=0}^{\nu} \sqrt{n}\cdot\sqrt{n}|a_n| + \frac{1}{\sqrt{\nu}}\sum_{n=\nu+1}^{\infty} \sqrt{n}|a_n| r_\nu^n \\ &\le (1 - r_\nu)\sqrt{\sum_{n=0}^{\nu} n \sum_{n=0}^{\nu} n|a_n|^2} + \frac{1}{\sqrt{\nu}}\sqrt{\sum_{n=\nu+1}^{\infty} n|a_n|^2 \sum_{n=\nu+1}^{\infty} r_\nu^{2n}} \\ &\le (1 - r_\nu)\sqrt{\nu^2\cdot\epsilon_0} + \frac{1}{\sqrt{\nu}}\sqrt{\frac{\epsilon_\nu}{1 - r_\nu}} = \sqrt{\epsilon_0}\sqrt{\epsilon_\nu} - \sqrt[4]{\epsilon_\nu} \to 0.\end{aligned}$$

The theorem is proved. □

From this, we deduce the following.

Theorem 11.42 *If*

$$\sum_{n=0}^{\infty} n|a_n|^2$$

converges and $f(z)$ is continuous in $|z| \le 1$ and regular in $|z| < 1$, then $\sum_{n=1}^{\infty} a_n z^n$ converges uniformly on the circumference of the circle.

Theorem 11.43 *If $f(z)$ is continuous in $|z| \le 1$ and univalent in $|z| < 1$, then its power series converges uniformly on the circle $|z| = 1$.*

Proof. It suffices to establish that

$$\sum_{n=0}^{\infty} n|a_n|^2 < \infty.$$

The transformation $w = f(z)$ maps the disc $|z| = r$ $(0 < r < 1)$ into a region with an area given by

$$\iint_{|z|\le r} |f'(x+iy)|^2\,dx\,dy = \int_0^r \rho\,d\rho \int_0^{2\pi} |f'(re^{i\theta})|^2\,d\theta$$

$$= \int_0^r \rho\,d\rho \int_0^{2\pi} \sum_{n=0}^{\infty} na_n\rho^{n-1}e^{i(n-1)\theta} \sum_{m=0}^{\infty} m\bar{a}_m\rho^{m-1}e^{-i(m-1)\theta}\,d\theta$$

$$= 2\pi \int_0^r \rho \sum_{n=0}^{\infty} n^2|a_n|^2\rho^{2n-2}\,d\rho = \pi \sum_{n=0}^{\infty} n^2|a_n|^2 r^{2n}.$$

The required result follows. □

11.15 The Borel polygon

The B'-sum of a power series

$$f(z) = \sum_{n=0}^{\infty} a_n z^n$$

is given by

$$\int_0^{\infty} e^{-t} \sum_{n=0}^{\infty} \frac{a_n(zt)^n}{n!}\,dt.$$

For example: Taking $f(z) = \sum z^n$, its B'-sum is given by

$$\int_0^{\infty} e^{-t}e^{zt}\,dt = \frac{1}{1-z}.$$

The equation is valid for $\Re z < 1$, so that the B'-sum for $\sum z^n$ spills over the disc of convergence $|z| < 1$ for the function into the half-plane $\Re z < 1$, giving an analytic expression there. Therefore Borel summability can be used as a tool for analytic continuation, and the purpose of this section is to investigate the extent for such continuation to be valid.

Theorem 11.44 *Let a power series be summable (B') at a point P. Then it is summable (B') at every point on the line segment OP. If Q is a point between O and P then the series is uniformly summable on QP.*

Proof. There being no assumption made on the circle of convergence, it does not matter if we assume that the point P is $z = 1$. Let

$$a(t) = \sum_{n=0}^{\infty} \frac{a_n t^n}{n!}$$

and

$$J(z) = \int_0^{\infty} e^{-t} a(zt) dt. \tag{11.36}$$

We proceed to prove that if (11.36) converges at $z = 1$ then it converges for $0 < z \leq 1$, and that, for any $\delta > 0$, the convergence is uniform in $\delta \leq z \leq 1$. Write

$$J(z) = \frac{1}{z} \int_0^{\infty} e^{-t/z} a(t) dt = \frac{K(z)}{z}, \quad 0 < z \leq 1. \tag{11.37}$$

Let

$$s = \frac{1}{z} - 1.$$

Then

$$K(z) = \int_0^{\infty} e^{-st} e^{-t} a(t) dt,$$

where the integral converges uniformly in $s \geq 0$. The theorem is proved. □

The full situation has not been disclosed in Theorem 11.44, because $J(z)$ actually converges uniformly in $0 \leq x \leq 1$. However, because of the presence of $1/z$ in (11.37) in the proof it is difficult to derive such a conclusion.

Theorem 11.45 *If $\sum a_n z^n$ is summable (B') at a point P then it is summable (B') uniformly on the line segment OP.*

Proof. The sum is the same as before, and we assume that the point P is $z = 1$, and we may also assume that a_n are real. We then have to prove the following: For any $\epsilon > 0$ there exists $H_0(\epsilon)$ such that if $H' > H \geq H_0(\epsilon)$ and $0 \leq z \leq 1$, then we have

$$|I| = |I(z, H, H')| = \left| \int_H^{H'} e^{-t} a(zt) dt \right| < \epsilon.$$

By Theorem 11.44 we already have uniform convergence in $(\frac{1}{2}, 1)$, so we may assume that $0 \leq z \leq \frac{1}{2}$. There are now three cases to consider: (i) $H'z \leq 1$; (ii) $Hz < 1 < H'z$, (iii) $Hz \geq 1$.

We deal with case (ii) only, because (i) and (iii) are simpler by comparison. It does not matter if we suppose that $H \geq 2$. Write

$$I = \int_H^{1/z} e^{-t} a(zt) dt + \int_{1/z}^{H'} e^{-t} a(zt) dt = I_1 + I_2,$$

and let

$$M = \max_{0 \le t \le 1} |a(t)|, \quad N = \max_{T > 1} \int_1^T e^{-t} a(t) dt,$$

so that

$$|I_1| \le M \int_H^\infty e^{-t}\, dt = Me^{-H},$$

$$I_2 = \frac{1}{z} \int_1^{H'z} e^{-t/z} a(t) dt = \frac{1}{z} \int_1^{H'z} e^{-st} e^{-t} a(t) dt$$

$$= \frac{e^{-s}}{z} \int_1^T e^{-t} a(t) dt,$$

where $s = 1/z - 1$ and $1 < T < H'z$. From $0 < z \le \frac{1}{2}$ and $2 \le H \le 1/z$, we have

$$|I_2| \le \frac{N}{z} e^{1-1/z} \le \frac{N}{z} e^{-1/(2z)} \le NHe^{-(1/2)H}$$

($ue^{-u/2}$ is decreasing for $u > 2$). Therefore, when $H \ge H_0(\epsilon)$,

$$|I| \le Me^{-H} + NHe^{-(1/2)H} < \epsilon.$$

□

Theorem 11.46 *If $\sum a_n z^n$ is summable (B') at a point P, then the sum is an analytic function of z on OP, which is regular inside the circle with OP as diameter.*

Proof. We continue to take $z = 1$ for the point P. From consideration of

$$J(z) = \int_0^\infty e^{-t} a(tz) dt = \frac{K(z)}{z},$$

we need to establish the following: The integral $J(z)$ is uniformly convergent in a region D formed by any two circular arcs through OP, making an acute angle η with OP. Write

$$z = re^{i\theta}, \quad s = 1/z - 1 = \rho e^{i\phi}.$$

Since $K(s)$ converges at $s = 0$, there is uniform convergence in $|\phi| < \eta$. The two sides subtending the angle correspond to the boundary arcs for D and the interior corresponds to that of D, so that $K(z)$ converges uniformly in D. The theorem is proved. □

The Borel sum of

$$f(z) = (c - z)^{-1} = \sum_{n=0}^\infty c^{-n-1} z^n,$$

being

$$J(z) = c^{-1} \int_0^\infty e^{-t(1-z/c)} dt,$$

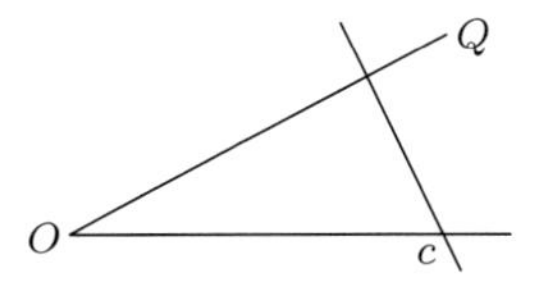

Figure 11.3.

is convergent for $\Re(z/c) < 1$. Taking a line L_c through $z = c$ perpendicular to the direction of c, we find that z and O lie on the same side of L_c. Such a consideration leads us to the following notion.

Let L_c pass through a singularity $z = c$ of $f(z)$. There is then a region D associated with all such L_c with the following property: Corresponding to each L_c, an interior point of D and O lies on the same side of L_c. Such a region is called the Borel polygon for $f(z)$.

It is not difficult to deduce the following theorem from Theorem 11.46.

Theorem 11.47 *The series for $f(z)$ is summable (B') at any point within the Borel polygon of $f(z)$, and not outside the polygon.*

There is no problem for the first part of the theorem, so we consider the remaining part. Let Q be a point lying outside the polygon (see Fig. 11.3). Then OQ must intersect with a certain L_c. If the series were summable (B') at Q then, by Theorem 11.46, it would be summable (B') inside the circle with OQ as diameter. Since c lies inside such a circle, the series would then be summable at c, and this contradicts c being a singular point.

12

Harmonic functions satisfying various types of boundary conditions

12.1 Introduction

Let $\mathcal{D}$ be a region with a closed curve C as its boundary. In this chapter we assume that $\mathcal{D}$ is simply connected and we let $\bar{\mathcal{D}}$ denote the closure of $\mathcal{D}$ (that is, $\mathcal{D} \cup C$). For the sake of simplicity, we also assume that the curve C has the parametric formulae

$$x = \phi(t), \quad y = \psi(t),$$

where ϕ, ψ are functions with continuous derivatives; we often use ζ to denote a variable point on C.

A function u satisfying Laplace's equation

$$\Delta u = \frac{\partial^2 u}{\partial x^2} + \frac{\partial^2 u}{\partial y^2} = 0 \tag{12.1}$$

in $\mathcal{D}$ is said to be harmonic in $\mathcal{D}$.

Concerning functions harmonic in a region satisfying boundary conditions, we have already discussed the following.

(I) Dirichlet's problem. In this problem we seek a solution $u(z)$ which is continuous in $\bar{\mathcal{D}}$, harmonic in $\mathcal{D}$ and takes on the values of a given function $\phi(\zeta)$ on C. We then simply write

$$\Delta u = 0, \quad u|_C = \phi(\zeta); \tag{12.2}$$

we shall use such a notation, always bearing in mind that $\chi|_C$ is meaningful only when χ is continuous in $\bar{\mathcal{D}}$ (we do not discuss the situation when there is a discontinuity on the boundary).

We have already dealt with this problem, and we know that Dirichlet's problem does have a solution, and is unique.

(II) Neumann's problem. This is the problem

$$\Delta u = 0, \quad \left.\frac{\partial u}{\partial n}\right|_C = \psi(\zeta). \tag{12.3}$$

A necessary and sufficient condition for this problem to have a solution is that

$$\int_C \psi(\zeta) d\zeta = 0; \tag{12.4}$$

because, from Green's formula,

$$\int_C \frac{\partial u}{\partial n} d\zeta = \iint_{\mathcal{D}} \Delta u \, dx \, dy = 0.$$

Although there is no unique solution, any two solutions to a problem only differ by a constant term.

In mathematical physics these two problems have the most extensive applications. However, there are times when we have to deal with the following types of mixed boundary problems.

(III) Certain parts of C have given values for u, and other parts have given values for $\partial u / \partial n$. More specifically, take $2n$ points

$$a_1, b_1, a_2, b_2, \ldots, a_n, b_n$$

along C, and we seek a harmonic function satisfying

$$u = \phi(\zeta) \quad \text{if } \zeta \in (a_k, b_k), \qquad \frac{\partial u}{\partial n} = \psi(\zeta) \quad \text{if } \zeta \in (b_k, a_{k+1}), \tag{12.5}$$

for $k = 1, 2, \ldots, n$ and $a_{n+1} = a_1$.

Another generalisation of (II) is as follows.

(IV) Mixed boundary problem.

$$\Delta u = 0, \quad a(\zeta) \frac{\partial u}{\partial x} + b(\zeta) \frac{\partial u}{\partial y}\bigg|_C = c(\zeta). \tag{12.6}$$

The second equation can be replaced by

$$\frac{\partial u}{\partial n} + h(\zeta)(u - p(\zeta))\big|_C = 0. \tag{12.7}$$

In this chapter we also deal with, in passing, the following.

(V) Doubly harmonic equation,

$$\Delta \Delta u = \frac{\partial^4 u}{\partial x^4} + 2 \frac{\partial^4 u}{\partial x^2 \partial y^2} + \frac{\partial^4 u}{\partial y^4} = 0, \tag{12.8}$$

$$u|_C = \phi(\zeta), \quad \frac{\partial u}{\partial n}\bigg|_C = \psi(\zeta), \tag{12.9}$$

and

(VI) Mixed forms of the partial differential equation

$$\frac{\partial^u}{\partial x^2} + \theta(y)\frac{\partial^2 u}{\partial y^2} = 0, \quad \theta(y) = \begin{cases} 1, & \text{if } y > 0, \\ -1, & \text{if } y < 0. \end{cases}$$

Since Laplace's equation is invariant under a conformal transformation, we only deal with the case when $\mathcal{D}$ is the unit disc or the upper half-plane in this chapter. In practice there is usually no real loss of generality.

The chapter is divided into three parts. In the first part, we try to deal with problems that can be solved without the introduction of new tools. We then use Cauchy's integral to tackle certain problems, and then make use of the Keldysh–Sedov formula to deal with certain other problems.

12.2 Poisson's equation

Poisson's equation

$$\Delta u = \rho(x, y) \tag{12.10}$$

may be considered as a generalisation of Laplace's equation, but in fact if (12.10) has a solution u_1, then the problem immediately reduces to Laplace's equation

$$\Delta(u - u_1) = 0,$$

together with whatever attendant boundary conditions. We now prove that

$$u(x, y) = \frac{1}{2\pi} \iint_{\mathcal{D}} \rho(\xi, \eta) \log \frac{1}{\sqrt{(x-\xi)^2 + (y-\eta)^2}} \, d\xi \, d\eta \tag{12.11}$$

is a solution to (12.10), and we shall not, henceforth, be particularly concerned with Poisson's equation.

Suppose that $\rho(\xi, \eta)$ is a continuous function of (ξ, η), so that $|\rho(\xi, \eta)| \leq C$.[1] For the sake of simplicity, we shall write $r = \sqrt{(x-\xi)^2 + (y-\eta)^2}$, noting that $\log 1/r$ is a function with a singularity at $x = \xi$, $y = \eta$. It is not difficult to show that, besides this singular point,

$$\frac{\partial}{\partial x} \log \frac{1}{r} = -\frac{x-\xi}{r^2}, \quad \frac{\partial}{\partial y} \log \frac{1}{r} = -\frac{y-\eta}{r^2},$$

$$\Delta \log \frac{1}{r} = 0. \tag{12.12}$$

Therefore, if (x, y) lies outside $\mathcal{D}$ then we can differentiate under the integral sign to give

$$\Delta u(x, y) = 0.$$

[1] Translator's footnote: The number C here, and in what follows, is not to be confused with the same symbol C use for the curve.

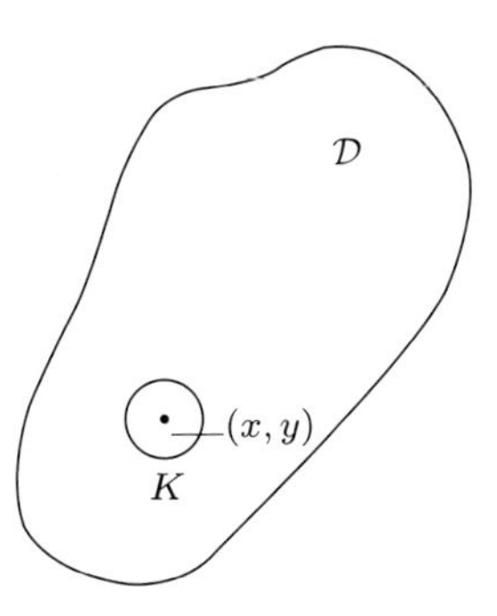

Figure 12.1.

When (x, y) is inside $\mathcal{D}$ the integral (12.11) is improper, and

$$\begin{cases} X = \dfrac{1}{2\pi} \displaystyle\iint_{\mathcal{D}} \rho(\xi, \eta) \frac{\partial}{\partial x} \log \frac{1}{r}\, d\xi\, d\eta = -\frac{1}{2\pi} \iint_{\mathcal{D}} \rho(\xi, \eta) \frac{x-\xi}{r^2}\, d\xi\, d\eta, \\ Y = \dfrac{1}{2\pi} \displaystyle\iint_{\mathcal{D}} \rho(\xi, \eta) \frac{\partial}{\partial y} \log \frac{1}{r}\, d\xi\, d\eta = -\frac{1}{2\pi} \iint_{\mathcal{D}} \rho(\xi, \eta) \frac{y-\eta}{r^2}\, d\xi\, d\eta, \end{cases} \tag{12.13}$$

are also improper, but it is not difficult to show that they are convergent.

(1) We first prove that

$$X = \frac{\partial u}{\partial x}, \quad Y = \frac{\partial u}{\partial y}. \tag{12.14}$$

That is, we prove the following: Given any $\epsilon > 0$, we may take $\delta > 0$ such that

$$\left| \frac{u(x + \Delta x, y) - u(x, y)}{\Delta x} - X \right| < \epsilon, \tag{12.15}$$

whenever $|\Delta x| < \delta$.

Take a disc K centre (x, y) and radius τ small enough so that it lies inside $\mathcal{D}$ (see Fig. 12.1). Divide the integral (12.11) into two parts: u_1 is taken over K, and u_2 is taken over $\mathcal{D} \setminus K$. The same meaning is to be given for X_1 and X_2.

From

$$\begin{aligned} \left| \frac{u(x + \Delta x, y) - u(x, y)}{\Delta x} - X \right| \leq{} & \left| \frac{u_2(x + \Delta x, y) - u_2(x, y)}{\Delta x} - X_2 \right| + |X_1| \\ & + \left| \frac{u_1(x + \Delta x, y) - u_1(x, y)}{\Delta x} \right|, \end{aligned}$$

where the three terms on the right-hand side will be denoted by J_1, J_2, J_3, we proceed to prove that δ can be so chosen that, for $|\Delta x| < \delta$, each such term is less than $\epsilon/3$.

Let us first consider $J_2 = |X_1|$. We have

$$|X_1| \leq \frac{1}{2\pi}\left|\iint_{r\leq\tau} \rho\frac{x-\xi}{r^2}\,d\xi\,d\eta\right| \leq \frac{C}{2\pi}\left|\iint_{r\leq\tau} \frac{x-\xi}{r^2}\,d\xi\,d\eta\right| \leq \frac{C}{2\pi}\iint_{r\leq\tau}\frac{d\xi\,d\eta}{r}$$
$$= \frac{C}{2\pi}\int_0^{2\pi}\int_0^{\tau} dr\,d\theta = \tau C,$$

and τ can be chosen so that $|X_1| < \epsilon/3$.

Next, we have

$$J_3 = \left|\frac{u_1(x+\Delta x, y) - u_1(x,y)}{\Delta x}\right|$$
$$= \frac{1}{|\Delta x| 2\pi}\left|\iint_K \rho \log\frac{\sqrt{(x-\xi)^2+(y-\eta)^2}}{\sqrt{(x+\Delta x-\xi)^2+(y-\eta)^2}}\,d\xi\,d\eta\right|,$$

and, by the triangle inequality,

$$\left|\sqrt{(x-\xi)^2+(y-\eta)^2} - \sqrt{(x+\Delta x-\xi)^2+(y-\eta)^2}\right| < |\Delta x|.$$

It then follows from $|\log(1+t)| \leq |t|$ that

$$\left|-\log\frac{\sqrt{(x-\xi)^2+(y-\eta)^2}}{\sqrt{(x+\Delta x-\xi)^2+(y-\eta)^2}}\right| \leq \frac{|\Delta x|}{r},$$

and hence

$$J_3 \leq \frac{C}{2\pi}\iint_K \frac{d\xi\,d\eta}{r},$$

which is again τC, and hence $J_3 < \epsilon/3$ for small τ.

Lastly, with τ fixed, since (x, y) lies inside K, and thus outside $\mathcal{D}\setminus K$, we have

$$\lim_{\Delta x\to 0}\frac{u_2(x+\Delta x, y) - u_2(x,y)}{\Delta x} = X_2,$$

so that δ can be so chosen that, for $|\delta x| < \delta$, we do have $J_3 < \epsilon/3$.

(2) In the proof set out in the above we can differentiate under the integral sign to obtain $\partial u/\partial x$, $\partial u/\partial y$. However, we cannot use the method to deliver the second derivatives because

$$\frac{1}{2\pi}\iint_{\mathcal{D}} \rho(\xi,\eta)\frac{\partial^2}{\partial x^2}\log r\,d\xi\,d\eta = \frac{1}{2\pi}\iint_{\mathcal{D}}\rho(\xi,\eta)\frac{(x-\xi)^2-(y-\eta)^2}{r^4}\,d\xi\,d\eta$$

does not converge.

Instead we now split u to become $u_1 + u_2$ with

$$u_1 = \frac{1}{2\pi}\iint_{r\leq\tau}\rho(\xi,\eta)\log\frac{1}{r}\,d\xi\,d\eta, \quad u_2 = \frac{1}{2\pi}\iint_{r>\tau}\rho(\xi,\eta)\log\frac{1}{r}\,d\xi\,d\eta.$$

After fixing τ, we can then differentiate u_2 under the integral sign and prove easily that $\Delta u_2 = 0$.

Let us now consider u_1. From (1) we already have

$$\frac{\partial u_1}{\partial x} = \frac{1}{2\pi}\iint_{r\le\tau} \rho(\xi,\eta)\frac{\partial}{\partial x}\log\frac{1}{r}\,d\xi\,d\eta = -\frac{1}{2\pi}\iint_{r\le\tau} \rho(\xi,\eta)\frac{\partial}{\partial \xi}\log\frac{1}{r}\,d\xi\,d\eta$$
$$= -\frac{1}{2\pi}\iint_{r\le\tau} \frac{\partial}{\partial\xi}\Big(\rho\log\frac{1}{r}\Big)d\xi\,d\eta + \frac{1}{2\pi}\iint_{r\le\tau}\log\frac{1}{r}\frac{\partial\rho}{\partial\xi}\,d\xi\,d\eta.$$

By Green's theorem we now have

$$\frac{\partial u_1}{\partial x} = -\frac{1}{2\pi}\int_{r\le\tau}\rho\log\frac{1}{r}\,d\eta + \frac{1}{2\pi}\iint_{r\le\tau}\log\frac{1}{r}\frac{\partial\rho}{\partial\xi}\,d\xi\,d\eta.$$

As in the proof in (1), we can differentiate this equation under the integral sign to give

$$\frac{\partial^2 u}{\partial x^2} = -\frac{1}{2\pi}\int_{r\le\tau}\rho\frac{\partial}{\partial x}\log\frac{1}{r}\,d\eta + \frac{1}{2\pi}\iint_{r\le\tau}\frac{\partial}{\partial x}\Big(\log\frac{1}{r}\Big)\frac{\partial\rho}{\partial\xi}\,d\xi\,d\eta = K_1 + K_2,$$

say. As $\tau \to 0$, we have

$$|K_2| \le C_1\iint_{r\le\tau}\frac{|x-\xi|}{r^2}\,d\xi\,d\eta \le C_1\iint_{r\le\tau}\frac{d\xi\,d\eta}{r} = o(\tau).$$

For K_1, we first examine the situation when $\rho = 1$: Let $\xi = x + r\cos\theta$, $\eta = y + r\sin\theta$. Then

$$-\frac{1}{2\pi}\int\frac{\partial}{\partial x}\log\frac{1}{r}\,d\eta = \frac{1}{2\pi}\int_0^{2\pi}\frac{x-\xi}{r^2}\,d\eta = \frac{1}{2\pi}\int_0^{2\pi}\cos^2\theta d\theta = \frac{1}{2}.$$

Thus, $K_1 \to \frac{1}{2}\rho(x,y)$ as $\tau \to 0$, and hence

$$\frac{\partial^2 u_1}{\partial x^2} + \frac{\partial^2 u_1}{\partial y^2} \to \rho(x,y).$$

Therefore the integral (12.11) satisfies

$$\frac{\partial^2 u}{\partial x^2} + \frac{\partial^2 u}{\partial y^2} = \rho(x,y).$$

12.3 Doubly harmonic equation

We seek $u(x, y)$ which is continuous in $\bar{\mathcal{D}}$ and with a continuous derivative $\partial u/\partial n$, satisfying

$$\Delta\Delta u = \frac{\partial^4 u}{\partial x^4} + 2\frac{\partial^4 u}{\partial x^\partial y^2} + \frac{\partial^4 u}{\partial y^4} = 0 \quad (\text{inside } \mathcal{D}), \tag{12.16}$$

$$u|_C = g(s), \quad \frac{\partial u}{\partial n}\Big|_C = h(s), \tag{12.17}$$

where $g(s)$ and $h(s)$ are functions on C, which has the curvilinear length s.

(1) Uniqueness of solution. Suppose that there is another solution u_1. Then $v = u - u_1$ satisfies

$$\Delta\Delta v = 0, \quad v|_C = 0, \quad \frac{\partial v}{\partial n}\Big|_C = 0. \tag{12.18}$$

Setting $\phi = v$, $\psi = \Delta v$ in Green's formula

$$\iint_{\mathcal{D}} (\psi\Delta\phi - \phi\Delta\psi) dx\, dy = \int_C \Big(\psi\frac{\partial\phi}{\partial n} - \phi\frac{\partial\psi}{\partial n}\Big) ds,$$

we have

$$\iint_{\mathcal{D}} (\Delta v)^2\, ds = 0,$$

and so $\Delta v = 0$. On applying $v|_C = 0$ we find that $v \equiv 0$, giving $u = u_1$.

(2) Using harmonic functions to represent doubly harmonic functions. If u_1, u_2 are two harmonic functions in $\mathcal{D}$ then $u = xu_1 + u_2$ must be a doubly harmonic function in $\mathcal{D}$.

We make use of

$$\Delta(\phi\psi) = \phi\Delta\psi + \psi\Delta\phi + 2\Big(\frac{\partial\phi}{\partial x}\frac{\partial\psi}{\partial x} + \frac{\partial\phi}{\partial y}\frac{\partial\psi}{\partial y}\Big)$$

to give

$$\Delta u = \Delta x u_1 = 2\frac{\partial u_1}{\partial x},$$

and from $\Delta u_1 = 0$ we have

$$\Delta\Delta u = 0.$$

(3) If any straight line parallel to the x-axis does not intersect C at more than two points then we have the following result as a converse: A function u which is doubly harmonic in $\mathcal{D}$ must be representable as $u = au_1 + u_2$, where u_1, u_2 are harmonic functions in $\mathcal{D}$.

Proof. Clearly it suffices to prove that there is a harmonic function u_1 such that $u - xu_1$ is also harmonic, that is

$$\Delta u_1 = 0, \quad \Delta u = \Delta(xu_1) = 2\frac{\partial u_1}{\partial x}. \tag{12.19}$$

The function

$$u_1^\star(x, y) = \int_{x_0}^{x} \frac{1}{2}\Delta u(\xi, y) d\xi$$

satisfies the second equation in (12.19), and

$$\frac{\partial}{\partial x}\Delta u_1^\star(x, y) = \Delta\frac{\partial}{\partial x}u_1^\star(x, y) = \frac{1}{2}\Delta\Delta u = 0,$$

so that $u_1^\star$ does not depend on x, and we may write

$$u_1^\star = v(y).$$

Obviously we can find a function $u_1^{\star\star}$, depending only on y, such that

$$\Delta u_1^{\star\star} = \frac{\partial^2}{\partial y^2} u_1^{\star\star} = -v(y).$$

Both equations in (12.19) are now satisfied when we set $u_1 = u_1^\star + u_1^{\star\star}$. □

(4) Let $\mathcal{D}$ be a star domain, so that there is an interior point, which we may assume to be the origin, with the property that any ray from the point intersects C at only one point. Then a doubly harmonic function in $\mathcal{D}$ must be representable as

$$u = (r^2 - r_0^2)u_1 + u_2, \quad r^2 = x^2 + y^2, \quad r_0 \text{ is a constant.}$$

Here, u_1, u_2 are harmonic functions, so that we have to prove that there is a harmonic function u_1 such that

$$\Delta(u - (r^2 - r_0^2)u_1) = 0.$$

We apply the identity

$$\Delta(\phi\psi) = \phi\Delta\psi + \psi\Delta\phi + 2\Big(\frac{\partial\phi}{\partial x}\frac{\partial\psi}{\partial x} + \frac{\partial\phi}{\partial y}\frac{\partial\psi}{\partial y}\Big)$$

and

$$\Delta r^2 = 4, \quad \frac{\partial u_1}{\partial r} = \frac{\partial u_1}{\partial x}\frac{\partial x}{\partial r} + \frac{\partial u_1}{\partial y}\frac{\partial y}{\partial r};$$

the proof is the same as that given earlier.

12.4 Formula for a doubly harmonic function in the unit disc

As far as the unit disc is concerned, we may assume that the doubly harmonic function being considered takes the form

$$u = (r^2 - 1)u_1 + u_2$$

with the boundary condition

$$u_2|_{r=1} = u|_{r=1} = g(\theta).$$

By Poisson's formula, we have

$$u_2 = \frac{1}{2\pi}\int_0^{2\pi} \frac{(1 - r^2)g(\psi)}{1 - 2r\cos(\theta - \psi) + r^2}\, d\psi, \tag{12.20}$$

and the second boundary condition takes the form

$$\frac{\partial u}{\partial n}\Big|_C = 2u_1 + \frac{\partial u_2}{\partial r}\Big|_{r=1} = h(\theta).$$

Since $2u_1 + \partial u_2/\partial r$ is also harmonic, we have

$$2u_1 + \frac{\partial u_2}{\partial r} = \frac{1}{2\pi}\int_0^{2\pi} \frac{(1-r^2)h(\psi)}{1-2r\cos(\theta-\psi)+r^2}\,d\psi. \tag{12.21}$$

On differentiating (12.20) with respect to r and substituting the result into (12.21), we arrive at

$$u = (r^2-1)u_1 + u_2 = \frac{1}{2\pi}(r^2-1)\Big[-\frac{1}{2}\int_0^{2\pi} \frac{h(\psi)d\psi}{1-2r\cos(\theta-\psi)+r^2}$$
$$+\int_0^{2\pi} \frac{1-r\cos(\theta-\psi)}{(1-2r\cos(\theta-\psi)+r^2)^2}g(\psi)d\psi\Big].$$

If the disc has radius r_0, then the solution is given by

$$u = \frac{1}{2\pi}(r^2-r_0^2)\Big[-\frac{1}{2}\int_0^{2\pi} \frac{h(\psi)d\psi}{r_0^2+r^2-2rr_0\cos(\theta-\psi)}$$
$$+\int_0^{2\pi} \frac{(r_0-r\cos(\theta-\psi))g(\psi)}{(1-2r\cos(\theta-\psi)+r^2)^2}\,d\psi\Big].$$

Exercise 12.1 Work out the solution for a doubly harmonic function when $\mathcal{D}$ is the upper half-plane.

12.5 The background to Cauchy's integral

Cauchy's integral has already been defined in Section 4.2. We now examine its meaning and, in Section 12.6, generalise it to any curve.

Given a function $\phi(\zeta)$ on the unit circle, with $\zeta = e^{i\theta}, 0 \le \theta \le 2\pi$, we may consider it as a function of θ with period 2π. We then call

$$F(z) = \frac{1}{2\pi i}\int_{|\zeta|=1} \frac{\phi(\zeta)}{\zeta - z}\,d\zeta = \frac{1}{2\pi}\int_0^{2\pi} \frac{\phi(e^{i\theta})e^{i\theta}d\theta}{e^{i\theta}-z} \tag{12.22}$$

Cauchy's integral.

Suppose that $\phi(\zeta)$ has the Fourier expansion

$$\phi(\zeta) = \sum_{n=-\infty}^{\infty} C_n e^{in\theta}. \tag{12.23}$$

For z lying inside the unit circle, we have

$$\frac{\zeta}{\zeta - z} = \sum_{m=0}^{\infty}\left(\frac{z}{\zeta}\right)^m = \sum_{m=0}^{\infty} z^m e^{-im\theta}.$$

Suppose that we can integrate term by term; then, by the orthogonal property associated with $e^{-im\theta}$, we have

$$F(z) = \frac{1}{2\pi}\int_0^{2\pi} \sum_{n=-\infty}^{\infty} C_n e^{in\theta} \sum_{m=0}^{\infty} z^m e^{-im\theta}\, d\theta = \sum_{n=0}^{\infty} C_n z^n. \tag{12.24}$$

This is a function analytic inside a circle and

$$\lim_{\rho\to 1_-} F(\rho e^{i\theta}) = \sum_{n=0}^{\infty} C_n e^{in\theta}, \tag{12.25}$$

which is the part of the original Fourier series expansion (12.23) with only the non-negative terms; we denote it by $F^+(\zeta)$.

If z lies outside the unit circle, then

$$\frac{\zeta}{\zeta - z} = -\frac{\zeta}{z}\left(1 - \frac{\zeta}{z}\right)^{-1} = \sum_{\ell=1}^{\infty}\left(\frac{\zeta}{z}\right)^{\ell},$$

and so

$$F(z) = -\frac{1}{2\pi}\int_0^{2\pi} \sum_{n=-\infty}^{\infty} C_n e^{in\theta} \sum_{\ell=1}^{\infty} z^{-\ell} e^{-i\ell\theta}\, d\theta = \sum_{\ell=1}^{\infty} \frac{C_{-\ell}}{z^{\ell}}. \tag{12.26}$$

This is a function analytic outside a circle (including ∞), and

$$\lim_{\rho\to 1_+} F(\rho e^{i\theta}) = -\sum_{\ell=1}^{\infty} C_{-\ell} e^{-i\ell\theta}, \tag{12.27}$$

which is the part of the original Fourier series expansion (12.23) with only the negative terms; we denote it by $F^-(\zeta)$.

Clearly

$$\phi(\zeta) = F^+(\zeta) - F^-(\zeta). \tag{12.28}$$

Now take a point $z = \zeta_0 = e^{i\theta_0}$ on the unit circle, and we first show that

$$\frac{1}{2\pi}\int_0^{2\pi} \frac{\zeta^{n+1}}{\zeta - \zeta_0}\, d\theta = \begin{cases} \frac{1}{2}\zeta_0^n, & \text{if } \ n \geq 0, \\ -\frac{1}{2}\zeta_0^n, & \text{if } \ n < 0. \end{cases} \tag{12.28$'$}$$

There is a singularity at $\theta = \theta_0$ in the integrand, so that the integral is improper; we take its Cauchy's principal value by splitting the range into $0 \leq \theta \leq \theta_0 - \epsilon$ and $\theta_0 + \epsilon \leq \theta \leq 2\pi$, and then let $\epsilon \to 0$.

We use mathematical induction. When $n = 0$, we have

$$\frac{1}{2\pi}\int_0^{2\pi} \frac{\zeta}{\zeta - \zeta_0}\, d\theta = \frac{1}{2\pi}\int_0^{2\pi} \frac{\zeta}{\zeta - 1}\, d\theta = \lim_{\epsilon\to 0} \frac{1}{2\pi}\int_{\epsilon}^{2\pi-\epsilon} \frac{e^{(1/2)i\theta}}{e^{(1/2)i\theta} - e^{-(1/2)i\theta}}\, d\theta$$
$$= \lim_{\epsilon\to 0} \frac{1}{2\pi}\left(\int_{\epsilon}^{2\pi-\epsilon} \frac{\cos\theta/2}{2i\sin\theta/2}\, d\theta + \frac{1}{2}\int_0^{2\pi} d\theta\right)$$
$$= \frac{1}{2} + \lim_{\epsilon\to 0} \frac{1}{2\pi} \log \frac{\sin(2\pi - \epsilon)/2}{\sin\epsilon/2} = \frac{1}{2}.$$

When $n > 0$, we have the inductive step

$$\frac{1}{2\pi}\int_0^{2\pi} \frac{\zeta^{n+1}}{\zeta - \zeta_0}\, d\theta = \frac{1}{2\pi}\int_0^{2\pi} \left(\zeta^n + \frac{\zeta_0\zeta^n}{\zeta - \zeta_0}\right) d\theta = \frac{1}{2}\zeta_0 \int_0^{2\pi} \frac{\zeta^n}{\zeta - \zeta_0}\, d\theta.$$

When $n = -\ell$, we consider

$$\overline{\frac{1}{2\pi}\int_0^{2\pi} \frac{\zeta^{-\ell+1}}{\zeta - \zeta_0}\, d\theta} = \frac{1}{2\pi}\int_0^{2\pi} \frac{\zeta^{\ell-1}}{\zeta^{-1} - \zeta_0^{-1}}\, d\theta$$
$$= \frac{\zeta_0}{2\pi}\int_0^{2\pi} \frac{\zeta^{\ell}}{\zeta_0 - \zeta}\, d\theta = -\frac{1}{2}\zeta_0^{\ell} = -\frac{1}{2}(\bar{\zeta}_0)^{-\ell},$$

and, by taking the imaginary parts, we also have (12.28′) for $n < 0$.

Let us now consider the meaning of the integral (12.22) when $z = \zeta_0$. This too is an improper integral, and Cauchy's principal value is again meant to be taken. Substituting (12.23) into (12.22), if term-by-term integration is allowed then

$$F(\zeta_0) = \sum_{n=-\infty}^{\infty} C_n \frac{1}{2\pi}\int_0^{2\pi} \frac{\zeta^{n+1}}{\zeta - \zeta_0}\, d\theta$$
$$= \frac{1}{2}\sum_{n=0}^{\infty} C_n\zeta_0^n - \frac{1}{2}\sum_{n=-\infty}^{-1} C_n\zeta_0^{-n} \tag{12.29}$$
$$= \frac{1}{2}\big(F^+(\zeta_0) + F^-(\zeta_0)\big).$$

Combined with (12.27) we arrive at

$$\begin{cases} F^+(\zeta) = F(\zeta) + \frac{1}{2}\phi(\zeta), \\ F^-(\zeta) = F(\zeta) - \frac{1}{2}\phi(\zeta). \end{cases} \tag{12.30}$$

This is the important formula for Cauchy's integral.

If $\phi(\zeta) = 0$ outside an arc r then (12.30) is valid at an interior point of r, so that if the integral in (12.22) is over a general circular arc, the conclusion (12.30) is still valid.

Moreover, by the fundamental theorem on conformal transformations, we hope that (12.30) is still valid for any arc. However, Cauchy's integral is not invariant under a conformal transformation, so that we still need to apply the method used in Chapter 4. We avoid all such trouble by starting from the beginning instead.

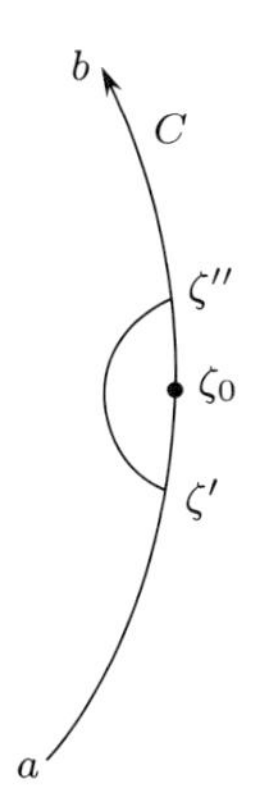

Figure 12.2.

12.6 Cauchy's integral

Let $\phi(\zeta)$ be a function defined on a curve C, which has the parametric equations $x = \psi(t)$, $y = \chi(t)$, with $\psi(t)$ and $\chi(t)$ having continuous derivatives. We then have the Cauchy integral

$$F(z) = \frac{1}{2\pi i} \int_C \frac{\phi(\zeta)}{\zeta - z} \, d\zeta, \tag{12.31}$$

and we also assume that $\phi(\zeta)$ satisfies Hölder's condition

$$|\phi(\zeta) - \phi(\zeta_0)| \le M|\zeta - \zeta_0|^{\mu}, \quad 0 < \mu \le 1.$$

If z is not on C then clearly (12.31) exists and is an analytic function of z. We now give a meaning to the integral (12.31) when $z = \zeta_0$ is a point on C. Specify the direction of C and take a circle centre ζ_0, radius ϵ, intersecting C at ζ', ζ'' (see Fig. 12.2). We proceed to prove that, as $\epsilon \to 0$,

$$\frac{1}{2\pi i}\left(\int_a^{\zeta'} + \int_{\zeta''}^{b}\right)\frac{\phi(\zeta)d\zeta}{\zeta - \zeta_0}$$

has a limit (noting that, going along C from a to b, we meet ζ' before ζ''). By Hölder's condition it is easy to see that

$$\frac{1}{2\pi}\int_a^b \frac{\phi(\zeta) - \phi(\zeta_0)}{\zeta - \zeta_0}\, d\zeta$$

exists. We now examine

$$\begin{aligned}\frac{\phi(\zeta_0)}{2\pi i}\left(\int_a^{\zeta'} + \int_{\zeta''}^{b}\right)\frac{d\zeta}{\zeta - \zeta_0} &= \frac{\phi(\zeta_0)}{2\pi i}\left(\log(\zeta - \zeta_0)|_a^{\zeta'} + \log(\zeta - \zeta_0)|_{\zeta''}^{b}\right) \\ &= \frac{\phi(\zeta_0)}{2\pi i}\left(\log\frac{b - \zeta_0}{a - \zeta_0} - \log\frac{\zeta'' - \zeta_0}{\zeta' - \zeta_0}\right),\end{aligned}$$

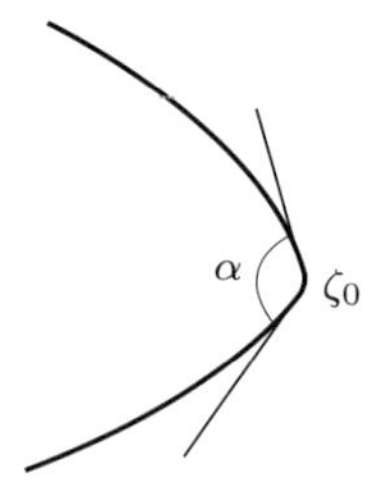

Figure 12.3.

in which the branches of the logarithm function on the arcs $\widehat{a\zeta'}$ and $\widehat{\zeta'' b}$ are specified as follows: Having specified the value of $\log(\zeta - \zeta_0)$ at $\zeta = a$, it changes continuously as ζ moves along C until it reaches $\zeta = \zeta'$, and the value changes continuously again as ζ moves along the circular arc $|z - \zeta_0| = \epsilon$ on the left-hand side of C to $\zeta = \zeta''$ (see Fig. 12.2); thus

$$|\zeta - \zeta_0| = |\zeta'' - \zeta_0|, \quad \lim_{\epsilon \to 0} \log \frac{\zeta'' - \zeta_0}{\zeta' - \zeta_0} = -i\pi. \tag{12.32}$$

We have made use of the condition that the curve has a tangent at ζ_0; if the tangents from the left and from the right are different, then obviously the formula will be different. Therefore

$$F(\zeta_0) = \frac{1}{2\pi i}\left[\int_C \frac{\phi(\zeta) - \phi(\zeta_0)}{\zeta - \zeta_0}\, d\zeta + \phi(\zeta_0) \log \frac{b - \zeta_0}{a - \zeta_0} + i\pi\phi(\zeta_0)\right], \tag{12.33}$$

which is called the principal value of the integral. If the curve is closed, that is $a = b$, then (12.33) becomes

$$F(\zeta_0) = \frac{1}{2\pi i}\int_C \frac{\phi(\zeta)}{\zeta - \zeta_0}\, d\zeta = \frac{1}{2\pi i}\int_C \frac{\phi(\zeta) - \phi(\zeta_0)}{\zeta - \zeta_0}\, d\zeta + \frac{1}{2}\phi(\zeta_0). \tag{12.34}$$

Note. It is not difficult to show that if the curve has a sharp bend at ζ_0, say with the left and the right tangents forming an angle α (see Fig. 12.3), then (12.32) becomes

$$\lim_{\epsilon \to 0} \log \frac{\zeta'' - \zeta_0}{\zeta' - \zeta_0} = -i\alpha,$$

and the corresponding formula (12.33) becomes

$$F(\zeta_0) = \frac{1}{2\pi i}\left[\int_C \frac{\phi(\zeta) - \phi(\zeta_0)}{\zeta - \zeta_0}\, d\zeta + \phi(\zeta_0) \log \frac{b - \zeta_0}{a - \zeta_0}\right] + \frac{\alpha}{2\pi}\phi(\zeta_0).$$

For this reason it is not that much more difficult to deal with such situations.

12.7 Sokhotsky's formula

Lemma 12.2 *Suppose that $\phi(\zeta)$ satisfies Hölder's condition at $\zeta = \zeta_0$. Then*

$$\lim_{z \to \zeta_0} \int_C \frac{\phi(\zeta) - \phi(\zeta_0)}{\zeta - z}\, d\zeta = \int_C \frac{\phi(\zeta) - \phi(\zeta_0)}{\zeta - \zeta_0}\, d\zeta.$$

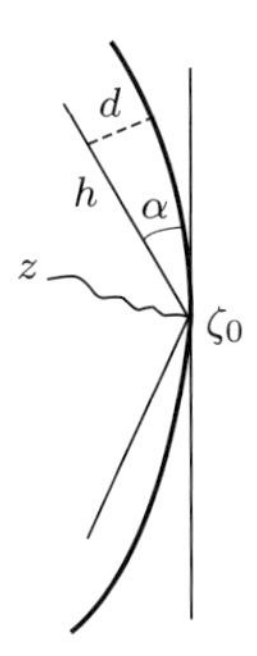

Figure 12.4.

Here z tends to ζ_0 under the following conditions: Let $h = |z - \zeta_0|$, and let d denote the shortest distance between z and C (see Fig. 12.4); then the ratio h/d has to be bounded.

Proof. We consider the difference Δ between the two integrals:

$$\Delta = \int_C (z - \zeta_0) \frac{\phi(\zeta) - \phi(\zeta_0)}{(\zeta - z)(\zeta - \zeta_0)} \, d\zeta.$$

Let $\delta > 0$ be suitably chosen later to correspond to any preassigned $\epsilon > 0$. Divide the integral as the sum of Δ_1 and Δ_2, with the former having the restriction $|\zeta - \zeta_0| < \delta$, and the exterior region being represented by C'.

For the first integral, we apply Hölder's condition and $|\zeta - z| \geq d$ to give

$$|\Delta_1| \leq \int_C \frac{h}{d} \frac{M|\zeta - \zeta_0|^{\mu}}{|\zeta - \zeta_0|} |d\zeta| = \frac{hM}{d} \int_C |\zeta - \zeta_0|^{\mu - 1} |d\zeta|.$$

With $|d\zeta| = \left|\sqrt{\psi'^2(t) + \chi'^2(t)}\, dt\right| \leq A\, dt$ we find that

$$|\Delta_1| \leq \frac{2h}{d} M A \int_0^{\delta} \frac{dt}{t^{--\mu}} = O(\delta^{\mu}),$$

which is less than $\epsilon/2$ when δ is sufficiently small.

With δ fixed, the integral over C' does not include ζ_0, so that the integral

$$\int_{C'} \frac{\phi(\zeta) - \phi(\zeta_0)}{\zeta - z} \, d\zeta,$$

as a function of z, is continuous at $z = \zeta_0$. Thus, corresponding to a sufficiently small $h = |z - \zeta_0|$, we also have $|\Delta_2|$ less than $\epsilon/2$, and hence $|\Delta| \leq |\Delta_1| + |\Delta_2| < \epsilon$. The lemma is proved. □

Note. The following is the underlying idea for h/d being bounded: Take two sides with ζ_0 as the vertex. The two sides form their own angles with the tangent at ζ_0 (such angles have already appeared in the study of the Abel–Tauber theorem, and are often called Stolz angles). In fact, it is not difficult to prove that, within a Stolz angle, there is uniformity in the limit formula in the lemma.

Theorem 12.3 (Sokhotsky) *Let ζ_0 be a point on C, but not an endpoint; let $\phi(\zeta)$ satisfy Hölder's condition at $\zeta = \zeta_0$, and, as $z \to \zeta_0$, let the ratio h/d be bounded. Then, as z approaches ζ_0 from the left, or the right, along C, Cauchy's integral*

$$F(z) = \frac{1}{2\pi i} \int_C \frac{\phi(\zeta)}{\zeta - z} \, d\zeta \tag{12.35}$$

separately approaches

$$F^+(\zeta_0) = F(\zeta_0) + \frac{1}{2}\phi(\zeta_0), \tag{12.36}$$

$$F^-(\zeta_0) = F(\zeta_0) - \frac{1}{2}\phi(\zeta_0). \tag{12.37}$$

In these formulae, $F(\zeta_0)$ represents the Cauchy principal value of the integral (12.35).

Proof. (1) Suppose that C is a closed curve, being traversed in the anti-clockwise direction. Then

$$\begin{aligned} F(z) &= \frac{1}{2\pi i} \int_C \frac{\phi(\zeta)}{\zeta - z} \, d\zeta \\ &= \frac{1}{2\pi i} \int_C \frac{\phi(\zeta) - \phi(\zeta_0)}{\zeta - z} \, d\zeta + \frac{\phi(\zeta_0)}{2\pi i} \int_C \frac{d\zeta}{\zeta - z}. \end{aligned}$$

By Lemma 12.2, as $z \to \zeta_0$, the first integral on the right-hand side approaches

$$\frac{1}{2\pi i} \int_C \frac{\phi(\zeta) - \phi(\zeta_0)}{\zeta - \zeta_0} \, d\zeta,$$

and the second integral is equal to either $\phi(\zeta_0)$ or 0, depending on whether z lies inside or outside C. We therefore have the limit formulae

$$\begin{aligned} F^+(z) &= \frac{1}{2\pi i} \int_C \frac{\phi(\zeta) - \phi(\zeta_0)}{\zeta - \zeta_0} \, d\zeta + \phi(\zeta_0) \\ F^-(z) &= \frac{1}{2\pi i} \int_C \frac{\phi(\zeta) - \phi(\zeta_0)}{\zeta - \zeta_0} \, d\zeta. \end{aligned}$$

The Sokhotsky formula now follows from (12.34).

(2) If C is not a closed curve then we may insert a section C' so that $C \cup C' = C_0$ is a closed curve and with $\phi(\zeta)$ being defined to be zero on C'. Note that the function is no longer continuous on C_0, but in our discussion in (1) the requirement is that it satisfies Hölder's condition in the neighbourhood of $\zeta = \zeta_0$. Therefore

$$F(z) = \frac{1}{2\pi i} \int_{C_0} \frac{\phi(\zeta)}{\zeta - z} \, d\zeta,$$

and if ζ_0 is not an endpoint of C we can deduce the theorem from (1). □

From Sokhotsky's formula we deduce the following: As a point moves across $\zeta = \zeta_0$ along C, the value of Cauchy's integral has the jump

$$F^+(\zeta_0) - F^-(\zeta_0) = \phi(\zeta_0). \tag{12.38}$$

Note. If ζ_0 is the vertex of a sharp bend on C, making an angle α ($0 < \alpha < 2\pi$, from the right), then Sokhotsky's formula should be modified to become

$$F^+(\zeta_0) = F(\zeta_0) + \frac{2\pi - \alpha}{2\pi}\phi(\zeta_0),$$

$$F^-(\zeta_0) = F(\zeta_0) - \frac{\alpha}{2\pi}\phi(\zeta_0).$$

Theorem 12.4 *Let C be a closed curve on which $\phi(\zeta)$ satisfies Hölder's condition. Then a necessary and sufficient condition for Cauchy's integral to become Cauchy's integral of the interior is that*

$$\int_C \zeta^n \phi(\zeta) d\zeta = 0, \quad n = 0, 1, 2, \dots; \tag{12.39}$$

and to become Cauchy's integral of the exterior is that

$$\int_C \zeta^{-m} \phi(\zeta) d\zeta = 0, \quad m = 1, 2, \dots. \tag{12.40}$$

By 'Cauchy's integral of the interior' we mean that $\phi(\zeta)$ delivers inside C an analytic function $F(z)$ which approaches the given boundary values on C. We only give the proof for the inside part; the outside part can be deduced from the inside part, or be proved similarly.

Proof. (1) Sufficiency. We have the expansion

$$\frac{1}{\zeta - z} = -\sum_{n=0}^{\infty} \frac{\zeta^n}{z^{n+1}}$$

at infinity so that, near $z = \infty$,

$$F(z) = \frac{1}{2\pi i}\int_C \phi(\zeta)\frac{d\zeta}{\zeta - z} = -\frac{1}{2\pi i}\sum_{n=0}^{\infty}\frac{1}{z^{n+1}}\int_C \zeta^n\phi(\zeta)d\zeta. \tag{12.41}$$

By condition (12.39) we know that $F(z) \equiv 0$ in the neighbourhood of $z = \infty$, and so $F(z) \equiv 0$ on the outside of C by virtue of being analytic; thus $F^-(\zeta) \equiv 0$. By Theorem 12.3 we find that $F(\zeta) = \frac{1}{2}\phi(\zeta)$ and $F^+(\zeta) = \phi(\zeta)$. This means that, inside C, Cauchy's integral does represent an analytic function $F(z)$. As z approaches the boundary, this function approaches the value of the function $F^+(\zeta) = \phi(\zeta)$ appearing in the integrand.

(2) Necessity. If $F(z)$ is a Cauchy's integral then, as far as z being outside of C is concerned, the expression $\phi(w)/(w - z)$ is an analytic function inside C and is continuous on C, so that, by Cauchy's theorem,

$$F(z) = \frac{1}{2\pi i}\int_C \frac{\phi(\zeta)}{\zeta - z}d\zeta = 0 \quad (z \text{ outside } C).$$

Thus, all the coefficients in (12.41) are equal to 0, and this is condition (12.40). □

Exercise 12.5 Generalise Sokhotsky's theorem to the square $|x| = \frac{1}{2}$, $|y| = \frac{1}{2}$, taking special care of the situation at the four corners.

Exercise 12.6 Suppose that $\phi(\zeta)$ has a finite number of discontinuities of the first kind. How should Theorem 12.4 be modified at these points of discontinuity? Consider the situation when $\phi(\theta) = (\pi - \theta)/2$ for $0 < \theta < 2\pi$ and C is the unit circle.

12.8 The Hilbert–Privalov problem

Problem 12.7 Let C be a closed curve, and let $a(\zeta) \neq 0$, $b(\zeta)$ be functions satisfying Hölder's condition. Find two functions $f^+(z)$ and $f^-(z)$, with one analytic inside and the other outside C, having the property that, when ζ is on C,

$$f^-(\zeta) = a(\zeta) f^+(\zeta) + b(\zeta). \tag{12.42}$$

There is a variety of important applications in mathematical physics associated with the problem.

We first consider the case when the curve is the unit circle: Let $a(\zeta)$ and $b(\zeta)$ have the Fourier series representations

$$\sum_{n=-\infty}^{\infty} a_n \zeta^n, \quad \sum_{n=-\infty}^{\infty} b_n \zeta^n, \tag{12.43}$$

respectively, and let

$$\sum_{n=-\infty}^{-1} c_n \zeta^n, \quad \sum_{n=1}^{\infty} d_n \zeta^n \tag{12.44}$$

be the Fourier expansions for $f^-(\zeta)$ and $f^+(\zeta)$, respectively, so that the problem becomes that of the determination of c_n and d_n.

A quick glance should tell us that sometimes there may not be a solution. Take, for example, the case when $a(\zeta)$ has only terms with positive exponents and $b(\zeta) = 0$. The right-hand side of (12.42) has only terms with positive exponents, and so cannot be matched with the left-hand side.

Also, the solution to such a problem cannot be unique, because on taking $a(\zeta) = 1/\zeta^n$, $b(\zeta) = 0$ there can be no uniqueness because of the identity

$$\left(a_0 + \frac{a_1}{\zeta} + \cdots + \frac{a_n}{\zeta^n}\right) = \frac{1}{\zeta^n}(a_0\zeta^n + \cdots + a_n)$$

for $a_0, a_1, \ldots, a_n$.

Privalov generalises the original Hilbert problem, which has $b(\zeta) = 0$. We now introduce the following solution given by Gakhov.

(1) $a(\zeta) = 1$. Although this is a particularly easy case, nevertheless the method does contain the main argument because, in Sokhotsky's formula, we may take $\phi(\zeta) = -b(\zeta)$ so that the Cauchy integral

$$f(z) = F(z) = -\frac{1}{2\pi i}\int_C \frac{b(\zeta)}{\zeta - z}\,d\zeta \tag{12.45}$$

is then a solution to the original problem.

If there is another solution

$$g^+(\zeta) - g^-(\zeta) = \phi(\zeta),$$

then

$$f^+(\zeta) - g^+(\zeta) = f^-(\zeta) - g^-(\zeta),$$

so that $f^+(z) - g^+(z)$ represents an analytic function inside C, and $f^-(z) - g^-(z)$ represents an analytic function outside C. Since they match completely on C the two constructions deliver a function which is analytic on the whole plane (including ∞). By Liouville's theorem such a function has to be constant. This means that the general solution is given by (12.45) with an additional constant, and there are no other solutions.

(2) $b(\zeta) \equiv 0$.

(I) Suppose that $\log a(\zeta)$ is a single-valued function on C. Then from

$$\log f^-(\zeta) = \log f^+(\zeta) + \log a(\zeta),$$

together with (1), we find that such a problem must have the following solution:

$$f^{\pm}(z) = Ae^{-F^{\pm}(z)}, \quad F(z) = \frac{1}{2\pi i}\int_C \frac{\log a(\zeta)d\zeta}{\zeta - z}, \tag{12.46}$$

where A is a constant.

(II) If $\log a(\zeta)$ is not single-valued then, because $a(\zeta)$ is single-valued, as ζ loops round C the value for $\log a(\zeta)$ increases by a multiple of 2π. Such a whole number m is called the index, and it is represented by

$$m = \frac{1}{2\pi}\Delta_C \arg a(\zeta) = \frac{1}{2\pi i}\int_C d\log a(\zeta).$$

It does not matter if we assume that the origin lies inside C, so that the index for

$$a_1(\zeta) = \zeta^{-m}a(\zeta)$$

has the value 0. By (I) we already have the function

$$g^{\pm}(z) = e^{-G^{\pm}(z)}, \quad G(z) = \frac{1}{2\pi i}\int_C \frac{\log a_1(\zeta)d\zeta}{\zeta - z}$$

satisfying

$$g^-(\zeta) = a_1(\zeta)g^+(\zeta).$$

Thus, the problem of finding $f^{\pm}$ to satisfy

$$f^{-}(\zeta) = a(\zeta) f^{+}(\zeta)$$

becomes one of finding $h^{\pm}$ so that

$$h^{-}(\zeta) = \zeta^{m} h^{+}(\zeta) \quad (f^{\pm} = g^{\pm} h^{\pm}). \tag{12.47}$$

(i) Suppose that $m = -n$ is a negative integer. Then

$$h^{+}(z), \quad z^{n} h^{-}(z)$$

are functions taking the same values on C, so that they are analytic continuations of each other. Since $h^{-}(z)$ is regular at ∞, the function concerned has a pole of order n at ∞, that is

$$h^{+}(z) = a_0 z^{n} + a_1 z^{n-1} + \cdots + a_n,$$

and

$$h^{-}(z) = a_0 + \frac{a_1}{z} + \cdots + \frac{a_n}{z^{n}}.$$

Thus

$$\begin{aligned} f^{+}(z) &= (a_0 z^{n} + a_1 z^{n-1} + \cdots + a_n) e^{-G^{+}(z)}, \\ f^{-}(z) &= \left(a_0 + \frac{a_1}{z} + \cdots + \frac{a_n}{z^{n}}\right) e^{-G^{-}(z)} \end{aligned} \tag{12.48}$$

and

$$G(z) = \frac{1}{2\pi i} \int_C \frac{\log\left(\zeta^{-m} a(\zeta)\right)}{\zeta - z} \, d\zeta. \tag{12.49}$$

Here $a_0, a_1, \ldots, a_n$ are constants, with a_0 being determined from $f^{-}(\infty)$. It is not difficult to show that there are no other solutions.

(ii) For the index $m > 0$ there is no solution to Hilbert's problem. From (12.47) we know that $h^{-}(z)$, $z^{m} h^{+}(z)$ deliver an analytic function, regular in the whole plane, and is thus a constant; but at $z = 0$ the function takes the value 0, which is then the value of the constant. Thus $h^{-} = h^{+} \equiv 0$.

Summarising our results we have the following.

Theorem 12.8 (Gakhov) *If the index $-n$ of the boundary value function $a(\zeta)$ is not positive then Hilbert's problem $f^{-}(\zeta) = a(\zeta) f^{+}(\zeta)$ has $n + 1$ linearly independent solutions. Otherwise, there is no solution.*

12.9 Continuation

We now consider the general Privalov problem

$$f^-(\zeta) = a(\zeta) f^+(\zeta) + b(\zeta). \tag{12.50}$$

Let $a_1(\zeta) = \zeta^{-m} a(\zeta)$, and set

$$\begin{aligned} &g^-(\zeta) = a_1(\zeta) g^+(\zeta), \\ &g^\pm(z) = e^{-G^\pm(z)}, \quad G(z) = \frac{1}{2\pi i}\int_C \frac{\log\left(\zeta^{-m}a(\zeta)\right)}{\zeta - z}\, d\zeta. \end{aligned} \tag{12.51}$$

Letting $f^\pm = g^\pm h^\pm$, we then have from (12.50) and (12.51) that

$$h^-(\zeta) = \zeta^m h^+(\zeta) + b(\zeta)/g^-(\zeta). \tag{12.52}$$

The case $m = 0$ is just (1) from Section 12.8, and so there is a solution:

$$\begin{aligned} &h^\pm(z) = A + H^\pm(z), \\ &H(z) = -\frac{1}{2\pi i}\int_C \frac{b(\zeta)e^{G^-(\zeta)}}{\zeta - z}\, d\zeta, \end{aligned} \tag{12.53}$$

where A is an arbitrary constant. Therefore

$$f^\pm(z) = e^{-G^\pm(z)}\left\{A + H^\pm(z)\right\}, \tag{12.54}$$

and it is not difficult to prove that there are no other solutions.

When $m = -n < 0$, we have

$$\begin{aligned} &h^-(z) = a_0 + \frac{a_1}{z} + \frac{a_2}{z^2} + \cdots + \frac{a_n}{z^n} + H^-(z), \\ &h^+(z) = a_0 z^n + a_1 z^{n-1} + \cdots + a_n + z^n H^+(z), \end{aligned}$$

that is

$$\begin{cases} f^-(z) = \left\{a_0 + \dfrac{a_1}{z} + \cdots + \dfrac{a_n}{z^n} + H^-(z)\right\} e^{-G^-(z)}, \\ f^-(z) = \left\{a_0 z^n + \cdots + a_n + z^n H^+(z)\right\} e^{-G^+(z)}; \end{cases} \tag{12.55}$$

it is also easy to prove that these are all of the solutions.

Finally, if the index $m > 0$ then

$$h^-(\zeta) = \zeta^m h^+(\zeta) + b(\zeta) e^{G^-(\zeta)},$$

and the equation yields

$$h^-(z) = A + H^-(z), \quad h^+(z) = \frac{A + H^+(z)}{z^m}; \tag{12.56}$$

it is easy to prove that only these two functions can satisfy (12.56).

We need to examine $H^+(z)$ to see how $h^+(z)$ can be made to be regular, and we do this through the expansion

$$\begin{aligned} H^+(z) &= -\frac{1}{2\pi i}\int_C b(\zeta)e^{G^-(\zeta)}\frac{d\zeta}{\zeta - z} \\ &= -\frac{1}{2\pi i}\int_C b(\zeta)e^{G^-(\zeta)}\sum_{n=0}^{\infty}\frac{z^n}{\zeta^{n+1}}\,d\zeta \\ &= \sum_{n=0}^{\infty}\left(-\frac{1}{2\pi i}\int_C \frac{b(\zeta)e^{G^-(\zeta)}}{\zeta^{n+1}}d\zeta\right)z^n. \end{aligned}$$

In order for $h^+(z)$ to be regular, the number A in the formula above may take a suitable value, and we need

$$\frac{1}{2\pi i}\int_C \frac{b(\zeta)e^{G^-(\zeta)}}{\zeta^{n+1}}d\zeta = 0, \qquad n = 1, 2, \ldots, m. \tag{12.57}$$

Therefore, when $m > 0$, Privalov's problem has a solution only when (12.57) is satisfied.

Summarising:

Theorem 12.9 (Gakhov) *If the index of the boundary function is $-n \le 0$ then Privalov's problem $f^-(\zeta) = a(\zeta)f^+(\zeta) + b(\zeta)$ has the solution (12.55) with $n+1$ parameters. If $a(\zeta)$ has index $m > 0$ then the problem has a solution only if (12.57) is satisfied.*

Exercise 12.10 Consider the situation when $a(\zeta)$, $b(\zeta)$ may have discontinuities.

Exercise 12.11 Consider the situation when the curve is not closed.

12.10 The Riemann–Hilbert problem

Let C be a closed curve, with the region $\mathcal{D}$ as its interior, and let $a(\zeta)$, $b(\zeta)$, $c(\zeta)$ be real functions defined on C. The problem is to find a function

$$f(z) = u(z) + iv(z)$$

which is analytic in $\mathcal{D}$, continuous in $\bar{\mathcal{D}}$ and satisfying

$$a(\zeta)u(\zeta) - b(\zeta)v(\zeta) = c(\zeta) \tag{12.58}$$

on C.

Muskhelishvili's method of solution: Suppose (without loss of generality) that $\mathcal{D}$ is the unit disc $|z| < 1$, that $a(\zeta)$, $b(\zeta)$, $c(\zeta)$ satisfy Hölder's condition, and that $a^2(\zeta) + b^2(\zeta) \neq 0$, so that (12.58) is rewritten as

$$2\Re((a+ib)f(\zeta)) = (a+bi)f(\zeta) + (a-bi)\overline{f(\zeta)} = 2c. \tag{12.59}$$

Going through Privalov's problem, we find two analytic functions $F^+(z)$, $F^-(z)$, one inside and the other outside the circle, such that

$$(a+bi)F^+(\zeta)+(a-bi)F^-(\zeta)=2c. \tag{12.60}$$

We set

$$F_*^+(z)=\overline{F^-(1/\bar{z})},\quad F_*^-(z)=\overline{F^+(1/\bar{z})},$$

which are also analytic functions, one inside and the other outside the circle, satisfying condition (12.60); the reason being that the conjugate of (12.60) is

$$(a-bi)\overline{F^+(1/\zeta)}+(a+bi)\overline{F^-(1/\zeta)}=2c,$$

that is

$$(a+bi)F_*^+(\zeta)+(a-bi)F_*^-(\zeta)=2c. \tag{12.61}$$

Let $f(z)=F^+(z)+F^-(z)$; then

$$\overline{f(z)}=\overline{F^+(1/\zeta)}+\overline{F_*^+(1/\zeta)}=F_*^-(\zeta)+F^-(\zeta),$$

and on adding (12.60) and (12.61) together we have

$$(a+bi)f(\zeta)+(a-bi)\overline{f(\zeta)}=2c.$$

Note. Let $f_1(z)=u_1+iv_1$ be the integral of $f(z)$, that is $\partial f_1/\partial z=f$. Then

$$\frac{\partial u_1}{\partial x}=u,\quad \frac{\partial v_1}{\partial x}=v=-\frac{\partial u_1}{\partial y},$$

and equation (12.58) can be rewritten as

$$a(\zeta)\frac{\partial u_1}{\partial x}+b(\zeta)\frac{\partial u_1}{\partial y}\bigg|_C=c(\zeta). \tag{12.62}$$

Therefore the Riemann–Hilbert problem can be viewed as follows: The search for a harmonic function u_1 satisfying the boundary condition (12.62), which can be considered as a generalisation of Neumann's problem. Conversely, we can also consider such a generalised Neumann's problem to be a Riemann-Hilbert problem.

12.11 The uniqueness of the solution to a mixed boundary value problem

Let C be a closed curve with the region $\mathcal{D}$ as its interior. Take $2n$ points $a_1, b_1, a_2, b_2, \ldots, a_n, b_n$ successively along C (for convenience we set $a_{n+1}=a_1$). For the mixed boundary value problem, we seek a function analytic in $\mathcal{D}$ with the property that, along the arc from a_k to b_k, its real part takes the values of a given function $\phi_k(\zeta)$ and, along the arc from b_k to a_{k+1}, its imaginary part takes the values of a given function $\psi_k(\zeta)$, for $k=1,2,\ldots,n$.

It goes without saying that we shall assume that, apart from the points a_k, b_k, the function concerned is continuous in $\bar{\mathcal{D}}$. We further suppose that $\phi_k(\zeta)$, $\psi_k(\zeta)$ satisfy Hölder's condition in the corresponding intervals. Such a problem can also be considered as a particular Riemann–Hilbert problem.

We now consider the situation when $\mathcal{D}$ is the upper half-plane and

$$a_1 < b_1 < a_2 < b_2 < \cdots < a_n < b_n.$$

Henceforth we use (b_n, a_{n+1}) to denote the union of the two intervals $b_n < x < \infty$, $-\infty < x < a_{n+1} = a_1$. Write $f(z) = u(z) + iv(z)$, so that the mixed boundary condition is given by

$$\begin{cases} u = \phi_k(x), & \text{when } a_k < x < b_k, \\ v = \psi_k(x), & \text{when } b_k < x < a_{k+1}, \end{cases} \tag{12.63}$$

for $k = 1, 2, \ldots, n$ and $v = 0$ in (b_n, a_{n+1}).

12.11.1 The important special case $\phi_k(x) = 0$, $\psi_k(x) = 0$

On the segment (b_{k-1}, a_k), we know from $v(x) = 0$ that $f(x)$ has to be real; that is, the function $W = f(z)$ maps the real line segment (b_{k-1}, a_k) into a real line segment. Therefore $f(z)$ has an analytic continuation across (b_{k-1}, a_k) into the lower half-plane, and with $f(x - iy) = u(z) - iv(z)$. Also, on (a_k, b_k), the expression $f(x)$ is purely imaginary, so that there is also an analytic continuation across (a_k, b_k) into the lower half-plane, and with $f(x - iy) = -u(z) + iv(z)$. Consequently, on looping round a point a_k, the value for $f(z)$ becomes $-f(z)$.

The functions

$$g(z) = \sqrt{\prod_{k=1}^{n} \frac{z - a_k}{z - b_k}} \tag{12.64}$$

and $f(z)$ both have the properties just described. When $x < a_1$, we take the positive branch of $g(z)$, so that the function has the following properties:

$$\begin{cases} \Re g(x) = 0, & \text{when } a_k < x < b_k, \\ \Im g(x) = 0, & \text{when } b_k < x < a_{k+1}, \end{cases} \tag{12.65}$$

and $g(z)$ becomes $-g(z)$ as z loops round either a_k or b_k.

The function

$$p(z) = \frac{f(z)}{g(z)} \tag{12.66}$$

is single-valued on the whole plane, and is analytic everywhere except a_k, b_k and ∞. Conversely, given a function $p(z)$ with such properties, we have

$$f(z) = g(z)p(z)$$

delivering a general solution to the problem; it is clear that the solution is not unique.

For uniqueness, we have to impose a condition so that $p(z)$ becomes a constant. We add the following condition:

(i) $$f(z) = O(|z - a_k|^{-\frac{1}{2}}) \quad \text{as } z \to a_k,$$

so that $p(z) = O(|z - a_k|^{-1})$ at the isolated singularity $z = a_k$, and hence is regular there.

Similarly, on imposing the condition

(ii) $$f(z) = O(|z - b_k|^{-\frac{3}{2}}) \quad \text{as } z \to b_k,$$

the function $p(z)$ will be regular at $z = b_k$.

Also, on supposing that

(iii) $$f(z) \to c \quad \text{as } z \to \infty,$$

the function $p(z)$ will also be regular at $z = \infty$, so that it is a constant[2] c. In other words, the mixed boundary value problem with the conditions (i), (ii) and (iii) has the following unique solution:

$$f(z) = cg(z) = c\sqrt{\prod_{k=1}^{n} \frac{z - a_k}{z - b_k}}.$$

12.11.2 Uniqueness for the general problem

Suppose that there are two functions $f(z)$, $f_1(z)$ both satisfying

$$u_1(x) = u(x) = \phi_k(x), \quad a_k < x < b_k,$$
$$v_1(x) = v(x) = \psi_k(x), \quad b_k < x < a_{k+1},$$

and also conditions (i), (ii) and (iii). By condition (iii), we have

$$\lim_{z\to\infty} \big(f(z) - f_1(z)\big) = 0,$$

so that $f(z) \equiv f_1(z)$.

12.12 The Keldysh–Sedov formula

Let $g(z)$ be given by (12.64), and consider Cauchy's integral for the function

$$2\frac{\phi_k(t)}{g(t)}$$

over the line segment $a_k < t < b_k$, that is

$$\frac{f_k(z)}{g(z)} = \frac{1}{\pi i}\int_{a_k}^{b_k} \frac{\phi_k(t)}{g(t)} \frac{dt}{t - z}. \tag{12.67}$$

[2] It is not difficult to prove that $p(z)$ is analytic in the whole plane.

For $b_\ell < t_1 < a_{\ell+1}$ ($\ell = 1, 2, \dots, n$), the expression $g(t_1)$ is real, while $g(t)$ is purely imaginary, so that $f_k(t_1)$ is real, that is

$$v_k(t_1) = \Im f_k(t_1) = 0. \tag{12.68}$$

Next, for $a_\ell < t_2 < b_\ell$ ($\ell = 1, 2, \dots, n,\ \ell \neq k$), both $g(t_2)$ and $g(t)$ are purely imaginary, so that $f_k(t_2)$ is purely imaginary, that is

$$u_k(t_2) = \Re f_k(t_2) = 0. \tag{12.69}$$

Now consider $a_k < t_0 < b_k$. As z approaches t_0 on the upper half-plane, the value of Cauchy's integral is given by (Sokhotsky's formula)

$$\frac{f_k^+(t_0)}{g(t_0)} = \frac{\phi_k(t_0)}{g(t_0)} + \frac{1}{\pi i}\int_{a_k}^{b_k} \frac{\phi_k(t)}{g(t)} \frac{dt}{t - t_0}.$$

Here we take Cauchy's principal value for the integral – by (12.33) it is

$$\frac{1}{\pi i}\int_{a_k}^{b_k} \left(\frac{\phi_k(t)}{g(t)} - \frac{\phi_k(t_0)}{g(t_0)}\right)\frac{dt}{t - t_0} + \frac{\phi_k(t_0)}{g(t_0)} + \frac{1}{\pi i}\frac{\phi_k(t_0)}{g(t_0)}\log\frac{b_k - t_0}{a_k - t_0}.$$

Since

$$\log\frac{b_k - t_0}{a_k - t_0} = \log\frac{b_k - t_0}{t_0 - a_k} - i\pi,$$

we have

$$\begin{aligned}\frac{f_k^+(t_0)}{g(t_0)} &= \frac{\phi_k(t_0)}{g(t_0)} + \frac{1}{\pi i}\int_{a_k}^{b_k} \left(\frac{\phi_k(t)}{g(t)} - \frac{\phi_k(t_0)}{g(t_0)}\right)\frac{dt}{t - t_0} \\ &\quad + \frac{1}{\pi i}\frac{\phi_k(t_0)}{g(t_0)}\log\frac{b_k - t_0}{t_0 - a_k},\end{aligned} \tag{12.70}$$

and on multiplication by $g(t_0)$ we find that

$$\Re f_k^+(t_0) = \phi_k(t_0). \tag{12.71}$$

Thus, the function $f_k(z)$ defined by Cauchy's integral (12.67) satisfies the mixed boundary conditions (12.68), (12.69) and (12.71).

We proceed to show that $f_k(z)$ satisfies also the additional conditions (i), (ii), (iii) ($c = 0$). It is clear that $f_k(z) \to 0$ as $z \to \infty$, and it is not difficult to prove that

$$\frac{f_k(z)}{g(z)} = O(1) \quad \text{as} \quad z \to a_\ell, \text{ or } b_\ell\ (\ell \neq k);$$

that is, $f_k(z) = O(|z - a_\ell|^{1/2})$ and $f_k(z) = O(|z - b_\ell|^{-1/2})$. Therefore conditions (i) and (ii) are satisfied in the neighbourhoods of $z = a_\ell, b_\ell$.

By (12.70) we see that

$$\frac{f_k^+(t_0)}{g(t_0)} = O\left(|t_0 - a_k|^{-1/2}\log\frac{1}{|t_0 - a_k|}\right), \quad t_0 \to a_k,$$

that is

$$f_k^+(t_0) = O\Big(\log \frac{1}{|t_0 - a_k|}\Big), \qquad t_0 \to a_k.$$

The problem is relatively simple when z approaches a_k from the upper half-plane; however, the lack of a formula similar to (12.70) means that there is a slight problem with the representation. Let $c_k = \frac{1}{2}(a_k + b_k)$; then clearly

$$\int_{c_k}^{b_k} \frac{\phi_k(t)}{g(t)} \frac{dt}{t - z} = O(1) \quad \text{as} \quad z \to a_k.$$

Note also that, when $\phi(t)$ is bounded,

$$\begin{aligned}\int_{c_k}^{b_k} \frac{\phi_k(t)}{g(t)} \frac{dt}{t - z} &= \int_0^{c_k - a_k} \frac{\phi(u + a_0)}{\sqrt{u}(u - w)}\, du \quad (w = z - a_k)\\ &= \frac{1}{w^{1/2}} \int_0^{c_k - a_k} \frac{\phi(\tau_w + a_k)}{\tau^{1/2}(\tau - 1)}\, d\tau \quad (u = \tau w).\end{aligned}$$

We remark that the integral $\int \tau^{-1/2}\, d\tau$ is convergent when τ is near 0, and that $\int \tau^{-3/2}\, d\tau$ is convergent when τ is near ∞. Since w is not real, the integral in the formula is $O(\log 1/|w|)$ in the neighbourhood of $\tau = 1$, so that

$$f_k(z) = O\Big(\log \frac{1}{|z - a_k|}\Big).$$

Therefore, as $z \to a_k$ or b_k, conditions (i) and (ii) are satisfied. (Note: This function may satisfy conditions which are even more restrictive than (i) and (ii); in other words, we may still have uniqueness under less restrictive conditions than (i) and (ii).)

In the same way we can show that Cauchy's integral

$$f_k^*(z) = \frac{g(t)}{\pi} \int_{b_k}^{a_{k+1}} \frac{\psi_k(t)}{g(t)} \frac{dt}{t - z}$$

satisfies conditions (i), (ii) and (iii) (with $c = 0$) and the following boundary conditions:

$$\begin{aligned}\Re f_k^*(z) &= 0, &&\text{when} \quad a_\ell < t < b_\ell,\ \ell = 1, 2, \ldots, n,\\ \Im f_k^*(z) &= \psi_k(t), &&\text{when} \quad b_k < t < a_{k+1},\\ &= 0, &&\text{when} \quad b_\ell < t < a_{\ell+1},\ \ell \neq k.\end{aligned}$$

When added together, we have the Keldysh–Sedov formula

$$f(z) = \sum_{k=1}^{n} f_k(z) + \sum_{k=1}^{n} f_k^*(z) + cg(z)$$

satisfying (i), (ii), (iii) and the accompanying boundary conditions.

We therefore have the following theorem giving the Keldysh–Sedov formula in a more explicit form.

Theorem 12.12 *The mixed boundary value problem on the upper half-plane does have a solution, and only the solution*

$$f(z) = \frac{g(z)}{\pi i}\left(\sum_{k=1}^{n}\int_{a_k}^{b_k}\frac{\phi_k(t)}{g(t)}\frac{dt}{t-z} + i\sum_{k=1}^{n}\int_{b_k}^{a_{k+1}}\frac{\phi_k(t)}{g(t)}\frac{dt}{t-z} + \pi i c\right)$$

when the following three additional conditions

(i) $f(z) = O(|z-a_k|^{-1/2})$ *as* $z \to a_k$,
(ii) $f(z) = O(|z-b_k|^{-3/2})$ *as* $z \to b_k$,
(iii) $f(z) \to c$, *a real constant, as* $z \to \infty$ *on the upper half-plane,*

are imposed on the solution.

12.13 The Keldysh–Sedov formula in other regions

The conformal transformation

$$z = \frac{-iw}{w-1},$$

which maps the upper half-plane $\Im z > 0$ onto the disc centre $\frac{1}{2}$, radius $\frac{1}{2}$, transforms a_k, b_k to α_k, β_k with

$$\frac{z-a_k}{z-b_k} = \left(\frac{w}{w-1} - \frac{\alpha_k}{\alpha_k-1}\right)\Big/\left(\frac{w}{w-1} - \frac{\beta_k}{\beta_k-1}\right) = \frac{w-\alpha_k}{w-\beta_k}\Big/\frac{1-\alpha_k}{1-\beta_k}.$$

Substituting this into the Keldysh–Sedov formula (for the sake of simplicity, we suppose that $f(\infty) = 0$) and replacing the symbols w, α_k, β_k by z, a_k, b_k, we have the formula

$$f(z) = \frac{g(z)}{\pi i}\int_C \frac{\tau(\zeta)}{g(\zeta)(\zeta-z)}\frac{z-1}{\zeta-1}\,d\zeta.$$

Here C represents the circle $|\zeta - \frac{1}{2}| = \frac{1}{2}$, with $\tau(\zeta)$ taking the values of $\phi_k(\zeta)$ on the arc (a_k, b_k) and the values of $i\psi_k(\zeta)$ on the arc (b_k, a_{k+1}).

We leave it to the reader to deduce the case when the curve is the unit circle. This time the Keldysh–Sedov formula takes the form

$$f(z) = \frac{1}{\pi i g(z)}\int_{|\zeta|=1} g(\zeta)\phi(\zeta)\left(\frac{1}{\zeta-z} - \frac{1}{2\zeta}\right)d\zeta$$
$$+ \sqrt{\prod_{k=1}^{n}\frac{a_k b_k}{(z-a_k)(z-b_k)}}(C_0 z^n + C_1 z^{n-1} + \cdots + C_n),$$

where $C_0, \ldots, C_n$ are complex constants.

For the purpose of applications in the following, we deal with the problem concerning the half-disc (Fig. 12.5)

$$\left|z - \frac{1}{2}\right| = \frac{1}{2}, \quad y > 0,$$

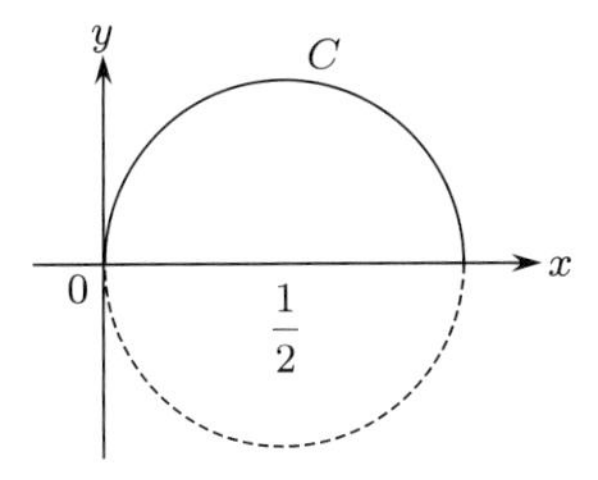

Figure 12.5.

and suppose that, on the perimeter of the half-disc,

$$u(z)\big|_C = \phi(\zeta),$$

and, on the diameter $y = 0$,

$$u(z) + v(z)\big|_{y=0} = \chi(x),$$

and that $\phi(\zeta)$, $\chi(\zeta)$ both satisfy Hölder's condition.

Such a problem can be dealt with by the method used in Sections 12.11 and 12.12, or by applying the result directly.

Let $f_1(z)$ be a solution satisfying

$$u_1(z)|_C = \phi(\zeta), \quad u_1(z) + v_1(z)|_{y=0} = 0.$$

The mapping $w = f(z)$ transforms the real axis to the line $u + v = 0$. By the reflection principle we know that $f_1(z)$ can be analytically continued into the lower half-disc C^*:

$$f_1(x - iy) = -v_1(x, y) - iu_1(x, y), \tag{12.72}$$

so that, on the upper semi-circle,

$$\Im f_1(z) = -\phi(\bar{\zeta}).$$

By (12.72) we have

$$f_1(z) = \frac{1}{\pi i}\left\{\int_C \sqrt{\frac{z(1-\zeta)}{\zeta(1-\zeta)}}\,\frac{\phi(\zeta)d\zeta}{\zeta - z} - i\int_{C^*}\sqrt{\frac{z(1-\zeta)}{\zeta(1-\zeta)}}\,\frac{\phi(\bar{\zeta})d\zeta}{\zeta - z}\right\}. \tag{12.73}$$

In polar coordinates the circle $x^2 + y^2 - x = 0$ has the representation $\rho = \cos\theta$, that is $\zeta = \cos\theta e^{i\theta}$, and the upper and lower semi-circles correspond to $0 < \theta < \pi$ and $-\pi < \theta < 0$, so that

$$1 - \zeta = -i\sin\theta e^{i\theta}, \quad d\zeta = ie^{2i\theta}\,d\theta.$$

With the substitution $\tau = -\theta$ (and w for $\bar{\zeta}$), we have

$$\int_{C^*} = i\int_C \sqrt{\frac{z(1-z)}{w(1-w)}}\,\frac{\phi(w)dw}{w - ze^{2i\tau}}.$$

Again, with $e^{2i\tau} = 2\cos^2\tau - 1 + 2i\sin\tau\cos\tau = 2\cos\tau e^{i\tau} - 1 = 2w - 1$, and replacing the dummy variable w by ζ, we find that the two integrals in (12.73) can be combined to give

$$f_1(z) = \frac{1}{\pi i}\int_C \sqrt{\frac{z(1-\zeta)}{\zeta(1-\zeta)}}\left\{\frac{1}{\zeta - z} - \frac{1}{\zeta + z - 2\zeta z}\right\}\phi(\zeta)d\zeta. \tag{12.74}$$

Let $f_2(z)$ satisfy

$$u_2(z)|_C = \phi(\zeta), \quad u_2(z) + v_2(z)|_{y=0} = \chi(x).$$

The mapping $w = f_2(z)$ transforms the upper semi-circle into the imaginary axis, so that $f_2(z)$ can be analytically continued across C onto the whole upper half-plane. By the reflection principle, the values taken by $f_2(z)$ at the two points $x, x/(2x-1)$ on the real axis along C are reflections with respect to the imaginary axis ($u_2 = 0$), that is

$$f_2(x) = -u_2\Big(\frac{x}{2x-1}, 0\Big) + iv_2\Big(\frac{x}{2x-1}, 0\Big).$$

We already know that, on $(0, 1)$,

$$\Re\{(1-i)f_2(x)\} = u_2(x, 0) + v_2(x, 0) = \chi(x),$$

and, on $(-\infty, 0)$, $(1, \infty)$,

$$\Im\{(1-i)f_2(x)\} = v_2\Big(\frac{x}{2x-1}, 0\Big) + u_2\Big(\frac{x}{2x-1}, 0\Big) = \chi\Big(\frac{x}{2x-1}\Big),$$

so that we can apply the Keldysh–Sedov formula on the upper half-plane to this problem. After a simple substitution, we arrive at

$$(1-i)f_2(z) = \frac{1}{\pi i}\int_0^1 \sqrt{\frac{z(1-z)}{t(1-t)}}\left\{\frac{1}{t-z} - \frac{1}{t+z-2tz}\right\}\chi(t)dt. \tag{12.75}$$

The solution to the general problem is then

$$f(z) = f_1(z) + f_2(z).$$

Note. Consider the case $\phi(\zeta) = 0$, $\chi(\zeta) = 0$. Taking a reflection once with respect to the real axis on the plane corresponds to taking a reflection once with respect to the circle; that is, as z takes a 180° turn about $z = 0$ (or $z = 1$), the mapping $w = f(z)$ makes a reflection with respect to $u + v = 0$ once, and a reflection with respect to the imaginary axis once, so that w makes a 90° turn on the z-plane. Therefore, as z loops round 0 (or 1) a whole turn, there is a change of sign for w, so that the function

$$p(z) = \frac{f(z)}{\sqrt{z(1-z)}}$$

is single-valued on the whole plane. By hypothesis we have $f(\frac{1}{2}) = \rho(1-i)$, where ρ is real and, by the relationship to the circle, $f(\infty) = -\rho(1+i)$, so that $p(z) \to 0$ as $z \to \infty$.

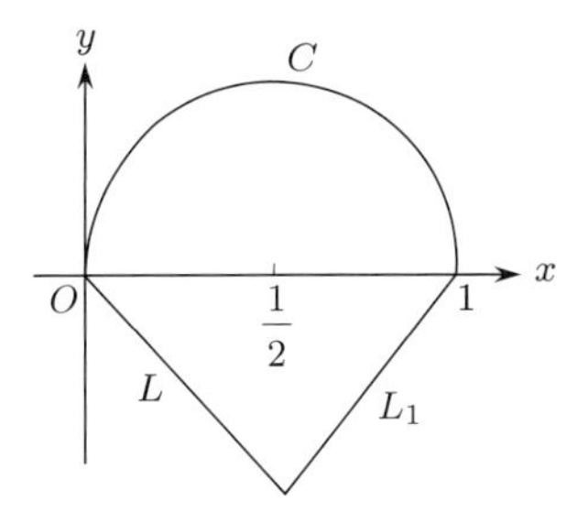

Figure 12.6.

If we assume that $f(z) = o(|z|^{-1/2})$ and $f(z) = o(|1 - z|^{-1/2})$, then $p(z) = o(|z|^{-1})$, which shows that $p(z)$ does not have any singularity apart from $z = 0$; thus $z = 1$ is not a singularity, so that $p(z)$ has to be zero. In other words, under such an assumption, the solution to the problem is unique.

If we assume that

$$\frac{\partial u}{\partial x} = o(|z|^{-3/2}), \quad \frac{\partial u}{\partial y} = o(|z|^{-3/2}),$$

then

$$f'(z) = o(|z|^{-3/2}).$$

If $u(0, 0) = u(1, 0) = 0$, then it can be deduced that $f(z) = o(|z|^{-1/2})$.

12.14 A mixed form of partial differential equations

We now consider Lavrentiev's partial differential equation

$$\frac{\partial^2 u}{\partial x^2} + \theta(y)\frac{\partial^2 u}{\partial y^2} = 0, \tag{12.76}$$

where $\theta(y) = \pm 1$ corresponds to $y \gtrless 0$.

The region $\mathcal{D}$ being considered lies inside the curve described as follows (see Fig. 12.6): The upper semi-circle

$$C : |z - \tfrac{1}{2}| = \tfrac{1}{2}, \quad y > 0,$$

centre $z = \frac{1}{2}$, radius $\frac{1}{2}$, together with the two line segments emitting as rays L, L_1 from 0 and 1 given by

$$L : x + y = 0, \quad L_1 : x - y = 1.$$

Let $\phi(\zeta)$ be a function defined on C, let $\psi(x)$ be a function defined on L, and suppose that they satisfy Hölder's condition, and that $\phi(0) = \psi(0) = 0$.

The problem is to find $u(x, y)$ satisfying the following conditions:

(i) When $y \neq 0$ in $\pm\mathcal{D}$, the function u satisfies equation (12.76);
(ii) the function is continuous in $\bar{\mathcal{D}}$, and $u|_C = \phi(\zeta)$, $u|_L = \psi(x)$;
(iii) the partial derivatives $\partial u/\partial x$, $\partial u/\partial y$ are continuous inside $\mathcal{D}$.
(iv) In the neighbourhoods of $z = 0$ and $z = 1$,

$$\frac{\partial u}{\partial x} = o(x^{-3/2}), \quad \frac{\partial u}{\partial y} = o(x^{-3/2}), \quad \frac{\partial u}{\partial x} = o(|1 - x|^{-3/2}), \quad \frac{\partial u}{\partial y} = o(|1 - x|^{-3/2}).$$

Here is the method of solution given by A. V. Bitsadze: The solution, in the lower half-plane, of the equation

$$\frac{\partial^2 u}{\partial x^2} - \frac{\partial^2 u}{\partial y^2} = \left(\frac{\partial}{\partial x} + \frac{\partial}{\partial y}\right)\left(\frac{\partial u}{\partial x} - \frac{\partial u}{\partial y}\right) = 0$$

can be represented as

$$u = \Phi(x + y) + \Psi(x - y). \tag{12.77}$$

Here, Φ, Ψ are arbitrary functions and, by condition (ii), we need to have

$$u|_L = \Phi(0) + \Psi(2x) = \psi(x).$$

Thus

$$u = \Phi(x + y) - \Phi(0) + \psi\left(\frac{x - y}{2}\right), \tag{12.78}$$

and, again by (ii),

$$u(x, 0) = \Phi(x) - \Phi(0) + \psi\left(\frac{x}{2}\right). \tag{12.79}$$

The function $u(x, y)$ is harmonic in the upper half-plane; let $v(x, y)$ denote its conjugate partner with $v(0, 0) = 0$. By (12.78) we find that, on the lower half-plane,

$$\frac{\partial u}{\partial y} = \Phi'(x + y) - \frac{1}{2}\psi'\left(\frac{x - y}{2}\right),$$

and, by (iii) and the continuity of $\partial u/\partial y$, we have, on the x-axis,

$$\frac{\partial v}{\partial x} = -\frac{\partial u}{\partial y} = -\Phi'(x) + \frac{1}{2}\psi'\left(\frac{x}{2}\right).$$

Integrating, we then have

$$v(x, 0) = -\Phi(x) + \Phi(0) + \psi\left(\frac{x}{2}\right), \tag{12.80}$$

and, together with (12.79),

$$u(x, 0) + v(x, 0) = 2\psi\left(\frac{x}{2}\right). \tag{12.81}$$

Consequently, the problem has been transformed into one that has already been discussed in Section 12.13 – that is the problem of finding a harmonic function satisfying

$$u|_C = \phi(\zeta), \quad u + v|_{y=0} = 2\chi(x) = 2\psi\left(\frac{x}{2}\right);$$

the existence and uniqueness of a solution have also been discussed in Section 12.13.

We repeat the proof of the uniqueness of the solution. We may assume without loss that u is zero on L and C. From (12.81) we then have

$$u(x,0) + v(x,0) = 0. \tag{12.81'}$$

Let K denote the disc centre $(\xi, 0)$, radius ϵ, lying entirely inside the circle $|z - \frac{1}{2}| = \frac{1}{2}$, and let C_K and $\bar{C}_K$ be the upper and lower semi-circles corresponding to K. Since the value for $F(z)$ inside the disc K is uniquely determined by the value of its real part on C_K, we need to determine $F(z)$, which is meromorphic inside K and continuous in the closed disc $\bar{K}$, satisfying the boundary conditions

$$\begin{cases} \Re F(z) = u(x,y), & z \in C_K, \\ \Im F(z) = -u(x,-y), & z \in \bar{C}_K. \end{cases} \tag{12.82}$$

By the Keldysh–Sedov formula we have

$$\begin{aligned} F(z) = {} & \frac{1}{\pi i}\int_{C_K} \sqrt{\frac{(z-\xi+\epsilon)(\xi+\epsilon-z)}{(t-\xi+\epsilon)(\xi+\epsilon-t)}}\,\frac{u(t)}{t-z}\,dt \\ & - \frac{1}{\pi}\int_{\bar{C}_K} \sqrt{\frac{(z-\xi+\epsilon)(\xi+\epsilon-z)}{(t-\xi+\epsilon)(\xi+\epsilon-t)}}\,\frac{u(\bar{t})}{t-z}\,dt. \end{aligned} \tag{12.83}$$

With $z = \xi$, and the substitution $t = \xi + \epsilon e^{i\theta}$, equation (12.83) then yields

$$F(\xi) = \frac{1}{\pi}\int_0^{\pi} \frac{u(\xi+\epsilon e^{i\theta})}{\sqrt{1-e^{2i\theta}}}\,d\theta - \frac{i}{\pi}\int_{\pi}^{2\pi} \frac{u(\xi+\epsilon e^{-i\theta})}{\sqrt{1-e^{2i\theta}}}\,d\theta.$$

Making the substitution $\theta = 2\pi - \phi$ in the second integral, and noting that

$$\frac{1}{\sqrt{1-e^{-2i\phi}}} = \frac{e^{i\phi}}{\sqrt{1-e^{2i\phi}}},$$

we then have

$$F(\xi) = \frac{1}{\pi}\int_0^{\pi} u(\xi+\epsilon e^{i\theta})\sqrt{\frac{1-e^{i\theta}}{1+e^{i\theta}}}\,d\theta = \frac{1}{\pi}\int_0^{\pi}\sqrt{\frac{1}{i}\tan\frac{\theta}{2}}\,u(\xi+\epsilon e^{i\theta})d\theta.$$

Thus

$$u(\xi,0) = \frac{1}{\pi}\int_0^{\pi}\sqrt{\frac{1}{2}\tan\frac{\theta}{2}}\,u(\theta)d\theta. \tag{12.84}$$

From (12.84) we can prove that $u(x, y)$ can assume neither the greatest positive value nor the least negative value on the line segment $(0, 1)$ along the real axis. For if u did take on the greatest value M at $(\xi, 0)$ then we could find a sufficiently small circle C_K centre at $(\xi, 0)$, radius ϵ with $|u(\xi + \epsilon e^{i\theta})| < M$, giving

$$M = u(\xi, 0) = \frac{1}{\pi} \int_0^{\pi} \sqrt{\frac{1}{2} \tan \frac{\theta}{2}} u(\xi + \epsilon e^{i\theta})\, d\theta < \frac{M}{\pi} \int_0^{\pi} \sqrt{\frac{1}{2} \tan \frac{\theta}{2}}\, d\theta = M.$$

Similarly, u cannot take on the least negative value. Also, from $u(0, 0) = u(1, 0) = 0$, we deduce that $u(x, 0)$ does take on the value 0 along the line segment $(0, 1)$, and therefore u has to be identically 0 in the region $\mathcal{D}$.

13
Weierstrass' elliptic function theory

13.1 Modules

Definition 13.1 By a module we mean a set M of numbers with the property that the sum and the difference of any two members of M also belong to M. We assume that M is an infinite set.

Example 13.2 The set of integers forms a module, and so does the set

$$nw, \quad n = 0, \pm 1, \pm 2, \ldots .$$

Example 13.3 The set of rational numbers also forms a module.

Example 13.4 The set of real numbers also forms a module.

Example 13.5 The set of complex numbers also forms a module.

Example 13.6 The set of numbers having the form

$$a + bi, \quad a, b = 0, \pm 1, \pm 2, \ldots$$

also forms a module.

Definition 13.7 If there are n numbers $w_1, \ldots, w_n$ in a module M such that any member of M has a unique representation

$$P_1 w_1 + \cdots + P_n w_n \quad (P_i \text{ are integers}),$$

then $\{w_1, \ldots, w_n\}$ forms a basis for M, and n is called the rank of M.

In Example 13.2, the number w is the basis, and the rank is 1.

In Example 13.6, $\{1, i\}$ is the basis, and the rank is 2.

If a module M has another basis $\{w'_1, \ldots, w'_m\}$ then, from Definition 13.7, we have

$$w_i = \sum_{j=1}^{m} a_{ij} w'_j, \quad w'_j = \sum_{k=1}^{n} b_{jk} w_k,$$

and hence

$$w_i = \sum_{j=1}^{m} \sum_{k=1}^{n} a_{ij} b_{jk} w_k.$$

The uniqueness of the representation then gives

$$\sum_{j=1}^{m} \sum_{k=1}^{n} a_{ij} b_{jk} = \begin{cases} 1, & \text{if} \quad i = k, \\ 0, & \text{if} \quad i \neq k. \end{cases}$$

Therefore $(a_{i,j})$ and (b_{jk}) are invertible square matrices, so that $m = n$.

Theorem 13.8 *If a module M has another basis $\{w'_1, \ldots, w'_m\}$ then $m = n$ and between the two bases there is a relationship*

$$w_i = \sum_{j=1}^{m} a_{ij} w'_j$$

in which the matrix (a_{ij}) has the values ± 1 for its determinant.

Theorem 13.9 *The rank of a real module with no finite limit point is 1, so that there exists w for which the module takes the form*

$$nw, \quad n = 0, \pm 1, \pm 2, \ldots .$$

To see this, we let w be the least positive member of M. If there is a number w' in M not having the form nw, then $w' = (n + \theta)w$ for some integer n and $0 < \theta < 1$, leading to $0 < w' - nw < w$, which contradicts the assumption on w.

More generally, we have the following.

Theorem 13.10 *Suppose that a module has no finite limit point, and that the ratio of any two of its members is real. Then the rank is 1.*

Theorem 13.11 *The rank of a complex module with no finite limit point is either 1 or 2.*

Proof. (1) Let members of the module be represented by points on the plane, with the member w having the least distance from the origin. Take the straight line L through $0, w$. On this line, by Theorem 13.10, only numbers of the form nw ($n = 0, \pm 1, \pm 2, \ldots$) can be members of M. If there are no other points then the rank for the module is 1.

(2) Suppose otherwise, so that there are members of M not on the line L. Take a circle centre the origin with the following properties: There are no points belonging to M inside the circle, except those lying on the line L, but there are members of M on the circle. Because M has no finite limit points, there can only be finitely many members on the circumference. There is now a point w' on the circle making the least angle subtended at the origin with w. We proceed to prove that members of M must have the form

$$nw + mw'.$$

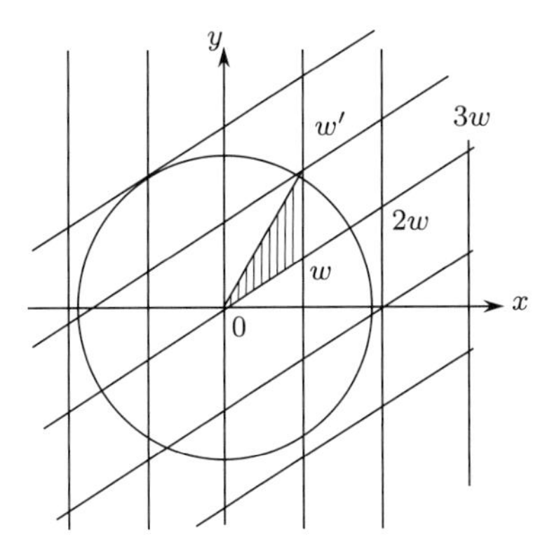

Figure 13.1.

Suppose otherwise, so that there is a member r with the representation

$$r = (n + \theta)w + (m + \theta')w', \quad 0 < \theta < 1, \ 0 < \theta' < 1.$$

With M being a module, the number

$$r_1 = \theta w + \theta' w'$$

is a member, and the point r_1 lies inside the parallelogram with the vertices 0, w, $w'w + w'$, and not on any of the vertices. By the definition of w', the shaded triangular region in Fig. 13.1 does not contain a point belonging to M. Thus r_1 has to lie in the remaining triangular part of the parallelogram; but now

$$w + w' - r_1 = (1 - \theta)w + (1 - \theta')w'$$

also belongs to M, and this point falls inside the shaded region. We therefore have a contradiction, and the theorem is proved. □

Exercise 13.12 Show that the module formed by n vectors has a rank $\leq n$.

13.2 Periodic functions

The trigonometric functions $\sin z$, $\cos z$, $\tan z$, ... etc. have the important property of being periodic. Thus, if $f(z)$ is any of these functions then

$$f(z + 2\pi) = f(z),$$

and hence, for any integer n, we always have

$$f(z + 2n\pi) = f(z).$$

Such a function is said to be periodic, and 2π is the period. In order to accommodate other types, we call these the singly periodic functions, and in this chapter we consider the doubly periodic functions.

Let w_1, w_2 be any two complex numbers with a ratio that is not a real number. A function $f(z)$ satisfying

$$f(z+2w_1)=f(z), \quad f(z+2w_2)=f(z),$$

is said to be doubly periodic, and $2w_1, 2w_2$ are the periods. A doubly periodic function which is meromorphic is called an elliptic function.

If the periods of a meromorphic function have a limit point then the function must be a constant. For if there is a limit point z_0 then there are infinitely many z_k in the neighbourhood of z_0 such that $f(z_k)=f(z_0)$, $k=1,2,\ldots$, and the function has to be constant according to Vitali's theorem.

Combined with the results of Section 13.1, we need only discuss meromorphic functions which are singly periodic or doubly periodic (with the ratio of the periods being complex), and there is not much point in considering doubly periodic functions with periods having a real ratio.

13.3 Expansions of periodic integral functions

Suppose that $f(z)$ is an integral function with period w, that is an integral function satisfying $f(z+w)=f(z)$. Let $W=2\pi z/w$ and write

$$f(z)=\phi(W),$$

so that

$$\phi(W+2\pi)=f(z+w)=f(z)=\phi(W),$$

which shows that $\phi(W)$ is periodic with period 2π. In order to study $\phi(W)$ on the whole plane it suffices to examine $\phi(W)$ in the infinite strip $0\le x<2\pi$, because the whole plane is the union of the strips

$$2\pi m\le x<2\pi(m+1), \quad m=0,\pm 1,\pm 2,\ldots,$$

and the behaviour of the function is the same in each of these strips.

Let $\phi(z)$ be expanded into its Fourier series

$$\phi(z)=\sum_{n=-\infty}^{\infty} C_n(y)e^{inx},$$

where

$$C_n(y)=\frac{1}{2\pi}\int_0^{2\pi}\phi(z)e^{-inx}\,dx.$$

Since $\phi(z)$ satisfies Laplace's equation we have

$$\begin{aligned} C_n''(y) &= \frac{1}{2\pi}\int_0^{2\pi} \frac{\partial^2}{\partial y^2}\phi(z)e^{-inx}\,dx \\ &= -\frac{1}{2\pi}\int_0^{2\pi} \frac{\partial^2}{\partial x^2}\phi(z)e^{-inx}\,dx \\ &= \frac{n^2}{2\pi}\int_0^{2\pi} \phi(z)e^{-inx}\,dx = n^2 C_n(y), \end{aligned}$$

having applied integration by parts twice and the periodicity of $\phi(z)$.

The solution to this differential equation is $a_n e^{-ny} + b_n e^{ny}$, so that

$$\phi(z) = \sum_{n=-\infty}^{\infty} a_n e^{inz} + \sum_{n=-\infty}^{\infty} b_n e^{in\bar{z}}.$$

This is the general form of a periodic harmonic function, with the analytic function satisfying $\partial\phi/\partial\bar{z} = 0$, so that it has the expansion

$$\phi(z) = \sum_{n=-\infty}^{\infty} a_n e^{inz},$$

delivering

$$f(z) = \sum_{n=-\infty}^{\infty} a_n e^{\frac{2\pi in}{w}z}.$$

The convergence aspects involved in this chapter can all be dealt with very easily, so that we have the following.

Theorem 13.13 *A periodic integral function $f(z)$ has a Fourier series representation which is convergent everywhere.*

Speaking generally for $f(z)$, the region between two parallel lines through 0 and w is called a periodic strip. Clearly the whole plane can be covered by shifting the periodic strip forward and backward. By Liouville's theorem we deduce the following.

Theorem 13.14 *If a periodic function is regular within a periodic strip then the function is constant.*

Theorem 13.15 *If, as z tends toward the two extremes of its periodic strip, $f(z)$ is of finite order, then $f(z)$ must be a trigonometric polynomial.*

More generally, we have Theorem 13.16.

Theorem 3.16 *Let $f(z)$ be a meromorphic periodic function. If $f(z)$ is of finite order as $z \to \infty$, then it is the ratio of two trigonometric polynomials.*

13.4 The fundamental region

Consider the group of substitutions

$$z' = z + mw + m'w', \qquad m, m' = 0, \pm 1, \pm 2, \ldots . \tag{13.1}$$

If z_0 becomes z under a substitution (13.1) then we say that the two points correspond, or are related, or are equivalent.

If w/w' is not a real number then, associated with the group, there is a special point z_0 with the property that the set of points related to z_0 has no limit points (apart from ∞).

The generating elements for the group are

$$z = z + w, \quad z = z + w'.$$

We may assume without loss that $\Im(w/w') > 0$. Corresponding to any integer m, take the straight line

$$z = mw + tw', \quad -\infty < t < \infty;$$

again, corresponding to any m', take the straight line

$$z = m'w' + tw \quad -\infty < t < \infty.$$

These straight lines divide the plane into infinitely many parallelograms which correspond to each other in the sense that any two parallelograms can be transformed to each other by means of a substitution (13.1).

Definition 13.17 A region in the plane is called a fundamental region if it satisfies the following conditions:

(i) any point must correspond to a point in the region;
(ii) no two points in the region correspond to each other.

Example 13.18 The parallelogram with the four vertices 0, w, w', $w + w'$ form a fundamental region, but, strictly speaking, from the four sides we should include only the two sides $0w$, $0w'$, and only 0 from the four vertices.

Example 13.19 The shape in Fig. 13.2 is obtained by removing a piece from one side of a fundamental region and attaching it to the opposite side; the result is also a fundamental region.

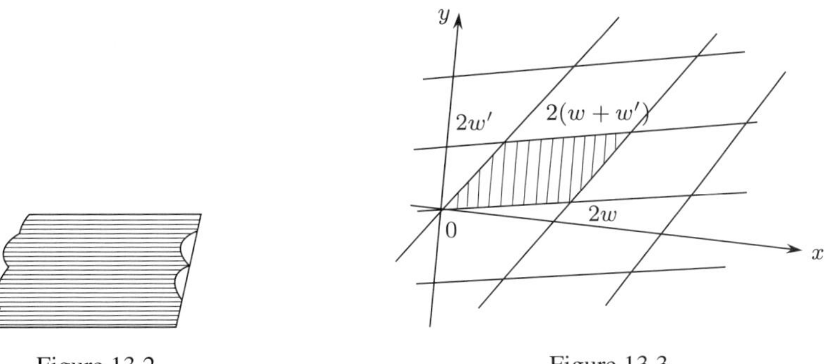

Figure 13.2. Figure 13.3.

13.5 General properties of an elliptic function

Let $f(z)$ be any elliptic function, that is a doubly periodic meromorphic function with periods $2w$ and $2w'$, with the ratio w/w' not being a real number; we may assume without loss that $\Im(w/w') > 0$. Therefore the parallelogram with the four vertices $0, 2w, 2w', 2(w + w')$ is not degenerate, and can be considered as a fundamental region (see Fig. 13.3). Numbers of the form $2(mw + m'w')$ (m, m' being integers) then deliver parallelograms on the plane which are similar to the fundamental region. The behaviour of $f(z)$ within a parallelogram is identical to that within the fundamental region.

Theorem 13.20 *If a doubly periodic function is an integral function then it must be a constant.*

The proof is exceedingly simple because $f(z)$ has to be bounded in the fundamental region, and therefore bounded in the whole plane, and the required result follows immediately from Liouville's theorem.

Therefore, within the fundamental region, or on its boundary, there must be at least one pole for $f(z)$. We use 'two-for-the-price-of-one' when we count the poles on the boundary; this is because if z is a pole then so are $z \pm w, z \pm w'$. If z is on the boundary then the four points $z \pm w, z \pm w'$ are also on the boundary; let z_0 be one of them. If we make a semi-circular loop round z outside the region then the corresponding semi-circular loop round z_0 will be inside the region. After such a modification within the fundamental region, the combined contribution from the poles z and z_0 may be considered as that from only one of them. Similarly, if there is a pole at one of the vertices, then all the vertices are poles, but only one of them should be counted. There are only finitely many poles within a fundamental region and, taking account of multiplicity, together with the above procedure, we now define the order of an elliptic function to be the number of poles in the fundamental region.

Theorem 13.21 *The sum of the residues over all the poles of an elliptic function in a fundamental region is equal to* 0.

Proof. It suffices to show that the value of the integral along the boundary curve C of the fundamental region is 0. (If there are poles on C then we need, of course, to make appropriate indentations, as described in the above; we shall not bother with such details.)

$$\begin{aligned}\int_C f(z)dz &= \left(\int_0^{2w} + \int_{2w}^{2(w+w')} + \int_{2(w+w')}^{2w'} + \int_{2w'}^{0}\right) f(z)dz \\ &= \int_0^{2w} f(z)dz + \int_0^{2w'} f(z+2w)dz + \int_{2w}^{0} f(z+2w')dz \\ &\quad + \int_{2w'}^{0} f(z)dz = 0.\end{aligned}$$

□

We deduce immediately the following theorems.

Theorem 13.22 *There is no elliptic function having order 1.*

Theorem 13.23 *The number of times a complex number a is taken by an elliptic function is equal to its order; in particular, the number of zeros is equal to the order.*

Proof. The difference between the number of solutions to $f(z) = a$ and the number of poles, within a fundamental region, is given by

$$\frac{1}{2\pi i}\int_C \frac{f'(z)}{f(z)-a}dz.$$

The integrand is an elliptic function, so that the value here is 0, as required. □

Theorem 13.24 *Let the solutions to $f(z) = a$ be denoted by $a_1, a_2, \ldots, a_n$, and let the solutions to $f(z) = \infty$ be $p_1, p_2, \ldots, p_n$. Then the two sums*

$$\sum_{i=1}^{n} a_i, \quad \sum_{i=1}^{n} p_i$$

have the same value.

Proof. We have

$$\begin{aligned}
\sum_{i=1}^{n} a_i - \sum_{i=1}^{n} p_i &= \frac{1}{2\pi i}\int_C z\frac{f'(z)}{f(z)-a}\,dz \\
&= \frac{1}{2\pi i}\left(\int_0^{2w} + \int_{2w}^{2(w+w')} + \int_{2(w+w')}^{2w'} + \int_{2w'}^{0}\right) z\frac{f'(z)}{f(z)-a}\,dz \\
&= \frac{1}{2\pi i}\left(\int_0^{2w} z\frac{f'(z)}{f(z)-a}\,dz + \int_0^{2w'} (z+2w)\frac{f'(z)}{f(z)-a}\,dz\right. \\
&\quad \left. + \int_{2w}^{0} (z+2w')\frac{f'(z)}{f(z)-a}\,dz + \int_{2w'}^{0} z\frac{f'(z)}{f(z)-a}\,dz\right) \\
&= \frac{1}{2\pi i}\left(-2w'\int_0^{2w} \frac{f'(z)}{f(z)-a}\,dz + 2w\int_0^{2w'} \frac{f'(z)}{f(z)-a}\,dz\right).
\end{aligned}$$

The value of the integral

$$\int_0^{2w} \frac{f'(z)}{f(z)-a}\,dz$$

is equal to the variation of $\log(f(z)-a)$ as z moves from 0 to $2w$, and because $f(2w)-a = f(0)-a$, the value concerned is $2m\pi i$, where m is an integer. The required result follows. □

13.6 Algebraic relationships

The set K of elliptic functions with periods $2w$, $2w'$ is closed with respect to the operations of addition, subtraction, multiplication and division (denominator $\neq 0$); the set K is said to form a field of algebraic functions.

Theorem 13.25 *Any two functions $f(z)$, $g(z)$ in a field of algebraic functions are algebraically related, in the sense that there is a polynomial $P(Z, W)$, with constant coefficients, such that*

$$P\big(f(z), g(z)\big) = 0.$$

Proof. Let $a_1, \ldots, a_m$ be the set of poles for $f(z)$ and $g(z)$, which have the orders $p_1, \ldots, p_m$ and $q_1, \ldots, q_m$, respectively, at $a_1, \ldots, a_m$, and write

$$A = \sum_{k=1}^{m} \max(p_k, q_k).$$

Let $Q(Z, W)$ be a certain polynomial of Z and W with degree n, which, together with the coefficients, are to be suitably chosen later. On setting $Z = f(z)$, $W = g(z)$, the resulting expression $F(z) = Q(f, g)$ is also a function in the field concerned, and we are done if we can show that $F(z)$ is a constant C, because then we may set $P = Q - C$. Indeed it suffices to show that all the principal parts for $F(z)$ at $z = a_k$ vanish.

The order of the pole for $F(z)$ at a_k is at most $n \max(p_k, q_k)$, which then delivers a bound for the number of terms in its principal part there. Thus the upper bound

$$n \sum_{k=1}^{m} \max(p_k, q_k) = nA$$

delivers a bound for the total number of terms in all the principal parts. Such a condition sets a bound for the linear relationship between the coefficients of the polynomial Q. There being $\frac{1}{2}(n+3)n$ coefficients to be chosen for Q (we have excluded the one corresponding to the constant term), it follows that if we choose n so that $n + 3 > 2A$, then there will be enough suitable coefficients to give a non-trivial solution to deliver our sought-after polynomial Q. The theorem is proved. □

Since the derivative $f'(z)$ of an elliptic function belongs to the same field of algebraic functions containing $f(z)$, we deduce Theorem 13.26.

Theorem 13.26 *Any elliptic function $f(z)$ must satisfy an algebraic differential equation*

$$P\big(f(z), f'(z)\big) = 0,$$

where $P(Z, W)$ is a polynomial.

Note. Let $f(z)$ be an odd elliptic function; we then have $f(w) = -f(-w) = -f(-w + 2w) = -f(w)$, which implies $f(w) = 0$. Thus, an odd elliptic function has either a zero or a pole at the half-period. Moreover, the order of this zero, or pole, has to be an odd number. Otherwise, if $f(z)$ has the order $2n$ at the zero then $f^{(2n)}(z)$ is an odd function which has no zero at $z = w$; if $f(z)$ has the order $2n$ at the pole then consideration of $\big(1/f(z)\big)^{(2n)}$ leads to a similar contradiction.

13.7 Two types of theories for elliptic functions

We have already remarked that there is no elliptic function of order 1. We therefore start with elliptic functions of order 2, hoping that, from such basic fundamentals, we may be able to construct all such elliptic functions. There is a clear dichotomy concerning elliptic functions of order 2: (i) there is a double pole for the elliptic function; (ii) there are two simple poles for the elliptic function. This is the initial departure point between Weierstrass' theory and Jacobi's theory.

More specifically, Weierstrass starts with the construction of an elliptic function with a double pole in the fundamental region and develops the general theory from there. By contrast, Jacobi's theory starts with the situation when there are two simple poles. Although the former theory seems to be more convenient to use, in practice Jacobi's functions appear more frequently.

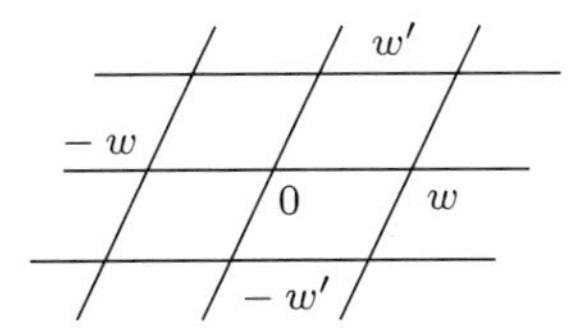

Figure 13.4.

13.8 Weierstrass' γ function

Perhaps the first remark to be made is that, in most texts, the function in this section is called Weierstrass' $\wp$ function.

We set the origin to be at the centre of the fundamental parallelogram (see Fig. 13.4) and examine the type of elliptic functions having a double pole at $z = 0$. By periodicity the points $z = 2mw + 2m'w'$ are also poles, and this suggests to us that we should consider the function

$$\sum_{m,m'=-\infty}^{\infty} \frac{1}{(z - 2mw - 2m'w')^2}.$$

In the following we use the symbol $\sum$ to denote summation over all m, m', and $\sum'$ for such summation with the term $m = m' = 0$ being omitted.

If the series is convergent then it will satisfy our requirements, so we must examine it carefully.

Lemma 13.27 *If $\Im(w/w') > 0$ then the series*

$$\sum_{m,m'=-\infty}^{\infty}{}' \frac{1}{(mw + m'w')^3}$$

is absolutely convergent.

Proof. Display the points $mw + m'w'$ on the plane as a lattice, and define π_n to be the set of $8n$ lattice points lying on the perimeter of the parallelogram with vertices at $n(w \pm w'), n(-w \pm w')$. Let ℓ denote the least distance of the eight points on π_1 to the origin, so that the distance of the points on π_n from the origin is at least $n\ell$, and therefore

$$\sum_{\pi_n} \frac{1}{|mw + m'w'|^3} \le \frac{8n}{(n\ell)^3} = \frac{8}{\ell^3} \frac{1}{n^2}.$$

Since the series $\sum 1/n^2$ is convergent, the lemma is proved. □

It follows from the lemma that the series

$$f(z) = \sum_{m,m'=-\infty}^{\infty} \frac{1}{(z - 2mw - 2m'w')^3} \tag{13.2}$$

is absolutely convergent for z not being one of the numbers $2mw + 2m'w'$. Indeed the convergence is 'locally uniform', that is uniform inside any circle not including a pole. It then follows that (13.2) represents a meromorphic function which is periodic with periods $2w, 2w'$; it is clear that the function is odd.

By integrating $f(z)$ we have the construction of an even elliptic function of order 2.

We perform integration along a curve leading from z_0 to z without passing through a pole:

$$\begin{aligned}\phi(z) &= C + \int_{z_0}^{z} f(\zeta)d\zeta \\ &= C - \frac{1}{2}\sum\left[\frac{1}{(z-2mw-2m'w')^2} - \frac{1}{(z_0-2mw-2m'w')^2}\right].\end{aligned}$$

On the removal of the term $m = m' = 0$ from the sum, we then have

$$\phi(z) + \frac{1}{2z^2} = C + \frac{1}{2z_0^2} - \frac{1}{2}\sum{}'\left[\frac{1}{(z-2mw-2m'w')^2} - \frac{1}{(z_0-2mw-2m'w')^2}\right].$$

The function on the right-hand side is regular at $z = 0$, and we may choose C so that its value is zero at $z = 0$, that is

$$0 = C + \frac{1}{2z_0^2} - \frac{1}{2}\sum{}'\left[\frac{1}{(2mw+2m'w')^2} - \frac{1}{(z_0-2mw-2m'w')^2}\right]$$

so that

$$\phi(z) = -\frac{1}{2}\left\{\frac{1}{z^2} + \sum{}'\left[\frac{1}{(z-2mw-2m'w')^2} - \frac{1}{(2mw+2m'w')^2}\right]\right\}.$$

The function inside the curly braces is the Weierstrass function, which is denoted by γ here, that is

$$\gamma(z) = \frac{1}{z^2} + \sum{}'\left[\frac{1}{(z-2mw-2m'w')^2} - \frac{1}{(2mw+2m'w')^2}\right]. \tag{13.3}$$

The series is absolutely convergent because, for sufficiently large T,

$$\left|\frac{1}{(z-T)^2} - \frac{1}{T^2}\right| = \left|\frac{(2T-z)z}{T^2(z-T)^2}\right| = O\left(\frac{1}{T^3}\right),$$

so that the series (13.3) is convergent according to the lemma.

It is also easy to see that $\gamma(z)$ is an even function, that is $\gamma(-z) = \gamma(z)$, and

$$\begin{aligned}\gamma'(z) &= -\frac{2}{z^3} - 2\sum{}'\frac{2}{(z-2mw-2m'w')^3} \\ &= -2\sum\frac{1}{(z-2mw-2m'w')^3} = -2f(z).\end{aligned}$$

Therefore $\gamma'(z)$ is an elliptic function with periods $2w, 2w'$, that is

$$\gamma'(z+2w)-\gamma'(z)=0,$$
$$\gamma'(z+2w')-\gamma'(z)=0.$$

On integration we then have

$$\gamma(z+2w)-\gamma(z)=C,$$
$$\gamma(z+2w')-\gamma(z)=C'.$$

Substituting $z=-w$ and $z=-w'$ separately into these equations, and making use of the fact that $\gamma(z)$ is even, we find that $C=C'=0$. Therefore $\gamma(z)$ is an elliptic function with a double pole inside a parallelogram, and the principal part at $2mw+2m'w'$ is the term $1/(z-2mw+-2m'w')^2$.

It is clear that $\gamma'(z)$ is an odd elliptic function of order 3 and, by the Note in Section 13.6, there are zeros at the half-periods $w, w', w+w'$, so that these are points at which $\gamma(z)$ takes its value twice; that is, if

$$e_1=\gamma(w),\quad e_2=\gamma(w+w'),\quad e_3=\gamma(w'),$$

then $\gamma(z)=e^1, e^2, e^3$ has double roots. If $e\neq e^1, e^2, e^3$ then $\gamma(z)=e$ definitely does not have a double root, the reason being that $\gamma'(z)=0$ cannot have an additional root.

The nth derivative of $\gamma(z)$ is given by

$$\gamma^{(n)}(z)=(-1)^n\frac{(n+1)!}{z^{n+2}}+(-1)^n(n+1)!\sum{}'\frac{1}{(z-2mw-2m'w')^{n+2}}.$$

13.9 The algebraic relationship between $\gamma(z)$ and $\gamma'(z)$

We first derive the Laurent expansion for $\gamma(z)$ at $z=0$. From

$$\frac{1}{(z-T)^2}-\frac{1}{T^2}=\frac{1}{T^2}\left[\left(1-\frac{z}{T}\right)^{-2}-1\right]=\sum_{n=1}^{\infty}\frac{n+1}{T^{n+2}}z^n,$$

and the fact that $\gamma(z)$ is an even function, it follows from (13.3) that

$$\gamma(z)=\frac{1}{z^2}+\frac{g_2}{20}z^2+\frac{g_3}{28}z^4+\cdots. \tag{13.4}$$

Here we adopt the common notation for the symbols

$$g_2=60\sum{}'\frac{1}{(2mw+2m'w')^4},\qquad g_3=140\sum{}'\frac{1}{(2mw+2m'w')^6}. \tag{13.5}$$

Differentiating (13.4) we have

$$\gamma'(z)=-\frac{2}{z^3}+\frac{g_2}{10}z+\frac{g_3}{7}z^3+\cdots. \tag{13.6}$$

By (13.4) and (13.6) we find that

$$\gamma'^2(z) = \frac{4}{z^6}\Big(1 - \frac{g_2}{10}z^4 - \frac{g_3}{7}z^6 + \cdots\Big),$$
$$\gamma^3(z) = \frac{1}{z^6}\Big(1 + \frac{3g_2}{20}z^4 + \frac{3g_3}{28}z^6 + \cdots\Big),$$

so that

$$(\gamma'(z))^2 - 4(\gamma(z))^3 + g_2\gamma(z) = -g_3 + C_2z^2 + C_3z^4 + \cdots. \tag{13.7}$$

The right-hand side is a doubly periodic function which is regular everywhere, so that it is constant; that is, $\gamma(z)$ and $\gamma'(z)$ satisfy the following algebraic relationship:

$$(\gamma'(z))^2 = 4(\gamma(z))^3 - g_2\gamma(z) - g_3. \tag{13.8}$$

We already know that $\gamma'(z)$ has zeros at $z = w, w', w + w'$, so that

$$(\gamma'(z))^2 = 4(\gamma(z) - e_1)(\gamma(z) - e_2)(\gamma(z) - e_3). \tag{13.9}$$

The numbers e_1, e_2, e_3 satisfy

$$e_1 + e_2 + e_3 = 0, \quad e_1e_2 + e_2e_3 + e_3e_1 = -\frac{g_2}{4}, \quad e_1e_2e_3 = \frac{g_3}{4},$$

and the value of the relevant determinant $(1/16)(g_2^3 - 27g_3^2)$ must be non-zero.

Set $W = \gamma(z)$, so that

$$\frac{dz}{dW} = \frac{1}{\sqrt{4W^3 - g_2W - g_3}}, \tag{13.10}$$

and hence

$$z - z_0 = \int_{W_0}^{W} \frac{dW}{\sqrt{4W^3 - g_2W - g_3}}, \qquad W_0 = \gamma(z_0).$$

Since $W \to \infty$ as $z \to z_0$, we have

$$z = \int_{\infty}^{W} \frac{dW}{\sqrt{4W^3 - g_2W - g_3}},$$

the inverse function to $\gamma(z)$.

The differential equation (13.8) also gives

$$2\gamma''(z) = 12\gamma^2(z) - g_2. \tag{13.11}$$

13.10 The function $\zeta(z)$

Weierstrass also introduced the function

$$\zeta(z) = \frac{1}{z} - \int_0^z \Big(\gamma(s) - \frac{1}{s^2}\Big)ds \quad (\zeta'(z) = -\gamma(z)). \tag{13.12}$$

From the series (13.3) for $\gamma(z)$, we have

$$\int_0^z \left(\gamma(s) - \frac{1}{s^2}\right) ds = -\sum{}' \left[\frac{1}{z - 2mw - 2m'w'} + \frac{1}{2mw + 2m'w'} + \frac{z}{(2mw + 2m'w')^2} \right]. \tag{13.13}$$

The series represents a meromorphic function, with simple poles at $2mw + 2m'w'$. From

$$(\zeta(z) + \zeta(-z))' = \zeta'(z) - \zeta'(-z) = \gamma(-z) - \gamma(z) = 0,$$

we find that $\zeta(z) + \zeta(-z)$ is a constant, the value of which can be seen to be 0 by letting $z \to 0$ in (13.12). Therefore $\zeta(z) = -\zeta(-z)$, an odd function.

There being only a simple pole for $\zeta(z)$ in a parallelogram, it cannot be an elliptic function. We now consider the situation at $\zeta(z + 2w)$ and $\zeta(z + 2w')$. From

$$\big(\zeta(z + 2w) - \zeta(z)\big)' = -\gamma(z + 2w) + \gamma(z) = 0,$$

we deduce that

$$\zeta(z + 2w) - \zeta(z) = 2\eta, \tag{13.14}$$

and similarly

$$\zeta(z + 2w') - \zeta(z) = 2\eta'. \tag{13.15}$$

There is a simple relationship between the values for w, w', η and η'. We evaluate the integral of $\zeta(z)$ over the four sides of the parallelogram with vertices $\pm w$, $\pm w'$. The function has only the simple pole, with the residual 1, at $z = 0$, so that

$$\begin{aligned}
2\pi i &= \left(\int_{w-w'}^{w+w'} + \int_{w+w'}^{-w+w'} + \int_{-w+w'}^{-w-w'} + \int_{-w-w'}^{w-w'} \right) \zeta(z) dz \\
&= \int_{-w-w'}^{-w+w'} \zeta(z + 2w) dz + \int_{w-w'}^{-w-w'} \zeta(z + 2w') dz \\
&\quad + \int_{-w+w'}^{-w-w'} \zeta(z) dz + \int_{-w-w'}^{w-w'} \zeta(z) dz \\
&= \int_{-w-w'}^{-w+w'} \big(\zeta(z + 2w) - \zeta(z)\big) dz + \int_{w-w'}^{-w-w'} \big(\zeta(z + 2w') - \zeta(z)\big) dz \\
&= 4\eta w' - 4\eta' w.
\end{aligned}$$

We therefore have the Legendre relationship

$$\eta w' - \eta' w = \frac{1}{2}\pi i. \tag{13.16}$$

13.11 The function $\sigma(z)$

Weierstrass' $\sigma(z)$ function can be defined by

$$\sigma(z) = z \exp\left(\int_0^z \left(\zeta(s) - \frac{1}{s}\right) ds\right), \tag{13.17}$$

that is

$$(\log \sigma(z))' = \zeta(z). \tag{13.18}$$

From the expansion (13.13) we have the following representation for $\zeta(z)$:

$$\zeta(z) - \frac{1}{z} = \sum{}' \left(\frac{1}{z-T} + \frac{1}{T} + \frac{z}{T^2}\right),$$

where T represents $2mw + 2m'w'$ for the summation. On exponentiation we then have an infinite product representation for $\sigma(z)$:

$$\begin{aligned} \sigma(z) &= z \exp\left(\sum{}' \left(\log\left(1 - \frac{z}{T}\right) + \frac{z}{T} + \frac{z^2}{2T^2}\right)\right) \\ &= z \prod{}' \left(1 - \frac{z}{T}\right) e^{\frac{z}{T} + \frac{z^2}{2T^2}}, \end{aligned} \tag{13.19}$$

which shows clearly that $\sigma(z)$ is an integral function with zeros at $z = T$.

From

$$\begin{aligned} \sigma(-z) &= -z \exp\left(\int_0^{-z} \left(\zeta(s) - \frac{1}{s}\right) ds\right) \\ &= -z \exp\left(-\int_0^z \left(\zeta(-t) + \frac{1}{t}\right) dt\right) \\ &= -z \exp\left(\int_0^z \left(\zeta(t) - \frac{1}{t}\right) dt\right) = -\sigma(z), \end{aligned}$$

we see that $\sigma(z)$ is an odd function.

By (13.18) we have

$$\frac{\sigma'(z+2w)}{\sigma(z+2w)} - \frac{\sigma'(z)}{\sigma(z)} = 2\eta,$$

and, on integrating and then exponentiating,

$$\sigma(z+2w) = \sigma(z) e^{2\eta z + \gamma}.$$

Substituting $z = -w$ to give $-1 = e^{-2\eta w + \gamma}$, we find that

$$\sigma(z+2w) = -\sigma(z) e^{2\eta(z+w)};$$

similarly,

$$\sigma(z + 2w') = -\sigma(z)e^{2\eta(z+w')}.$$

Besides $\sigma(z)$ we still have three more σ functions:

$$\sigma_1(z) = -\frac{e^{2\eta z}}{\sigma(w)}\sigma(z - w),$$

$$\sigma_2(z) = -\frac{e^{2\eta' z}}{\sigma(w')}\sigma(z - w'),$$

$$\sigma_3(z) = -\frac{e^{2\eta'' z}}{\sigma(w'')}\sigma(z - w''),$$

where $w'' = w + w'$, and η'' is given by $\zeta(ww'') - \zeta(z) = 2\eta''$. The negative signs are introduced so that $\sigma_k(0) = 1$, although it is also common to do without them.

13.12 General representations for elliptic functions

A general elliptic function can be represented in one of the following ways:

(i) representation from $\sigma(z)$ (factorisation into factors method);
(ii) representation from $\zeta(z)$ (partial fractions method);
(iii) representation from $\gamma(z)$ and $\gamma'(z)$.

(i) Let $f(z)$ have, in the fundamental region, zeros at $a_1, \ldots, a_n$ and poles at $b_1, \ldots, b_n$ (repeated according to multiplicity, and dealing with those that are on the boundary as before); we have already seen in Theorem 13.24 that

$$\sum_{i=1}^{n} a_i - \sum_{i=1}^{n} b_i = 2\Omega$$

is a period.

Take the function

$$\phi(z) = \frac{\sigma(z - a_1)\cdots\sigma(z - a_n)}{\sigma(z - b_1)\cdots\sigma(z - b_{n-1})\sigma(z - b_n - 2\Omega)},$$

which has the same zeros and poles as those for $f(z)$, and

$$\phi(z + 2w) = e^{2\eta(b_1+\cdots+b_n-a_1-\cdots-a_n+2\Omega)}\phi(z) = \phi(z),$$

so that $f(z)/\phi(z)$ is an elliptic function with no poles, and hence

$$f(z) = C\phi(z).$$

(ii) Let $f(z)$ have the principal part

$$g_k(z) = \frac{C_{k1}}{z - b_k} + \frac{C_{k2}}{(z - b_k)^2} + \cdots + \frac{C_{kn_k}}{(z - b_k)^{n_k}}$$

at the pole $z = b_k$ $(= 1, 2, \ldots, m)$. Then, the difference

$$f(z) - \sum_{k=1}^{m} \left\{ C_{k1}\zeta(z - b_k) - C_{k2}\zeta'(z - b_k) + \cdots \right.$$
$$\left. + (-1)^{n_k - 1} \frac{C_{kn_k}}{(n_k - 1)!} \zeta^{(n_k - 1)}(z - b_k) \right\}$$

is a function which is analytic throughout the plane. Since it is elliptic and, as z becomes $z + 2w$, its value increases by

$$-2\eta \sum_{k=1}^{m} C_{k1},$$

we deduce that the value of the sum here has to be 0. The function concerned is constant, and we have Hermite's formula:

$$f(z) = C + \sum_{k=1}^{m} \left\{ C_{k1}\zeta(z - b_k) - C_{k2}\zeta'(z - b_k) + \cdots \right.$$
$$\left. + (-1)^{n_k - 1} \frac{C_{kn_k}}{(n_k - 1)!} \zeta^{(n_k - 1)}(z - b_k) \right\}.$$

(iii) We first establish the identity

$$\zeta(u + v) - \zeta(u) - \zeta(v) = \frac{1}{2} \frac{\gamma'(u) - \gamma'(v)}{\gamma(u)\gamma(v)}. \tag{13.20}$$

The expression

$$\frac{\sigma(u + v)\sigma(u - v)}{\sigma^2(u)},$$

as a function of u, has the same zeros and poles as those of $\gamma(u) - \gamma(v)$, so that

$$\gamma(u) - \gamma(v) = C \frac{\sigma(u + v)\sigma(u - v)}{\sigma^2(u)}.$$

On multiplication by $\sigma^2(u)$ and letting $u \to 0$, we have $1 = -C\sigma^2(v)$, so that

$$\gamma(u) - \gamma(v) = -\frac{\sigma(u + v)\sigma(u - v)}{\sigma^2(u)\sigma^2(v)}. \tag{13.21}$$

Differentiating with respect to u and v separately, we find that

$$\frac{\gamma'(u)}{\gamma(u) - \gamma(v)} = \zeta(u + v) + \zeta(u - v) - 2\zeta(u),$$
$$\frac{-\gamma'(v)}{\gamma(u) - \gamma(v)} = \zeta(u + v) + \zeta(u - v) - 2\zeta(v),$$

from which (13.20) follows.

In (13.20) we take $u = z - a$, $v = a - b$ to give

$$\zeta(z-b) - \zeta(z-a) = \zeta(a-b) + \frac{1}{2}\frac{\gamma'(z-a) - \gamma'(a-b)}{\gamma(z-a)\gamma(z-b)}.$$

Consequently, if $\sum_{k=1}^{m} C_{k1} = 0$ then

$$\sum_{k=1}^{m} C_{k1}\zeta(z-b_k)$$

can be represented as functions of γ and γ', and so $\zeta', \zeta'', \ldots$ can also be represented as functions of γ, γ'. Therefore, any elliptic function must be representable as a function of γ, γ'.

We already know that if $a \neq w, w', w + w'$ then the equation

$$\gamma(z) - \gamma(a) = 0$$

has two distinct roots $z = \pm a$, and the equation

$$\gamma(z) - \gamma(w) = 0$$

has a repeated root at $z = w$. With $f(z)$ being even, the order of the zero (or pole) has to be even, since otherwise $f'(z)$ would be an odd function with an even order at a repeated root, which is impossible (see the Note at the the end of Section 13.6).

Let the zeros of $f(z)$ in the parallelogram with vertices $\pm w, \pm w'$ be

$$\pm a_1, \pm a_2, \ldots, \pm a_k,$$

being listed in accordance with their orders, but being halved if the zero is one of w, w', $w + w'$. The expression

$$(\gamma(z) - \gamma(a_1))(\gamma(z) - \gamma(a_2)) \cdots (\gamma(z) - \gamma(a_n))$$

has the same zeros as those of $f(z)$; similarly, the expression

$$\frac{1}{(\gamma(z) - \gamma(b_1)) \cdots (\gamma(z) - \gamma(b_n))}$$

has the same poles as those of $f(z)$, so that

$$\phi(z) = \frac{(\gamma(z) - \gamma(a_1))(\gamma(z) - \gamma(a_2)) \cdots (\gamma(z) - \gamma(a_n))}{(\gamma(z) - \gamma(b_1))(\gamma(z) - \gamma(b_2)) \cdots (\gamma(z) - \gamma(b_n))}$$

has the same zeros and poles as those of $f(z)$, and no new zero or pole has been introduced. Therefore

$$f(z) = C\phi(z).$$

Any function $F(z)$ can be decomposed as the sum of its even and odd parts as follows:

$$F(z) = \frac{F(z) + F(-z)}{2} + \frac{F(z) - F(-z)}{2},$$

so that, for our function, we have

$$f(z) = R(\gamma(z)) + \gamma'(z)R_1(\gamma(z)),$$

where $R(Z)$ and $R_1(Z)$ are rational functions of Z.

13.13 Addition formulae

Consider the function

$$\gamma'(z) - A\gamma(z) - B.$$

It has a triple pole at $z = 0$, so that there are three roots, the sum of which is a period (Theorem 13.24). Thus, if u, v are roots, then $-u - v$ is also a root, so that

$$\begin{aligned} \gamma'(u) - A\gamma(u) - B &= 0, \\ \gamma'(v) - A\gamma(v) - B &= 0, \\ \gamma'(-u - v) - A\gamma(u + v) - B &= 0. \end{aligned} \tag{13.22}$$

Eliminating A, b we have

$$\begin{vmatrix} \gamma(u) & \gamma'(u) & 1 \\ \gamma(v) & \gamma'(v) & 1 \\ \gamma(u+v) & -\gamma'(u+v) & 1 \end{vmatrix} = 0. \tag{13.23}$$

The function $\gamma'(z)^2 - \{A\gamma(z) + B\}^2$ also has a zero at $z = u, v, u + v$, and it can be rewritten as

$$4\gamma^3(z) - A^2\gamma^2(z) - (2AB + g_2)\gamma(z) - (B^2 + g_2).$$

For general u, v, the cubic equation for $\gamma(z)$ is not the same as that for $\gamma(u)$, $\gamma(v)$, $\gamma(u + v)$, which are the three roots of the cubic equation

$$4Z^3 - A^2Z^2 - (2AB + g_2)Z - (B^2 + g_2) = 0,$$

so that

$$\gamma(u) + \gamma(v) + \gamma(u + v) = \frac{1}{4}A^2.$$

From (13.22) we have

$$A = \frac{\gamma'(u) - \gamma'(v)}{\gamma(u) - \gamma(v)},$$

so that

$$\gamma(u + v) = \frac{1}{4}\left(\frac{\gamma'(u) - \gamma'(v)}{\gamma(u) - \gamma(v)}\right)^2 - \gamma(u) - \gamma(v). \tag{13.24}$$

Letting $v \to u$ we then have

$$\gamma(2u) = \frac{1}{4}\lim_{h\to 0}\Big(\frac{\gamma'(u)-\gamma'(u-h)}{\gamma(u)-\gamma(u-h)}\Big)^2 - 2\gamma(u)$$
$$= \frac{1}{4}\Big(\frac{\gamma''(u)}{\gamma'(u)}\Big)^2 - 2\gamma(u). \tag{13.25}$$

13.14 The integral of an elliptic function

From Hermite's formula we immediately derive the following formula for the integral:

$$\int f(z)dz = Cz + \sum_{k=1}^{m}\Big\{C_{k1}\log[\sigma(z-b_k)] - C_{k2}\zeta(z-b_k)$$
$$+\cdots(-1)^{n_k-1}\frac{C_{kn_k}}{(n_k-1)!}\zeta^{(n_k-2)}(z-b_k)\Big\}.$$

Thus the integral of an elliptic function can be represented as a function of σ, ζ, γ, albeit with σ appearing under the logarithm.

For the integral of an elliptic function to be also elliptic we need to ensure that there are no branches arising from the logarithm, that is the residues C_{k1} should be 0, and also

$$2Cw - 2\eta\sum_k C_{k2} = 0, \quad 2Cw' - 2\eta'\sum_k C_{k2} = 0,$$

which means that $C = 0$ and $\sum_k C_{k2} = 0$.

Conversely, if these conditions are satisfied then the integral of $f(z)$ is an elliptic function.

In elementary calculus we know how to integrate

$$\int R\big(x, \sqrt{ax^2+2bx+c}\big)dx,$$

where $R(u, v)$ is a rational function of u, v. However, advancing one step further, the integral

$$\int R\big(x, \sqrt{4x^3-g_2x-g_3}\big)dx$$

is no longer representable in terms of elementary functions. On the other hand, on setting $x = \gamma(z)$, such an integral becomes a rational function of

$$\gamma(z), \quad \gamma'(z).$$

We know that rational functions of $\gamma(z)$, $\gamma'(z)$ can be represented as

$$R(\gamma(z)) + R_1(\gamma(z))\gamma'(z),$$

where $R(Z)$ and $R_1(Z)$ are rational functions of Z. The integral of $R_1(\wp(z))\wp'(z)$ being easy, it remains to deal with

$$\int R(\wp(z))dz.$$

We first examine

$$I_n = \int (\wp(z))^n dz$$

and the relevant derivative

$$\frac{d}{dz}(\wp^{n-1}(z)\wp'(z)) = (n-1)\wp^{n-2}(z)\wp'^2(z) + \wp^{n-1}(z)\wp''(z).$$

Replacing $\wp'^2(z)$ and $\wp''(z)$ by $4\wp^3(z) - g_2\wp(z) - g_3$ and $6\wp^2(z) - \frac{1}{2}g_2$, we have

$$\begin{aligned}\frac{d}{dz}(\wp^{n-1}(z)\wp'(z)) &= (4n+2)\wp^{n+1}(z)\\ &\quad - \left(n-\tfrac{1}{2}\right)g_2(\wp(z))^{n-1} - (n-1)g_3(\wp(z))^{n-2},\end{aligned}$$

so that, on integration,

$$\wp^{n-1}(z)\wp'(z) = (4n+2)I_{n+1} - \left(n-\tfrac{1}{2}\right)g_2 I_{n-1} - (n-1)g_3 I_{n-2}.$$

This is a reduction formula from which we can evaluate I_n from I_0, I_1; we already have

$$I_0 = \int dz = z, \quad I_1 = \int \wp(z)dz = -\zeta(z),$$

so that I_n can be represented in terms of $\wp(z)$, $\wp'(z)$, $\zeta(z)$.

If $R(z)$ is a proper fraction $Q(z)/P(z)$ then we can apply the method of partial fractions, but we only explain how we can evaluate

$$\int \frac{du}{\wp(u) - \wp(v)};$$

here $v \neq w, w', w''$ and $\wp(v) \neq e_1, e_2, e_3$.

In Section 13.12 we established that

$$\frac{-\wp'(v)}{\wp(u) - \wp(v)} = \zeta(u+v) - \zeta(u-v) - 2\zeta(v),$$

so that

$$\int \frac{du}{\wp(u) - \wp(v)} = \frac{-1}{\wp'(v)}\big(\log\sigma(u+v) - \log\sigma(u-v) - 2u\zeta(v)\big) + C.$$

13.15 The field of algebraic functions

In Sections 13.9 and 13.12 we resolved the following important problem.

Theorem 13.28 *In the field K of periodic functions with period $2w$, $2w'$ we can find two functions $\gamma(z)$, $\gamma'(z)$ with the property that*

$$\gamma'^2(z) = 4\gamma^3(z) - g_2\gamma(z) - g_3,$$

and any member of K can be represented as a rational function of $\gamma(z)$, $\gamma'(z)$.

Given any two numbers g_2, g_3 with $g_2^3 - 27g_3^2 \neq 0$, we can say that such an elliptic function $\gamma(z)$ definitely exists, so that we have the following conclusion: Rational functions of the form $R(z, W)$ $(W^2 = 4z^2 - g_2z - g_3)$ are closed with respect to the usual arithmetic operations. If we use the parametric representation $z = \gamma(u)$, $W = \gamma'(z)$, then $R(z, W)$ is an individual elliptic function of u.

Stepping one pace back, let us consider $R(z, \sqrt{az^2 + bz + c}\,)$, or an even more general form of $R(z, W)$ in which z and W satisfy the relationship

$$az^2 + 2bzW + cW^2 + 2dz + 2cW + f = 0.$$

We saw in Volume I, Section 9.7 that there is a parameter t such that

$$z = R_1(t), \qquad W = R_2(t);$$

here R_1, R_2 are rational functions of t, and when substituted into $R(z, W)$, the expression becomes an aggregate of rational functions of t. Later, we prove that it is not possible to have such a parameter t resolving $R(z, \sqrt{4z^3 - g_2z - g_3})$ as an aggregate of rational functions of t.

Having seen this example, perhaps we should consider $R(z, W)$ in which z, W satisfy some cubic equation. It may appear to be only a small error of judgement, but actually it would be a rather large one. The simplest special example is $W^2 = 4z^3 - g_2z - g_3$, which is precisely the theory of elliptic functions, and a general cubic relationship would lead us beyond such a theory.

Nevertheless, $R(z, \sqrt{a_0z^4 + a_1z^3 + a_2z^2 + a_3z + a_4}\,)$, with $a_0 \neq 0$, may not go beyond elliptic function theory. Here is the explanation: Let z_0 be a root of $a_0z^4 + a_1z^3 + a_2z + a_3 = 0$; with the substitution $z = u + z_0$, we may assume that $a_4 = 0$. The substitution $z = 1/v$ then gives

$$\sqrt{a_0z^4 + a_1z^3 + a_2z^2 + a_3z} = \frac{1}{v^2}\sqrt{a_3v^3 + a_2v^2 + a_1v + a_0},$$

which reduces to the previous situation.

Generally speaking, the study of $R(x, y)$ in which x, y satisfy some relationship

$$f(x, y) = 0$$

belongs to problems in algebra and algebraic geometry. Functions from such a theory are said to be self-similar, and they naturally include elliptic functions as an example.

13.16 The inverse problem

Let w, w' be a pair of complex numbers satisfying $\Im(w'/w) > 0$. We already know that there is an elliptic function $\gamma(z; w, w')$ with the two invariants g_2, g_3 defined by

$$\begin{cases} g_2(w, w') = 60 \sum{}' \dfrac{1}{(mw + m'w')^4}, \\ g_3(w, w') = 140 \sum{}' \dfrac{1}{(mw + m'w')^6}, \end{cases} \tag{13.26}$$

where $\sum'$ denotes summation over all pairs of integers (m, m'), with the pair $(0, 0)$ omitted. We also know that $g_2^3 - 27g_3^2 \neq 0$.

We now consider the inverse problem: Given any two numbers a, b satisfying $a^3 - 27b^2 \neq 0$, can we find w, w' so that $g_2(w, w') = a$, $g_2(w, w') = b$?

In order to resolve this important problem, we introduce a function $J(\tau)$ which is defined as follows: Let $\tau = w'/w$, and let $\Delta(w, w') = g_2^3 - 27g_3^2$. Then

$$g_2(w, w') = w^{-4} g_2(1, \tau), \quad g_3(w, w') = w^{-6} g_3(1, \tau), \quad \Delta(w, w') = w^{-12} \Delta(1, \tau).$$

Therefore, if we define

$$J(\tau) = \frac{g_2^3(1, \tau)}{\Delta(1, \tau)} = \frac{g_2^3(w, w')}{\Delta(w, w')},$$

then

$$J(\tau) - 1 = \frac{27g_3^2(1, \tau)}{\Delta(1, \tau)}. \tag{13.27}$$

This function $J(\tau)$ has the properties as discussed in Theorem 13.29.

Theorem 13.29 *The function $J(\tau)$ is analytic in the upper half-plane $\Im(\tau) > 0$. Moreover, for any integers satisfying*

$$ad - bc = 1,$$

we always have

$$J\left(\frac{a\tau + b}{c\tau + d}\right) = J(\tau). \tag{13.28}$$

Proof. (1) We first show that the series

$$g_2(1, \tau) = 60 \sum{}' \frac{1}{(m + n\tau)^4}$$

represents an analytic function in the upper half-plane $\Im\tau > 0$. Since each summand is an analytic function, it suffices to show that the series is uniformly convergent in a vertical strip S of the form

$$-a \leq \Re\tau \leq a, \quad \Im\tau \geq b,$$

where a, b are arbitrary positive real numbers.

Let h be the smaller of the two heights of the parallelogram $(0, 1, 1+\tau, \tau)$. If τ lies inside S then there exists ϵ such that $0 < \epsilon \le h$. The method used in Section 13.8 now shows that the series converges uniformly in S, so that $g_2(1, \tau)$ is an analytic function of τ in the upper half-plane.

The same method shows that $g_3(1, \tau)$ and $\Delta(1, \tau)$ are analytic in the upper half-plane and, since $\Delta(1, \tau) \neq 0$, so is $J(\tau)$.

(2) Let a, b, c, d be integers satisfying $ad - bc = 1$, and let $W' = aw' + bw$, $W = cw' + dw$. Then

$$\frac{W'}{W} = \frac{aw' + bw}{cw' + dw} = \frac{a\tau + b}{c\tau + d}.$$

From $ad - bc > 0$ we find that the upper half-plane for τ has been transformed into the upper half-plane for $T = W'/W$, and

$$g_2(w, w') = g_2(W, W'), \quad g_3(w, w') = g_3(W, W'),$$

so that

$$J(\tau) = \frac{g_2^3(w, w')}{\Delta(w, w')} = \frac{g_2^3(W, W')}{\Delta(W, W')} = J\Big(\frac{a\tau + b}{c\tau + d}\Big).$$

□

Functions satisfying (13.28) are known as modular functions. The group of transformations $\tau' = (a\tau + b)/(c\tau + d)$ is called the modular group. In the investigation of $J(\tau)$ we first discuss some general properties associated with the modular group and modular functions.

13.17 Modular substitutions

We consider the modular group of substitutions of the form

$$z' = \frac{az + b}{cz + d},$$

where a, b, c, d are integers satisfying $ad - bc = 1$. In general, every such substitution has two distinct fixed points, or invariant points (that is $z' = z$), which are roots of the quadratic equation

$$cz^2 + (d - a)z - b = 0.$$

If z_1, z_2 are the two roots then the transformation can be written in the standard form

$$\frac{z' - z_1}{z' - z_2} = \lambda \frac{z - z_1}{z - z_2}.$$

Taking $z = \infty$ we have $z' = a/c$, so that

$$\lambda = \frac{a - cz_1}{a - cz_2}.$$

- If $|\lambda| = 1$, $\lambda \neq 1$ then the transformation is said to be elliptic.
- If λ is real and $\neq \pm 1$ then it is hyperbolic.
- If λ is complex and $|\lambda| \neq 1$ then it is loxodromic.
- If the two fixed points coincide then it is parabolic.

If the repeated application of a transformation becomes the unit (identity) transformation E then the transformation is said to be of finite order, and the period is the least number of applications taken to arrive at E. Only elliptic transformations with $\lambda^n = 1$ can have a period, which is then the least such n. If $n = 2$, then $\lambda = -1$, the period is 2, and the transformation is called an involution.

It is easy to see that λ satisfies the quadratic equation

$$\lambda + \lambda^{-1} = a^2 + d^2 + 2bc = (a + d)^2 - 2,$$

the discriminant of which is

$$[(a + d)^2 - 2]^2 - 4 = (a + d)^2[(a + d)^2 - 4].$$

For our discussion, we may assume that $a + d \geq 0$, since otherwise we may replace a, b, c, d by $-a, -b, -c, -d$.

(1) If $a + d > 2$ then the transformation is hyperbolic, and the two fixed points are the real roots of the quadratic equation

$$cz^2 + (d - a)z - b = 0.$$

The condition for the quadratic equation to have rational roots is

$$(d - a)^2 + 4bc = (a + d)^2 - 4 = u^2,$$

where u is an integer. Since the only integer solutions to $x^2 - y^2 = 4$ are $x = \pm 2$, $y = 0$, we find that, for a hyperbolic modular transformation, the fixed points have to be the irrational roots of a quadratic equation with rational coefficients. Such numbers are called algebraic numbers of degree 2.

(2) If $a + d = 2$ then $\lambda = 1$, and the parabolic transformation is given by

$$\frac{1}{z' - (a - 1)/c} = \frac{1}{z - (a - 1)/c} + c.$$

If $c = 0$, then $a = d = 1$, giving

$$z' = z + b.$$

The fixed point is the rational point $(a - 1)/c$ for the former case, and ∞ for the latter.

(3) If $a + d = 1$ then

$$\lambda^2 + \lambda + 1 = 0,$$

so that λ is either $\rho = e^{2\pi i/3} = (-1 + \sqrt{-3})/2$ or ρ^2. The fixed points are

$$z_1 = \frac{a + \rho^2}{c}, \quad z_2 = \frac{a + \rho}{c},$$

and the standard form of the transformation is

$$\frac{z' - (a+\rho^2)/c}{z' - (a+\rho)/c} = \rho \frac{z - (a+\rho^2)/c}{z - (a+\rho)/c}.$$

This is an elliptic transformation, with period 3. If $\lambda = \rho$ is replaced by ρ^2, then we have another elliptic transformation with period 3.

(4) If $a + d = 0$ then $(\lambda + 1)^2 = 0$, so that $\lambda = -1$, and the fixed points are the roots of

$$cz^2 - 2az - b = 0,$$

that is

$$z = \frac{a \pm i}{c}.$$

The standard form of the transformation is

$$\frac{z' - (a+i)/c}{z' - (a-i)/c} = -\frac{z - (a+i)/c}{z - (a-i)/c},$$

an elliptic transformation with period 2.

Summarising, if $a + d = 0$ then the modular transformation is an involution; if $a + d = \pm 1$ then it has the period 3; if $a + d = \pm 2$ then it is parabolic, and its fixed point is either a rational number or infinity; if $|a + d| > 2$ then it is hyperbolic, and the fixed points are real algebraic numbers of degree 2.

13.18 The fundamental region

Definition 13.30 Two points z, z' on the upper half-plane are said to be equivalent if there is a modular transformation mapping z to z', and we write $z \sim z'$ to denote such a relationship.

Clearly we have:

(i) $z \sim z$;
(ii) if $z \sim z'$ then $z' \sim z$;
(iii) if $z \sim z'$ and $z' \sim z''$, then $z \sim z''$.

We take the following region (see Fig. 13.5) in the upper half-plane:

$$D : \begin{cases} -\frac{1}{2} \le x < \frac{1}{2}, \\ x^2 + y^2 > 1, & \text{when } x > 0, \\ x^2 + y^2 \ge 1, & \text{when } x \le 0. \end{cases}$$

Definition 13.31 The set D is called the fundamental region (or domain), forming a triangle with angles having the values $(0, \pi/3, \pi/3)$. Members of D are called reduced points.

Theorem 13.32 *No two reduced points are equivalent.*

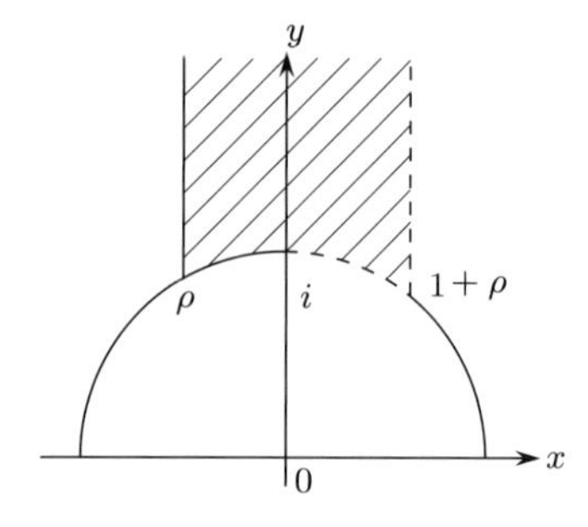

Figure 13.5.

Proof. Let z, z' be two distinct reduced points, and suppose that

$$z' = \frac{az+b}{cz+d}.$$

If $z = z + iy$ and $z' = x' + iy'$ then

$$y' = \frac{y}{|cz+d|^2}.$$

We have

$$\begin{aligned}|cz+d|^2 &= c^2 z\bar{z} + cd(z+\bar{z}) + d^2 \\ &= c^2(x^2+y^2) + 2cdx + d^2 \\ &\geq c^2 - |cd| + d^2 > 1,\end{aligned}$$

where we must exclude the following exceptional cases: $c = \pm 1, d = 0$, or $c = 0, d = \pm 1$, or $c = d = 1$. Therefore, apart from these exceptional cases, we always have

$$y' < y.$$

When $c = d = 1$, we can have $|cz+d|^2 = 1$ only when $z = \rho$. From $a - b = 1$ and $\rho^2 + \rho + 1 = 0$, we have

$$z' = \frac{a\rho + b}{\rho + 1} = -\frac{a\rho + b}{\rho^2} = -a\rho^2 - b\rho = -\rho^2 + b,$$

so that $\Im(z') = \sqrt{3}/2$. If $z' \in D$ then $z' = \rho$, which contradicts z, z' being distinct points.

We also have

$$z = \frac{dz' - b}{-cz' + d},$$

so that

$$y < y';$$

again we must exclude the following cases: $c = \pm 1, a = 0$, or $c = 0, a = \pm 1$. Since we cannot have both $y > y'$ and $y < y'$, we need only examine the following two cases: (i) $c = 0, a = d = 1$; (ii) $c = 1, a = d = 0$.

In the first case we have

$$z' = z + b, \quad b \neq 0,$$

so that $x' = x + b$, and hence $|x' - x| \geq 1$, and therefore z, z' cannot be both in D.

In the second case we have $b = -1$, that is $z' = -1/z$, and hence $|z'z| = 1$. Thus, if $|z| > 1$ then $|z'| < 1$, and hence z' cannot be a reduced point; similarly, z cannot be a reduced point if $|z'| > 1$. If $|z| = 1$ then z has to lie on the circular arc from ρ to i, and z' $(= -1/z)$ must lie on the arc from $\rho + 1$ to i. If $z \neq i$ then z' cannot be a reduced point; if $z = i$ then $z' = i = z$, contradicting the assumption that $z \neq z'$. The theorem is proved. □

Theorem 13.33 *The number of points in the rectangular strip* $-\frac{1}{2} \leq x < \frac{1}{2}, y \geq \gamma$ $(\gamma > 0)$ *that are equivalent to a fixed point is finite. That is, if we partition the strip into sets of mutually equivalent points, then each set has only finitely many points.*

Proof. Let $z = x + iy$ and

$$z' = \frac{az + b}{cz + d}.$$

Then we have

$$y' = \frac{y}{|cz + d|^2} = \frac{y}{c^2(x^2 + y^2) + 2cdx + d^2}.$$

If $y' \geq \gamma$ then

$$c^2(x^2 + y^2) + 2cdx + d^2 \leq \frac{y}{\gamma},$$

that is

$$(cx + d)^2 + c^2 y^2 \leq \frac{y}{\gamma},$$

and clearly there can only be a finite number of integers c, d satisfying this.

Let (c', d') be any such pair of integers with $\gcd(c', d') = 1$. Then all the solutions (a, b) of the equation $ad' - bc' = 1$ can be represented by

$$a = a' + mc', \quad b = b' + md',$$

where a', b' is a fixed solution (that is $a'd' - b'c' = 1$), and m is any integer. Thus

$$z' = \frac{az + b}{cz + d} = \frac{a'z + b'}{c'z + d'} + m.$$

There can only be one m such that $-\frac{1}{2} \leq x' < \frac{1}{2}$. Therefore corresponding to each pair (c', d') with $\gcd(c', d') = 1$ there is only one set a, b such that $-\frac{1}{2} \leq x' < \frac{1}{2}$. Therefore the number of points in the strip which are equivalent to z is finite. □

Theorem 13.34 *Every point on the upper half-plane is equivalent to a unique reduced point.*

Proof. Let $z = x_0 + iy_0$, $y_0 > 0$. We take the unique integer m in

$$-\frac{1}{2} \leq x_0 + m < \frac{1}{2}$$

and let

$$z' = z + m.$$

If $|z'| > 1$ then z' is a reduced point, and there is nothing to prove. If $|z'| = 1$ and z' lies on the arc from ρ to i, then it is a reduced point, and if it lies on the arc from $1 + \rho$ to i, then the transformation $-1/z$ will give the former situation. If $|z'| < 1$, then we let

$$z'' = -\frac{1}{z'} \quad \text{and} \quad y'' = \frac{y_0}{|z'|^2} > y_0.$$

Choose m' such that

$$z''' = z'' + m', \quad -\frac{1}{2} \leq x''' < \frac{1}{2}.$$

If z''' is still not a reduced point, then we use the same method and construct $z^{\text{iv}} = -1/z'''$. In this way we obtain $z', z''', \ldots$ all lying in the strip

$$-\frac{1}{2} \leq x < \frac{1}{2}, \quad y > y_0.$$

By Theorem 13.33 there can only be finitely many such points.

Therefore every point must be equivalent to a reduced point. Also, by Theorem 13.32, there cannot be two equivalent reduced points. The theorem is proved. □

Exercise 13.35 Show that all the points

$$z = \frac{a + i}{c}, \quad a^2 + bc + 1 = 0,$$

are equivalent to i.

Exercise 13.36 Show that all the points

$$z = \frac{a + \rho}{c}, \quad a(1 - a) - bc = 1,$$

are equivalent to ρ.

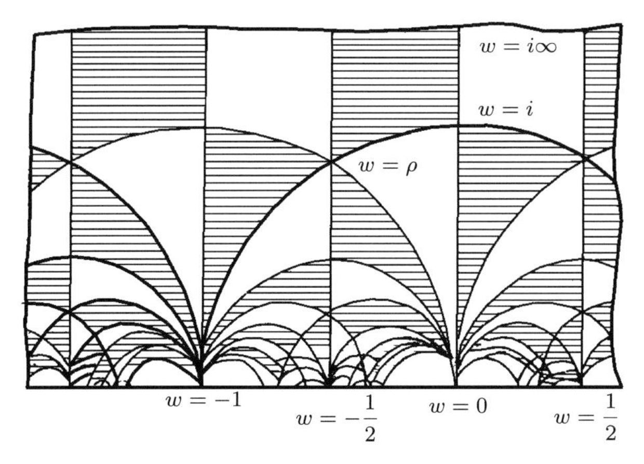

Figure 13.6.

13.19 The net of the fundamental region

Theorem 13.37 *Suppose that z is not a fixed point of any (non-unit) modular transformation. Let U, V be two distinct modular transformations, and write Uz for the image of z under U. Then*

$$Uz \neq Vz.$$

Proof. If $Uz = Vz$ then $z = U^{-1}Vz$ is a fixed point of a modular transformation. □

Theorem 13.38 *The set of all triangular images of the fundamental region form a covering of the upper half-plane with no overlaps.*

Proof. The first part of the theorem follows from Theorem 13.34. If U and V are two distinct modular transformations whose triangular images of D overlap, then the transformation $U^{-1}V$ must map D into a triangular region which overlaps with D. Let z be a point in this overlap. Then there must be a point in D which is equivalent to z, and this is impossible if z is in D. □

We can interpret this theorem in terms of covering by tiles. In ordinary space we can cover regions without overlaps using equal size square tiles, and by this we mean that each tile can be 'translated' from one place to another which is occupied by another tile. Here the fundamental region is the shape of our new tile, and 'translation' is now a modular transformation. The above theorem then tells us that, in the language of non-Euclidean geometry, such a tile can be used to cover the upper half-plane with no overlaps (see Fig. 13.6).

We can alter the definition of a fundamental region as follows: Any region in the upper half-plane is called a fundamental region if

(i) any point is equivalent to a point in the region;
(ii) no two points in the region are equivalent.

Take any point z in the upper half-plane which is not a fixed point of any modular transformation. Construct the points $z_1, z_2, \ldots$ which are equivalent to z, and then construct the perpendicular bisectors of (z, z_i), that is those points which have two equal non-Euclidean distances from z and z_i. Discard the part on the side of z_i. Then the remaining part is a fundamental region. (The reader should supply the proof for this, and also determine the fundamental region corresponding to $z = z_i$.)

This shows clearly that Lobachevskian geometry has useful applications in function theory.

We note the following: The fixed points of an elliptic transformation with period 2 lie on the lines joining vertices with angles $\pi/3$. The fixed points of an elliptic function with period 3 are the common vertices of six triangles. The fixed points of a parabolic transformation are those points with infinitely many lines through them. The fixed points of a hyperbolic transformation cannot be vertices of any triangle (and it is even clearer that they cannot lie on the sides).

13.20 The structure of the modular group

Let us denote by S and T the transformations $z' = z + 1$ and $z' = -1/z$, respectively. Then S^{-1} denotes the transformation $z' = z - 1$. The three substitutions transform a fundamental region into the three neighbouring regions, and conversely the transformation which maps a neighbouring region into a fundamental region must be one of S, T or S^{-1}.

Let M be any modular transformation, and let z be any point in the fundamental region D. We join z to M_z by a curve not passing through any vertices. Suppose that the various regions that this curve crosses are

$$D,\ D_1,\ D_2,\ \ldots,\ D_n\ (= MD).$$

Also, denote by M_i the modular transformation which maps D into D_i. Now $M_1 = S, T$ or S^{-1}. Suppose that M_k can be represented as a product of S and T. Since M_k^{-1} maps D_k into D, and D_{k+1} is a neighbouring region to D_k, it follows that M_k^{-1} maps D_{k+1} into D'_{k+1}, a neighbouring region of D. But D'_{k+1} can be mapped into D via a transformation M'^{-1} ($= S, T$ or S^{-1}). That is,

$$M'^{-1} M_k^{-1} D_{k+1} = M'^{-1} D'_{k+1} = D,$$

or

$$M_k M' D = D_{k+1}.$$

Therefore $M_{k+1} = M_k M'$ can be represented as a product of S and T, and hence so can M itself. We have therefore proved the following theorem.

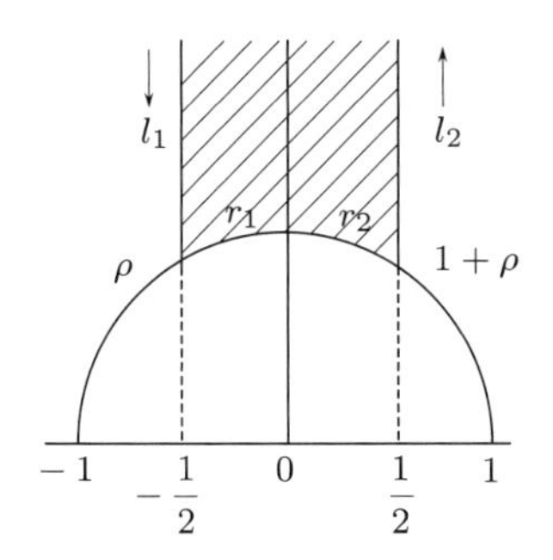

Figure 13.7.

Theorem 13.39 *Any modular transformation is representable as a product of S and T.*

Theorem 13.39 has the following explicit interpretation: If

$$M = S^{m_1} T S^{m_2} T S^{m_3} \cdots T S^{m_\nu},$$

then

$$z' = m_1 - \frac{1}{m_2 -} \frac{1}{m_3 -} \cdots \frac{1}{m_\nu + z}.$$

This clearly shows the relationship between modular transformations and continued fractions.

It is easy to see that

$$T^2 = E, \quad (ST)^3 = E.$$

If we extend the definition of a modular transformation to

$$z' = \frac{az+b}{cz+d}, \quad ad - bc = \pm 1,$$

then we can have the result

$$z' = m_1 + \frac{1}{m_2 +} \frac{1}{m_3 +} \cdots \frac{1}{m_\nu + z}.$$

13.21 The definition and properties of modular functions

Definition 13.40 A meromorphic function which is invariant under a modular transformation is called a modular function.

From the properties of the modular group, the study of the properties of a modular function can be confined to the fundamental region (Fig. 13.7). We call the point ∞ the parabolic vertex of the fundamental region, and i, ρ, $1+\rho$ are called elliptic vertices. The two sides defined by $\tau = \pm\frac{1}{2}$ are the matching sides; the two circular arcs from ρ to i and from i to $1+\rho$ are also called matching sides. The parabolic vertex is invariant under the modular transformation $\tau' = \tau + 1$, and i is invariant under $\tau' = -1/\tau$. The two remaining vertices ρ and $1+\rho$ are invariant under $\tau' = -1/(\tau+1)$ and $\tau' = 1 - 1/\tau$, respectively.

Theorem 13.41 *The order of a zero, or a pole, of a modular function $f(\tau)$ at the point $\tau = i$ is even.*

Proof. Suppose that $f(\tau)$ has a zero of order k at $\tau = i$. Then

$$f(\tau) = (\tau - i)^k \phi(\tau), \quad \text{where} \quad \phi(i) \neq 0,$$

and $\phi(\tau)$ is also modular. From

$$f(\tau) = f\Big(-\frac{1}{\tau}\Big) = \Big(-\frac{1}{\tau} - i\Big)^k \phi\Big(-\frac{1}{\tau}\Big) = (\tau - i)^k \phi(\tau) = f(\tau),$$

we have

$$\phi(\tau) = \Big(-\frac{i}{\tau}\Big)^k \phi\Big(-\frac{1}{\tau}\Big),$$

and hence

$$\phi(i) = (-1)^k \phi(i),$$

which implies that k is even.

The same argument applied to $1/f(\tau)$ gives the result for a pole. □

Theorem 13.42 *The order of a zero, or a pole, of a modular function $f(\tau)$ at the points $\tau = \rho$ and $\tau = 1 + \rho$ is a multiple of 3.*

The proof is similar to that of the previous theorem, but now we use the relationship

$$f\Big(-\frac{1}{\tau+1}\Big) = f(\tau).$$

We omit the details.

Theorem 13.43 *A modular function $f(\tau)$ has the expansion*

$$\sum_{n=-\infty}^{\infty} a_n e^{2\pi i n\tau}; \tag{13.29}$$

in other words, in terms of the variable $t = 2\pi i\tau$, the function $f(\tau)$ has an isolated singularity at $t = 0$.

Proof. From $f(\tau+1) = f(\tau)$ we have $f(x+1+iy) = f(x+iy)$, so that f is a periodic function with period 1. Therefore, for any fixed y, as long as there is no pole in $x + iy$ $(-\infty < x < \infty)$, the function $f(\tau)$ has the Fourier expansion

$$f(\tau) = \sum_{n=-\infty}^{\infty} P_n(y) e^{2\pi i n x}, \quad P_n(y) = \int_0^1 f(\tau) e^{-2\pi i n x}\, dx,$$

and the series is uniformly convergent when y lies in a bounded region.

Being analytic, we have

$$\frac{d}{dx}f(\tau) = -i\frac{d}{dy}f(\tau),$$

so that

$$2\pi i n P_n(y) = -i P_n'(y), \quad \text{and hence} \quad P_n(y) = a_n e^{-2\pi n y}.$$

The theorem is proved. □

Definition 13.44 Modular functions with finitely many terms with negative n in (13.29) are said to be simple.

Definition 13.45 The zeros and poles of a simple modular function in the fundamental region are counted according to the following method.

(i) Within the interior of the region, they are counted according to their multiplicities.
(ii) For a pair of matching sides, only those on the left-hand side are counted.
(iii) At $\tau = i$, we take only half of the multiplicity.
(iv) At $\tau = \rho, 1+\rho$, we take only one-third of the multiplicity.
(v) At $\tau = \infty$, we take the order.

Theorem 13.46 *The number of zeros and the number of poles for a simple modular function $f(\tau)$ in the fundamental region is the same.*

Proof. (1) Suppose first that there are neither zeros nor poles on the perimeter. Consider the integral

$$\frac{1}{2\pi i}\left(\int_{r_1} + \int_{r_2} + \int_{\ell_2} + \int_{\ell_1}\right) d(\log f(\tau))$$

round the fundamental region (see Fig. 13.7), the value for which is the difference between the number of zeros and the number of poles for $f(\tau)$ inside the region. The equation $\tau' = \tau + 1$ reverses ℓ_1 and ℓ_2, so that

$$\left(\int_{\ell_2} + \int_{\ell_1}\right) d(\log f(\tau)) = 0.$$

Also, $\tau' = -1/\tau$ reverses r_1 and r_2, so that the required result follows.

(2) Suppose that there is a zero, or a pole, on a side. For example, there is a zero at z_0 on the side $z = -\frac{1}{2}$; then there is also a zero at $1 + z_0$, which lies on the side $z = \frac{1}{2}$. We construct a small semi-circle round z_0, lying outside the region, and a matching semi-circle round $1 + z_0$, lying inside the region. Such a technique then side-steps the difficulty of having zeros on the sides.

(3) Suppose that there is a zero, with order $2h$, at $\tau = i$. Construct a circular arc inside the region (see Fig. 13.8) centre i, radius ϵ, leading from A on r_1 to B on r_2. Write

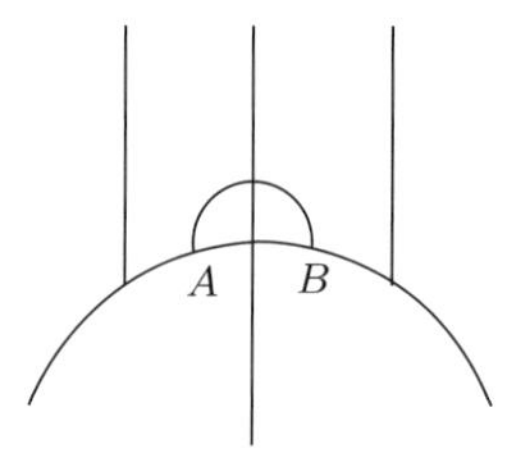

Figure 13.8.

$f(\tau) = (\tau - i)^{2h}\phi(\tau)$ and consider the integral

$$\int_A^B d\log f(\tau) = 2h\int_A^B d\log(\tau - i) + \int_A^B \frac{\phi'(\tau)}{\phi(\tau)}\,d\tau$$

along the arc. As $\epsilon \to 0$, the first integral on the right-hand side tends to πi, and the second integral tends to 0, so that the contribution is half the number of zeros there.

The same argument applies at ρ and $1 + \rho$.

(4) Finally, consider $\tau = \infty$. When y_0 is sufficiently large, we integrate along the straight line $x + iy_0$, $|x| \leq \frac{1}{2}$, which amounts to integrating round $t = 0$, and we again have the required result. □

13.22 The function $J(\tau)$

Let us now return to the study of the function $J(\tau)$ given in (13.27).

Theorem 13.47 *The function $J(\tau)$ has a simple pole at the parabolic vertex; that is, if $t = e^{2\pi i\tau}$ then $J(\tau)$, as a function of t, has a simple pole at $t = 0$.*

Proof. (1) We start with the expansion for $\cot \pi\tau$:

$$\pi \cot \pi\tau = \frac{1}{\tau} + \sum_{n=-\infty}^{\infty}{}' \left(\frac{1}{\tau - n} + \frac{1}{n}\right), \tag{13.30}$$

and, with $t = e^{2\pi i\tau}$,

$$\pi \cot \pi\tau = -\pi i(1 + 2t + 2t^2 + \cdots).$$

Differentiating τ three and five times, and noting that $dt/d\tau = 2\pi i t$, we have

$$\begin{cases} -6\displaystyle\sum_{m=-\infty}^{\infty} \frac{1}{(m+\tau)^4} = -16\pi^4(t + 8t^2 + \cdots), \\ -120\displaystyle\sum_{m=-\infty}^{\infty} \frac{1}{(m+\tau)^6} = 64\pi^6(t + 32t^2 + \cdots), \end{cases} \tag{13.31}$$

and by differentiating (13.30) we deduce that

$$\sum_{m=-\infty}^{\infty}{}' \frac{1}{m^4} = \frac{\pi^4}{45}, \qquad \sum_{m=-\infty}^{\infty}{}' \frac{1}{m^6} = \frac{2\pi^6}{945}. \tag{13.32}$$

(2) Consequently we find that

$$\begin{aligned} g_2(1,\tau) &= 60\left(\sum_{m=-\infty}^{\infty}{}' \frac{1}{m^4} + 2\sum_{n=1}^{\infty}\sum_{m=-\infty}^{\infty} \frac{1}{(m+n\tau)^4} \right) \\ &= 60\left(\frac{\pi^4}{45} + \frac{16\pi^4}{3} \sum_{n=1}^{\infty} (t^n + 8t^{2n} + \cdots) \right), \end{aligned}$$

noting that, as τ is replaced by $n\tau$, we need to replace t by t^n. Therefore

$$g_2(1,\tau) = \pi^4\left(\frac{4}{3} + 320t + \cdots\right).$$

Similarly we find that

$$\begin{aligned} g_3(1,\tau) &= 140\left(\sum_{m=-\infty}^{\infty}{}' \frac{1}{m^6} + 2\sum_{n=1}^{\infty}\sum_{m=-\infty}^{\infty} \frac{1}{(m+n\tau)^6} \right) \\ &= 140\left(\frac{2\pi^6}{945} - \frac{16\pi^6}{15} \sum_{n=1}^{\infty} (t^n + 32t^{2n} + \cdots) \right) \\ &= \pi^6\left(\frac{8}{27} - \frac{448}{3}t + \cdots \right). \end{aligned}$$

Thus

$$\Delta(1,\tau) = g_2^3(1,\tau) - 27g_3^2(1,\tau) = \pi^{12}(4096t + \cdots),$$

giving

$$J(\tau) = \frac{(4/3 + 320t + \cdots)^3}{4096t + \cdots} = \frac{1}{1728t} + C_0 + C_1 t + \cdots .$$

The theorem is proved, and we note that $J(\tau) \to \infty$ as $t \to +\infty$. □

Theorem 13.48 *Given any C, the equation*

$$J(\tau) = C$$

has a unique solution in the fundamental region.

This follows immediately as a corollary to Theorem 13.47. From

$$\frac{1}{140} g_3(1,i) = \sum_{m,n=-\infty}^{\infty}{}' \frac{1}{(m+in)^6} = -\sum_{m,n=-\infty}^{\infty}{}' \frac{1}{(n-im)^6} = -\frac{1}{140} g_3(1,i),$$

we find that $g_3(1,i) = 0$, and hence, by (13.27), $J(i) = 1$.

Also, from

$$\frac{1}{60}g_2(1,\rho) = \sum_{m,n=-\infty}^{\infty}{}' \frac{1}{(m+\rho n)^4} = \frac{1}{\rho}\sum_{m,n=-\infty}^{\infty}{}' \frac{1}{(m\rho^2+n)^4}$$
$$= \frac{1}{\rho}\sum_{m,n=-\infty}^{\infty}{}' \frac{1}{(m-n)-m\rho)^4} = \frac{1}{\rho}\frac{1}{60}g_2(1,\rho),$$

we find that $g_2(1,\rho)=0$, and hence $J(\rho)=0$. We summarise these in the following theorem.

Theorem 13.49 *We have* $J(\infty)=\infty$, $J(i)=1$, $J(\rho)=0$.

When will $J(\tau)$ be a real number? The transformation $\tau' = -\tau$ is a reflection with respect to the imaginary axis, and it is easy to see that

$$g_2(1,\tau') = 60\sum_{m,n=-\infty}^{\infty}{}' \frac{1}{(m+n\tau')^4} = 60\sum_{m,n=-\infty}^{\infty}{}' \frac{1}{(m+n\bar{\tau})^4} = \overline{g_2(1,\tau)},$$

and similarly $g_3(1,\tau') = \overline{g_3(1,\tau)}$. Therefore

$$J(\tau') = \bar{J}(\tau). \tag{13.33}$$

First, points satisfying $\tau' = \tau$ lie on the imaginary axis, and for such points $J(\tau)=\bar{J}(\tau)$, so that $J(\tau)$ is real. Next, $J(\tau'+1) = J(\tau') = \bar{J}(\tau)$, and $\tau'+1=\tau$, meaning that $x=\frac{1}{2}$, so that $J(\tau)$ is also real for such τ. Finally, from $J(-1/\tau') = \bar{J}(\tau)$, we see that when $\tau = e^{i\theta}$, $-1/\tau' = e^{i\theta}$, which means that $J(\tau)$ is also real on the circumference of the unit circle. Thus, in the fundamental region, $J(\tau)$ takes real values on the boundary and the imaginary axis.

There are no other points at which $J(\tau)$ can be real. Otherwise there would be points τ_0 and $-1/\tau_0$ at which $J(-\bar{\tau}_0) = \bar{J}(\tau_0) = J(\tau_0)$, which contradicts Theorem 13.48.

We see from the above that, in the two halves of the fundamental region, $J(\tau) = u+iv$ has different fixed signs for v, since otherwise there would be two τ_1, τ_2 in the same half with $v(\tau_1)v(\tau_2) < 0$, and so there would be a point τ within the half-region at which $v(\tau)=0$, which is impossible.

Let us investigate which half of the region is mapped to $v>0$. As τ moves from ρ to i, so that the region lies to the left-hand side, $J(\tau)$ moves from 0 to 1, so that the image region should also lie on the left-hand side. This means that the left half of the fundamental region is mapped to the upper half-plane.

To summarise: The transformation $z = J(\tau)$ defined in Theorem 13.47 is conformal, mapping the fundamental region onto the whole plane, the left-hand half of the region being mapped onto the upper half-plane. The boundary and the imaginary axis correspond to the real axis. As τ moves from i to ∞ along the imaginary axis, z moves from 1 to $+\infty$ along the real axis. As z moves from 1 to $-\infty$ along the real axis, the corresponding points of the fundamental region lie on the 'seam' with its neighbouring regions.

Theorem 13.50 *The inverse function to J is an infinitely many-valued analytic function J^{-1}, with singularities at* 0, 1, ∞, *and with image values lying on the upper half-plane.*

If $F(z)$ is a meromorphic function with three values a, b, c missing from its range, then

$$G(z) = \frac{F(z) - a}{F(z) - c} \Big/ \frac{b - a}{b - c}$$

is a meromorphic function which does not take the values $0, 1, \infty$. The function $J^{-1}(G(z))$ is then analytic everywhere, and is thus a constant. Therefore we have the following.

Theorem 13.51 *A meromorphic function which omits three values is a constant.*

Theorem 13.52 *An integral function which omits two values is a constant.*

13.23 Solutions to $g_2(w, w') = a$, $g_3(w, w') = b$

Theorem 13.53 *Let a, b be any two finite numbers satisfying $a^3 - 27b^2 \neq 0$. Then there is a pair of complex numbers w, w', with a non-real ratio, such that*

$$g_2(w, w') = a, \quad g_3(w, w') = b. \tag{13.34}$$

Suppose that $ab \neq 0$. Then, from (13.34),

$$\frac{g_2^3(w, w')}{g_2^3(w, w') - 27g_3^2(w, w')} = \frac{a^3}{a^3 - 27b^2}, \quad \frac{g_2(w, w')}{g_3(w, w')} = \frac{a}{b}. \tag{13.35}$$

Conversely, if g_2, g_3 satisfy (13.35) then (13.34) follows.

Let $\tau = w'/w$. The first equation in (13.35) is then $J(\tau) = a^3/(a^3 - 27b^2)$ and, by Theorem 13.48, there is a solution τ in the upper half-plane. Knowing τ, the second equation in (13.35) becomes

$$\frac{w^2 g_2(1, \tau)}{g_2(1, \tau)} = \frac{a}{b},$$

from which we can obtain w and $w' = \tau w$.

If $a = 0$, the problem is equivalent to solving $J(\tau) = 0$, $w^{-6}\, g_3(1, \tau) = b$, which have the solutions $\tau = \rho$ and $w = (g_3(1, \rho)/b)^{1/6}$. The same method can be used to deal with $b = 0$.

13.24 Any modular function is a rational function of $J(\tau)$

Any simple modular function $f(\tau)$ can have only finitely many zeros and poles in the fundamental region (otherwise the point at infinity would be an isolated singularity). Let $a_1, a_2, \ldots, a_n$ be the zeros and $b_1, b_2, \ldots, b_n$ the poles. Take the function

$$F(\tau) = \frac{\big(J(\tau) - J(a_1)\big) \cdots \big(J(\tau) - J(a_n)\big)}{\big(J(\tau) - J(b_1)\big) \cdots \big(J(\tau) - J(b_n)\big)}$$

so that $J(\tau)/F(\tau)$ has neither zero nor pole in the fundamental region, and is therefore a constant. Note, however, that if i is a zero then it must be a zero with order $2k$, and the factor we use has to be $(J(i)-1)^k$; we also have to deal with $\tau=\rho$ in a similar way.

13.25 Appendix: Important formulae[1] (Weierstrass)

13.25.1 Background

$$\sin \pi u = \pi u \prod_n{}' \left\{ \left(1-\frac{u}{n}\right) e^{\frac{u}{n}} \right\}; \tag{13.36}$$

$$\pi \cot \pi u = \frac{1}{u} + \sum_n{}' \left\{ \frac{1}{u-n} + \frac{1}{n} \right\}; \tag{13.37}$$

$$\frac{\pi^2}{\sin^2 \pi u} = \sum_n \frac{1}{(u-n)^2}; \tag{13.38}$$

$$\prod_n \left\{ \left(1-\frac{u}{n-a}\right) e^{\frac{u}{n-a}} \right\} = \frac{\sin \pi(u+a)}{\sin \pi a} e^{-u\pi \cot \pi a}; \tag{13.39}$$

$$\sum_n \left\{ \frac{1}{u+a-n} + \frac{1}{n-a} \right\} = \pi[\cot \pi(u+a) - \cot \pi a]. \tag{13.40}$$

13.25.2

$$\sigma(u) = u \prod_S{}' \left\{ \left(1-\frac{u}{S}\right) e^{\frac{u}{S}+\frac{u^2}{2S^2}} \right\}. \tag{13.41}$$

Here, and in the following, $S = 2mu + 2m'w'$, with $\Im(w/w') > 0$, and $\prod'_S$ represents the product over all integer pairs (m, m'), omitting the pair $(0, 0)$.

$$\zeta(u)\frac{d}{du} = \log \sigma(u) = \frac{1}{u} + \sum_S{}' \left\{ \frac{1}{u-S} + \frac{1}{S} + \frac{u}{S^2} \right\}; \tag{13.42}$$

$$\gamma(u) = -\zeta'(u) = -\frac{d^2}{du^2} \log \sigma(u) = \frac{1}{u^2} + \sum_S{}' \left\{ \frac{1}{(u-S)^2} - \frac{1}{S^2} \right\}; \tag{13.43}$$

$$-\frac{1}{2}\gamma(u) = \sum_S \frac{1}{(u-S)^3}; \tag{13.44}$$

[1] Some of the formulae have been established in the text, and others are not difficult – the reader should attempt them.

$$\begin{cases} \sigma(-u) = -\sigma(u), & \gamma(-u) = \gamma(u), \\ \zeta(-u) = -\zeta(u), & \gamma^{(n)}(-u) = (-1)^n \gamma^{(n)}(u). \end{cases} \tag{13.45}$$

13.25.3

$$\sigma(\lambda u|\lambda w, \lambda w') = \lambda\sigma(u|w, w'); \tag{13.46}$$

$$\zeta(\lambda u|\lambda w, \lambda w') = \frac{1}{\lambda}\zeta(u|w, w'); \tag{13.47}$$

$$\gamma(\lambda u|\lambda w, \lambda w') = \frac{1}{\lambda^2}\gamma(u|w, w'); \tag{13.48}$$

$$\gamma^{(n)}(\lambda u|\lambda w, \lambda w') = \frac{1}{\lambda^{n+2}}\gamma^{(n)}(u|w, w'). \tag{13.49}$$

13.25.4

$$\begin{cases} g_2 = 60 \sum_S{}' \dfrac{1}{S^4}, \quad g_3 = 140 \sum_S{}' \dfrac{1}{S^6}, \\ g_2(\lambda w, \lambda w') = \lambda^{-4} g_2(w, w'), \\ g_3(\lambda w, \lambda w') = \lambda^{-6} g_3(w, w'); \end{cases} \tag{13.50}$$

$$\sigma(u) = u - \frac{g_2 u^5}{2^4{\cdot}3{\cdot}5} - \frac{g_3 u^7}{2^3{\cdot}3{\cdot}5{\cdot}7} - \frac{g_2^2 u^9}{2^9{\cdot}3^2{\cdot}5{\cdot}7} + \cdots; \tag{13.51}$$

$$\zeta(u) = \frac{1}{u} - \frac{g_2 u^3}{2^2{\cdot}3{\cdot}5} - \frac{g_3 u^5}{2^2{\cdot}5{\cdot}7} - \frac{g_2^2 u^7}{2^4{\cdot}3{\cdot}5^2{\cdot}7} + \cdots; \tag{13.52}$$

$$\gamma(u) = \frac{1}{u^2} + \frac{g_2 u^2}{2^2{\cdot}5} + \frac{g_3 u^4}{2^2{\cdot}7} + \frac{g_2^2 u^6}{2^4{\cdot}3{\cdot}5^2} + \cdots. \tag{13.53}$$

13.25.5

$$\frac{\sigma(u+a)}{\sigma(a)} e^{-u\zeta(a) + \frac{1}{2}u^2\gamma(a)} = \prod_s \left\{ \left(1 - \frac{u}{S-a}\right) e^{\frac{u}{s-a} + \frac{u^2}{2(S-a)^2}} \right\}; \tag{13.54}$$

$$9pt]\zeta(u+a) - \zeta(a) - u\gamma(u) = \sum_s \left\{ \frac{1}{u+a-S} - \frac{1}{a-s} + \frac{u}{(a-S)^2} \right\}; \tag{13.55}$$

$$\gamma(u+a) - \gamma(a) = \sum_s \left\{ \frac{1}{(u+a-S)^2} - \frac{1}{(a-S)^2} \right\}. \tag{13.56}$$

13.25.6

$$\sigma(u+2w) = -e^{2\eta(u+w)}\sigma(u); \tag{13.57}$$

$$\sigma(u+2w') = -e^{2\eta'(u+w)}\sigma(u); \tag{13.58}$$

$$\eta w' - \eta' w = \frac{\pi i}{2}; \tag{13.59}$$

$$\sigma(u+2mw+2nw') = (-1)^{mn+m+n} e^{(2m\eta+2n\eta')(\eta+mw+nw')}\sigma(u); \tag{13.60}$$

$$\zeta(u+2mw+2nw') = \zeta(u) + 2m\eta + 2n\eta'; \tag{13.61}$$

$$\zeta(w) = \eta, \quad \zeta(w') = \eta', \quad \zeta(w+w') = \eta+\eta'; \tag{13.62}$$

$$\gamma(u+2mw+2nw') = \gamma(u); \tag{13.63}$$

$$\gamma'(u+2mw+2nw') = \gamma'(u); \tag{13.64}$$

$$\gamma'(w) = \gamma'(w') = \gamma'(w+w') = 0. \tag{13.65}$$

13.25.7

$$\gamma(u) - \gamma(v) = -\frac{\sigma(u+v)\sigma(u-v)}{\sigma^2(u)\sigma^2(v)}; \tag{13.66}$$

$$\begin{cases} \gamma(w_1) = e_1, \quad \gamma(w_2) = e_2, \quad \gamma(w_1+w_2) = e_3, \\ e_1+e_2+e_3 = 0; \end{cases} \tag{13.67}$$

$$\begin{cases} \gamma'^2(u) = 4(\gamma(u)-e_1)(\gamma(u)-e_2)(\gamma(u)-e_3), \\ g_2 = -4(e_1e_2+e_2e_3+e_3e_1) = 2(e_1^2+e_2^2+e_3^2), \\ g_3 = 4e_1e_2e_3; \end{cases} \tag{13.68}$$

$$\gamma''(u) = 6\gamma^2(u) - \frac{1}{2}g_2; \tag{13.69}$$

$$\gamma'''(u) = 16\gamma(u)\gamma'(u). \tag{13.70}$$

13.25.8

$$J(\tau) = \frac{27g_2^3}{g_2^3 - 27g_3^2} + 1, \quad \tau = w'/w; \tag{13.71}$$

$$J(\tau+1) = J(\tau); \tag{13.72}$$

$$J\left(-\frac{1}{\tau}\right) = J(\tau). \tag{13.73}$$

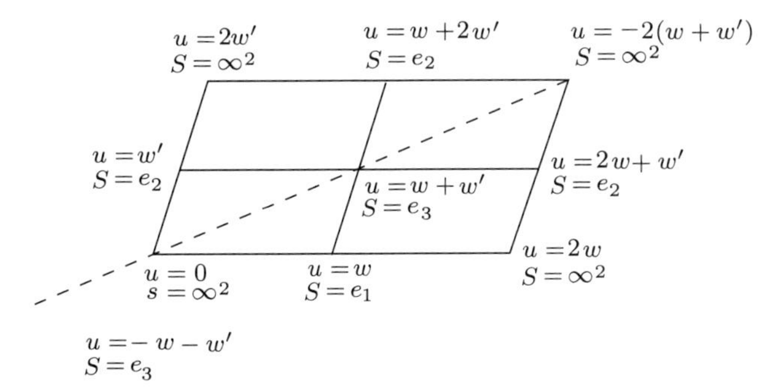

Figure 13.9.

In Fig. 13.9 there are four adjacent fundamental regions, with $S = \gamma(u)$.

14

Jacobian elliptic function theory

14.1 The theta functions $\theta(z, q)$

The elliptic functions in the previous chapter are given in terms of series and products which are slowly convergent, and are thus rather inconvenient when actual computations are required. Such a defect can be remedied by means of Jacobi's theta functions, which converge rapidly and can be used to represent elliptic functions.

We now introduce the symbol

$$q = e^{\pi i \tau}, \quad \Im\tau > 0,$$

so that $|q| < 1$.

Definition 14.1 We define

$$\theta(z, q) = \sum_{n=-\infty}^{\infty} (-1)^n q^{n^2} e^{2niz}. \tag{14.1}$$

If $|z| \leq A$ then $|q^{n^2} e^{2niz}| \leq |q|^{n^2} e^{2nA}$, and from

$$\frac{|q|^{(n+1)^2} e^{2(n+1)A}}{|q|^{n^2} e^{2nA}} = |q|^{2n+1} e^{2A} \to 0, \quad \text{as} \quad n \to \infty,$$

we see that the series

$$\sum_{n=-\infty}^{\infty} |q|^{n^2} e^{2nA}$$

is convergent. Therefore the series in (14.1) is uniformly convergent in any finite region, and hence $\theta(z, q)$ is an integral function.

We have, obviously,

$$\theta(z, q) = 1 + 2\sum_{n=1}^{\infty} (-1)^n q^{n^2} \cos 2nz, \tag{14.2}$$

from which we deduce at once that

$$\theta(z + \pi, q) = \theta(z, q). \tag{14.3}$$

Also, from

$$\theta(z+\pi\tau, q) = \sum_{n=-\infty}^{\infty} (-1)^n q^{n^2} q^{2n} e^{2niz}$$
$$= -q^{-1}e^{-2iz} \sum_{n=-\infty}^{\infty} (-1)^{n+1} q^{(n+1)^2} e^{2(n+1)iz},$$

we have

$$\theta(z+\pi\tau, q) = -q^{-1}e^{-2iz}\theta(z, q). \tag{14.4}$$

We call $-q^{-1}e^{-2iz}$ the multiplicative factor associated with the period $\pi\tau$, so that the multiplicative factors for $\theta(z, q)$ associated with the periods π and $\pi\tau$ are 1 and $-q^{-1}e^{-2iz}$, respectively.

Following established tradition, we write $\theta_4(z, q)$ for this function $\theta(z, q)$, and the other three θ functions are defined in the following:

$$\theta_3(z, q) = \theta_4(z+\pi/2, q) = 1 + 2\sum_{n=1}^{\infty} q^{n^2} \cos 2nz \tag{14.5}$$

$$\begin{aligned} \theta_1(z, q) &= -ie^{iz+\pi i\tau/4}\theta_4(z+\pi\tau/2, q) \\ &= -i \sum_{n=-\infty}^{\infty} (-1)^n q^{(n+1/2)^2} e^{(2n+1)iz} \\ &= 2\sum_{n=1}^{\infty}(-1)^n q^{(n+1/2)^2} \sin(2n+1)z, \end{aligned} \tag{14.6}$$

$$\begin{aligned} \theta_2(z, q) &= \theta_1(z+\pi/2, q) \\ &= 2\sum_{n=1}^{\infty} q^{(n+1/2)^2} \cos(2n+1)z. \end{aligned} \tag{14.7}$$

Thus $\theta_1(z, q)$ is an odd function, and the others are even functions.

For the sake of simplicity, we shall henceforth write $\theta_\nu(z)$ for $\theta_\nu(z, q)$. Occasionally we need to emphasise the appearance of τ, in which case we write $\theta_\nu(z|\tau)$; we also use θ_ν and θ'_ν to denote the values for $\theta_\nu(0)$ and $\theta'_\nu(0)$, respectively.

It is easy to verify that the θ functions take the values as given in Table 14.1 for their multiplicative factors at the periods π and $\pi\tau$; we write $N = q^{-1}e^{-2iz}$. It follows at once

Table 14.1.

	$\theta_1(z)$	$\theta_2(z)$	$\theta_3(z)$	$\theta_4(z)$
π	-1	-1	1	1
$\pi\tau$	$-N$	N	N	$-N$

(14.8)

that, for any θ, we always have

$$\frac{\theta'(z+\pi)}{\theta(z+\pi)} = \frac{\theta'(z)}{\theta(z)}, \tag{14.9}$$

$$\frac{\theta'(z+\pi\tau)}{\theta(z+\pi\tau)} = -2i + \frac{\theta'(z)}{\theta(z)}. \tag{14.10}$$

Exercise 14.2 Show that

$$\theta_3(z,q) = \theta_3(2z,q^4) + \theta_2(2z,q^4),$$
$$\theta_4(z,q) = \theta_3(2z,q^4) - \theta_2(2z,q^4).$$

Exercise 14.3 Let $M = q^{1/4}e^{iz}$. Show that

$$\theta_1(z,q) = -\theta_2(z+\tfrac{\pi}{2}) = -iM\theta_3(z+\tfrac{\pi(1+\tau)}{2}) = -iM\theta_4(z+\tfrac{\pi\tau}{2}),$$
$$\theta_2(z,q) = M\theta_3(z+\tfrac{\pi\tau}{2}) = M\theta_4(z+\tfrac{\pi(1+\tau)}{2}) = \theta_1(z+\tfrac{\pi}{2}),$$
$$\theta_3(z,q) = \theta_4(z+\tfrac{\pi}{2}) = M\theta_1(z+\tfrac{\pi(1+\tau)}{2}) = M\theta_2(z+\tfrac{\pi\tau}{2}),$$
$$\theta_4(z,q) = -iM\theta_1(z+\tfrac{\pi\tau}{2}) = iM\theta_2(z+\tfrac{\pi(1+\tau)}{2}) = \theta_3(z+\tfrac{\pi}{2}).$$

14.2 The zeros of, and infinite products for, θ functions

If z_0 is a zero of $\theta(z)$, then so are

$$z_0 + m\pi + n\pi\tau, \qquad m, n \text{ integers};$$

they are called points associated with z_0.

Theorem 14.4 *Corresponding to any t, there is only a single zero of $\theta(z)$ inside the parallelogram C with the four vertices*

$$t, \quad t+\pi, \quad t+\pi+\pi\tau, \quad t+\pi\tau.$$

Proof. With $\theta(z)$ being an integral function, the number N of zeros inside C is given by

$$\begin{aligned} N &= \frac{1}{2\pi i}\int_C \frac{\theta'(z)}{\theta(z)}\,dz \\ &= \frac{1}{2\pi i}\int_t^{t+\pi}\left[\frac{\theta'(z)}{\theta(z)} - \frac{\theta'(z+\pi\tau)}{\theta(z+\pi\tau)}\right]dz - \frac{1}{2\pi i}\int_t^{t+\pi}\left[\frac{\theta'(z)}{\theta(z)} - \frac{\theta'(z+\pi)}{\theta(z+\pi)}\right]dz, \end{aligned}$$

and, by (14.9) and (14.10), we find that

$$N = \frac{1}{2\pi i}\int_t^{t+\pi} 2i\,dt = 1.$$

Clearly 0 is a zero of $\theta_1(z)$, so that $\theta_1(z), \theta_2(z), \theta_3(z), \theta_4(z)$ all have $0, \pi/2, \pi(1+\tau)/2, \pi\tau/2$ and their associated points, as zeros. □

Theorem 14.5 *We have the following infinite product representations for the θ functions:*

$$\theta_1(z) = 2Gq^{1/4} \sin z \prod_{n=1}^{\infty}(1 - 2q^{2n} \cos 2z + q^{4n}),$$

$$\theta_2(z) = 2Gq^{1/4} \cos z \prod_{n=1}^{\infty}(1 + 2q^{2n} \cos 2z + q^{4n}),$$

$$\theta_3(z) = G \prod_{n=1}^{\infty}(1 + 2q^{2n-1} \cos 2z + q^{4n-2}),$$

$$\theta_4(z) = G \prod_{n=1}^{\infty}(1 - 2q^{2n-1} \cos 2z + q^{4n-2}),$$

where G is a constant (which will be shown to be $G = \prod_{n=1}^{\infty}(1 - q^{2n})$ in Section 14.3).

Proof. Starting with $\theta_4(z)$, we know that it has zeros at

$$m\pi + (n + \tfrac{1}{2})\pi\tau.$$

Clearly

$$1 - 2q^{2n-1} \cos 2z + q^{4n-2} = (1 - q^{2n-1}e^{2iz})(1 - q^{2n-1}e^{-2iz})$$

has zeros at $z = -\frac{1}{2}\pi\tau - \pi m + \pi\tau n$ for $-\infty < m < \infty$ and $z = \frac{1}{2}\pi\tau - \pi m - \pi\tau n$, so that the product

$$f(z) = \prod_{n=1}^{\infty}(1 - 2q^{2n-1} \cos 2z + q^{4n-2})$$

has the same zeros as those of $\theta_4(z)$.

It is also clear that $f(z + \pi) = f(z)$ and that

$$\begin{aligned} f(z + \pi\tau) &= \prod_{n=1}^{\infty}(1 - q^{2n+1}e^{2iz})(1 - q^{2n-3}e^{-2iz}), \\ &= f(z)(1 - q^{-1}e^{-2iz})/(1 - qe^{2iz}) \\ &= -q^{-1}e^{-2iz} f(z). \end{aligned}$$

Therefore $\theta_4(z)/f(z)$ is a function with no zero, and it is periodic with periods $\pi, \pi\tau$, so that it has to be a constant, which is denoted by G.

Replacing z by $z + \pi/2$ we have the product formula for $\theta_3(z)$. Again, from $\theta_1(z) = -iq^{1/4}e^{iz}\theta(z + \pi\tau/2)$ and $\theta_2(z) = \theta_1(z + \pi/2)$, the remaining two formulae also follow. □

14.3 $G = \prod_{n=1}^{\infty}(1 - q^{2n})$

Theorem 14.6 *We have* $\theta_1'(0) = \theta_2(0)\theta_3(0)\theta_4(0)$.

Proof. (1) Taking the logarithmic derivative with respect to z of the product formula for $\theta_3(z)$, we have

$$\theta_3'(z) = \theta_3(z)\left[\sum_{n=1}^{\infty} \frac{2iq^{2n-1}e^{2iz}}{1+q^{2n-1}e^{2iz}} - \sum_{n=1}^{\infty} \frac{2iq^{2n-1}e^{-2iz}}{1+q^{2n-1}e^{-2iz}}\right].$$

Differentiating again, we have

$$\theta_3''(z) = \theta_3'(z)\left[\sum_{n=1}^{\infty} \frac{2iq^{2n-1}e^{2iz}}{1+q^{2n-1}e^{2iz}} - \sum_{n=1}^{\infty} \frac{2iq^{2n-1}e^{-2iz}}{1+q^{2n-1}e^{-2iz}}\right]$$
$$+ \theta_3(z)\left[\sum_{n=1}^{\infty} \frac{(2i)^2q^{2n-1}e^{2iz}}{(1+q^{2n-1}e^{2iz})^2} + \sum_{n=1}^{\infty} \frac{(2i)^2q^{2n-1}e^{-2iz}}{(1+q^{2n-1}e^{-2iz})^2}\right].$$

Letting $z \to 0$ we find that

$$\theta_3'(0) = 0, \quad \theta_3''(0) = -8\theta_3(0)\sum_{n=1}^{\infty} \frac{q^{2n-1}}{(1+q^{2n-1})^2}.$$

Similarly, we have

$$\theta_4'(0) = 0, \quad \theta_4''(0) = 8\theta_4(0)\sum_{n=1}^{\infty} \frac{q^{2n-1}}{(1-q^{2n-1})^2},$$

$$\theta_2'(0) = 0, \quad \theta_2''(0) = \theta_2(0)\left[-1 - 8\sum_{n=1}^{\infty} \frac{q^{2n}}{(1+q^{2n-1})^2}\right];$$

and, on writing $\theta_1(z) = \sin z\phi(z)$, we also have

$$\phi'(0) = 0, \quad \phi''(0) = 8\phi(0)\sum_{n=1}^{\infty} \frac{q^{2n}}{(1-q^{2n})^2}.$$

Differentiating $\theta_1(z) = \sin z\phi(z)$ thrice we have

$$\theta_1'(0) = \phi(0), \quad \theta_1'''(0) = 3\phi(0) - \phi(0),$$

so that

$$\frac{\theta_1'''(0)}{\theta_1'(0)} = 24\sum_{n=1}^{\infty} \frac{q^{2n}}{(1-q^{2n})^2} - 1.$$

Therefore

$$
\begin{aligned}
1 &+ \frac{\theta_2''(0)}{\theta_2(0)} + \frac{\theta_3''(0)}{\theta_3(0)} + \frac{\theta_4''(0)}{\theta_4(0)} \\
&= 8\left[-\sum_{n=1}^{\infty} \frac{q^{2n}}{(1+q^{2n})^2} - \sum_{n=1}^{\infty} \frac{q^{2n-1}}{(1+q^{2n-1})^2} + \sum_{n=1}^{\infty} \frac{q^{2n-1}}{(1-q^{2n-1})^2} \right] \\
&= 8\left[-\sum_{n=1}^{\infty} \frac{q^{n}}{(1+q^{n})^2} - \sum_{n=1}^{\infty} \frac{q^{n}}{(1-q^{n})^2} + \sum_{n=1}^{\infty} \frac{q^{2n}}{(1-q^{2n})^2} \right] \\
&= 8\left[-\sum_{n=1}^{\infty} \frac{q^n(1-q^n)^2 - q^n(1+q^n)^2}{(1+q^n)^2} - \sum_{n=1}^{\infty} \frac{q^{2n}}{(1-q^{2n})^2} \right] \\
&= 24\sum_{n=1}^{\infty} \frac{q^{2n}}{(1-q^{2n})^2} = 1 + \frac{\theta_1'''(0)}{\theta_1(0)},
\end{aligned}
$$

giving the equation

$$
\frac{\theta_1'''(0)}{\theta_1(0)} = \frac{\theta_2''(0)}{\theta_2(0)} + \frac{\theta_3''(0)}{\theta_3(0)} + \frac{\theta_4''(0)}{\theta_4(0)}. \tag{14.11}
$$

(2) From the differentiation of the series for the θ functions we have a lemma.

Lemma 14.7 *The functions $\Phi = \theta_i(z|\tau)$ $(i = 1, 2, 3, 4)$ satisfy the following differential equation:*

$$
\frac{\pi i}{4}\frac{\partial^2 \Phi}{\partial z^2} + \frac{\partial \Phi}{\partial \tau} = 0. \tag{14.12}
$$

For example,

$$
\begin{aligned}
\frac{\partial^2 \theta_3(z|\tau)}{\partial z^2} &= \frac{\partial^2}{\partial z^2}\left(1 + 2\sum_{n=1}^{\infty} e^{n^2\pi i\tau}\cos 2nz\right) \\
&= -8\sum_{n=1}^{\infty} n^2 e^{n^2\pi i\tau}\cos 2nz = -\frac{4}{\pi\tau}\frac{\partial}{\partial\tau}\theta_3(z|\tau).
\end{aligned}
$$

(3) Making use of the differential equation (14.12) we rewrite (14.11) as

$$
\begin{aligned}
\frac{1}{\theta_1'(0|\tau)}\frac{d}{d\tau}\theta_1'(0|\tau) = {}& \frac{1}{\theta_2(0|\tau)}\frac{d\theta_2(0|\tau)}{d\tau} \\
&+ \frac{1}{\theta_3(0|d\tau)}\frac{d\theta_3(0|\tau)}{d\tau} + \frac{1}{\theta_4(0|\tau)}\frac{d\theta_4(0|\tau)}{d\tau}.
\end{aligned}
$$

Integrating with respect to τ yields

$$
\theta_1'(0|\tau) = C\theta_2(0|\tau)\theta_3(0|\tau)\theta_4(0|\tau),
$$

where C is a constant, which has the value 1 because

$$
q^{-1/4}\theta_1' \to 2, \quad q^{-1/4}\theta_2 \to 2, \quad \theta_3 \to 1, \quad \theta_4 \to 1 \quad \text{as} \quad q \to 0.
$$

We have arrived at the required result

$$\theta_1' = \theta_2\theta_3\theta_4.$$

□

Theorem 14.8 *The constant in Theorem 14.5 is* $G = \prod_{n=1}^{\infty}(1 - q^{2n})$.

Proof. From the product formulae in Section 14.2, we have

$$\theta_1' = \phi(0) = 2q^{1/4}G\prod_{n=1}^{\infty}(1 - q^{2n})^2,$$

$$\theta_2 = 2q^{1/4}G\prod_{n=1}^{\infty}(1 + q^{2n})^2,$$

$$\theta_3 = G\prod_{n=1}^{\infty}(1 + q^{2n-1})^2,$$

$$\theta_4 = G\prod_{n=1}^{\infty}(1 - q^{2n-1})^2.$$

Substituting into Theorem 14.6, we have

$$\prod_{n=1}^{\infty}(1 - q^{2n})^2 = G^2\prod_{n=1}^{\infty}(1 + q^{2n})^2\prod_{n=1}^{\infty}(1 + q^{2n-1})^2\prod_{n=1}^{\infty}(1 - q^{2n-1})^2.$$

These infinite products are absolutely convergent, so that we may rearrange the terms in any way we wish. From

$$\left(\prod_{n=1}^{\infty}(1 + q^{2n})^2\prod_{n=1}^{\infty}(1 + q^{2n-1})^2\right)\left(\prod_{n=1}^{\infty}(1 - q^{2n})^2\prod_{n=1}^{\infty}(1 - q^{2n-1})^2\right)$$
$$= \prod_{n=1}^{\infty}(1 + q^n)^2\prod_{n=1}^{\infty}(1 - q^n)^2 = \prod_{n=1}^{\infty}(1 - q^{2n})^2,$$

we deduce that

$$\prod_{n=1}^{\infty}(1 - q^{2n})^2 = G^2$$

so that

$$G = \pm\prod_{n=1}^{\infty}(1 - q^{2n}).$$

As $q \to 0$, we note from the series and the product representations for $\theta_3(z)$ that we need to have $G \to 1$. The theorem is proved. □

Note. We have established, in passing, the identity

$$\prod_{n=1}^{\infty}(1+q^{2n})\prod_{n=1}^{\infty}(1+q^{2n-1})\prod_{n=1}^{\infty}(1-q^{2n-1})=1.$$

The reader may wish to prove this from

$$\prod_{n=1}^{\infty}(1+q^{2n})\prod_{n=1}^{\infty}(1-q^{4n-2})=\prod_{n=1}^{\infty}(1+q^{4n})\prod_{n=1}^{\infty}(1-q^{8n-4})=\cdots.$$

14.4 Using θ functions to represent elliptic functions

Let $f(z)$ be an elliptic function with periods $2w, 2w'$, and let $\alpha_1, \alpha_2, \ldots, \alpha_n$ be its zeros, and $\beta_1, \beta_2, \ldots, \beta_n$ its poles, in the fundamental region, where we already know (by Theorem 13.24) that

$$\sum_{r=1}^{n}\alpha_r=\sum_{r=1}^{n}\beta_r.$$

We proceed to show that the function

$$\prod_{r=1}^{n}\frac{\theta_1\left(\frac{\pi(z-\alpha_r)}{2w}\middle|\tau\right)}{\theta_1\left(\frac{\pi(z-\beta_r)}{2w}\middle|\tau\right)}$$

has the same zeros and poles as those of $f(z)$, and that it also has the same periods $2w, 2w'$. It then follows that any elliptic function $f(z)$ can be represented as

$$f(z)=A\prod_{r=1}^{n}\frac{\theta_1\big(\pi(z-\alpha_r)/2w\,|\,\tau\big)}{\theta_1\big(\pi(z-\beta_r)/2w\,|\,\tau\big)}.$$

The advantages of having such a representation over that of representations in terms of the σ function are: (i) the formulae for the θ functions converge much more rapidly and are therefore more suitable for computation; (ii) in the application of elliptic functions to deal with problems in applied mathematics, the real periods often have some particularly important properties, and θ functions offer clarity concerning these properties of such periods.

Returning to our present problem, if the principal part of $f(z)$ at β_r is

$$\sum_{m=1}^{m_r}A_{m,r}(z-\beta_r)^{-m},$$

then

$$f(z)=A+\sum_{r=1}^{n}\left\{\sum_{m=1}^{m_r}\frac{(-1)^{m-1}A_{m,r}}{(m-1)!}\frac{d^m}{dz^m}\log\theta_1\left(\frac{\pi(z-\beta_r)}{2w}\middle|\tau\right)\right\}.$$

We then have to prove that

$$\sigma(z)=\frac{2w}{\pi}\exp\left(\frac{\eta z^2}{2w}\right)\cdot\frac{1}{2}q^{-1/4}\prod_{n=1}^{\infty}(1-q^{2n})^3\theta_1\left(\frac{\pi z}{2w}\middle|\tau\right). \tag{14.13}$$

We first compare

$$\sigma(z)\quad\text{with}\quad f(z)=\exp\left(\frac{\eta z^2}{2w}\right)\theta_1\left(\frac{\pi z}{2w}\middle|\tau\right).$$

Well, they both have zeros at $2mw+2m'w'$, where m, m' are integers; next,

$$f(z+2w)=\exp\left(\frac{\eta z^2}{2w}+2\eta(z+w)\right)\cdot\theta_1\left(\frac{\pi z}{2w}+\pi\middle|\tau\right)=-e^{2\eta(z+w)}f(z),$$

and finally

$$\begin{aligned}f(z+2w')&=\exp\left(\frac{\eta z^2}{2w}+2\tau\eta(z+w')\right)\cdot\theta_1\left(\frac{\pi z}{2w}+\pi\tau\middle|\tau\right)\\&=-e^{2\tau\eta(z+w')}\exp\left(\frac{\eta z^2}{2w}\right)\cdot q^{-1}e^{-i\pi z/w}\theta_1\left(\frac{\pi z}{2w}\middle|\tau\right)\\&=-e^{2\eta'(z+w')}f(z),\end{aligned}$$

where we have made use of $\eta w'-\eta' w=\frac{1}{2}\pi i$, that is $\eta\tau-\frac{1}{2}(\pi i/w)=\eta'$. Thus $\sigma(z)/f(z)$ is an elliptic function with no zero, and is therefore a constant C. For the determination of the value for C we make use of

$$\lim_{z\to 0}\frac{\sigma(z)}{z}=1$$

and

$$\begin{aligned}\lim_{z\to 0}\frac{f(z)}{z}=\lim_{z\to 0}\frac{\theta_1\left(\frac{\pi z}{2w}\middle|\tau\right)}{z}&=\frac{\pi}{2w}\cdot 2Gq^{1/4}\prod_{n=1}^{\infty}(1-q^{2n})^2\\&=\frac{\pi}{2w}\cdot 2q^{1/4}\prod_{n=1}^{\infty}(1-q^{2n})^3,\end{aligned}$$

and the required formula is proved.

In formula (14.13), η can also be represented by a θ function – by taking the logarithm and differentiating twice we have

$$-\gamma(z)=\frac{\eta}{w}-\left(\frac{\pi}{2w}\right)^2\operatorname{cosec}^2\left(\frac{\pi z}{2w}\right)+\left(\frac{\pi}{2w}\right)^2\left[\frac{\phi''(v)}{\phi(v)}-\frac{\phi'(v)}{\phi(v)}\right],$$

where $v=\frac{1}{2}\pi z/w$ and $\phi(z)=\theta_1(z)/\sin z$.

On expanding in powers of z and comparing coefficients we find that

$$0=\frac{\eta}{w}-\frac{1}{3}\left(\frac{\pi}{2w}\right)^2+\left(\frac{\pi}{2w}\right)^2\frac{\phi''(0)}{\phi(0)},$$

so that

$$\eta = -\frac{\pi^2}{16w}\frac{\theta_1''}{\theta_1'}.$$

Finally,

$$\sigma(z) = \frac{2w}{\pi\theta_1'}\exp\Big(-\frac{v^2\theta_1'''}{6\theta_1'}\Big)\theta_1(v|\tau), \quad v = \frac{\pi z}{2w}.$$

Exercise 14.9 Show that

$$\eta' = -\Big(\frac{\pi^2 w'\theta_1'''}{12w^2\theta_1'} + \frac{\pi i}{2w'}\Big).$$

14.5 Various quadratic relationships between θ functions

Theorem 14.10 *We have*

$$\theta_1^2(z)\theta_3^2 = \theta_4^2(z)\theta_2^2 - \theta_2^2(z)\theta_4^2, \tag{14.14}$$
$$\theta_1^2(z)\theta_2^2 = \theta_4^2(z)\theta_3^2 - \theta_3^2(z)\theta_4^2, \tag{14.15}$$
$$\theta_1^2(z)\theta_4^2 = \theta_3^2(z)\theta_2^2 - \theta_2^2(z)\theta_3^2, \tag{14.16}$$
$$\theta_4^2(z)\theta_4^2 = \theta_3^2(z)\theta_3^2 - \theta_2^2(z)\theta_2^2. \tag{14.17}$$

Proof. The functions $\theta_1^2(z), \theta_2^2(z), \theta_3^2(z), \theta_4^2(z)$ are all periodic functions with period $\pi, \pi\tau$ and multiplicative factors $1, q^{-2}e^{-4iz}$, so that, corresponding to all a, b, a', b', the functions

$$\frac{a\theta_1^2(z) + b\theta_4^2(z)}{\theta_2(z)}, \qquad \frac{a'\theta_1^2(z) + b'\theta_4^2(z)}{\theta_3(z)}$$

are elliptic functions with periods $\pi, \pi\tau$. Generally speaking, they have a double pole, but we can choose a, b, a', b' so that the numerators are zero at the appropriate places, resulting in a simple pole instead. It then follows that they are constants, so that there are the following relationships:

$$\theta_2^2(z) = a\theta_1^2(z) + b\theta_4^2(z),$$
$$\theta_3^2(z) = a'\theta_1^2(z) + b'\theta_4^2(z).$$

Taking $z = \pi\tau/2$ and 0, together with

$$\theta_2(\pi\tau/2) = q^{-1/4}\theta_3, \quad \theta_4(\pi\tau/2) = 0, \quad \theta_1(\pi\tau/2) = iq^{-1/4}\theta_4,$$

we find that

$$\theta_3^2 = -a\theta_4^2, \quad \theta_2^2 = b\theta_4^2, \quad \theta_2^2 = a'\theta_4^2, \quad \theta_3^2 = b'\theta_4^2,$$

which then deliver (14.14) and (14.15).

Replacing z by $z + \pi/2$, we have (14.16) and (14.17). □

Taking $z = 0$ in (14.17) we have

$$\theta_2^4 + \theta_4^4 = \theta_3^4,$$

and if we write the equation in terms of the series for their terms then

$$\begin{aligned} &16q(1 + q^{1\cdot 2} + q^{2\cdot 3} + q^{3\cdot 4} + \cdots)^4 + (1 - 2q + 2q^4 - 2q^9 + \cdots)^4 \\ &\quad = (1 + 2q + 2q^4 + 2q^9 + \cdots)^4. \end{aligned} \tag{14.18}$$

In terms of products, we have

$$\left\{\prod_{n=1}^{\infty}(1 - q^{2n-1})\right\}^8 + 16q\left\{\prod_{n=1}^{\infty}(1 + q^{2n})\right\}^8 = \left\{\prod_{n=1}^{\infty}(1 + q^{2n-1})\right\}^8. \tag{14.19}$$

Exercise 14.11 Show that

$$\theta_1^4(z) + \theta_3^4(z) = \theta_2^4(z) + \theta_4^4(z). \tag{14.20}$$

14.6 Formulae for sums and differences

We derive a collection of formulae for $\theta_k(x + y)$ and $\theta_k(x - y)$; the method for the proofs does not deviate much from that given in Section 14.5.

For example, suppose that we wish to find a formula to represent $\theta_1(x + y)\theta_1(x - y)$. We fix x, and consider

$$D_{11}(y) = \theta_1(x + y)\theta_1(x - y)$$

as a function of y. By (14.8) we have

$$\begin{aligned} D_{11}(y + \pi) &= \theta_1(x + y + \pi)\theta_1(x - y - \pi) \\ &= \theta_1(x + y)\theta_1(x - y) = D_{11}(y) \end{aligned}$$

and

$$\begin{aligned} D_{11}(y + \pi\tau) &= \theta_1(x + y + \pi\tau)\theta_1(x - y - \pi\tau) \\ &= -q^{-1}e^{-2i(x+y)}\theta_1(x + y)\cdot(-q^{-1})e^{-2i(x-y)}\theta_1(x - y) \\ &= q^{-2}e^{-4iy}D_{11}(y), \end{aligned}$$

which show that we have a periodic function with multiplicative factors $1, q^{-2}e^{-4iy}$ for the periods $\pi, \pi\tau$.

The function has the same transformation formula as that for

$$A\theta_1^2(y) + B\theta_2^2(y), \tag{14.21}$$

so that

$$\frac{A\theta_1^2(y) + B\theta_2^2(y)}{\theta_1(x + y)\theta_1(x - y)}$$

is an elliptic function. The function $D_{11}(y)$ has two simple poles at $y = \pm x$. If we take $A = \theta_2^2(x)$, $B = -\theta_1^2(x)$ then there is a zero in the numerator at $y = x$. Thus this elliptic function can only have at most a simple pole at $y = -x$, and is therefore a constant, giving the relationship

$$\theta_2^2(x)\theta_1^2(y) - \theta_1^2(x)\theta_2^2(y) = C\theta(x+y)\theta(x-y).$$

Here C may depend on x, but it is independent of y. Taking $y = 0$ we have

$$-\theta_1^2(x)\theta_2^2 = C\theta_1^2(x), \quad C = -\theta_2^2,$$

giving the formula

$$\theta_2^2\theta_1(x+y)\theta_1(x-y) = \theta_1^2(x)\theta_2^2(y) - \theta_2^2(x)\theta_1^2(y). \tag{14.22}$$

Making use of the results in Theorem 14.10 we can represent $\theta_1(x+y)\theta_1(x-y)$ as a quadratic form in other θ functions. For example, using

$$\theta_4^2\theta_1^2(z) + \theta_3^2\theta_2^2(z) = \theta_2^2\theta_3^2(z),$$

we see from (14.22) that

$$\begin{aligned}\theta_4^2\theta_2^2\theta_1(x+y)\theta_1(x-y) &= \theta_2^2(y)\big(\theta_2^2\theta_3^2(x) - \theta_3^2\theta_2^2(x)\big) - \theta_2^2(x)\big(\theta_2^2\theta_3^2(y) - \theta_3^2\theta_2^2(y)\big)\\ &= \theta_2^2\theta_2^2(y)\theta_3^2(x) - \theta_2^2\theta_2^2(x)\theta_3^2(y),\end{aligned}$$

giving

$$\theta_4^2\theta_1(x+y)\theta_1(x-y) = \theta_2^2(y)\theta_3^2(x) - \theta_2^2(x)\theta_3^2(y). \tag{14.23}$$

Naturally, this can also be proved directly by the method of proving (14.22).

Collecting all such results we have the following identities:

$$\begin{aligned}\theta_1(x+y)\theta_1(x-y) &= \theta_2^{-2}\big(\theta_1^2(x)\theta_2^2(y) - \theta_2^2(x)\theta_1^2(y)\big)\\ &= \theta_2^{-2}\big(\theta_4^2(x)\theta_3^2(y) - \theta_3^2(x)\theta_4^2(y)\big)\\ &= \theta_3^{-2}\big(\theta_1^2(x)\theta_3^2(y) - \theta_3^2(x)\theta_1^2(y)\big)\\ &= \theta_3^{-2}\big(\theta_4^2(x)\theta_2^2(y) - \theta_2^2(x)\theta_4^2(y)\big)\\ &= \theta_4^{-2}\big(\theta_1^2(x)\theta_4^2(y) - \theta_4^2(x)\theta_1^2(y)\big)\\ &= \theta_4^{-2}\big(\theta_3^2(x)\theta_2^2(y) - \theta_2^2(x)\theta_3^2(y)\big).\end{aligned}$$

From $\theta_1(z+\pi/2)=\theta_2(z)$ and $\theta_1(z+\pi\tau/2)=iq^{-1/4}e^{-iz}\theta_4(z)$, we also deduce the following three groups of identities:

$$\begin{aligned}
\theta_2(x+y)\theta_2(x-y) &= \theta_2^{-1}\big(\theta_2^2(x)\theta_2^2(y)-\theta_1^2(x)\theta_1^2(y)\big)\\
&= \theta_2^{-1}\big(\theta_3^2(x)\theta_3^2(y)-\theta_4^2(x)\theta_4^2(y)\big)\\
&= \theta_3^{-2}\big(\theta_2^2(x)\theta_3^2(y)-\theta_4^2(x)\theta_1^2(y)\big)\\
&= \theta_3^{-2}\big(\theta_3^2(x)\theta_2^2(y)-\theta_1^2(x)\theta_4^2(y)\big)\\
&= \theta_4^{-2}\big(\theta_2^2(x)\theta_4^2(y)-\theta_3^2(x)\theta_1^2(y)\big)\\
&= \theta_4^{-2}\big(\theta_4^2(x)\theta_2^2(y)-\theta_1^2(x)\theta_3^2(y)\big);\\
\theta_3(x+y)\theta_3(x-y) &= \theta_2^{-2}\big(\theta_3^2(x)\theta_2^2(y)-\theta_4^2(x)\theta_1^2(y)\big)\\
&= \theta_2^{-2}\big(\theta_2^2(x)\theta_3^2(y)-\theta_1^2(x)\theta_4^2(y)\big)\\
&= \theta_3^{-2}\big(\theta_3^2(x)\theta_3^2(y)-\theta_1^2(x)\theta_1^2(y)\big)\\
&= \theta_3^{-2}\big(\theta_2^2(x)\theta_2^2(y)-\theta_4^2(x)\theta_4^2(y)\big)\\
&= \theta_4^{-2}\big(\theta_3^2(x)\theta_4^2(y)-\theta_2^2(x)\theta_1^2(y)\big)\\
&= \theta_4^{-2}\big(\theta_4^2(x)\theta_3^2(y)-\theta_1^2(x)\theta_2^2(y)\big);\\
\theta_4(x+y)\theta_4(x-y) &= \theta_2^{-2}\big(\theta_4^2(x)\theta_2^2(y)-\theta_3^2(x)\theta_1^2(y)\big)\\
&= \theta_2^{-2}\big(\theta_1^2(x)\theta_3^2(y)-\theta_2^2(x)\theta_4^2(y)\big)\\
&= \theta_3^{-2}\big(\theta_4^2(x)\theta_3^2(y)-\theta_2^2(x)\theta_1^2(y)\big)\\
&= \theta_3^{-2}\big(\theta_1^2(x)\theta_2^2(y)-\theta_3^2(x)\theta_4^2(y)\big)\\
&= \theta_4^{-2}\big(\theta_4^2(x)\theta_4^2(y)-\theta_1^2(x)\theta_1^2(y)\big)\\
&= \theta_4^{-2}\big(\theta_3^2(x)\theta_3^2(y)-\theta_2^2(x)\theta_2^2(y)\big).
\end{aligned}$$

There is a similar collection of identities for $D_{ij}(y)=\theta_i(x+y)\theta_j(x-y)$. For example, with $D_{14}(y)=\theta_1(x+y)\theta_4(x-y)$ we have

$$D_{14}(y+\pi)=-D_{14}(y),\quad D_{14}(y+\pi\tau)=q^{-2}e^{-4iy}D_{14}(y),$$

so that

$$\frac{A\theta_2(y)\theta_2(y)+B\theta_1(y)\theta_4(y)}{\theta_1(x+y)\theta_4(x-y)}$$

is an elliptic function of y with periods $\pi,\pi\tau$. There are simple zeros at $x+y=0$ and $x-y=\pi\tau/2$ in the denominator, so that on taking $A=\theta_1(x)\theta_4(x)$ and $B=\theta_2(x)\theta_3(x)$, the numerator has a zero at $y=-x$. Consequently,

$$\theta_1(x)\theta_2(y)\theta_3(y)\theta_4(x)+\theta_1(y)\theta_2(x)\theta_3(x)\theta_4(y)=C\theta_1(x+y)\theta_4(x-y),$$

and we find that $C = \theta_2\theta_3$ by setting $y = 0$. We then have an identity, which is displayed with the others as follows:

$$\theta_3\theta_4\theta_1(x+y)\theta_2(x-y) = \theta_1(x)\theta_2(x)\theta_3(y)\theta_4(y) + \theta_1(y)\theta_2(y)\theta_3(x)\theta_4(x),$$
$$\theta_2\theta_4\theta_1(x+y)\theta_3(x-y) = \theta_1(x)\theta_2(y)\theta_3(x)\theta_4(y) + \theta_1(y)\theta_2(x)\theta_3(y)\theta_4(x),$$
$$\theta_2\theta_3\theta_1(x+y)\theta_4(x-y) = \theta_1(x)\theta_2(y)\theta_3(y)\theta_4(x) + \theta_1(y)\theta_2(x)\theta_3(x)\theta_4(y),$$
$$\theta_2\theta_3\theta_2(x+y)\theta_3(x-y) = \theta_2(x)\theta_3(x)\theta_2(y)\theta_3(y) - \theta_1(x)\theta_4(x)\theta_1(y)\theta_4(y),$$
$$\theta_2\theta_4\theta_2(x+y)\theta_4(x-y) = \theta_2(x)\theta_4(x)\theta_2(y)\theta_4(y) - \theta_1(x)\theta_3(x)\theta_1(y)\theta_3(y),$$
$$\theta_3\theta_4\theta_3(x+y)\theta_4(x-y) = \theta_3(x)\theta_4(x)\theta_3(y)\theta_4(y) - \theta_1(x)\theta_2(x)\theta_1(y)\theta_2(y).$$

From these it is not difficult to derive the double-valued formulae:

$$\theta_2^3\theta_2(2x) = \theta_2^4(x) - \theta_1^4(x) = \theta_3^4(x) - \theta_4^4(x),$$
$$\theta_3^3\theta_3(2x) = \theta_1^4(x) + \theta_3^4(x) = \theta_2^4(x) + \theta_4^4(x),$$
$$\theta_4^3\theta_4(2x) = \theta_4^4(x) - \theta_1^4(x) = \theta_3^4(x) - \theta_2^4(x),$$
$$\theta_3^2\theta_2\theta_2(2x) = \theta_2^2(x)\theta_3^2(x) - \theta_1^2(x)\theta_4^2(x),$$
$$\theta_4^2\theta_2\theta_2(2x) = \theta_2^2(x)\theta_4^2(x) - \theta_1^2(x)\theta_3^2(x),$$
$$\theta_2^2\theta_3\theta_3(2x) = \theta_2^2(x)\theta_3^2(x) + \theta_1^2(x)\theta_4^2(x),$$
$$\theta_4^2\theta_3\theta_3(2x) = \theta_3^2(x)\theta_4^2(x) - \theta_1^2(x)\theta_2^2(x),$$
$$\theta_2^2\theta_4\theta_4(2x) = \theta_1^2(x)\theta_3^2(x) + \theta_2^2(x)\theta_4^2(x),$$
$$\theta_3^2\theta_4\theta_4(2x) = \theta_1^2(x)\theta_2^2(x) + \theta_3^2(x)\theta_4^2(x).$$

14.7 Differential equations satisfied by ratios of θ functions

The function

$$\theta_1(z)/\theta_4(z)$$

is periodic with multiplicative factors $-1, +1$ at the periods $\pi, \pi\tau$, so that its derivative

$$\{\theta_1'(z)\theta_4(z) - \theta_4'(z)\theta_1(z)\}/\theta_4^2(z)$$

also has the same properties. Again, the function

$$\theta_2(z)\theta_3(z)/\theta_4^2(z)$$

also has these properties, so that

$$\phi(z) = \frac{\theta_1'(z)\theta_4(z) - \theta_4'(z)\theta_1(z)}{\theta_2(z)\theta_3(z)}$$

is periodic with periods $\pi, \pi\tau$; moreover, $\phi(z)$ can only have (simple) poles at

$$z = \frac{\pi}{2}, \frac{\pi(1+\tau)}{2}$$

and their associated points.

Now consider $\phi(z+\pi\tau/2)$. By Exercise 14.3, we know that

$$\theta_1(z+\pi\tau/2) = iq^{-1/4}e^{-iz}\theta_4(z),$$
$$\theta_4(z+\pi\tau/2) = iq^{-1/4}e^{-iz}\theta_1(z),$$
$$\theta_2(z+\pi\tau/2) = q^{-1/4}e^{-iz}\theta_3(z),$$
$$\theta_3(z+\pi\tau/2) = q^{-1/4}e^{-iz}\theta_2(z),$$

so that

$$\phi(z+\pi\tau/2) = \frac{-\theta_4'(z)\theta_1(z)+\theta_1'(z)\theta_4(z)}{\theta_3(z)\theta_2(z)} = \phi(z).$$

Thus $\phi(z)$ is an elliptic function, with periods π and $\pi\tau/2$, and with only a simple pole at $\pi/2$ within the fundamental region. The function has to be a constant, the value of which is found to be

$$\theta_1'\theta_4/\theta_2\theta_3 = \theta_4^2$$

by letting $z \to 0$. We have therefore derived the important differential equation

$$\frac{d}{dz}\left\{\frac{\theta_1(z)}{\theta_4(z)}\right\} = \theta_4^2\frac{\theta_2(z)}{\theta_4(z)}\frac{\theta_3(z)}{\theta_4(z)}. \tag{14.25}$$

Let $\xi = \theta_1(z)/\theta_4(z)$; by (14.14) and (14.15) we have

$$\left(\frac{d\xi}{dz}\right)^2 = (\theta_2^2-\xi^2\theta_3^2)(\theta_3^2-\xi^2\theta_2^2).$$

This differential equation has the particular solution $\theta_1(z)/\theta_4(z)$, and the general solution is $\pm\theta_1(z+c)/\theta_4(z+c)$.

Similarly to (14.25) we also have

$$\frac{d}{dz}\left\{\frac{\theta_2(z)}{\theta_4(z)}\right\} = -\theta_3^2\frac{\theta_1(z)}{\theta_4(z)}\frac{\theta_3(z)}{\theta_4(z)},$$
$$\frac{d}{dz}\left\{\frac{\theta_3(z)}{\theta_4(z)}\right\} = -\theta_2^2\frac{\theta_1(z)}{\theta_4(z)}\frac{\theta_2(z)}{\theta_4(z)}.$$

14.8 Jacobian elliptic functions

The three important Jacobian elliptic functions can be defined in terms of the θ functions:

$$\operatorname{sn} u = \frac{\theta_3}{\theta_2}\frac{\theta_1(u\theta_3^{-2})}{\theta_4(u\theta_3^{-2})}, \tag{14.26}$$

$$\operatorname{cn} u = \frac{\theta_4}{\theta_2}\frac{\theta_2(u\theta_3^{-2})}{\theta_4(u\theta_3^{-2})}, \tag{14.27}$$

$$\operatorname{dn} u = \frac{\theta_4}{\theta_3}\frac{\theta_3(u\theta_3^{-2})}{\theta_4(u\theta_3^{-2})}. \tag{14.28}$$

By Theorem 14.10 we have

$$\mathrm{sn}^2 u + \mathrm{cn}^2 u = 1, \tag{14.29}$$
$$k^2 \mathrm{sn}^2 u + \mathrm{dn}^2 u = 1, \tag{14.30}$$

where

$$k^{1/2} = \frac{\theta_2}{\theta_3}. \tag{14.31}$$

From (14.27), (14.28) and (14.29) we have

$$\frac{d\,\mathrm{sn}\,u}{du} = \mathrm{cn}\,u\,\mathrm{dn}\,u, \tag{14.32}$$
$$\frac{d\,\mathrm{cn}\,u}{du} = -\mathrm{sn}\,u\,\mathrm{dn}\,u, \tag{14.33}$$
$$\frac{d\,\mathrm{dn}\,u}{du} = -k^2 \mathrm{sn}\,u\,\mathrm{cn}\,u. \tag{14.34}$$

From (14.29) and (14.30) we see that $y = \mathrm{sn}\,u$ satisfies the differential equation

$$\left(\frac{dy}{du}\right)^2 = (1 - y^2)(1 - k^2 y^2). \tag{14.35}$$

Similarly, $y = \mathrm{cn}\,u$ satisfies

$$\left(\frac{dy}{du}\right)^2 = (1 - y^2)(k'^2 + k^2 \mathrm{cn}^2 u), \tag{14.36}$$

where

$$k'^2 = 1 - k^2; \tag{14.37}$$

$y = \mathrm{cn}\,u$ satisfies

$$\left(\frac{dy}{du}\right)^2 = (1 - y^2)(y^2 - k'^2). \tag{14.38}$$

The number k defined by (14.31) is called the (elliptic) modulus of the elliptic function sn, and we may need to emphasise it by writing $\mathrm{sn}(u, k)$ for $\mathrm{sn}\,u$. The number k' defined by (14.37) is called the complementary modulus; actually (14.37) does not specify k'; we define it instead by

$$k'^{1/2} = \theta_4/\theta_3, \tag{14.39}$$

and equation (14.37) then follows from $\theta_2^4 + \theta_4^4 = \theta_3^4$.

14.9 Periodicity

From the formulae in Exercise 14.3 we find that

$$\begin{aligned}
\theta_1(z+\pi/2) &= \theta_2(z), & \theta_1(z+\pi\tau/2) &= iq^{-1/4}e^{-iz}\theta_4(z),\\
\theta_2(z+\pi/2) &= -\theta_1(z), & \theta_2(z+\pi\tau/2) &= q^{-1/4}e^{-iz}\theta_3(z),\\
\theta_3(z+\pi/2) &= \theta_4(z), & \theta_3(z+\pi\tau/2) &= q^{-1/4}e^{-iz}\theta_2(z),\\
\theta_4(z+\pi/2) &= \theta_3(z), & \theta_4(z+\pi\tau/2) &= iq^{-1/4}e^{-iz}\theta_1(z).
\end{aligned}$$

Let $K=\pi\theta_3^2/2$ and $K'=-i\pi\tau\theta_3^2/2$; we then deduce that

$$\operatorname{sn}(u+K)=\frac{\theta_3}{\theta_2}\frac{\theta_1(u\theta_3^{-2}+\pi/2)}{\theta_4(u\theta_3^{-2}+\pi/2)}=\frac{\theta_3}{\theta_2}\frac{\theta_2(u\theta_3^{-2})}{\theta_3(u\theta_3^{-2})}=\frac{\operatorname{cn}u}{\operatorname{dn}u} \tag{14.40}$$

and

$$\operatorname{sn}(u+iK')=\frac{\theta_3}{\theta_2}\frac{\theta_1(u\theta_3^{-2}+\pi\tau/2)}{\theta_4(u\theta_3^{-2}+\pi\tau/2)}=\frac{\theta_3}{\theta_2}\frac{\theta_4(u\theta_3^{-2})}{\theta_1(u\theta_3^{-2})}=\frac{1}{k\operatorname{sn}u}. \tag{14.41}$$

Similarly, we can show that

$$\operatorname{cn}(u+K)=-k'\frac{\operatorname{sn}u}{\operatorname{dn}u},\qquad \operatorname{cn}(u+iK')=-\frac{i}{k}\frac{\operatorname{dn}u}{\operatorname{sn}u}, \tag{14.42}$$

$$\operatorname{dn}(u+K)=k'\frac{\operatorname{sn}u}{\operatorname{dn}u},\qquad \operatorname{dn}(u+iK')=-i\frac{\operatorname{cn}u}{\operatorname{sn}u}. \tag{14.43}$$

We can then deduce that

$$\operatorname{sn}(u+2K)=-\operatorname{sn}u,\qquad \operatorname{sn}(u+2iK')=\operatorname{sn}u, \tag{14.44}$$

$$\operatorname{cn}(u+2K)=-\operatorname{cn}u,\qquad \operatorname{cn}(u+2iK')=-\operatorname{cn}u, \tag{14.45}$$

$$\operatorname{dn}(u+2K)=\operatorname{dn}u,\qquad \operatorname{dn}(u+2iK')=-\operatorname{dn}u. \tag{14.46}$$

Thus we arrive at the periodicity properties

$$\operatorname{sn}(u+4K)=\operatorname{sn}(u+2iK')=\operatorname{sn}u, \tag{14.47}$$

$$\operatorname{cn}(u+4K)=\operatorname{cn}(u+2iK')=\operatorname{cn}u, \tag{14.48}$$

$$\operatorname{dn}(u+2K)=\operatorname{dn}(u+4iK')=\operatorname{dn}u. \tag{14.49}$$

From $\theta_1(0)=0$, $\theta_4(0)\neq 0$ and (14.40) we find that $\operatorname{sn}0=0$. Also, from (14.41) and (14.42) we have $\operatorname{cn}0=\operatorname{dn}0=1$. From (14.40) we now have

$$\operatorname{sn}K=1,\quad \operatorname{cn}K=0,\quad \operatorname{dn}K=k', \tag{14.50}$$

and

$$\operatorname{sn}K'=\infty,\quad \operatorname{cn}K'=\infty,\quad \operatorname{dn}K'=\infty. \tag{14.51}$$

14.10 Being analytic

Since $\theta_1(z)$ has zeros at $m\pi + n\tau\pi$ (m, n being integers), the function $\operatorname{sn} u$ has zeros at $(m\pi + n\tau\pi)\theta_3^2 = 2mK + 2niK'$. Being periodic with periods $4K, 2iK'$, there are now two zeros within the fundamental region, namely

$$u = 0,\ 2K.$$

Again, because $\theta_4(z)$ has zeros at $m\pi + \pi\tau(n + \frac{1}{2})$, the function $\operatorname{sn} u$ has poles at $2mK + (2n+1)iK'$, and the two poles within the fundamental region are at

$$u = iK',\ 2K + iK'.$$

Since $\operatorname{sn} u$ is an odd function, and

$$\frac{d}{du}\operatorname{sn} u = \operatorname{cn} u \operatorname{dn} u,$$
$$\frac{d^3}{du^3}\operatorname{sn} u = 4k^2 \operatorname{sn}^2 u \operatorname{cn} u \operatorname{dn} u - \operatorname{cn} u \operatorname{dn} u(\operatorname{dn}^2 u + k^2 \operatorname{cn}^2 u),$$

we find that

$$\operatorname{sn} u = u - \frac{1}{6}(1+k^2)u^3 + O(|u|^5) \quad \text{as} \quad u \to 0,$$

and the same method shows that

$$\operatorname{cn} u = 1 - \frac{1}{2}u^2 + O(|u|^4), \quad \operatorname{dn} u = 1 - \frac{1}{2}k^2u^2 + O(|u|^4).$$

It then follows that, as $u \to 0$,

$$\begin{aligned}\operatorname{sn}(u + iK') &= \frac{1}{k \operatorname{sn} u} = \frac{1}{ku}\left(1 - \frac{1}{6}(1+k^2)u^2 + O(|u|^4)\right)^{-1} \\ &= \frac{1}{ku} + \frac{1+k^2}{6k}u + O(|u|^3),\end{aligned}$$

and the same method shows that

$$\operatorname{cn}(u + iK') = \frac{-i}{ku} + \frac{2k^2-1}{6k}iu + O(|u|^3),$$
$$\operatorname{dn}(u + iK') = \frac{-i}{u} + \frac{2-k^2}{6k}iu + O(|u|^3).$$

Thus $\operatorname{sn} u$ has the residue $1/k$ at the pole iK'. Since the sum of the two residues has to be zero, we conclude the following: The function $\operatorname{sn} u$ is periodic with periods $4K$, $2iK'$, and is analytic everywhere except for the simple poles at

$$u \equiv iK',\ 2K + iK' \quad (\operatorname{mod} 4K,\ 2iK');$$

the residues at these (sets of) poles are $1/k$ and $-1/k$, respectively.

Similarly, $\operatorname{cn} u$ is periodic with periods $4K, 2K + 2iK'$, and is analytic everywhere except for the simple poles at

$$u \equiv iK',\ 2K + iK' \quad (\text{mod } 4K,\ 2K + 2iK');$$

the residues at these (sets of) poles are $-i/k$ and i/k, respectively; the zeros are at $u \equiv K$ $(\text{mod } 4K,\ 2iK')$.

Also, $\operatorname{dn} u$ is periodic with periods $2K, 4iK'$, and is analytic everywhere except for the simple poles at

$$u \equiv iK',\ 3iK' \quad (\text{mod } 2K,\ 4iK');$$

the residues at these (sets of) poles are $-i$ and i, respectively, and the simple zeros are at $u \equiv K + iK' (\text{mod } 2K,\ 2iK')$.

14.11 The relationship between Weierstrass and Jacobian functions

Let e_1, e_2, e_3 be three numbers such that $e_1 + e_2 + e_3 = 0$. Then the transformation

$$y = e_3 + \frac{e_1 - e_2}{\operatorname{sn}^2(\lambda u, k)}$$

satisfies

$$\begin{aligned}\left(\frac{dy}{du}\right)^2 &= 4(e_1 - e_3)^2\lambda^2 \frac{1}{\operatorname{sn}^2\lambda u} \frac{\operatorname{cn}^2\lambda u}{\operatorname{sn}^2\lambda u} \frac{\operatorname{dn}^2\lambda u}{\operatorname{sn}^2\lambda u} \\ &= 4(e_1 - e_3)^2\lambda^2 \frac{1}{\operatorname{sn}^2\lambda u}\left(\frac{1}{\operatorname{sn}^2\lambda u} - 1\right)\left(\frac{1}{\operatorname{sn}^2\lambda u} - k^2\right) \\ &= 4\lambda^2(e_1 - e_3)^{-1}(y - e_3)(y - e_1)\{y - k^2(e_1 - e_2) - e_3\}.\end{aligned}$$

Take $\lambda^2 = e_1 - e_3$ and

$$k^2 = \frac{e_2 - e_3}{e_1 - e_3},$$

so that we have

$$\left(\frac{dy}{du}\right)^2 = 4y^3 - g_2 y - g_3,$$

and hence

$$\frac{e_3 + (e_1 - e_3)}{\operatorname{sn}^2\left\{u(e_1 - e_2)^{1/2}, \sqrt{\frac{e_2 - e_3}{e_1 - e_3}}\right\}} = \wp(u + \alpha; g_2, g_3).$$

Here α is a constant, the value of which is a period, as can be seen by letting $u \to 0$. We then have the relationship

$$\wp(u; g_2, g_3) = \frac{e_3 + (e_1 - e_3)}{\operatorname{sn}^2\{u(e_1 - e_2)^{1/2}\}},$$

with the modulus for the Jacobian function being given in the above.

14.12 Addition formulae

Theorem 14.12 *We have*

$$\operatorname{sn}(u+v) = \frac{\operatorname{sn} u \operatorname{cn} v \operatorname{dn} v + \operatorname{sn} v \operatorname{cn} u \operatorname{dn} u}{1 - k \operatorname{sn}^2 u \operatorname{sn}^2 v},$$
$$\operatorname{cn}(u+v) = \frac{\operatorname{cn} u \operatorname{cn} v - \operatorname{sn} u \operatorname{sn} v \operatorname{dn} u \operatorname{dn} v}{1 - k \operatorname{sn}^2 u \operatorname{sn}^2 v},$$
$$\operatorname{dn}(u+v) = \frac{\operatorname{dn} u \operatorname{dn} v - k \operatorname{sn} u \operatorname{sn} v \operatorname{cn} u \operatorname{cn} v}{1 - k \operatorname{sn}^2 u \operatorname{sn}^2 v}.$$

Proof. In Section 14.6 we established the formulae

$$\theta_2\theta_3\theta_1(x+y)\theta_4(x-y) = \theta_1(x)\theta_2(y)\theta_3(y)\theta_4(x) + \theta_1(y)\theta_2(x)\theta_3(x)\theta_4(y),$$
$$\theta_4^2\theta_4(x+y)\theta_4(x-y) = \theta_4^2(x)\theta_4^2(y) - \theta_1^2(x)\theta_1^2(y).$$

Their ratio is then

$$\frac{\theta_2\theta_3}{\theta_4^2}\frac{\theta_1(x+y)}{\theta_4(x+y)} = \frac{\theta_1(x)\theta_2(y)\theta_3(y)\theta_4(x) + \theta_1(y)\theta_2(x)\theta_3(x)\theta_4(y)}{\theta_4^2(x)\theta_4^2(y) - \theta_1^2(x)\theta_1^2(y)}$$
$$= \frac{\dfrac{\theta_1(x)}{\theta_4(x)}\dfrac{\theta_2(y)}{\theta_4(y)}\dfrac{\theta_3(y)}{\theta_4(y)} + \dfrac{\theta_1(y)}{\theta_4(y)}\dfrac{\theta_2(x)}{\theta_4(x)}\dfrac{\theta_3(x)}{\theta_4(x)}}{1 - \left(\dfrac{\theta_1(x)}{\theta_4(x)}\right)^2\left(\dfrac{\theta_1(y)}{\theta_4(y)}\right)^2}.$$

If we set $x = \theta_3^{-2}u$, $y = \theta_3^{-2}v$, and recall that $k = (\theta_2/\theta_3)^2$, then the first formula in the theorem now follows. A similar argument delivers the remaining two formulae. □

14.13 Representations of K, K' as functions of k, k'

Let τ be purely imaginary, so that $q = \pi i \tau$ is real and $0 < q < 1$. The numbers

$$k^{1/2} = \frac{\theta_2}{\theta_3}, \quad k'^{1/2} = \frac{\theta_4}{\theta_3}$$

are also real, and $k^2 + k'^2 = (\theta_2^4 + \theta_4^4)/\theta_3^4 = 1$, so that $0 < k < 1$ and $0 < k' < 1$. Again, the numbers

$$K = \frac{1}{2}\pi\theta_3^2, \quad K' = -\frac{1}{2}i\pi\tau\theta_3^2$$

are also real.

From (14.35) we deduce that $y = \operatorname{sn} u$ is the inverse function of

$$u = \int_0^y (1-t^2)^{-1/2}(1-k^2t^2)^{-1/2}\,dt, \tag{14.52}$$

and from $\operatorname{sn} K = 1$, we find that

$$K \equiv \int_0^1 (1-t^2)^{-1/2}(1-k^2t^2)^{-1/2}\,dt \pmod{4K, 2iK'}. \tag{14.53}$$

On assuming that the path of integration is along the real axis from 0 to 1, we can then only have

$$(1+4m)K = \int_0^1 (1-t^2)^{-1/2}(1-k^2t^2)^{-1/2}\,dt.$$

Here, m is an integer, which has to be non-negative because the right-hand side is positive. If $m \geq 1$, then, by the increasing nature of the integral, there must be an $\alpha < 1$ such that

$$K = \int_0^\alpha (1-t^2)^{-1/2}(1-k^2t^2)^{-1/2}\,dt,$$

so that $\operatorname{sn} K = \alpha < 1$, which is impossible. Therefore we must have

$$K = \int_0^1 (1-t^2)^{-1/2}(1-k^2t^2)^{-1/2}\,dt. \tag{14.54}$$

In (14.41) we take $u = K$ to give

$$\operatorname{sn}(K+iK') = \frac{1}{k}.$$

Therefore

$$K+iK' \equiv \int_0^{\frac{1}{k}} (1-t^2)^{-1/2}(1-k^2t^2)^{-1/2}\,dt \quad (\text{mod } 4K, 2iK'),$$

and, by (14.54),

$$iK' \equiv -i\int_1^{\frac{1}{k}} (t^2-1)^{-1/2}(1-k^2t^2)^{-1/2}\,dt \quad (\text{mod } 4K, 2iK'),$$

that is

$$K' = -\int_1^{\frac{1}{k}} (t^2-1)^{-1/2}(1-k^2t^2)^{-1/2}\,dt. \tag{14.55}$$

The substitution

$$t = (1-k'^2s^2)^{-1/2}$$

yields

$$t^2-1 = \frac{k'^2s^2}{1-k'^2s^2}, \qquad 1-k^2t^2 = \frac{k'^2(1-s^2)}{1-k'^2s^2}, \qquad dt = (1-k'^2s^2)^{-3/2}k'^2s\,ds,$$

so that

$$K' = -\int_0^1 (1-s^2)^{-1/2}(1-k'^2s^2)^{-1/2}\,ds.$$

When $|k| < 1$, the factor $(1 - k^2t^2)^{-1/2}$ can be expanded into a uniformly convergent power series with respect to k^2, so that

$$
\begin{aligned}
K &= \int_0^1 (1-t^2)^{-1/2}\Big(1 + \sum_{n=1}^{\infty} \frac{1\cdot 3\cdots(2n-1)}{2\cdot 4\cdots 2n} k^{2n} t^{2n}\Big) dt \\
&= \frac{1}{2}\sum_{n=0}^{\infty} \frac{\Gamma(n+1/2)}{\Gamma(1/2) n!} \int_0^1 (1-u)^{-1/2} u^{n-1/2}\, du \\
&= \frac{1}{2}\sum_{n=0}^{\infty} \left(\frac{\Gamma(n+1/2)}{n!}\right)^2 k^{2n}.
\end{aligned}
$$

If $k^2 = c$ is taken to be complex, then K is an analytic function of c inside the unit disc $|c| < 1$. We can then prove that the function can be analytically continued into the whole complex c-plane, and K can be made into a single-valued function on the cut plane with a slit from 1 to ∞.

14.14 Various formulae associated with the Jacobian elliptic functions

From the results for the θ functions we deduce immediately the following infinite product formulae, in which $u = 2Kx/\pi$:

$$
\operatorname{sn} u = 2q^{1/4} k^{-1/2} \sin x \prod_{n=1}^{\infty} \frac{1 - 2q^{2n}\cos 2x + q^{4n}}{1 - 2q^{2n-1}\cos 2x + q^{4n-2}}, \tag{14.56}
$$

$$
\operatorname{cn} u = 2q^{1/4} k'^{-1/2} \cos x \prod_{n=1}^{\infty} \frac{1 + 2q^{2n}\cos 2x + q^{4n}}{1 - 2q^{2n-1}\cos 2x + q^{4n-2}}, \tag{14.57}
$$

$$
\operatorname{dn} u = k'^{-1/2} \sin x \prod_{n=1}^{\infty} \frac{1 + 2q^{2n-1}\cos 2x + q^{4n-2}}{1 - 2q^{2n-1}\cos 2x + q^{4n-2}}. \tag{14.58}
$$

By considering $\operatorname{sn} u$ to be a function of x, it is an odd periodic function with period 2π, so that it has the Fourier expansion

$$
\operatorname{sn} u = \sum_{n=1}^{\infty} b_n \sin nx,
$$

with

$$
b_n = \frac{1}{\pi}\int_{-\pi}^{\pi} \operatorname{sn} u \sin nx\, dx = \frac{1}{\pi i}\int_{-\pi}^{\pi} \operatorname{sn} u\, e^{inx}\, dx.
$$

Consider the contour integral

$$
\int \operatorname{sn} u\, e^{inx}\, dx
$$

round the parallelogram with vertices $-\pi, \pi, \pi\tau, -2\pi + \pi\tau$. By periodicity of the integrand, the parts of the integral from π to $\pi\tau$ and from $-2\pi + \pi\tau$ to $-\pi$ cancel. There are only two poles at $-\pi + \frac{1}{2}\pi\tau$ and $\frac{1}{2}\pi\tau$ inside the contour, with the residues

$$-k^{-1}\left(\frac{\pi}{2K}\right)\exp\left(-ni\pi + \frac{1}{2}n\pi i\tau\right), \quad k^{-1}\left(\frac{\pi}{2K}\right)\exp\left(\frac{1}{2}n\pi i\tau\right),$$

so that

$$\left(\int_{-\pi}^{\pi} - \int_{-2\pi+\pi\tau}^{\pi\tau}\right)\operatorname{sn} u\, e^{inx}\, dx = \frac{\pi^2 i}{Kk}q^{n/2}\{1-(-1)^n\}.$$

Replacing x by $x - \pi + \pi\tau$ in the second integral we find that

$$\{1-(-1)^n q^n\}\int_{-\pi}^{\pi} \operatorname{sn} u\, e^{inx}\, dx = \frac{\pi^2 i}{Kk}q^{n/2}\{1-(-1)^n\}.$$

Thus, if n is even $b_n = 0$, and if n is odd

$$b_n = \frac{2\pi}{Kk}\frac{q^{in}}{1-q^n},$$

so that

$$\operatorname{sn} u = \frac{2\pi}{Kk}\sum_{n=0}^{\infty}\frac{q^{n+1/2}\sin(2n+1)x}{1-q^{2n+1}}. \tag{14.59}$$

When x is real we have $q^{n/2}e^{\pm nix} \to 0$ as $n \to \infty$, and it is not difficult to show that the formula holds for $|\theta x| < \frac{1}{2}\pi\theta\tau$.

We leave it to the reader to show that, for $|\theta x| < \frac{1}{2}\pi\theta\tau$,

$$\operatorname{cn} u = \frac{2\pi}{Kk}\sum_{n=0}^{\infty}\frac{q^{n+1/2}\cos(2n+1)x}{1+q^{2n+1}}, \tag{14.60}$$

$$\operatorname{dn} u = \frac{\pi}{2K} + \frac{2\pi}{K}\sum_{n=1}^{\infty}\frac{q^n\cos 2nx}{1+q^{2n}}. \tag{14.61}$$

14.15 Notes

We have not considered the inverse problem associated with the elliptic function $\operatorname{sn}(u, k)$ in the manner that we did for γ in Chapter 13. Thus, given $c = k^2$, we may ask if there is a value for τ such that

$$c = \frac{\theta_4^2(0|\tau)}{\theta_3^2(0|\tau)}.$$

The solution, by means of the theory of modular functions, is the same as that given in Chapter 13; we shall omit its presentation here.

14.16 Important formulae (Jacobi)

14.16.1

$$\theta_1(z,q) = 2\sum_{n=0}^{\infty}(-1)^n q^{(n+1/2)^2}\sin(2n+1)z, \tag{14.62}$$

$$\theta_2(z,q) = 2\sum_{n=0}^{\infty} q^{(n+1/2)^2}\cos(2n+1)z, \tag{14.63}$$

$$\theta_3(z,q) = 1 + 2\sum_{n=1}^{\infty} q^{n^2}\cos 2nz, \tag{14.64}$$

$$\theta_4(z,q) = 1 + 2\sum_{n=1}^{\infty}(-1)^n q^{n^2}\cos 2nz; \tag{14.65}$$

$$\theta_1(z,q) = 2q_0 q^{1/4}\sin\pi z\prod_{n=1}^{\infty}(1 - 2q^{2n}\cos 2z + q^{4n}), \tag{14.66}$$

$$\theta_2(z,q) = 2q_0 q^{1/4}\cos\pi z\prod_{n=1}^{\infty}(1 + 2q^{2n}\cos 2z + q^{4n}), \tag{14.67}$$

$$\theta_3(z,q) = q_0\prod_{n=1}^{\infty}(1 + 2q^{2n-1}\cos 2z + q^{4n-2}), \tag{14.68}$$

$$\theta_4(z,q) = q_0\prod_{n=1}^{\infty}(1 - 2q^{2n-1}\cos 2z + q^{4n-2}); \tag{14.69}$$

$$q_0 = \prod_{n=1}^{\infty}(1 - q^{2n}).$$

14.16.2

Zeros (mod $\pi, \tau\pi$)

θ_1	θ_2	θ_3	θ_4
0	$\frac{1}{2}\pi$	$\frac{1+\tau}{2}\pi$	$\frac{\tau}{2}\pi$

(14.70)

$$\theta_1(-z) = -\theta_1(z), \quad \theta_{\nu+1}(-z) = \theta_{\nu+1}(z), \quad \nu = 1, 2, 3; \tag{14.71}$$

$$\begin{cases} \theta_1(z+\pi) = -\theta_1(z), & \theta_2(z+\pi) = -\theta_2(z), \\ \theta_3(z+\pi) = \theta_3(z), & \theta_4(z+\pi) = \theta_4(z); \end{cases} \tag{14.72}$$

$$\begin{cases} \theta_1(z+\pi/2) = -\theta_2(z), & \theta_2(z+\pi/2) = -\theta_1(z), \\ \theta_3(z+\pi/2) = \theta_4(z), & \theta_4(z+\pi/2) = \theta_3(z); \end{cases} \tag{14.73}$$

$$\begin{cases} \theta_1(z+\tau\pi) = -A\theta_1(z), & \theta_2(z+\tau\pi) = A\theta_2(z), \\ \theta_3(z+\tau\pi) = A\theta_3(z), & \theta_4(z+\tau\pi) = -A\theta_4(z) \end{cases} \tag{14.74}$$
$$(A = q^{-1}e^{-2iz});$$

$$\begin{cases} \theta_1(z+\tau\pi/2) = iB\theta_4(z), & \theta_2(z+\tau\pi/2) = B\theta_3(z), \\ \theta_3(z+\tau\pi/2) = B\theta_2(z), & \theta_4(z+\tau\pi/2) = iB\theta_1(z) \end{cases} \tag{14.75}$$
$$(B = q^{-1/4}e^{-iz}).$$

14.16.3

$$\theta_1' = \theta_2\theta_3\theta_4; \tag{14.76}$$
$$\theta_2^4 + \theta_4^4 = \theta_3^4. \tag{14.77}$$

14.16.4

$$k^{1/2} = \frac{\theta_2(0)}{\theta_3(0)} = \frac{2q^{1/4} + 2q^{9/4} + 2q^{25/4} + \cdots}{1 + 2q + 2q^4 + 2q^9 + \cdots}, \tag{14.78}$$
$$k'^{1/2} = \frac{\theta_4(0)}{\theta_3(0)} = \frac{1 - 2q + 2q^4 - 2q^9 + \cdots}{1 + 2q + 2q^4 + 2q^9 + \cdots}; \tag{14.79}$$
$$k^2 + k'^2 = 1. \tag{14.80}$$
$$K = \tfrac{1}{2}\pi\theta_3^2, \tag{14.81}$$
$$K' = -\tfrac{1}{2}i\pi\tau\theta_3^2; \tag{14.82}$$
$$K = \int_0^1 (1-t^2)^{-1/2}(1-k^2t^2)^{-1/2}\,dt, \tag{14.83}$$
$$K' = \int_0^1 (1-t^2)^{-1/2}(1-k'^2t^2)^{-1/2}\,dt. \tag{14.84}$$

14.16.5

$$\operatorname{sn} u = \frac{\theta_3}{\theta_2}\frac{\theta_2(u\theta_3^{-2})}{\theta_4(u\theta_3^{-2})} = \frac{1}{\sqrt{k}}\frac{\theta_1\big(\pi u/2K\big)}{\theta_4\big(\pi u/2K\big)} = -\operatorname{sn}(-u), \tag{14.85}$$
$$\operatorname{cn} u = \frac{\theta_4}{\theta_2}\frac{\theta_2(u\theta_3^{-2})}{\theta_4(u\theta_3^{-2})} = \sqrt{\frac{k'}{k}}\frac{\theta_2\big(\pi u/2K\big)}{\theta_4\big(\pi u/2K\big)} = \operatorname{cn}(-u), \tag{14.86}$$
$$\operatorname{dn} u = \frac{\theta_4}{\theta_3}\frac{\theta_3(u\theta_3^{-2})}{\theta_4(u\theta_3^{-2})} = \sqrt{k'}\frac{\theta_3\big(\pi u/2K\big)}{\theta_4\big(\pi u/2K\big)} = \operatorname{dn}(-u); \tag{14.87}$$
$$q = e^{\pi i\tau} = \exp(-\pi K'/K). \tag{14.88}$$

14.16.6

$$\mathrm{sn}^2 u + \mathrm{cn}^2 u = 1, \tag{14.89}$$

$$\mathrm{dn}^2 u + k^2 \mathrm{sn}^2 u = 1. \tag{14.90}$$

14.16.7

$$\mathrm{sn}' u = \mathrm{cn}\, u\, \mathrm{dn}\, u, \tag{14.91}$$

$$\mathrm{cn}' u = -\mathrm{sn}\, u\, \mathrm{dn}\, u, \tag{14.92}$$

$$\mathrm{dn}' u = -k^2 \mathrm{sn}\, u\, \mathrm{cn}\, u; \tag{14.93}$$

$$\mathrm{sn}'^2 u = (1 - \mathrm{sn}^2 u)(1 - k^2 \mathrm{sn}^2 u), \tag{14.94}$$

$$\mathrm{cn}'^2 u = (1 - \mathrm{cn}^2 u)(k'^2 + k^2 \mathrm{cn}^2 u), \tag{14.95}$$

$$\mathrm{dn}'^2 u = (1 - \mathrm{dn}^2 u)(\mathrm{dn}^2 u - k^2). \tag{14.96}$$

14.16.8

$$\gamma(u) = e_3 + \frac{e_1 - e_2}{\mathrm{sn}^2(u\sqrt{e_1 - e_2})}; \tag{14.97}$$

$$\mathrm{sn}(u\sqrt{e_1 - e_2}) = \frac{\sqrt{e_1 - e_3}}{\sqrt{\gamma(u) - e_3}}, \tag{14.98}$$

$$\mathrm{cn}(u\sqrt{e_1 - e_2}) = \frac{\sqrt{\gamma(u) - e_1}}{\sqrt{\gamma(u) - e_2}}, \tag{14.99}$$

$$\mathrm{dn}(u\sqrt{e_1 - e_2}) = \frac{\sqrt{\gamma(u) - e_3}}{\sqrt{\gamma(u) - e_2}}. \tag{14.100}$$

14.16.9

$$\mathrm{sn}(u + K) = \frac{\mathrm{cn}\, u}{\mathrm{dn}\, u}, \tag{14.101}$$

$$\mathrm{cn}(u + K) = -k' \frac{\mathrm{sn}\, u}{\mathrm{dn}\, u}, \tag{14.102}$$

$$\mathrm{dn}(u + K) = k' \frac{1}{\mathrm{dn}\, u}; \tag{14.103}$$

$$\mathrm{sn}(u + iK') = \frac{1}{k} \frac{1}{\mathrm{sn}\, u}, \tag{14.104}$$

$$\mathrm{cn}(u + iK') = -\frac{i}{k} \frac{\mathrm{dn}\, u}{\mathrm{sn}\, u}, \tag{14.105}$$

$$\mathrm{dn}(u + iK') = -i\frac{\mathrm{cn}\,u}{\mathrm{sn}\,u}; \tag{14.106}$$

$$\mathrm{sn}(u + K + iK') = \frac{1}{k}\frac{\mathrm{dn}\,u}{\mathrm{cn}\,u}, \tag{14.107}$$

$$\mathrm{cn}(u + K + iK') = -i\frac{k'}{k}\frac{1}{\mathrm{cn}\,u}, \tag{14.108}$$

$$\mathrm{dn}(u + K + iK') = ik'\frac{\mathrm{sn}\,u}{\mathrm{cn}\,u}; \tag{14.109}$$

$$\mathrm{sn}(u + 2K) = -\mathrm{sn}\,u, \tag{14.110}$$

$$\mathrm{cn}(u + 2K) = -\mathrm{cn}\,u, \tag{14.111}$$

$$\mathrm{dn}(u + 2K) = \mathrm{dn}\,u; \tag{14.112}$$

$$\mathrm{sn}(u + 2iK') = \mathrm{sn}\,u, \tag{14.113}$$

$$\mathrm{cn}(u + 2iK') = -\mathrm{cn}\,u, \tag{14.114}$$

$$\mathrm{dn}(u + 2iK') = -\mathrm{dn}\,u; \tag{14.115}$$

$$\mathrm{sn}(u + 2K + 2iK') = -\mathrm{sn}\,u, \tag{14.116}$$

$$\mathrm{cn}(u + 2K + 2iK') = \mathrm{cn}\,u, \tag{14.117}$$

$$\mathrm{dn}(u + 2K + 2iK') = -\mathrm{dn}\,u. \tag{14.118}$$

14.16.10

$$\mathrm{sn}(u+v) = \frac{\mathrm{sn}\,u\,\mathrm{cn}\,v\,\mathrm{dn}\,v + \mathrm{sn}\,v\,\mathrm{cn}\,u\,\mathrm{dn}\,u}{1 - k^2\mathrm{sn}^2 u\,\mathrm{sn}^2 v}, \tag{14.119}$$

$$\mathrm{cn}(u+v) = \frac{\mathrm{cn}\,u\,\mathrm{cn}\,v + \mathrm{sn}\,u\,\mathrm{dn}\,u\,\mathrm{sn}\,v\,\mathrm{dn}\,v}{1 - k^2\mathrm{sn}^2 u\,\mathrm{sn}^2 v}, \tag{14.120}$$

$$\mathrm{dn}(u+v) = \frac{\mathrm{dn}\,u\,\mathrm{dn}\,v - k^2\mathrm{sn}\,u\,\mathrm{cn}\,u\,\mathrm{sn}\,v\,\mathrm{cn}\,v}{1 - k^2\mathrm{sn}^2 u\,\mathrm{sn}^2 v}; \tag{14.121}$$

$$\mathrm{sn}(u+v) + \mathrm{sn}(u-v) = \frac{2\mathrm{sn}\,u\,\mathrm{cn}\,v\,\mathrm{dn}\,v}{1 - k^2\mathrm{sn}^2 u\,\mathrm{sn}^2 v}; \tag{14.122}$$

$$\mathrm{sn}(u+v) - \mathrm{sn}(u-v) = \frac{2\mathrm{sn}\,v\,\mathrm{cn}\,u\,\mathrm{dn}\,u}{1 - k^2\mathrm{sn}^2 u\,\mathrm{sn}^2 v}; \tag{14.123}$$

$$\mathrm{cn}(u+v) + \mathrm{cn}(u-v) = \frac{2\mathrm{cn}\,u\,\mathrm{cn}\,v}{1 - k^2\mathrm{sn}^2 u\,\mathrm{sn}^2 v}; \tag{14.124}$$

$$\mathrm{cn}(u+v) - \mathrm{cn}(u-v) = \frac{-2\mathrm{sn}\,u\,\mathrm{dn}\,u\,\mathrm{sn}\,v\,\mathrm{dn}\,v}{1 - k^2\mathrm{sn}^2 u\,\mathrm{sn}^2 v}; \tag{14.125}$$

$$\mathrm{dn}(u+v) + \mathrm{dn}(u-v) = \frac{2\mathrm{dn}\,u\,\mathrm{dn}\,v}{1 - k^2\mathrm{sn}^2 u\,\mathrm{sn}^2 v}; \tag{14.126}$$

$$\mathrm{dn}(u+v) - \mathrm{dn}(u-v) = \frac{-2k^2\mathrm{sn}\,u\,\mathrm{cn}\,u\,\mathrm{sn}\,v\,\mathrm{cn}\,v}{1 - k^2\mathrm{sn}^2 u\,\mathrm{sn}^2 v}. \tag{14.127}$$

14.16.11

$$\mathrm{sn}(u+v)\mathrm{sn}(u-v) = \frac{\mathrm{sn}^2 u - \mathrm{sn}^2 v}{1 - k^2 \mathrm{sn}^2 u\, \mathrm{sn}^2 v}, \tag{14.128}$$

$$\mathrm{cn}(u+v)\mathrm{cn}(u-v) = \frac{\mathrm{cn}^2 v - \mathrm{dn}^2 v\, \mathrm{sn}^2 u}{1 - k^2 \mathrm{sn}^2 u\, \mathrm{sn}^2 v}, \tag{14.129}$$

$$\mathrm{dn}(u+v)\mathrm{dn}(u-v) = \frac{\mathrm{dn}^2 v - k^2 \mathrm{cn}^2 v\, \mathrm{sn}^2 u}{1 - k^2 \mathrm{sn}^2 u\, \mathrm{sn}^2 v}. \tag{14.130}$$

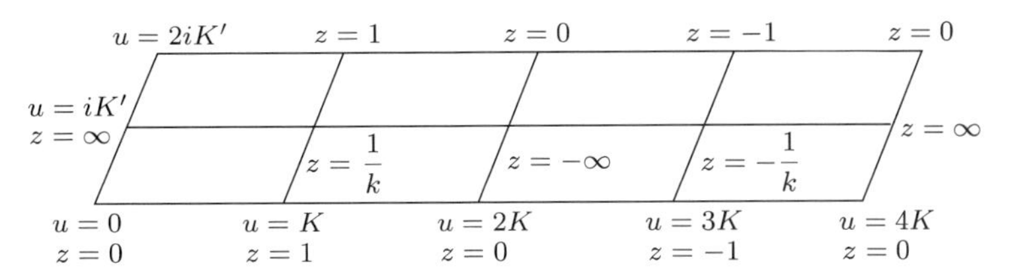

Figure 14.1. $z = \mathrm{sn}(u, k)$.

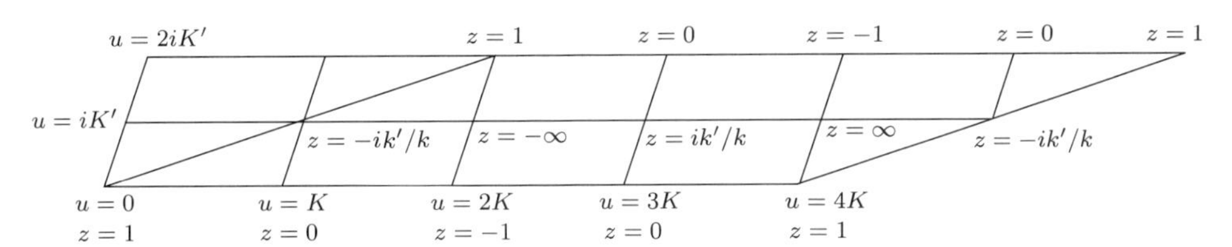

Figure 14.2. $z = \mathrm{cn}(u, k)$.

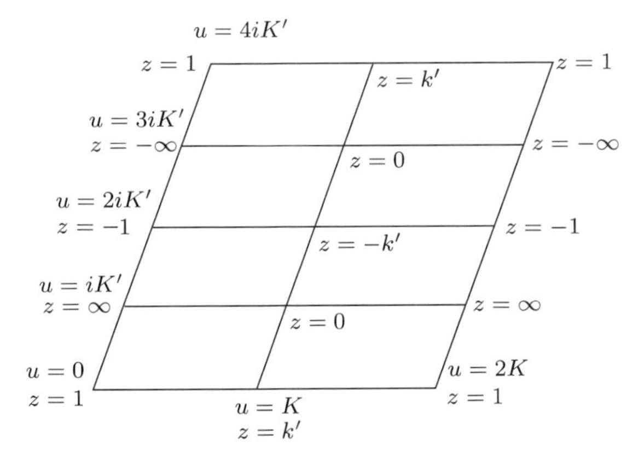

Figure 14.3. $z = \mathrm{dn}(u, k)$.

15

Linear systems and determinants (review)

15.1 Linear systems

Consider the homogeneous system

$$\sum_{j=1}^{n} a_{ij}x_j = 0, \qquad i = 1, \dots, n, \tag{15.1}$$

where a_{ij} are complex (or real) numbers and x_i are the unknowns. Obviously the system has the trivial solution

$$x_1 = \cdots = x_n = 0. \tag{15.2}$$

The fundamental problems concerning a homogeneous system are whether there are other solutions besides the trivial one, and whether it is possible to determine all the solutions.

Consider next the inhomogeneous system

$$\sum_{j=1}^{n} a_{ij}x_j = b_i, \qquad i = 1, \dots, n. \tag{15.3}$$

Suppose that (15.3) has a solution $(x_1^{(0)}, \dots, x_n^{(0)})$, that is

$$\sum_{j=1}^{n} a_{ij}x_j^{(0)} = b_i \quad \text{so that} \quad \sum_{j=1}^{n} a_{ij}(x_j - x_j^{(0)}) = 0.$$

Then $y_j = x_j - x_j^{(0)}$ is a solution to (15.1). Consequently, after a substitution, the fundamental problem for an inhomogeneous system becomes whether there is a solution at all in the first place. If there is indeed a solution, then all the solutions can be obtained from the solutions of the homogeneous system (15.1).

Concerning the existence of a solution, we have the following important results:

- If (15.1) has a non-trivial solution, then there are some $b_1, \dots, b_n$ for which the system (15.3) is not soluble.
- If (15.1) has only the trivial solution then (15.3) is always soluble for any $b_1, \dots, b_n$.

15.2 Method of eliminations

We now give a revision of the Gaussian elimination method for the solution of the linear system (15.3). Consider, for example, the following system of four linear equations with four unknowns:

$$\begin{aligned} a_{11}x_1 + a_{12}x_2 + a_{13}x_3 + a_{14}x_4 &= a_{15}, \\ a_{21}x_1 + a_{22}x_2 + a_{23}x_3 + a_{24}x_4 &= a_{25}, \\ a_{31}x_1 + a_{32}x_2 + a_{33}x_3 + a_{34}x_4 &= a_{35}, \\ a_{41}x_1 + a_{42}x_2 + a_{43}x_3 + a_{44}x_4 &= a_{45}. \end{aligned} \tag{15.4}$$

Take the leading coefficient a_{11} from the first equation, and set

$$b_{1j} = \frac{a_{1j}}{a_{11}} \quad (j > 1), \tag{15.5}$$

which then leads to the new equation

$$x_1 + b_{12}x_2 + b_{13}x_3 + b_{14}x_4 = b_{15}. \tag{15.6}$$

Using (15.6) to eliminate x_1 in the final three equations in (15.4) we now have a supplementary system of only three equations in three unknowns. The procedure can be carried out systematically by multiplying (15.6) by a_{21}, a_{31}, a_{41}, which are, respectively, the leading coefficients for the second, third and fourth equations in (15.4), and then subtracting the corresponding terms. This then eliminates one of the unknowns for the new system, with the coefficients $a_{ij,1}$ being given by

$$a_{ij,1} = a_{ij} - a_{i1}b_{1j} \quad (i, j \geq 2). \tag{15.7}$$

We next divide the first equation of the new system by the new leading coefficient $a_{22,1}$ to obtain

$$x_2 + b_{23,1}x_3 + b_{24,1}x_4 = b_{25,1}, \tag{15.8}$$

where

$$b_{2j,1} = \frac{a_{2j,1}}{a_{22,1}} \quad (j > 2). \tag{15.9}$$

Using the same procedure we arrive at a system of two equations with two unknowns and the coefficients

$$a_{ij,2} = a_{ij,1} - a_{i2,1}b_{2j,1} \quad (i, j \geq 3). \tag{15.10}$$

We next take $a_{33,2}$ to be the leading coefficient of the first equation in the new system, and set

$$b_{3j,2} = \frac{a_{3j,2}}{a_{33,2}} \quad (j > 3) \tag{15.11}$$

to yield

$$x_3 + b_{34,2}x_4 = b_{35,2}.$$

Table 15.1.

x_1	x_2	x_3	x_4		Σ						Σ
a_{11}	a_{12}	a_{13}	a_{14}	a_{15}	a_{16}	1.00	0.42	0.54	0.66	0.33	2.92
a_{21}	a_{22}	a_{23}	a_{24}	a_{25}	a_{26}	0.42	1.00	0.32	0.44	0.5	2.68
a_{31}	a_{32}	a_{33}	a_{34}	a_{35}	a_{36}	0.54	0.32	1.00	0.22	0.7	2.78
a_{41}	a_{42}	a_{43}	a_{44}	a_{45}	a_{46}	0.66	0.44	0.22	1.00	0.9	3.22
1	b_{12}	b_{13}	b_{14}	b_{15}	b_{16}	1	0.42	0.54	0.66	0.3	2.92
	$a_{22,1}$	$a_{23,1}$	$a_{24,1}$	$a_{25,1}$	$a_{26,1}$		0.82360	0.09320	0.16280	0.37400	1.45360
	$a_{32,1}$	$a_{33,1}$	$a_{34,1}$	$a_{35,1}$	$a_{36,1}$		0.09320	0.70840	−0.13640	0.53800	1.20320
	$a_{42,1}$	$a_{43,1}$	$a_{44,1}$	$a_{45,1}$	$a_{46,1}$		0.16280	−0.13640	0.56440	0.70200	1.29280
	1	$b_{23,1}$	$b_{24,1}$	$b_{25,1}$	$b_{26,1}$		1	0.11316	0.19767	0.45410	1.76493
		$a_{33,2}$	$a_{34,2}$	$a_{35,2}$	$a_{36,2}$			0.69785	−0.15482	0.49568	1.03871
		$a_{43,2}$	$a_{44,2}$	$a_{45,2}$	$a_{46,2}$			−0.15482	0.53222	0.62807	1.00547
		1	$b_{34,2}$	$b_{35,2}$	$b_{36,2}$			1	−0.22185	0.71030	1.48844
			$a_{44,3}$	$a_{45,3}$	$a_{46,3}$				0.49787	0.73804	1.23591
			1	x_4	$\bar{x}_4$				1	1.48240	2.48240
		1		x_3	$\bar{x}_3$			1		1.03917	2.03916
	1			x_2	$\bar{x}_2$		1			0.04348	1.04348
1				x_1	$\bar{x}_1$	1				−1.25780	−0.25779

Finally we arrive at an equation with a single unknown to give, on division by the coefficient $a_{44,3}$,

$$x_4 = b_{45,3}.$$

Using the coefficients $b_{ij,i-1}\,(j > i)$ to form the equations, we have a triangular system which is equivalent to the original system in the sense that they have the same solutions. We note, however, that the process described above can only be carried out if all the leading coefficients being used do not take the value zero.

The above procedure of arriving at the triangular system is called the *forward substitution* process, and the solutions are then obtained from the *backward substitution* process (see Table 15.1).

Consider now a checking procedure, using the substitution $\bar{x}_i = x_i + 1$. This gives also a system with $\bar{x}_i$ as unknowns, with the same coefficients as before, and the new constants as the sum of the coefficients together with the corresponding old constants. We may then perform the calculations for both systems and see if $\bar{x}_i$ agree with $x_i + 1$ as a check.

Here is a brief description of the entries in Table 15.1: The coefficients, together with the constants and also the constants for checking, are first entered into the table as a matrix. We then divide the first row by its leading coefficient and enter the result as the next row at the bottom above the line. Next, the coefficients $a_{ij,1}\,(i, j \geq 2)$ are obtained from the corresponding terms of the known matrix by subtracting the product obtained from the leading coefficient described in the above, and we do this for each of the rows. The whole process is repeated until there is only one row in the matrix.

For the backward substitution process we make use of the rows containing the entry 1, starting from the last row. More precisely, from the last row we obtain the value of an unknown from the constant term, and also that for the check value. The values for the remaining unknowns can now be successively evaluated in the same manner. We only need to apply subtraction in terms of the b values multiplied by the evaluated unknowns with the appropriate suffix i, and the last coefficient being 1 enables us to complete the task from the various coefficients for x. For example, we have

$$\begin{aligned} x_2 &= b_{25,1} - b_{23,1}x_3 - b_{24,1}x_4 \\ &= 0.45410 - 0.11316\times 1.03917 - 0.19767\times 1.48240 = 0.04348. \end{aligned}$$

Finally, we remark that, for such a linear system with n unknowns, the required number of multiplications and divisions is $\frac{n}{3}(n^2 + 3n - 1)$.

15.3 Geometric interpretation of the method

Consider first the situation when there are only two unknowns:

$$\ell : \quad ax + by = c, \qquad \ell' : \quad a'x + b'y = c',$$

which represent two straight lines on the plane with a point of intersection. On eliminating y, we have an equation for x only, giving the abscissa for the point of intersection, that is the image of the projection of the point onto the x-axis.

We can also view this in another way: The two equations represent the two straight lines ℓ and ℓ', and the equation

$$\lambda(ax + by - c) + \mu(a'x + b'y - c') = 0$$

defines a family of straight lines, which will be denoted by $\lambda\ell + \mu\ell'$. These lines have an important common property, namely that they all pass through the point of intersection between ℓ and ℓ' (see Fig. 15.1). It is also easy to establish the converse: every straight line that passes through the point of intersection between ℓ and ℓ' belongs to the family. Within this family, there is one line which is parallel to the y-axis; the equation for this line is that for which y has been eliminated.

Consider next the situation when there are three unknowns:

$$\begin{aligned} \ell &: ax + by + cz = d, \\ \ell' &: a'x + b'y + c'z = d', \\ \ell'' &: a''x + b''y + c''z = d''. \end{aligned}$$

There are three plane surfaces here, and the equation

$$\lambda\ell + \mu\ell' = 0$$

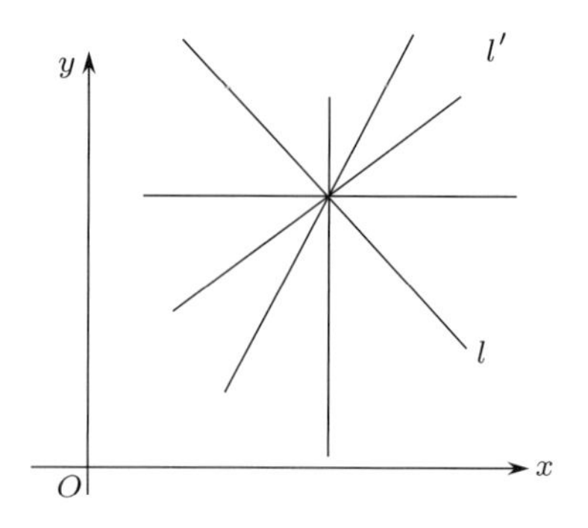

Figure 15.1.

now represents the family of planes passing through both the planes ℓ and ℓ'. On eliminating x from ℓ and ℓ', we obtain an equation with the following interpretations: On one hand, it represents a plane passing through the line of intersection of two planes, and is parallel to the x-axis; on the other hand, it represents the projection of the line of intersection onto the yz-plane, the line obtained from the elimination of x from ℓ and ℓ' to arrive at a line on the plane; similarly, on eliminating x from ℓ, ℓ'', we arrive at a straight line on the yz-plane.

Thus, on the elimination of x, the problem of the intersection of three planes in three dimensional space now becomes one of the intersection of two lines on the yz-plane. These two straight lines are just the projected images of the lines (intersection of planes) in space onto the plane.

In general, we can consider the system of equations

$$a_{i1}x_1 + \cdots + a_{in}x_n = b_i, \quad i = 1, \ldots, m,$$

which represents m hyperplanes intersecting in n-dimensional space. From the method of elimination, the problem becomes one of the intersection of one less hyperplane in the space with one less dimension.

15.4 Mechanical interpretation of the method

On the n points $P_1, \ldots, P_n$ of a chord, with both endpoints being fixed, we attach various weights exerting the downward forces $F_1, \ldots, F_n$. We now consider the values of the corresponding vertical drops $y_1, \ldots, y_n$ along the chord (see Fig. 15.2).

It is assumed that these weights on the chord satisfy the principle of 'linear superposition':

(1) When two sets of weights are combined together, the corresponding effect on the drops will also be added together.

(2) If any one of the weights is multiplied by a constant, then the corresponding effect on the drops will also be multiplied by the same constant.

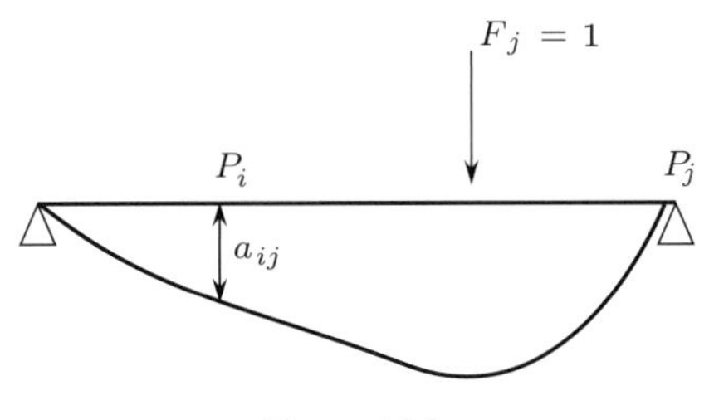

Figure 15.2.

Let a_{ij} denote the drop of P_i corresponding to a unit force applied at the point P_j. Then the combined forces $F_1, \ldots, F_n$ yield the drops $y_1, \ldots, y_n$ given by

$$\sum_{j=1}^{n} a_{ij} F_j = y_i \quad (i = 1, 2, \ldots, n). \tag{15.12}$$

As a linear system, the problem becomes one of finding the values for the forces $F_1, \ldots, F_n$ from the specified values for the drops $y_1, \ldots, y_n$.

Let P_1 be given a reacting force R. Then adding a unit force on P_j will give the corresponding value for the drop of P_j,

$$b_{ij} = a_{ij} + Ra_{i1}.$$

On examining the situation when P_1 becomes fixed, that is $b_{1j} = 0$, $a_{1j} + Ra_{11} = 0$, we have $R = -a_{1j}/a_{11}$. So if a unit force has been added to P_j, and a force is needed to be applied at P_1 for it to be fixed, then this force has to be $-a_{1j}/a_{11}$, giving $b_{ij} = a_{ij} - a_{1j}a_{i1}/a_{11}$. Eliminating F_1 from (15.12) we obtain

$$\sum_{j=2}^{n} (a_{ij} - a_{1j}a_{i1}/a_{11}) F_j = y - a_{i1}y_1/a_{11}. \tag{15.13}$$

This is the equation of equilibrium after fixing the point P_1. We may continue to apply a unit force to P_j, with the points P_i taking on the drops b_{ij}. Successive eliminations can thus be interpreted as the successive insertions of weights to points.

15.5 Economic equilibrium

Suppose there are n products $P_1, \ldots, P_n$ being produced, with the production of P_j requiring a_{ij} units of P_i. If $x_1, \ldots, x_n$ denote the amounts of such products, then the total cost for the production of P_i is given by

$$\sum_{j=1}^{n} a_{ij} x_j, \quad 1 \leq i \leq n.$$

In order to satisfy the market we need to have

$$x_i - \sum_{j=1}^{n} a_{ij} x_j = b_i.$$

Thus, knowing that the market demand is $b_1, \ldots, b_n$, we now have to compute the production indices $x_1, \ldots, x_n$. This is also a linear system.

Obviously one can just apply the usual elimination method to deal with the problem. However, there is a much better iterative method of solution which will be discussed later.

15.6 Linear regression analysis

A certain variable ξ is determined by n parameters

$$\eta_1, \ldots, \eta_n.$$

Suppose that we have already performed N experiments with the results

$$\xi^{(i)}, \eta_1^{(i)}, \eta_2^{(i)}, \ldots, \eta_n^{(i)}, \quad i = 1, \ldots, N.$$

We now examine the linear relation

$$\xi = \sum_{j=1}^{n} a_j \eta_j,$$

and the problem is to find out what kind of linear relation will give the least 'mean-square deviation'. That is, if we let

$$\zeta^{(i)} = \sum_{j=1}^{n} a_j \eta_j^{(i)},$$

then we wish to determine a_j so that

$$\sum_{i=1}^{N} (\xi^{(i)} - \zeta^{(i)})^2$$

is least; in other words, we need to find the least value for

$$F(a_1, \ldots, a_n) = \sum_{i=1}^{N} \left(\xi^{(i)} - \sum_{j=1}^{n} a_j \eta_j^{(i)} \right)^2. \tag{15.14}$$

Let

$$a_{jk} = \sum_{i=1}^{N} \eta_j^{(i)} \eta_k^{(i)}, \qquad b_k = \sum_{i=1}^{N} \xi^{(i)} \eta_k^{(i)},$$

and we proceed to show that the solution x_j to

$$\sum_{j=1}^{n} a_{jk} x_j = b_k, \quad k = 1, 2, \ldots, n, \tag{15.15}$$

will deliver the sought-after least value in (15.14) when we set $x_j = a_j$.

We first establish the following point: Suppose that $a_1, \ldots, a_n$ do not satisfy (15.15), and say that there is a value k for which

$$\sum_{j=1}^{n} a_{jk} a_j - b_k = -\alpha_k \neq 0.$$

We then examine

$$\begin{aligned}
&F(a_1, \ldots, a_{k-1}, a_k - \epsilon, a_{k+1}, \ldots, a_n)\\
&= \sum_{i=1}^{N} \Big(\xi^{(i)} - \sum_{j=1}^{n} a_j \eta_j^{(i)} + \epsilon \eta_k^{(i)} \Big)^2\\
&= \sum_{i=1}^{N} \Big(\xi^{(i)} - \sum_{j=1}^{n} a_j \eta_j^{(i)} \Big)^2 + 2\epsilon \sum_{i=1}^{N} \Big(\xi^{(i)} - \sum_{j=1}^{n} a_j \eta_j^{(i)} \Big) \eta_k^{(i)} + \epsilon^2 \sum_{i=1}^{N} \Big(\eta_k^{(i)} \Big)^2\\
&= \sum_{i=1}^{N} \Big(\xi^{(i)} - \sum_{j=1}^{n} a_j \eta_j^{(i)} \Big)^2 + 2\epsilon \Big(b_k - \sum_{j=1}^{n} a_{jk} a_j \Big) + \epsilon^2 a_{kk}\\
&= F(a_1, \ldots, a_n) + 2\epsilon \alpha_k + \epsilon^2 a_{kk}.
\end{aligned}$$

On completing the square, we find that

$$\begin{aligned}
&F(a_1, \ldots, a_{k-1}, a_k - \epsilon, a_{k+1}, \ldots, a_n)\\
&= F(a_1, \ldots, a_n) + a_{kk} \Big(\epsilon + \frac{\alpha_k}{a_{kk}} \Big)^2 - \frac{\alpha_k^2}{a_{kk}}.
\end{aligned} \tag{15.16}$$

If $\alpha_k \neq 0$ then $F(a_1, \ldots, a_n)$ cannot be the least value, because on setting $\epsilon = -\alpha/a_{kk}$ in (15.16) we find that $F(a_1, \ldots, a_{k-1}, a_k - \epsilon, a_{k+1}, \ldots, a_n)$ now takes on a smaller value than $F(a_1, \ldots, a_n)$ does. Therefore, the problem for least-square regression can be transformed to one for a linear system.

We still need to show that a solution to (15.15) yields the least value for F, but this is not difficult. If there is only one solution to (15.15) then there is no more problem, because we already saw from (15.16) that any input not satisfying (15.15) cannot deliver the least value. (We leave it to the reader to show that (15.15) is definitely soluble, and also to deal with the situation when there is more than one solution.)

Obviously we can use the elimination method to deal with (15.15). However, there is symmetry in the system because $a_{jk} = a_{kj}$, and for this we have a far better method for the solution.

An alert reader may have observed a rather good point in the above, namely that we avoided the use of calculus, using instead the more elementary argument of 'completing the square'. An even better point is that we have just introduced an important concept in numerical analysis, namely the 'relaxation method'.

The relaxation method is particularly valuable when applied to regression analysis. The method is as follows:

(1) First select any set $a_1, \ldots, a_n$.
(2) Compute

$$\alpha_k = \sum_{j-1}^{n} a_{jk} a_j - b_k;$$

if $\alpha_k \neq 0$, then take

$$a_1, \ldots, a_{k-1}, a_k - \epsilon, \ldots, a_n$$

as an initial input; if $\alpha_k = 0$, then take another k.
(3) In general, we apply the method for $k = 1, 2, \ldots, n, 1, 2, \ldots, n, \ldots$ repeatedly, and the optimal solution will eventually emerge.

The name 'relaxation method' comes from the following: If there is an error within the calculation at a point, it does not matter if we still carry on with the calculation; we do not have to start from the beginning all over again, and the correct result will emerge (starting from an erroneous $(a_1, \ldots, a_n)$, the result will still come through). Obviously we do not mean that there are always errors – just that when there are occasional errors it does not really matter.

Although we call the method 'relaxation', we can tighten up its execution a little. For example, we can first compare α_k^2/a_{kk} for $k = 1, 2, \ldots, n$ to find which k yields the largest value. It is for this k that the method is optimum, as we see from (15.16) that the value for F will be reduced by the most, thereby speeding up the whole process.

15.7 Determinants

The material in Section 15.1 leads us to the introduction of determinants, which can be defined inductively by

$$|a_{ij}| = \begin{vmatrix} a_{11} & \cdots & a_{1n} \\ \cdots & \cdots & \cdots \\ a_{n1} & \cdots & a_{nn} \end{vmatrix} = a_{11}A_{11} + a_{12}A_{12} + \cdots + a_{1n}A_{1n}.$$

Here A_{ij} is the determinant of the matrix of order $n-1$ when the ith row and the jth column have been deleted, and the sign $(-1)^{i+j}$ has also been attached to it – we call it the *cofactor* corresponding to a_{ij}.

In the following we list the most important properties for determinants:

(1) If a row (column) is multiplied by k, then the value of the determinant also changes by the factor k.
(2) The addition of a multiple of a row (column) to another row (column) will not alter the value of the determinant.
(3) If two rows (columns) are interchanged then the sign of the determinant will change. Consequently if two rows (columns) are identical then the value of the determinant has to be zero.

By Cramer's rule the system

$$\sum_{j=1}^{n} a_{ij}x_j = b_i$$

has the solution

$$x_1 = \frac{\begin{vmatrix} b_1 & a_{12} & \cdots & a_{1n} \\ b_2 & a_{22} & \cdots & a_{2n} \\ \cdots & \cdots & \cdots & \cdots \\ b_n & a_{n2} & \cdots & a_{nn} \end{vmatrix}}{|a_{ij}|};$$

in the numerator the first column for $|a_{ij}|$ has been replaced by 'b'; a similar formula exists for x_i, except that it will be the ith column being replaced by 'b'.

Fundamental theorem for a homogeneous linear system *A necessary and sufficient condition for the system*

$$\sum_{j=1}^{n} a_{ij}x_j = 0, \quad i = 1, \ldots, n,$$

to have a non-trivial solution is that

$$|a_{ij}| = 0.$$

Simple deduction 15.1 *If $n > m$, then the linear system*

$$\sum_{j=1}^{n} a_{ij}x_j = 0, \quad i = 1, \ldots, m,$$

must have a non-trivial solution, because we can artificially augment the system by setting

$$\sum_{j=1}^{n} a_{ij}x_j = 0, \quad i = m+1, \ldots, n,$$

where $a_{ij} = 0$ *for* $m+1 \le i \le n$, $1 \le j \le n$.

Simple deduction 15.2 *If* $n \le m$ *and the linear system*

$$\sum_{j=1}^{n} a_{ij}x_j = 0, \quad i = 1, \ldots, m, \tag{15.17}$$

has a non-trivial solution then any of the determinants of order n *must have the value* 0.

Here is a slight generalisation of the fundamental theorem: If all the relevant determinants of order n have the value 0 then (15.17) has a non-trivial solution.

Proof. (1) By hypothesis we have

$$\begin{vmatrix} a_{11} & \cdots & a_{1n} \\ a_{21} & \cdots & a_{2n} \\ \cdots & \cdots & \cdots \\ a_{n-1,1} & \cdots & a_{n-1,n} \\ a_{i1} & \cdots & a_{in} \end{vmatrix} = 0.$$

On expanding this we find that

$$a_{i1}A_1 + \cdots + a_{in}A_n = 0, \quad A_n = \begin{vmatrix} a_{11} & \cdots & a_{1,n-1} \\ \cdots & \cdots & \cdots \\ a_{n-1,1} & \cdots & a_{n-1,n-1} \end{vmatrix}.$$

If $A_1, \ldots, A_n$ are not all zero then the system (15.17) obviously has the solution $x_i = A_i$.

(2) If a reordering of the determinant, or a reordering of x_i, results in

$$\begin{vmatrix} a_{11} & \cdots & a_{1,n-1} \\ \cdots & \cdots & \cdots \\ a_{n-1,1} & \cdots & a_{n-1,n-1} \end{vmatrix} \neq 0,$$

then we see from (1) that the theorem holds.

(3) Consider the situation not covered by (2). In examining $x_n = 0$, we now find that

$$\begin{vmatrix} a_{11} & \cdots & a_{1,n-1} \\ \cdots & \cdots & \cdots \\ a_{n-2,1} & \cdots & a_{n-2,n-1} \\ a_{i1} & \cdots & a_{i,n-1} \end{vmatrix} = 0, \quad a_{i1}B_1 + \cdots + a_{i,n-1}B_{n-1} = 0,$$

and we may take $x_1 = B_1, \ldots, x_{n-1} = B_{n-1}, x_n = 0$ as a solution.

The theorem can then be established by mathematical induction. $\square$

We now consider the relationship between the homogeneous and the inhomogeneous systems introduced in Section 15.1.

First, if

$$\sum_{j=1}^{n} a_{ij}x_j = 0, \quad i = 1, \dots, n,$$

has a non-trivial solution, then

$$\sum_{i=1}^{n} a_{ij}x_i = 0, \quad j = 1, \dots, n, \tag{15.18}$$

also has a non-trivial solution. This is obvious, since the determinant of the transpose of a matrix will have the same value. Now let

$$\sum_{i=1}^{n} \xi_i a_{ij} = 0,$$

and we may assume that $\xi_1 \neq 0$; then from

$$\sum_{j=1}^{n} a_{ij}x_j = b_i, \tag{15.19}$$

we find that

$$\sum_{i=1}^{n} b_i\xi_i = \sum_{i=1}^{n} \xi_i\Big(\sum_{j=1}^{n} a_{ij}x_j\Big) = \sum_{j=1}^{n}\Big(\sum_{i=1}^{n} a_{ij}\xi_i\Big)x_j = 0.$$

Clearly, when $b_1 = 1, b_2 = \cdots = b_n = 0$, the system (15.19) has no solution.

If (15.18) has only the trivial solution, then $|a_{ij}| \neq 0$, and (15.19) can now be solved by Cramer's rule.

Note that, beautifully simple though Cramer's rule is, in practice we hardly ever use it; this is because the amount of arithmetic involved in applying the rule becomes prohibitively large even for a modest size n.

15.8 The Vandermonde determinant

Theorem 15.3 *We have*

$$\begin{vmatrix} 1 & 1 & \cdots & 1 \\ x_1 & x_2 & \cdots & x_n \\ \cdots & \cdots & \cdots & \cdots \\ x_1^{n-1} & x_2^{n-1} & \cdots & x_n^{n-1} \end{vmatrix} = \prod_{1 \le j \le i \le n} (x_i - x_j).$$

Proof. (1) Use mathematical induction, the case $n = 2$ being obvious.

(2) Write $\Delta = \Delta(x_1, x_2, \dots, x_n)$ for the value of the determinant, and subtract from the rows $2, 3, \dots, n$ the previous rows with the factor x_1 to give

$$\Delta = \begin{vmatrix} x_2 - x_1 & \cdots & x_n - x_1 \\ x_2(x_2 - x_1) & \cdots & x_n(x_n - x_1) \\ \cdots & \cdots & \cdots \\ x_2^{n-2}(x_2 - x_1) & \cdots & x_n^{n-2}(x_n - x_1) \end{vmatrix}$$

$$= (x_2 - x_1) \cdots (x_n - x_1) \begin{vmatrix} 1 & \cdots & 1 \\ x_2 & \cdots & x_n \\ \cdots & \cdots & \cdots \\ x_2^{n-2} & \cdots & x_n^{n-2} \end{vmatrix};$$

the required result follows by mathematical induction. □

Another proof. If $x_i = x_j$ then $\Delta = 0$, so that $x_i - x_j$ is a factor for Δ, and hence the product $\prod_{1 \le j \le i \le n}(x_i - x_j)$ is also a factor. The required result now follows by comparing the coefficients for $x_n^{n-1} x_{n-1}^{n-2} \cdots x_2 \cdot 1$. □

The theorem has the following corollary.

Theorem 15.4 *Let $P_i(x)$ be a polynomial with degree i, with the coefficient a_j for x^j. Then*

$$\begin{vmatrix} P_0(x_1) & \cdots & P_0(x_n) \\ P_1(x_1) & \cdots & P_1(x_n) \\ \cdots & \cdots & \cdots \\ P_{n-1}(x_1) & \cdots & P_{n-1}(x_n) \end{vmatrix} = a_0 \cdots a_n \prod_{1 \le j \le i \le n} (x_i - x_j). \tag{15.20}$$

To see this we note that the entries in the first row are all a_0, which become 1 on division by a_0. Let $P_1(x) = a_1 x + a_1'$, and subtract from the second row a_1' times the first row, and then divide by a_1, so that the second row becomes $x_1, \dots, x_n$. Let $P_2(x) = a_2 x^2 + a_2' x + a_2''$, and subtract from the third row a_2'' times the first row, and also a_2' times the second row, and then divide by a_2, so that the third row becomes $x_1^2, \dots, x_n^2$. Continuing in this way, the required result is obtained.

Theorem 15.5 *We have*

$$\begin{vmatrix} 1 & \cdots & 1 \\ \cos\theta_1 & \cdots & \cos\theta_n \\ \cdots & \cdots & \cdots \\ \cos(n-1)\theta_1 & \cdots & \cos(n-1)\theta_n \end{vmatrix} = 2^{(1/2)(n-2)(n-1)} \prod_{1 \le j \le i \le n} (\cos\theta_i - \cos\theta_j),$$

and also

$$\begin{vmatrix} \sin\theta_1 & \cdots & \sin\theta_n \\ \sin 2\theta_1 & \cdots & \sin 2\theta_n \\ \cdots & \cdots & \cdots \\ \sin n\theta_1 & \cdots & \sin n\theta_n \end{vmatrix} = 2^{(1/2)n(n-1)} \sin\theta_1 \cdots \sin\theta_n \prod_{1\le j\le i\le n} (\cos\theta_i - \cos\theta_j).$$

Proof. From $(\cos\theta + i\sin\theta)^n = \cos n\theta + i\sin n\theta$, we find that

$$\begin{aligned} 2\cos n\theta &= (\cos\theta + i\sin\theta)^n + (\cos\theta - i\sin\theta)^n \\ &= \sum_{\ell=0}^{n} \binom{n}{\ell} \big((i\sin\theta)^\ell + (-i\sin\theta)^\ell\big) \cos^{n-\ell}\theta \\ &= 2\sum_{k=0}^{[n/2]} \binom{n}{2k} (-1)^k \sin^{2k}\theta, \end{aligned}$$

that is

$$\cos n\theta = \sum_{k=0}^{[n/2]} (-1)^k \binom{n}{2k} (1-\cos^2\theta)^k \cos^{n-2k}\theta = P_n(\cos\theta),$$

where $[x]$ is the integer part of x and P_n is a polynomial of degree n in $\cos\theta$, with the leading coefficient

$$\sum_{k=0}^{[n/2]} (-1)^k(-1)^k \binom{n}{2k} = \frac{1}{2}\sum_{\ell=0}^{n} \big(1+(-1)^\ell\big)\binom{n}{\ell} = 2^{n-1}.$$

Substituting $P_n(\cos\theta)$ for $\cos n\theta$ into the original equation, we obtain

$$\begin{aligned} \begin{vmatrix} 1 & \cdots & 1 \\ \cos\theta_1 & \cdots & \cos\theta_n \\ \cdots & \cdots & \cdots \\ \cos(n-1)\theta_1 & \cdots & \cos(n-1)\theta_n \end{vmatrix} &= \begin{vmatrix} P_0(\cos\theta_1) & \cdots & P_0(\cos\theta_n) \\ P_1(\cos\theta_1) & \cdots & P_0(\cos\theta_n) \\ \cdots & \cdots & \cdots \\ P_{n-1}(\cos\theta_1) & \cdots & P_{n-1}(\cos\theta_n) \end{vmatrix} \\ &= 2^{0+1+2+\cdots+(n-2)} \prod_{1\le j\le i\le n} (\cos\theta_i - \cos\theta_j) \end{aligned}$$

and

$$\begin{aligned} 2i\sin n\theta &= (\cos\theta + i\sin\theta)^n - (\cos\theta - i\sin\theta)^n \\ &= \sum_{\ell=0}^{n} \binom{n}{\ell} [(i\sin\theta)^\ell - (-i\sin\theta)^\ell] \cos^{n-\ell}\theta \\ &= 2i \sum_{0\le 2k+1\le n} \binom{n}{2k+1} (-1)^k \sin^{2k+1}\theta \cos^{n-2k-1}\theta, \end{aligned}$$

that is

$$\sin n\theta = \sin\theta \sum_{0\le 2k+1\le n} \binom{n}{2k+1}(-1)^k(1-\cos^2\theta)^k \cos^{n-2k-1}\theta$$
$$= \sin\theta\, Q_{n-1}(\cos\theta),$$

where $Q_{n-1}(\cos\theta)$ is a polynomial of $\cos\theta$ with degree $n-1$, and the leading coefficient

$$\sum_{0\le 2k+1\le n} \binom{n}{2k+1} = \sum_{\ell=0}^{n}\binom{n}{\ell} - \sum_{0\le 2k\le n}^{n}\binom{n}{2k} = 2^n - 2^{n-1} = 2^{n-1}.$$

Therefore

$$\begin{vmatrix} \sin\theta_1 & \cdots & \sin\theta_n \\ \sin 2\theta_1 & \cdots & \sin 2\theta_n \\ \cdots & \cdots & \cdots \\ \sin n\theta_1 & \cdots & \sin n\theta_n \end{vmatrix} = \sin\theta_1\cdots\sin\theta_n \begin{vmatrix} P_0(\cos\theta_1) & \cdots & P_0(\cos\theta_n) \\ P_1(\cos\theta_1) & \cdots & P_0(\cos\theta_n) \\ \cdots & \cdots & \cdots \\ P_{n-1}(\cos\theta_1) & \cdots & P_{n-1}(\cos\theta_n) \end{vmatrix}$$
$$= 2^{(1/2)n(n-1)} \sin\theta_1\cdots\sin\theta_n \prod_{1\le j\le i\le n}(\cos\theta_i - \cos\theta_j).$$

□

Theorem 15.6 *We have*

$$\begin{vmatrix} \sin\frac{1}{2}\theta_1 & \cdots & \sin\frac{1}{2}\theta_n \\ \sin\frac{3}{2}\theta_1 & \cdots & \sin\frac{3}{2}\theta_n \\ \cdots & \cdots & \cdots \\ \sin\left(n-\frac{1}{2}\right)\theta_1 & \cdots & \sin\left(n-\frac{1}{2}\right)\theta_n \end{vmatrix} = 2^{(1/2)n(n-1)}\prod_{i=1}^{n}\sin\frac{1}{2}\theta_i \prod_{1\le j\le i\le n}(\cos\theta_i - \cos\theta_j).$$

Proof. We multiply each column by $\cos\frac{1}{2}\theta_1, \ldots, \cos\frac{1}{2}\theta_n$; from

$$\sin\left(v-\frac{1}{2}\right)\theta_i \cos\frac{1}{2}\theta_i = \frac{1}{2}\left(\sin v\theta_i + \sin(v-1)\theta_i\right),$$

we see that the original determinant becomes

$$\frac{1}{2^n\prod_{i=1}^{n}\cos\frac{1}{2}\theta_i}\begin{vmatrix} \sin\theta_1 & \cdots & \sin\theta_n \\ \sin 2\theta_1 - \sin\theta_1 & \cdots & \sin 2\theta_n - \sin\theta_n \\ \cdots & \cdots & \cdots \\ \sin n\theta_1 - \sin(n-1)\theta_1 & \cdots & \sin n\theta_n - \sin(n-1)\theta_n \end{vmatrix}$$
$$= \frac{1}{2^n\prod_{i=1}^{n}\cos\frac{1}{2}\theta_i}\begin{vmatrix} \sin\theta_1 & \sin\theta_2 & \cdots & \sin\theta_n \\ \sin 2\theta_1 & \sin 2\theta_2 & \cdots & \sin 2\theta_n \\ \cdots & \cdots & \cdots & \cdots \\ \sin n\theta_1 & \cdots & \cdots & \sin n\theta_n \end{vmatrix}$$
$$= 2^{(1/2)n(n-1)}\sin\frac{1}{2}\theta_1\cdots\sin\frac{1}{2}\theta_n \prod_{1\le j\le i\le n}(\cos\theta_i - \cos\theta_j).$$

□

Theorem 15.7 *We have*

$$\begin{vmatrix} \cos\frac{1}{2}\theta_1 & \cdots & \cos\frac{1}{2}\theta_n \\ \cos\frac{3}{2}\theta_1 & \cdots & \cos\frac{3}{2}\theta_n \\ \cdots & \cdots & \cdots \\ \cos\left(n-\frac{1}{2}\right)\theta_1 & \cdots & \cos\left(n-\frac{1}{2}\right)\theta_n \end{vmatrix} = 2^{(1/2)n(n-1)}\prod_{i=1}^{n}\cos\frac{1}{2}\theta_i \prod_{1\le j\le i\le n}(\cos\theta_i - \cos\theta_j).$$

Proof. We multiply each column by $\sin\frac{1}{2}\theta_1, \ldots, \cos\frac{1}{2}\theta_n$; from

$$\sin\frac{1}{2}\theta_i\cos\left(v+\frac{1}{2}\right)\theta_i = \frac{1}{2}\left(\sin(v+1)\theta_i - \sin v\theta_i\right),$$

we see that the original determinant becomes

$$\frac{1}{2^n\prod_{i=1}^{n}\sin\frac{1}{2}\theta_i}\begin{vmatrix} \sin\theta_1 & \cdots & \sin\theta_n \\ \sin 2\theta_1 - \sin\theta_1 & \cdots & \sin 2\theta_n - \sin\theta_n \\ \cdots & \cdots & \cdots \\ \sin n\theta_1 - \sin(n-1)\theta_1 & \cdots & \sin n\theta_n - \sin(n-1)\theta_n \end{vmatrix}$$

$$= \frac{1}{2^n\prod_{i=1}^{n}\sin\frac{1}{2}\theta_i}\begin{vmatrix} \sin\theta_1 & \sin\theta_2 & \cdots & \sin\theta_n \\ \sin 2\theta_1 & \sin 2\theta_2 & \cdots & \sin 2\theta_n \\ \cdots & \cdots & \cdots & \cdots \\ \sin n\theta_1 & \cdots & \cdots & \sin n\theta_n \end{vmatrix}$$

$$= 2^{(1/2)n(n-1)}\cos\frac{1}{2}\theta_1\cdots\sin\frac{1}{2}\theta_n\prod_{1\le j\le i\le n}(\cos\theta_i - \cos\theta_j).$$

□

Exercise 15.8 Let the vertices of a polyhedron with n vertices be labelled in a certain way. Form a determinant $|a_{ij}|$ so that $a_{ij} = a_{ji} = 1$ if the vertices with labels i and j are linked by a side and $a_{ij} = 0$ otherwise; in particular, $a_{ii} = 0$. Show that the value of the determinant is independent of the labelling, and evaluate the determinants for the tetrahedron, the cube and the octahedron. For example, for the tetrahedron, we have

$$\begin{vmatrix} 0 & 1 & 1 & 1 \\ 1 & 0 & 1 & 1 \\ 1 & 1 & 0 & 1 \\ 1 & 1 & 1 & 0 \end{vmatrix} = -3.$$

Hint: On interchanging the labels i and j, the effect on the determinant is the interchange of the ith and jth rows, and also the columns.

Answer: For the cube: 9; for the octahedron: 0.

Exercise 15.9 Evaluate

$$\begin{vmatrix} 1 & 1 & 1 & \cdots & 1 \\ b_1 & a_1 & a_1 & \cdots & a_1 \\ b_1 & b_2 & a_2 & \cdots & a_2 \\ \cdots & \cdots & \cdots & \cdots & \cdots \\ b_1 & b_2 & b_3 & \cdots & a_n \end{vmatrix}.$$

Exercise 15.10 Show that, for $\ell mn \neq 0$,

$$\begin{vmatrix} \rho & \frac{\ell}{m}+\frac{m}{\ell} & \frac{n}{\ell}+\frac{\ell}{n} \\ \frac{\ell}{m}+\frac{m}{\ell} & \rho & \frac{m}{n}+\frac{n}{m} \\ \frac{n}{\ell}+\frac{\ell}{n} & \frac{m}{n}+\frac{n}{m} & \rho \end{vmatrix} = (\rho-2)(\rho+1+\sqrt{P})(\rho+1-\sqrt{P}),$$

where

$$P = \left(\frac{1}{\ell^2}+\frac{1}{m^2}+\frac{1}{n^2}\right)(\ell^2+m^2+n^2).$$

15.9 Symmetric functions

On expanding the product

$$f(x) = (x-x_1)\cdots(x-x_n) \tag{15.21}$$

to give

$$f(x) = x^n - \sigma_1 x^{n-1} + \sigma_2 x^{n-2} - \sigma_3 x^{n-3} + \cdots, \tag{15.22}$$

we find that

$$\sigma_1 = x_1 + x_2 + \cdots + x_n,$$
$$\sigma_2 = x_1x_2 + x_1x_3 + \cdots + x_1x_n + x_2x_3 + \cdots + x_2x_n + \cdots + x_{n-1}x_n;$$

here σ_i is the sum of all the products of $x_1, \ldots, x_n$ taking i terms at a time, so that there are $\binom{n}{i}$ such products in the sum.

These polynomials $\sigma_1, \ldots, \sigma_n$ are called the elementary symmetric functions of $x_1, \ldots, x_n$.

Definition 15.11 By a symmetric function F of n variables we mean one such that

$$F(x_1, \ldots, x_n) = F(x_{i_1}, \ldots, x_{i_n}),$$

where $x_{i_1}, \ldots, x_{i_n}$ is any permutation of $x_1, \ldots, x_n$.

Obviously the elementary symmetric functions are symmetric; another important set of symmetric functions comprises the sums of symmetric powers:

$$s_i = x_1^i + \cdots + x_n^i.$$

We note that $\sigma_0 = 1$ and $s_0 = n$.

Definition 15.12 We use

$$\sum x_{i_1}^{\ell_1} x_{i_2}^{\ell_2} \cdots x_{i_m}^{\ell_m}$$

to denote the sum over all possible combinations of the terms $x_{i_1}^{\ell_1} x_{i_2}^{\ell_2} \cdots x_{i_m}^{\ell_m}$.

For example, we have

$$\sigma_2 = \sum x_i x_j;$$

note that this is not the same as

$$\sum_{j=1}^{n}\sum_{i=1}^{n} x_i x_j$$

because x_i^2 is absent from the former sum and present in the latter sum.

On setting $x = x_i$ in (15.22) we have

$$x_i^n - \sigma_1 x_i^{n-1} + \sigma_2 x_i^{n-2} - \cdots + (-1)^n \sigma_n = 0, \quad i = 1, 2, \ldots, n. \tag{15.23}$$

This can be considered as a linear system with $-\sigma_1, \sigma_2, \ldots, (-1)^n\sigma_n$ as the n unknowns. By Cramer's rule, we have the solution $(-1)^{n-\ell}\sigma_{n-\ell}$ given by

$$\begin{vmatrix} x_1^{n-1} \cdots x_1^{\ell+1} & -x_1^n & x_1^{\ell-1} \cdots x_1 & 1 \\ x_2^{n-1} \cdots x_2^{\ell+1} & -x_2^n & x_2^{\ell-1} \cdots x_2 & 1 \\ \vdots \qquad \vdots & \vdots & \vdots \qquad \vdots & \vdots \\ x_n^{n-1} \cdots x_n^{\ell+1} & -x_n^n & x_n^{\ell-1} \cdots x_n & 1 \end{vmatrix} \Bigg/ \begin{vmatrix} x_1^{n-1} x_1^{n-2} \cdots x_1 & 1 \\ x_1^{n-1} x_1^{n-2} \cdots x_1 & 1 \\ \vdots \quad \vdots \qquad \vdots & \vdots \\ x_1^{n-1} x_1^{n-2} \cdots x_1 & 1 \end{vmatrix}. \tag{15.24}$$

We therefore have the following theorem.

Theorem 15.13 *Let*

$$\Delta(\ell_1, \ldots, \ell_n) = \begin{vmatrix} x_1^{\ell_1} \cdots x_n^{\ell_1} \\ \vdots \qquad \vdots \\ x_1^{\ell_n} \cdots x_n^{\ell_n} \end{vmatrix};$$

then

$$\sigma_{n-\ell} = \frac{\Delta(n, n-1, \ldots, \ell+1, \ell-1, \ldots, 0)}{\Delta(n-1, n-2, \ldots, 1, 0)}.$$

Exercise 15.14 Prove that $\Delta(n-1, n-2, \ldots, 1, 0)$ divides every $\Delta(\ell_1, \ldots, \ell_n)$, and that the quotient is a symmetric function.

Exercise 15.15 Evaluate

$$\Delta(n+1, n-2, \ldots, 1, 0) \quad \text{and} \quad \Delta(n+1, n, n-2, n-3 \ldots, 1, 0).$$

Hint: First show that

$$\Delta(n+1, n-2, \ldots, 1, 0) = \Delta(n-1, n-2, \ldots, 1, 0)\Big(A\sum x_i^2 + B\sum x_i x_j\Big),$$

and then find A and B.

We now investigate the relationship between

$$\sigma_1, \sigma_2, \ldots, \sigma_n \quad \text{and} \quad s_1, s_2, \ldots, s_n.$$

Taking the logarithmic derivative of (15.21) we have

$$f'(x) = \frac{f(x)}{x - x_1} + \frac{f(x)}{x - x_2} + \cdots + \frac{f(x)}{x - x_n},$$

and from

$$\frac{x^\ell - x_1^\ell}{x - x_1} = x^{\ell-1} + x^{\ell-2}x_1 + \cdots + x_1^{\ell-1}$$

we find that

$$\begin{aligned}\frac{f(x)}{x - x_i} &= \frac{f(x) - f(x_i)}{x - x_i} = \sum_{\ell=0}^{n}(-1)^\ell \sigma_\ell \frac{x^{n-\ell} - x_i^{n-\ell}}{x - x_i} \\ &= \sum_{\ell=0}^{n-1}(-1)^\ell \sigma_\ell \sum_{t=1}^{n-\ell-1} x^t x_i^{n-\ell-t-1} \\ &= \sum_{t=0}^{n-1}\left(\sum_{\ell=0}^{n-t-1}(-1)^\ell \sigma_\ell x_i^{n-\ell-t-1}\right)x^t.\end{aligned}$$

Summing over i we then have

$$f'(x) = \sum_{t=0}^{n-1}\left(\sum_{\ell=0}^{n-t-1}(-1)^\ell \sigma_\ell s_{n-\ell-t-1}\right)x^t.$$

On the other hand, differentiating (15.22) we have

$$f'(x) = \sum_{t=0}^{n}(-1)^{n-t}\sigma_{n-t}(x^t)' = \sum_{t=1}^{n}(-1)^{n-t} t \sigma_{n-t} x^{t-1}.$$

Comparing coefficients then yields

$$\sum_{\ell=0}^{n-t-1}(-1)^\ell \sigma_\ell s_{n-\ell-t-1} = (-1)^{n-t-1}(t+1)\sigma_{n-t-1}, \quad t = 0, 1, 2, \ldots, n-1,$$

which, on replacing t by $n - \tau$, becomes

$$\sum_{\ell=0}^{\tau-1}(-1)^\ell \sigma_\ell s_{\tau-\ell-1} = (-1)^{\tau-1}(n - \tau + 1)\sigma_{\tau-1}, \quad \tau = 1, 2, \ldots, n,$$

that is

$$\begin{aligned} s_0 &= n\sigma_0, \\ \sigma_0 s_1 - \sigma_1 s_0 &= -(n-1)\sigma_1, \\ \sigma_0 s_2 - \sigma_1 s_1 + \sigma_2 s_0 &= (n-2)\sigma_2, \\ &\cdots \\ \sigma_0 s_m - \sigma_1 s_{m-1} + \cdots + (-1)^m \sigma_m s_0 &= (-1)^m (n-m)\sigma_m. \end{aligned}$$

With $\sigma_0 = 1$ and $s_0 = n$ this becomes

$$\begin{aligned} s_1 &= \sigma_1, \\ \sigma_1 s_1 - s_2 &= 2\sigma_2, \\ &\cdots \\ \sigma_{m-1}s_1 - \sigma_{m-2}s_2 + \cdots + (-1)^{m-1}s_m &= m\sigma_m. \end{aligned}$$

By considering $s_1, \ldots, s_m$ to be unknowns, solving the system then gives the following theorem.

Theorem 15.16 (Newton) *We have*

$$s_1 = \sigma_1, \quad s_2 = \begin{vmatrix} \sigma_1 & 1 \\ 2\sigma_2 & \sigma_1 \end{vmatrix}, \quad s_3 = \begin{vmatrix} \sigma_1 & 1 & 0 \\ 2\sigma_2 & \sigma_1 & 1 \\ 3\sigma_3 & \sigma_2 & \sigma_1 \end{vmatrix}, \quad \ldots,$$

$$s_m = \begin{vmatrix} \sigma_1 & 1 & 0 & \cdots & 0 \\ 2\sigma_2 & \sigma_1 & 1 & \cdots & 0 \\ 3\sigma_3 & \sigma_2 & \sigma_1 & \cdots & 0 \\ \vdots & \vdots & \vdots & & \vdots \\ m\sigma_m & \sigma_{m-1} & \sigma_{m-2} & \cdots & \sigma_1 \end{vmatrix}, \quad m = 1, 2, \ldots, n.$$

Again, by considering $\sigma_1, \ldots, \sigma_m$ to be unknowns, solving the system then gives the following theorem.

Theorem 15.17 *We have*

$$\sigma_2 = \frac{1}{2}\begin{vmatrix} s_1 & 1 \\ s_2 & s_1 \end{vmatrix}, \quad \sigma_3 = \frac{1}{3!}\begin{vmatrix} s_1 & 1 & 0 \\ s_2 & s_1 & 2 \\ s_3 & s_2 & s_1 \end{vmatrix}, \quad \sigma_4 = \frac{1}{4!}\begin{vmatrix} s_1 & 1 & 0 & 0 \\ s_2 & s_1 & 2 & 0 \\ s_3 & s_2 & s_1 & 3 \\ s_4 & s_3 & s_2 & s_1 \end{vmatrix}, \quad \ldots.$$

If $m > n$, then, on multiplying x_i^{m-n} into

$$x_i^n - \sigma_1 x_i^{n-1} + \cdots + (-1)^n \sigma_n = 0$$

and then summing over $i = 1, 2, \ldots, n$, we arrive at

$$s_m - \sigma_1 s_{m-1} + \cdots + (-1)^n \sigma_n s_{m-n},$$

so that, as a reduction formula, s_m can be obtained from $\sigma_1, \ldots, \sigma_n$ and $s_{m-1}, \ldots, s_{m-n}$.

15.10 The fundamental theorem on symmetric functions

Theorem 15.18 *Any symmetric (polynomial) function can be expressed as a polynomial in the elementary symmetric functions.*

Proof. (1) By the results in Section 15.9 we need only show that a symmetric polynomial is representable as a polynomial in $s_1, s_2, \ldots, s_n$.

(2) From

$$s_m = \sum_{i=1}^{n} x_i^m, \quad s_p = \sum_{i=1}^{n} x_i^p,$$

we have

$$s_m s_p = \sum x_i^{m+p} + \sum_{i \neq j} x_i^m x_j^p,$$

which then delivers the following representations:

$$\sum x_i^m x_j^p = s_m s_p - s_{m+p} \ (m \neq p), \quad \sum (x_i x_j)^m = \frac{1}{2}(s_m^2 - s_{2m}).$$

(3) We next consider symmetric polynomials of the form $\sum x_i^m x_j^p x_k^q$, with three distinct factors.

If m, p, q are distinct and $m + q \neq p$, $p + q \neq m$, then

$$s_q \sum x_i^m x_j^p = \sum x_i^{m+q} x_j^p + \sum x_i^m x_j^{p+q} + \sum x_i^m x_j^p x_k^q.$$

By (2), we now have

$$s_q(s_m s_p - s_{m+p}) = s_{m+q} s_p - s_{m+p+q} + s_m s_{p+q} - s_{m+p+q} + \sum x_i^m x_j^p x_k^q,$$

that is

$$\sum x_i^m x_j^p x_k^q = s_m s_p s_q - s_{m+p} s_q - s_{m+q} s_p - s_{p+q} s_m + 2 s_{m+p+q}.$$

If m, p, q are distinct, but $m + q = p$, then

$$\sum x_i^m x_j^p x_k^q = s_m s_p s_q - s_q s_{m+p} - \frac{1}{2} s_p^2 - s_{p+q} s_m + \frac{3}{2} s_{2p}.$$

If m, p, q are distinct, but $p + q = m$, then

$$\sum x_i^m x_j^p x_k^q = s_m s_p s_q - s_q s_{m+p} - \frac{1}{2} s_m^2 - s_{m+q} s_p + \frac{3}{2} s_{2m}.$$

If $m = p$ then

$$2 \sum (x_i x_j)^m x_k^q = s_m^2 s_q - s_{2m} s_q - -2 s_{m+q} s_m + 2 s_{2m+q}.$$

If $m = p = q$ then

$$6 \sum (x_i x_j x_k)^m = s_m^3 - 3 s_{2m} s_m + 2 s_{3m}.$$

Thus the symmetric polynomial $\sum x_i^m x_j^p x_k^q$ is representable as a polynomial in the elementary symmetric functions.

The theorem can be proved by continuing successively in the same way. □

15.11 Common roots for two algebraic equations

Earlier we considered an important result: If there is a sequence of numbers $x_1, \ldots, x_n$, not all 0, such that

$$\sum_{j=1}^{n} a_{ij} x_j = 0, \quad i = 1, 2, \ldots, n,$$

then the determinant

$$|a_{ij}| = 0.$$

This very important fundamental result is also an often used tool, and we give a couple of examples to illustrate its application.

Example 15.19 Suppose that the two equations

$$a_0 x^4 + a_1 x^3 + a_2 x^2 + a_3 x + a_4 = 0, \tag{15.25}$$

$$b_0 x^3 + b_1 x^2 + b_2 x + b_3 = 0 \tag{15.26}$$

have a common root ξ. Then

$$\begin{aligned}
a_0\xi^6 + a_1\xi^5 + a_2\xi^4 + a_3\xi^3 + a_4\xi^2 &= 0,\\
a_0\xi^5 + a_1\xi^4 + a_2\xi^3 + a_3\xi^2 + a_4\xi &= 0,\\
a_0\xi^4 + a_1\xi^3 + a_2\xi^2 + a_3\xi + a_4 &= 0,\\
b_0\xi^6 + b_1\xi^5 + b_2\xi^4 + b_3\xi^3 &= 0,\\
b_0\xi^5 + b_1\xi^4 + b_2\xi^3 + b_3\xi^2 &= 0,\\
b_0\xi^4 + b_1\xi^3 + b_2\xi^2 + b_3\xi &= 0,\\
b_0\xi^3 + b_1\xi^2 + b_2\xi + b_3 &= 0,
\end{aligned}$$

which can be considered as a system of seven linear equations (in seven unknowns $x_1, x_2, \ldots, x_7$, say) having the solution

$$x_1 = \xi^6,\ x_2 = \xi^5,\ x_3 = \xi^4,\ x_4 = \xi^3,\ x_5 = \xi^2,\ x_6 = \xi,\ x_7 = 1.$$

The solution being non-trivial we deduce that the determinant

$$\Delta = \begin{vmatrix}
a_0 & a_1 & a_2 & a_3 & a_4 & 0 & 0\\
0 & a_0 & a_1 & a_2 & a_3 & a_4 & 0\\
0 & 0 & a_0 & a_1 & a_2 & a_3 & a_4\\
b_0 & b_1 & b_2 & b_3 & 0 & 0 & 0\\
0 & b_0 & b_1 & b_2 & b_3 & 0 & 0\\
0 & 0 & b_0 & b_1 & b_2 & b_3 & 0\\
0 & 0 & 0 & b_0 & b_1 & b_2 & b_3
\end{vmatrix} = 0.$$

This then is the condition for equations (15.25) and (15.26) to have a common root.

In general, the condition for the two equations

$$f(x) = a_0x^n + a_1x^{n-1} + \cdots + a_n = 0,$$
$$g(x) = b_0x^m + b_1x^{m-1} + \cdots + b_m = 0$$

to have a common root can be obtained from the elimination of $x^{m+n-1}, \ldots, x$ from

$$x^{m-1}f(x) = 0, \ \ldots, \quad xf(x) = 0, \quad f(x) = 0,$$
$$x^{n-1}g(x) = 0, \ \ldots, \quad xg(x) = 0, \quad g(x) = 0,$$

to arrive at the determinant

$$\Delta = 0.$$

If $\alpha_1, \ldots, \alpha_n$ and $\beta_1, \ldots, \beta_m$ are the roots of the equations $f(x) = 0$ and $g(x) = 0$, respectively, then

$$\Delta = a_0^m b_0^n \prod_{i=1}^{n}\prod_{j=1}^{m}(\alpha_i - \beta_j).$$

We omit the proof of such a result.

Example 15.20 Let us consider the condition for

$$f(x) = x^3 + px + q = 0$$

to have a repeated root, which is the same as $f(x) = 0$ and $f'(x) = 0$ having a common root. The relevant determinant is

$$\begin{vmatrix} 1 & 0 & p & q & 0 \\ 0 & 1 & 0 & p & q \\ 3 & 0 & p & 0 & 0 \\ 0 & 3 & 0 & p & 0 \\ 0 & 0 & 3 & 0 & p \end{vmatrix} = \begin{vmatrix} -2p & -3q & 0 \\ 0 & -2p & -3q \\ 3 & 0 & p \end{vmatrix} = 4p^3 + 27q^2,$$

so that $f(x)$ has repeated roots when $4p^3 + 27q^2 = 0$.

Actually, there is no need to go through this, because the required result follows at once from substituting the roots for $f'(x) = 3x^2 + p = 0$, namely $x = \pm\sqrt{-p/3}$, into $f(x) = 0$.

Exercise 15.21 Determine the condition for

$$x^5 + 5px^3 + 5p^2x + q = 0$$

to have repeated roots.

15.12 Intersections of algebraic curves

Let $f(x, y)$, $g(x, y)$ be polynomials in x, y so that

$$f(x, y) = 0, \quad g(x, y) = 0$$

each represents an algebraic curve. The problem is the determination of the intersections between these two algebraic curves.

Here then is a method. Consider $f(x, y)$ and $g(x, y)$ as polynomials in x, with coefficients being functions of y. Use the method in Section 15.11 to eliminate y in order to obtain a certain determinant, the expansion of which becomes a polynomial in y. Equating the said determinant to 0 then delivers a solution for y, from which x can then also be obtained.

Example 15.22 Determine the intersections between the ellipse and the rectangular hyperbola

$$\frac{x^2}{a^2} + \frac{y^2}{b^2} = 1, \quad xy = 1.$$

We eliminate x^2, x from

$$\frac{x^2}{a^2} + \frac{y^2}{b^2} - 1 = 0,$$
$$yx^2 - x = 0,$$
$$yx - 1 = 0,$$

to obtain

$$\begin{vmatrix} \frac{1}{a^2} & 0 & \frac{y^2}{b^2} - 1 \\ y & -1 & 0 \\ 0 & y & -1 \end{vmatrix} = 0.$$

We can now solve this for y, and then $x = 1/y$.

Note. Although we use determinants in our demonstrations, in practice the method of eliminations is usually better.

15.13 Power series associated with determinants

The following is another definition for a determinant:

$$\begin{vmatrix} a_{11} & \cdots & a_{1n} \\ \vdots & & \vdots \\ a_{n1} & \cdots & a_{nn} \end{vmatrix} = \sum_{\binom{1\ 2\ \cdots\ n}{i_1\ i_2\ \cdots\ i_n}} \delta^{1\,2\cdots n}_{i_1 i_2 \cdots i_n} a_{1i_1} \cdots a_{ni_n}.$$

Here $\delta^{1\,2\cdots n}_{i_1 i_2\cdots i_n} = \pm 1$, depending on whether $\binom{1\ 2\ \cdots\ n}{i_1\ i_2\ \cdots\ i_n}$ is an even or an odd substitution.

Such a definition appears to be rather abstract, and is rather complicated to use in any actual evaluation. Nevertheless, it is very important and indeed useful to have such a definition for theoretical purposes, and we demonstrate this with an example.

Theorem 15.23 *Let*

$$f_i(z) = \sum_{t=0}^{\infty} a_t^{(i)} z^t, \quad i = 1, 2, \ldots, n,$$

represent n power series which are all convergent in $|z| < \rho$*. Then, for* $|z_1| < \rho, \ldots, |z_n| < \rho$*, we have the identity*

$$\begin{vmatrix} f_1(z_1) & \cdots & f_1(z_n) \\ \vdots & & \vdots \\ f_n(z_1) & \cdots & f_n(z_n) \end{vmatrix} = \sum_{\ell_1 > \ell_2 > \cdots \ell_n \geq 0} \begin{vmatrix} a_{\ell_1}^{(1)} & \cdots & a_{\ell_n}^{(1)} \\ \vdots & & \vdots \\ a_{\ell_1}^{(n)} & \cdots & a_{\ell_n}^{(n)} \end{vmatrix} \begin{vmatrix} z_1^{\ell_1} & \cdots & z_1^{\ell_n} \\ \vdots & & \vdots \\ z_n^{\ell_1} & \cdots & z_n^{\ell_n} \end{vmatrix}.$$

Proof. On expanding a determinant from its definition, we find that the left-hand side is given by

$$\begin{aligned}
&\sum_{\binom{1\ 2\ \cdots\ n}{i_1\ i_2 \cdots\ i_n}} \delta_{i_1 i_2 \cdots i_n}^{1 2 \cdots n} f_{i_1}(z_1) \cdots f_{i_n}(z_n) \\
&= \sum_{\binom{1\ 2\ \cdots\ n}{i_1\ i_2 \cdots\ i_n}} \delta_{i_1 i_2 \cdots i_n}^{1 2 \cdots n} \sum_{t_1=0}^{\infty} \cdots \sum_{t_n=0}^{\infty} a_{t_i}^{(i_1)} a_{t_2}^{(i_2)} \cdots a_{t_n}^{(i_n)} z_1^{t_1} z_2^{t_2} \cdots z_n^{t_n} \\
&= \sum_{t_1=0}^{\infty} \cdots \sum_{t_n=0}^{\infty} \left(\sum_{\binom{1\ 2\ \cdots\ n}{i_1\ i_2 \cdots\ i_n}} \delta_{i_1 i_2 \cdots i_n}^{1 2 \cdots n} a_{t_i}^{(i_1)} a_{t_2}^{(i_2)} \cdots a_{t_n}^{(i_n)} \right) z_1^{t_1} \cdots z_n^{t_n} \\
&= \sum_{t_1=0}^{\infty} \cdots \sum_{t_n=0}^{\infty} \begin{vmatrix} a_{t_1}^{(1)} & \cdots & a_{t_n}^{(1)} \\ \vdots & & \vdots \\ a_{t_1}^{(n)} & \cdots & a_{t_n}^{(n)} \end{vmatrix} z_1^{t_1} \cdots z_n^{t_n} \\
&= \sum_{\ell_1 > \ell_2 > \cdots \ell_n \geq 0} \begin{vmatrix} a_{\ell_1}^{(1)} & \cdots & a_{\ell_n}^{(1)} \\ \vdots & & \vdots \\ a_{\ell_1}^{(n)} & \cdots & a_{\ell_n}^{(n)} \end{vmatrix} \sum_{\binom{1\ 2\ \cdots\ n}{i_1\ i_2 \cdots\ i_n}} \delta_{i_1 i_2 \cdots i_n}^{1 2 \cdots n} z_1^{\ell_{i_1}} \cdots z_n^{\ell_{i_n}},
\end{aligned}$$

and hence we obtain the required result. □

In particular, taking

$$f_i(z) = f(x_i z)$$

and

$$f(z) = a_0 + a_1 z + a_2 z^2 + \cdots,$$

that is

$$a_i^{(i)} = a_t x_i^t,$$

we then have

$$\begin{vmatrix} a_{\ell_1}^{(1)} & \cdots & a_{\ell_n}^{(1)} \\ \vdots & & \vdots \\ a_{\ell_1}^{(n)} & \cdots & a_{\ell_n}^{(n)} \end{vmatrix} = \begin{vmatrix} a_{\ell_1} x_1^{\ell_1} & \cdots & a_{\ell_n} x_1^{\ell_n} \\ \vdots & & \vdots \\ a_{\ell_1} x_n^{\ell_1} & \cdots & a_{\ell_n} x_n^{\ell_n} \end{vmatrix} = a_{\ell_1} a_{\ell_2} \cdots a_{\ell_n} \begin{vmatrix} x_1^{\ell_1} & \cdots & x_1^{\ell_n} \\ \vdots & & \vdots \\ x_n^{\ell_1} & \cdots & x_n^{\ell_n} \end{vmatrix},$$

and consequently the following theorem.

Theorem 15.24 *Let*

$$f(z) = a_0 + a_1 z + \cdots$$

be a power series which converges in $|z| < \rho$. *Then, for* $|x_i y_j| < \rho$, *we have*

$$\begin{vmatrix} f(x_1 y_1) & \cdots & f(x_1 y_n) \\ \vdots & & \vdots \\ f(x_n y_1) & \cdots & f(x_n y_n) \end{vmatrix} = \sum_{\ell_1 > \cdots > \ell_n \geq 0} a_{\ell_1} \cdots a_{\ell_n} \begin{vmatrix} x_1^{\ell_1} & \cdots & x_1^{\ell_n} \\ \vdots & & \vdots \\ x_n^{\ell_1} & \cdots & x_n^{\ell_n} \end{vmatrix} \begin{vmatrix} y_1^{\ell_1} & \cdots & y_1^{\ell_n} \\ \vdots & & \vdots \\ y_n^{\ell_1} & \cdots & y_n^{\ell_n} \end{vmatrix}.$$

Again, on taking, in particular,

$$f(z) = (1 - z)^{-1},$$

we have the following theorem.

Theorem 15.25 *For* $|x_i y_j| < 1$ *we have*

$$\sum_{\ell_1 > \cdots > \ell_n \geq 0} a_{\ell_1} \cdots a_{\ell_n} \begin{vmatrix} x_1^{\ell_1} & \cdots & x_1^{\ell_n} \\ \vdots & & \vdots \\ x_n^{\ell_1} & \cdots & x_n^{\ell_n} \end{vmatrix} \begin{vmatrix} y_1^{\ell_1} & \cdots & y_1^{\ell_n} \\ \vdots & & \vdots \\ y_n^{\ell_1} & \cdots & y_n^{\ell_n} \end{vmatrix} = \frac{\Delta(x_1, \ldots, x_n)\Delta(y_1, \ldots, y_n)}{\prod_{i=1}^n \prod_{j=1}^n (1 - x_i y_j)},$$

where $\Delta(x_1, \ldots, x_n) = \prod_{n \geq i > j \geq 1} (x_i - x_j)$.

Proof. By Theorem 15.24 we need only show that

$$\begin{vmatrix} \frac{1}{1-x_1 y_1} & \cdots & \frac{1}{1-x_1 y_n} \\ \vdots & & \vdots \\ \frac{1}{1-x_n y_1} & \cdots & \frac{1}{1-x_n y_n} \end{vmatrix} = \frac{\Delta(x_1, \ldots, x_n)\Delta(y_1, \ldots, y_n)}{\prod_{i=1}^n \prod_{j=1}^n (1 - x_i y_j)},$$

which, by an obvious substitution, is the same as the following theorem. □

Theorem 15.26 (Cauchy) *We have*

$$\begin{vmatrix} \frac{1}{x_1+y_1} & \cdots & \frac{1}{x_1+y_n} \\ \vdots & & \vdots \\ \frac{1}{x_n+y_1} & \cdots & \frac{1}{x_n+y_n} \end{vmatrix} = \frac{\Delta(x_1, \ldots, x_n)\Delta(y_1, \ldots, y_n)}{\prod_{i=1}^n \prod_{j=1}^n (x_i + y_j)}.$$

Proof. On subtracting the first row from subsequent rows, and making use of

$$\frac{1}{x_\ell + y_h} - \frac{1}{x_1 + y_h} = \frac{x_1 - x_\ell}{(x_1 + y_h)(x_\ell + y_h)}, \qquad \ell, h = 1, 2, \ldots, n,$$

the value of the determinant concerned becomes

$$\frac{(x_1 - x_2)(x_1 - x_3) \cdots (x_1 - x_n)}{\prod_{h=1}^n (x_1 + y_h)} \begin{vmatrix} 1 & 1 & \cdots & 1 \\ \frac{1}{x_2+y_1} & \frac{1}{x_2+y_2} & \cdots & \frac{1}{x_2+y_n} \\ \vdots & \vdots & & \vdots \\ \frac{1}{x_n+y_1} & \frac{1}{x_n+y_2} & \cdots & \frac{1}{x_n+y_n} \end{vmatrix}.$$

On subtracting the first column from subsequent columns, the determinant has the value

$$\frac{(y_1 - y_2) \cdots (y_1 - y_n)}{\prod_{j=2}^n (x_j + y_1)} \begin{vmatrix} \frac{1}{x_2+y_2} & \cdots & \frac{1}{x_2+y_n} \\ \vdots & & \vdots \\ \frac{1}{x_n+y_2} & \cdots & \frac{1}{x_n+y_n} \end{vmatrix}.$$

The theorem now follows from an inductive argument. □

15.14 Power series expansion for the Wronskian

We have

$$\begin{vmatrix} z_1^{\ell_1} & \cdots & z_1^{\ell_n} \\ \vdots & & \vdots \\ z_n^{\ell_1} & \cdots & z_n^{\ell_n} \end{vmatrix} = \Delta(z_1, \ldots, z_n) P(z_1, \ldots, z_n),$$

where P is a homogeneous function of $z_1, \ldots, z_n$, with degree

$$\ell_1 + \cdots + \ell_n - (n-1) - (n-2) - \cdots - 1 = \ell_1 + \cdots + \ell_n - \frac{1}{2}n(n-1).$$

Since $\ell_1 > \ell_2 > \cdots > \ell_n \geq 0$, apart from

$$\ell_1 = n-1, \; \ell_2 = n-2, \; \ldots, \; \ell_{n-1} = 1, \; \ell_n = 0,$$

the degree for P is positive, so that $P(z_1, \ldots, z_n) \to 0$ as $z_1 \to 0, \ldots, z_n \to 0$. Therefore, under the hypothesis of Theorem 15.23,

$$\lim_{\substack{z_1 \to 0 \\ \cdots \\ z_n \to 0}} \frac{1}{\Delta(z_1, \ldots, z_n)} \begin{vmatrix} f_1(z_1) & \cdots & f_1(z_n) \\ \vdots & & \vdots \\ f_n(z_1) & \cdots & f_n(z_n) \end{vmatrix} = \begin{vmatrix} a_{n-1}^{(1)} & a_{n-2}^{(1)} & \cdots & a_1^{(1)} & a_0^{(1)} \\ \vdots & \vdots & & \vdots & \vdots \\ a_{n-1}^{(n)} & a_{n-2}^{(n)} & \cdots & a_1^{(n)} & a_0^{(n)} \end{vmatrix}$$

$$= \frac{(-1)^{(1/2)n(n-1)}}{1!\,2! \cdots (n-1)!} \begin{vmatrix} f_1(0) & \cdots & f_n(0) \\ f_1'(0) & \cdots & f_n'(0) \\ \vdots & & \vdots \\ f_1^{(n-1)}(0) & \cdots & f_n^{(n-1)}(0) \end{vmatrix}.$$

With an obvious substitution, we have the following theorem.

Theorem 15.27 *Let $f_1(z), \ldots, f_n(z)$ be analytic at $z = z_0$. Then*

$$\lim_{\substack{z_1 \to z_0 \\ \cdots \\ z_n \to z_0}} \frac{1}{\Delta(z_1, \ldots, z_n)} \begin{vmatrix} f_1(z_1) & \cdots & f_1(z_n) \\ \vdots & & \vdots \\ f_n(z_1) & \cdots & f_n(z_n) \end{vmatrix}$$

$$= \frac{(-1)^{(1/2)n(n-1)}}{1!\,2! \cdots (n-1)!} \begin{vmatrix} f_1(z_0) & \cdots & f_n(z_0) \\ f_1'(z_0) & \cdots & f_n'(z_0) \\ \vdots & & \vdots \\ f_1^{(n-1)}(z_0) & \cdots & f_n^{(n-1)}(z_0) \end{vmatrix}.$$

Exercise 15.28 Apply Taylor's formula with a remainder term to deal with the theorem, thus weakening the hypothesis being imposed on $f_1(z), \ldots, f_n(z)$. Take, in particular,

$$f_i(z) = z^{\ell_i}, \quad \ell_i > 0,$$

to obtain

$$\lim_{\substack{z_1 \to 1 \\ \cdots \\ z_n \to 1}} \frac{1}{\Delta(z_1, \ldots, z_n)} \begin{vmatrix} z_1^{\ell_1} & \cdots & z_n^{\ell_1} \\ \vdots & & \vdots \\ z_1^{\ell_n} & \cdots & z_n^{\ell_n} \end{vmatrix}$$

$$= \frac{(-1)^{(1/2)n(n-1)}}{1!\,2! \cdots (n-1)!} \begin{vmatrix} 1 & \cdots & 1 \\ \ell_1 & \cdots & \ell_n \\ \vdots & & \vdots \\ \ell_1(\ell_1-1)\cdots(\ell_1-n+1) & \cdots & \ell_n(\ell_n-1)\cdots(\ell_n-n+1) \end{vmatrix}$$

$$= \frac{(-1)^{(1/2)n(n-1)} \Delta(\ell_1, \ldots, \ell_n)}{1!\,2! \cdots (n-1)!}.$$

Dividing the formula in Theorem 15.23 by $\Delta(z_1, \ldots, z_n)$ and letting $z_1 \to z, \ldots, z_n \to z$, we arrive at the following theorem.

Theorem 15.29 *Under the hypothesis of Theorem 15.23, we have*

$$
\begin{vmatrix}
f_1(z) & \cdots & f_n(z) \\
f_1'(z) & \cdots & f_n'(z) \\
\vdots & & \vdots \\
f_1^{(n-1)}(z) & \cdots & f_n^{(n-1)}(z)
\end{vmatrix}
= \sum_{\ell_1 > \ell_2 > \cdots \ell_n \geq 0} \Delta(\ell_1, \ldots, \ell_n)
\begin{vmatrix}
a_{\ell_1}^{(1)} & \cdots & a_{\ell_n}^{(1)} \\
\vdots & & \vdots \\
a_{\ell_1}^{(n)} & \cdots & a_{\ell_n}^{(n)}
\end{vmatrix}
z^{\ell_1 + \cdots + \ell_n - (1/2)n(n-1)}.
$$

The determinant on the left-hand side of the equation is called the Wronskian for the functions $f_1, \ldots, f_n$.

16
Equivalence of matrices

16.1 Notations

We let $A = A^{(m,n)}$ represent a matrix with m rows and n columns, and $A = A^{(n)} = A^{(n,n)}$ is then a square matrix. A matrix with a single row ($m = 1$) is called a vector, or a row vector, and that with a single column ($n = 1$) a column vector. Thus, on setting $x = x^{(1,m)}$, $b = b^{(1,n)}$, $A = A^{(m,n)}$, the matrix equation

$$xA = b$$

represents the linear system

$$\sum_{i=1}^{m} x_i a_{ij} = b_j, \quad j = 1, 2, \ldots, n.$$

We write $[d_1, \ldots, d_n]$ for the diagonal matrix, with the entries $d_1, \ldots, d_n$ for the main diagonal of the square matrix; $I = I^{(n)}$ is the unit square matrix of order n, and $0 = 0^{(m,n)}$ is the zero $m \times n$ matrix. The transpose of A, the matrix with the rows and columns interchanged, is denoted by A'.

We do not repeat the definitions for addition and multiplication, and the accompanying simple properties, associated with matrices. We use

$$A\begin{pmatrix} s_1 \cdots s_r \\ t_1 \cdots t_r \end{pmatrix}$$

to denote the determinant with order r by taking the rows $s_1, \ldots, s_r$ and the columns $t_1, \ldots, t_r$ from A.

Theorem 16.1 *Let*

$$A = A^{(m,n)} = (a_{ij}), \quad B = B^{(n,\ell)} = (b_{jk}), \quad C = AB(= C^{(m,\ell)}).$$

If $p \leq n$ then

$$|(c_{ij})|_{1 \leq i,j \leq p} = \sum_{r_1 < r_2 < \cdots < r_p} A\begin{pmatrix} 1\ 2 \cdots p \\ r_1 r_2 \cdots r_p \end{pmatrix} B\begin{pmatrix} r_1 r_2 \cdots r_p \\ 1\ 2 \cdots p \end{pmatrix},$$

where $r_1, \ldots, r_p$ *are taken from* $1, 2, \ldots, n$. *If* $p > n$, *then the value of the determinant is* 0.

Proof. Suppose that $p \leq n$. Then from

$$c_{ij} = \sum_{h=1}^{n} a_{ih} b_{hj}$$

we have

$$
\begin{aligned}
|(c_{ij})|_{1 \leq i,j \leq p} &= \begin{vmatrix} \sum_{s_1=1}^{n} a_{1s_1} b_{s_1 1} & \cdots & \sum_{s_p=1}^{n} a_{1s_p} b_{s_p p} \\ \vdots & & \vdots \\ \sum_{s_1=1}^{n} a_{ps_1} b_{s_1 1} & \cdots & \sum_{s_p=1}^{n} a_{1ps_p} b_{s_p p} \end{vmatrix} \\
&= \sum_{s_1=1}^{n} b_{s_1 1} \begin{vmatrix} a_{1s_1} & \sum_{s_2=1}^{n} a_{1s_2} b_{s_2 2} & \cdots & \sum_{s_p=1}^{n} a_{1s_p} b_{s_p p} \\ \vdots & \vdots & & \vdots \\ a_{ps_1} & \sum_{s_2=1}^{n} a_{ps_2} b_{s_2 2} & \cdots & \sum_{s_p=1}^{n} a_{pps_p} b_{s_p p} \end{vmatrix} \\
&= \sum_{s_1=1}^{n} \cdots \sum_{s_p=1}^{n} b_{s_1 1} \cdots b_{s_p p} A \begin{pmatrix} 1\,2 \cdots p \\ s_1 s_2 \cdots s_p \end{pmatrix}. \qquad (16.1)
\end{aligned}
$$

In this sum, terms with two equal values for s_h are 0, so that we need only consider those terms in which $s_1, \ldots, s_p$ are pairwise distinct. Take any sequence

$$r_1 < r_2 < \cdots < r_p$$

from $1, 2, \ldots, n$. There are $p!$ choices for $s_1, \ldots, s_p$, so that

$$|(c_{ij})|_{1 \leq i,j \leq p} = \sum_{r_1 < r_2 < \cdots < r_p} \sum_{\begin{pmatrix} r_1 r_2 \cdots r_p \\ t_1 t_2 \cdots t_p \end{pmatrix}} A \begin{pmatrix} 1\,2 \cdots p \\ t_1 t_2 \cdots t_p \end{pmatrix} b_{t_1 1} \cdots b_{t_p p},$$

where $t_1, \ldots, t_p$ run over all possible sequences from $r_1, \ldots, r_p$. Since

$$A \begin{pmatrix} 1\,2 \cdots p \\ t_1 t_2 \cdots t_p \end{pmatrix} = A \begin{pmatrix} 1\,2 \cdots p \\ r_1 r_2 \cdots r_p \end{pmatrix} \delta_{t_1 t_2 \cdots t_p}^{r_1 r_2 \cdots r_p},$$

we find that

$$|(c_{ij})|_{1\le i,j\le p} = \sum_{r_1<r_2<\cdots<r_p} A\begin{pmatrix}1\,2\,\cdots\,p\\ r_1 r_2\cdots r_p\end{pmatrix} \sum_{\begin{pmatrix}r_1r_2\cdots r_p\\ t_1t_2\cdots t_p\end{pmatrix}} \delta^{r_1r_2\cdots r_p}_{t_1t_2\cdots t_p} b_{t_1 1}\cdots b_{t_p p}$$

$$= \sum_{r_1<r_2<\cdots<r_p} A\begin{pmatrix}1\,2\,\cdots\,p\\ r_1 r_2\cdots r_p\end{pmatrix} B\begin{pmatrix}r_1\cdots r_p\\ 1\,\cdots\,p\end{pmatrix},$$

as required.

If $p > n$ then the terms in the sum (16.1) must have two s with the same value, so that the sum is 0. The theorem is proved. □

Remarks

(1) The same method gives the more general identity

$$C\begin{pmatrix}s_1\cdots s_p\\ t_1\cdots t_p\end{pmatrix} = \sum_{r_1<r_2<\cdots<r_p} A\begin{pmatrix}s_1\cdots s_p\\ r_1\cdots r_p\end{pmatrix} B\begin{pmatrix}r_1\cdots r_p\\ t_1\cdots t_p\end{pmatrix},$$

where $r_1, \ldots, r_p$ run over $1, 2, \ldots, n$.

(2) The determinant of the product of two matrices A, B is the product of the determinants of A and B.

16.2 Rank

If the matrix A has at least one non-zero sub-determinant of order r, and every sub-determinant of order $r + 1$ has the value 0, then r is called the rank of A. In the following we let r_A, r_B, r_C denote the ranks of the matrices A, B, C.

Theorem 16.2 *If* $C = AB$ *then* $r_C \le \min(r_A, r_B)$.

The theorem follows from the results of Section 16.1. Indeed, if A is an invertible square matrix then $r_C = r_B$. For, by Theorem 16.1, we have $r_C \le r_B$, and from $B = A^{-1}C$ we also have $r_B \le r_C$. Similarly, if B is an invertible square matrix then $r_C = r_A$.

Remarks

(1) We may have equality, for example when

$$A = B = C = \begin{pmatrix}1 & 0\\ 0 & 0\end{pmatrix}.$$

(2) We may have inequality, for example when

$$A = \begin{pmatrix}1 & 0\\ 0 & 0\end{pmatrix}, \quad B = \begin{pmatrix}0 & 0\\ 0 & 1\end{pmatrix}, \quad C = 0.$$

Theorem 16.3 *If $C = A + B$ then $r_C \le r_A + r_B$.*

Proof. Let $p = r_A + r_B + 1$, and consider any sub-determinant of order p for C:

$$\begin{vmatrix} a_{i_1h_1}+b_{i_1h_1} & \cdots & a_{i_1h_p}+b_{i_1h_p} \\ \vdots & & \vdots \\ a_{i_ph_1}+b_{i_ph_1} & \cdots & a_{i_ph_p}+b_{i_ph_p} \end{vmatrix} = \begin{vmatrix} a_{i_1h_1} & a_{i_1h_2}+b_{i_1h_2} & \cdots & a_{i_1h_p}+b_{i_1h_p} \\ \vdots & \vdots & & \vdots \\ a_{i_ph_1} & a_{i_ph_1}+b_{i_ph_1} & \cdots & a_{i_ph_p}+b_{i_ph_p} \end{vmatrix}$$
$$+ \begin{vmatrix} b_{i_1h_1} & a_{i_1h_2}+b_{i_1h_2} & \cdots & a_{i_1h_p}+b_{i_1h_p} \\ \vdots & \vdots & & \vdots \\ b_{i_ph_1} & a_{i_ph_1}+b_{i_ph_1} & \cdots & a_{i_ph_p}+b_{i_ph_p} \end{vmatrix}.$$

Continuing to break the expression down in this way, we end up with the sum of 2^p determinants having the form

$$\begin{vmatrix} a_{i_1h_1} & \cdots & a_{i_1h_s} & b_{i_1h_{s+1}} & \cdots & b_{i_1h_p} \\ \vdots & & \vdots & \vdots & & \vdots \\ a_{i_ph_1} & \cdots & a_{i_ph_s} & b_{i_ph_{s+1}} & \cdots & b_{i_ph_p} \end{vmatrix}.$$

We proceed to show that all such determinants have the value 0. In such a determinant, either there are r_A columns of 'a', or r_B columns of 'b', so that either $s > r_A$ or $p - s > r_B$. Suppose the latter. Expand along the first column, and then the second column, until the sth column, to end up with

$$\sum \pm a_{j_1h_1} a_{j_2h_2} \cdots a_{j_sh_s} \begin{vmatrix} b_{j_{s+1}h_{s+1}} & \cdots & b_{j_{s+1}h_p} \\ \vdots & & \vdots \\ b_{j_ph_{s+1}} & \cdots & b_{j_ph_p} \end{vmatrix}.$$

Here the sum is over $j_1, \ldots, j_p$ from $i_1, \ldots, i_p$, but with $j_s \ne j_t$, and from $p - s > r_B$ the determinant has to have the value 0. The same argument applies when $s > r_A$, so that the theorem is proved. □

Remarks

(1) For $A = B = \begin{pmatrix} 1 & 0 \\ 0 & 0 \end{pmatrix}$, we have $r_C < r_A + r_B$.

(2) For $A = \begin{pmatrix} 1 & 0 \\ 0 & 0 \end{pmatrix}$, $B = \begin{pmatrix} 0 & 0 \\ 0 & 1 \end{pmatrix}$, we have $r_C = r_A + r_B$.

16.3 Elementary operations

When dealing with a linear system, we apply the following three types of operations:

(1) multiply the ith equation by a number q;
(2) subtract t times the jth equation from the ith equation;
(3) interchange the ith and the jth equations.

We examine the meaning, or effect, of such types of operations when the system is represented by the matrix

$$P = \begin{pmatrix} a_{11} & \cdots & a_{1n} & b_1 \\ \vdots & & \vdots & \vdots \\ a_{n1} & \cdots & a_{nn} & b_n \end{pmatrix}.$$

(1) The effect of multiplying the ith row of the matrix P by q is the same as multiplication from the left by the diagonal matrix

$$[1, \ldots, 1, q, 1, \ldots, 1],$$

where q is at the ith position. For convenience, we let E_{ij} denote the square matrix of order n in which every entry is 0, except for the entry $a_{ij} = 1$. We then have

$$[1, \ldots, 1, q, 1, \ldots, 1] = E_{11} + \cdots + E_{i-1,i-1} + qE_{ii} + E_{i+1,i+1} + \cdots + E_{nn}.$$

(2) Subtraction of t times the jth row from the ith row is the same as multiplication of P from the left by

$$I - tE_{ij}.$$

(3) Interchanging the ith and jth rows is the same as multiplication of P from the left by

$$I - E_{ii} - E_{jj} + E_{ij} + E_{ji}.$$

Therefore, all the operations involved in Gaussian eliminations can be effected by multiplications of P from the left by these three types of square matrices. Actually, a matrix of the third type is also a product of matrices of the first two types: Thus, for $n = 2$, we have

$$\begin{pmatrix} 1 & 1 \\ 0 & 1 \end{pmatrix}\begin{pmatrix} 1 & 0 \\ -1 & 1 \end{pmatrix}\begin{pmatrix} 1 & 1 \\ 0 & 1 \end{pmatrix} = \begin{pmatrix} 0 & 1 \\ -1 & 1 \end{pmatrix}\begin{pmatrix} 1 & 1 \\ 0 & 1 \end{pmatrix} = \begin{pmatrix} 0 & 1 \\ -1 & 0 \end{pmatrix}$$

and

$$\begin{pmatrix} 1 & 1 \\ 0 & -1 \end{pmatrix}\begin{pmatrix} 0 & 1 \\ -1 & 0 \end{pmatrix} = \begin{pmatrix} 0 & 1 \\ 1 & 0 \end{pmatrix}.$$

Matrices of the second type are called transvections, or shears, and they occupy an important place in the study of square matrices.

Theorem 16.4 *Any square matrix with full rank can be represented as a product of matrices of the first two types.*

Proof. There must be a non-zero entry in the first column of the matrix

$$A = (a_{ij}),$$

and on multiplication from the left by a matrix of the third type if necessary we may assume that $a_{11} \neq 0$. Indeed, we may assume that $a_{11} = 1$, since otherwise we may multiply A from the left by $[a_{11}^{-1}, 1, \ldots, 1]$. On multiplication from the left by $I - a_{21}E_{21}$, the resulting matrix then has $a_{21} = 0$; similarly we can have $a_{31} = \cdots = a_{n1} = 0$.

By dealing with

$$\begin{pmatrix} a_{22} & \cdots & a_{2n} \\ \vdots & & \vdots \\ a_{n2} & \cdots & a_{nn} \end{pmatrix}$$

in the same way, we see that, on multiplication of appropriate matrices of types (1), (2) and (3), from the left, a matrix with full rank then takes the form of a triangular matrix:

$$\begin{pmatrix} 1 & a_{12} & a_{13} & \cdots & a_{1n} \\ 0 & 1 & a_{23} & \cdots & a_{2n} \\ \vdots & \vdots & \vdots & & \vdots \\ 0 & 0 & 0 & \cdots & 1 \end{pmatrix}.$$

On multiplication from the left by

$$\begin{pmatrix} 1 & -a_{12} & 0 & \cdots & 0 \\ 0 & 1 & 0 & \cdots & 0 \\ \vdots & \vdots & \vdots & & \vdots \\ 0 & 0 & 0 & \cdots & 1 \end{pmatrix},$$

the entry a_{12} becomes 0, and continuing in this way we finally arrive at the identity matrix I, that is

$$L_1 L_2 \cdots L_r A = I,$$

where L_i are square matrices of types (1), (2) or (3).

We note that the inverse of a matrix in a type still belongs to the type:

$$\begin{aligned} ([1, \ldots, 1, q, 1, \ldots, 1])^{-1} &= [1, \ldots, 1, q^{-1}, 1, \ldots, 1], \\ (I - tE_{ij})^{-1} &= I + tE_{ij}, \\ (I - E_{ii} - E_{jj} + E_{ij} + E_{ji})^{-1} &= I - E_{ii} - E_{jj} + E_{ij} + E_{ji}. \end{aligned}$$

The theorem is proved. □

Note. The method can be used to find the inverse of a square matrix, and we need only use some of the 'rough' part of the process. For example, we multiply

$$\begin{pmatrix} 1 & a_{12} & \cdots & a_{1n} \\ a_{21} & a_{22} & \cdots & a_{2n} \\ \vdots & \vdots & & \vdots \\ a_{n1} & a_{n2} & \cdots & a_{nn} \end{pmatrix}$$

from the left by

$$\begin{pmatrix} 1 & 0 & 0 & \cdots & 0 \\ -a_{21} & 1 & 0 & \cdots & 0 \\ -a_{31} & 0 & 1 & \cdots & 0 \\ \vdots & \vdots & \vdots & & \vdots \\ -a_{n1} & 0 & 0 & \cdots & 1 \end{pmatrix},$$

which then results in having the whole first column, except the first entry, taking the value 0; moreover the inverse of this square matrix is

$$\begin{pmatrix} 1 & 0 & 0 & \cdots & 0 \\ a_{21} & 1 & 0 & \cdots & 0 \\ a_{31} & 0 & 1 & \cdots & 0 \\ \vdots & \vdots & \vdots & & \vdots \\ a_{n1} & 0 & 0 & \cdots & 1 \end{pmatrix}.$$

Theorem 16.5 *Any matrix with a determinant having the value 1 must be a product of shears.*

We only mention the essential points involved in the proof.

(1) From

$$\begin{pmatrix} 1 & t \\ 0 & 1 \end{pmatrix}\begin{pmatrix} 1 & 0 \\ s & 1 \end{pmatrix} = \begin{pmatrix} 1+ts & t \\ s & 1 \end{pmatrix},$$

$$\begin{pmatrix} 1 & 0 \\ -s(1+ts)^{-1} & 1 \end{pmatrix}\begin{pmatrix} 1+ts & t \\ s & 1 \end{pmatrix} = \begin{pmatrix} 1+ts & t \\ 0 & (1+ts)^{-1} \end{pmatrix},$$

$$\begin{pmatrix} 1+ts & t \\ 0 & (1+ts)^{-1} \end{pmatrix}\begin{pmatrix} 1 & -(1+ts)^{-1}t \\ 0 & 1 \end{pmatrix} = \begin{pmatrix} 1+ts & 0 \\ 0 & (1+ts)^{-1} \end{pmatrix},$$

we see that $\begin{pmatrix} \lambda & 0 \\ 0 & \lambda^{-1} \end{pmatrix}$ is representable as a product of shears.

(2) Do use $[a_{11}^{-1}, a_{11}, 1, \ldots, 1]$ instead of $[a_{11}^{-1}, \ldots, 1]$.

(3) Do use $\begin{pmatrix} 0 & 1 \\ -1 & 0 \end{pmatrix}$ instead of $\begin{pmatrix} 0 & 1 \\ 1 & 0 \end{pmatrix}$.

16.4 Equivalence

Definition 16.6 Two $m \times n$ matrices A, B are said to be equivalent if there are matrices $P(= P^{(m)})$, $Q = (Q^{(n)})$ with full ranks such that

$$A = PBQ,$$

and we denote this by writing

$$A \stackrel{E}{=} B.$$

It is clear that being equivalent is an equivalence relation in the sense that the relation is (1) reflexive, that is $A \stackrel{E}{=} A$, (2) symmetric, that is if $A \stackrel{E}{=} B$ then $B \stackrel{E}{=} A$, and (3) transitive, that is if $A \stackrel{E}{=} B$ and $B \stackrel{E}{=} C$ then $A \stackrel{E}{=} C$.

Theorem 16.7 *A necessary and sufficient condition for two matrices of the same order to be equivalent is that they should have the same rank. Any matrix with rank r must be equivalent to*

$$\begin{pmatrix} I^{(r)} & 0 \\ 0 & 0 \end{pmatrix}.$$

Proof. After the interchange of rows and columns (that is, the multiplication from the left and from the right by matrices of the third type), we may assume that

$$A = \begin{pmatrix} A_1 & A_2 \\ A_3 & A_4 \end{pmatrix}, \quad A_1 = A_1^{(r)}, \quad |A_1| \neq 0,$$

and on multiplication from the left by

$$P = \begin{pmatrix} A_1^{-1} & 0 \\ -A_3 A_1^{-1} & I \end{pmatrix},$$

we have

$$\begin{pmatrix} I & A_1^{-1} A_2 \\ 0 & \star \end{pmatrix}.$$

Since the matrix here has rank r, all the entries for '$\star$' have to be 0. The required result follows by setting

$$Q = \begin{pmatrix} I & -A_1^{-1} A_2 \\ 0 & I \end{pmatrix}.$$

□

The reader should apply Theorem 16.7 to show that a matrix with rank 1 has the representation

$$\begin{pmatrix} a_1 b_1 & a_1 b_2 & \cdots & a_1 b_n \\ \vdots & \vdots & & \vdots \\ a_m b_1 & a_m b_2 & \cdots & a_m b_n \end{pmatrix} = \begin{pmatrix} a_1 \\ \vdots \\ a_m \end{pmatrix} (b_1, \ldots, b_n).$$

If A has rank m, then we can attach $n-m$ rows to it so that

$$\begin{pmatrix} A \\ C \end{pmatrix}$$

is an invertible square matrix.

Proof. From

$$A = P(I\ 0)Q,$$

we insert

$$C = (0\ I)Q,$$

so that

$$\begin{pmatrix} A \\ C \end{pmatrix} = \begin{pmatrix} P & 0 \\ 0 & I \end{pmatrix} Q,$$

as required. □

16.5 The n-dimensional vector space

We let R_n denote the set of vectors

$$x = (x_1, \dots, x_n),$$

and we call it the n-dimensional vector space.

We omit the revision concerning linear dependence and independence and their associated properties, apart from emphasising the following point: The m vectors

$$x^{(j)} = \left(x_1^{(j)}, \cdots, x_n^{(j)}\right), \quad j = 1, 2, \dots, m,$$

are linearly independent, or dependent, according to whether the matrix

$$X = \begin{pmatrix} x^{(1)} \\ \vdots \\ x^{(m)} \end{pmatrix} = \begin{pmatrix} x_1^{(1)} & \cdots & x_n^{(1)} \\ \vdots & & \vdots \\ x_1^{(m)} & \cdots & x_n^{(m)} \end{pmatrix}$$

has rank equal to m, or not. Thus, if X has rank m then $x^{(1)}, \dots, x^{(m)}$ are linearly independent, otherwise they are linearly dependent. Suppose that the rank is r; then there are r linearly independent vectors, and the remaining vectors can be represented as a linear combination of such vectors.

Suppose that

$$x^{(1)}, \dots, x^{(m)}$$

are m linearly independent vectors; then the collection of vectors

$$x = c_1 x^{(1)} + \cdots + c_m x^{(m)}$$

forms an m-dimensional subspace. The m vectors are said to be base vectors spanning the subspace, and X is called the representation matrix for the subspace. If there are another ℓ vectors

$$y^{(1)}, \ldots, y^{(\ell)}$$

also spanning the same subspace, then

$$x^{(i)} = \sum_{j=1}^{\ell} c_{ij} y^{(j)}, \quad 1 \leq i \leq m;$$

that is,

$$X = CY.$$

Thus $r(X) \leq r(Y)$, and similarly $r(Y) \leq r(X)$, so that $\ell = m$. Consequently: If X, Y both represent the same subspace, then there is a square matrix Q with full rank such that

$$X = QY. \tag{16.2}$$

Example 16.8 The vector $(1, 0, \ldots, 0)$ represents a one dimensional subspace, the collection of vectors of the form $(x, 0, \ldots, 0)$. Indeed, corresponding to any fixed $a_1, \ldots, a_n$, the vectors $(a_1 t, \ldots, a_n t)$ also form a one dimensional subspace.

We may describe (16.2) as X, Y being 'left-equivalent'; it is also an equivalence relation.

16.6 Transformations in a vector space

The transformation

$$y = xA, \quad A = A^{(n)}, \tag{16.3}$$

maps a vector x to a vector y in R_n.

If A is of full rank, we can solve the equation to give

$$x = yA^{-1}, \tag{16.4}$$

the inverse transformation. Such a transformation then gives a one-to-one correspondence between the vectors in R_n.

If we continue to let

$$z = yB, \tag{16.5}$$

then

$$z = x(AB),$$

which shows that the matrix corresponding to two successive transformations is the product of the two matrices.

Let the m vectors

$$x^{(1)}, \dots, x^{(m)}$$

in R_n be transformed, by (16.3), to the m vectors

$$y^{(i)} = x^{(i)} A, \quad i = 1, 2, \dots, m;$$

we write this as a matrix equation

$$Y = XA.$$

The relationship of being equivalent can be given the following geometric interpretation: After the transformation (16.3), the representation X of an m-dimensional subspace has the representation

$$X = QXA$$

for the same m-dimensional subspace.

From the theorem concerning equivalence we deduce at once the following.

Theorem 16.9 *Any m-dimensional linear subspace can be transformed by (16.3) to the subspace spanned by*

$$e_1 = (1, 0, 0, \dots, 0), \quad e_2 = (0, 1, 0, \dots, 0), \quad \dots, \quad e_m = (0, \dots, 0, 1, 0, \dots, 0).$$

We may interpret this by saying that any m-dimensional subspace forms a 'reachable' set under the 'group' of transformations (16.3); in particular, any vector can be transformed to $(1, 0, \dots, 0)$, and any two vectors can be transformed into each other by such a transformation.

We do not assume that A is invertible. Suppose that A has rank r. Then the result of (16.3) is the projection of R_n onto an r-dimensional subspace. This is because

$$PAQ = \begin{pmatrix} I^{(r)} & 0 \\ 0 & 0 \end{pmatrix}, \quad |P| \neq 0, \quad |Q| \neq 0,$$

and

$$y = xA$$

becomes

$$yQ = xP^{-1} \begin{pmatrix} I^{(r)} & 0 \\ 0 & 0 \end{pmatrix}.$$

Let $y^\star = yQ, x^\star = xP^{-1}$; then $y^\star = (\overbrace{\star, \ldots, \star}^{r \text{ lots}}, 0, \ldots, 0)$ form an r-dimensional subspace, so that all the vectors y also form such a subspace.

We next consider those vectors x that are mapped to 0 under (16.3). We need to solve

$$0 = xA,$$

that is

$$0 = xP^{-1}\begin{pmatrix} I^{(r)} & 0 \\ 0 & 0 \end{pmatrix},$$

so that

$$x^\star = (\overbrace{0, \ldots, 0}^{r \text{ lots}}, \star, \ldots, \star).$$

Thus (16.3) maps an $(n-r)$ dimensional subspace to 0.

16.7 Length, angle, area and volume

We now consider real vectors.

Definition 16.10 The inner product between two vectors a, b is defined by

$$ab' = \sum_{i=1}^{n} a_i b_i = ba'.$$

The length of a vector a is $\sqrt{aa'} = \sqrt{a_1^2 + \cdots + a_n^2}$, and the lengths of the three sides of the triangle formed by the two vectors a, b are

$$\sqrt{aa'}, \quad \sqrt{bb'}, \quad \sqrt{(a-b)(a-b)'};$$

by the cosine rule

$$(a-b)(a-b)' = aa' + bb' - 2\sqrt{aa'bb'}\cos\theta, \tag{16.6}$$

where θ is the angle between the two vectors a, b. Thus

$$\cos\theta = \frac{ab'}{\sqrt{aa'bb'}},$$

and we therefore say that a, b are orthogonal, or perpendicular, to each other if $ab' = 0$.

The area of the triangle is given by

$$\Delta = \frac{1}{2}\sqrt{aa'bb'}\sin\theta = \frac{1}{2}\sqrt{aa'bb' - (ab')^2};$$

the number under the square-root sign is the determinant of the square matrix

$$\begin{pmatrix} aa' & ab' \\ ba' & bb' \end{pmatrix} = \begin{pmatrix} a \\ b \end{pmatrix} \begin{pmatrix} a \\ b \end{pmatrix}';$$

also,

$$aa'bb' - (ab')^2 = \sum_{i \le i < j \le n} (a_i b_j - a_j b_i)^2$$

is the sum of the squares of all 2×2 determinants of sub-matrices obtained from $\begin{pmatrix} a \\ b \end{pmatrix}$.

We consider (without giving proofs) the situation when there are m vectors,

$$a^{(1)}, \dots, a^{(m)},$$

forming what is called an 'm-dimensional simplex'. By such a simplex we mean the following: Consider the points 0 and $a^{(i)}$ $(i = 1, \dots, m)$ as vertices. Any fixed m vertices now yield an $(m-1)$ dimensional hyperplane, and the region lying between these $m+1$ hyperplanes is called an m-dimensional simplex. On setting the matrix

$$A = \begin{pmatrix} a^{(1)} \\ \vdots \\ a^{(m)} \end{pmatrix},$$

the volume of the simplex is then given by

$$\frac{1}{m!}|AA'|^{1/2}.$$

The simplest example of a simplex is that formed by $e_1 = (1, 0 \dots, 0)$, $e_2 = (0, 1, 0 \dots, 0)$, ..., $e_n = (0, 0 \dots, 1)$, which is the set of points $(x_1, \dots, x_n)$ satisfying

$$x_1 \ge 0, \ \dots, \ x_n \ge 0, \quad x_1 + \cdots + x_n \le 1.$$

Next, the set $(t_1, \dots, t_n)$ satisfying

$$0 \le t_1 \le t_2 \le \cdots \le t_n \le 1$$

is also a simplex, and on setting $t_1 = x_1, t_2 = x_1 + x_2, \dots, t_n = x_1 + \cdots + x_n$, the two are transformed into each other. It is clear that the second simplex has volume $1/n!$ because, for each $i = 1, \dots, n$, the region $0 \le x_i \le 1$ forms a simplex, and there are $n!$ simplexes, each of which becomes, after the transformation, one described in the above.

When considering volume, we need to take direction into account. For example, on the plane we take the anti-clockwise direction to be positive, and in three dimensional space three vectors ordered to form a right-handed system give a positive direction. In n-dimensional space, if we specify the direction of the ordering of

$$e_1, e_2, \dots, e_n,$$

and any even permutation thereof, to be positive, whereas any odd permutation of the ordering is deemed to be negative, then, corresponding to n vectors

$$a^{(1)}, \ldots, a^{(n)},$$

the sign of the direction is the sign of the determinant of the matrix

$$\begin{pmatrix} a^{(1)} \\ \vdots \\ a^{(n)} \end{pmatrix}.$$

16.8 The determinant of a transformation (Jacobian)

We have already considered the linear transformation

$$y_i = \sum_{j=1}^{n} a_{ij} x_j, \quad i = 1, 2, \ldots, m.$$

We now consider a common general transformation

$$y_i = \phi_i(x_1, \ldots, x_n), \quad i = 1, 2, \ldots, m. \tag{16.7}$$

Suppose that these $\phi_1, \ldots, \phi_m$ all have partial derivatives, so that

$$dy_i = \sum_{j=1}^{n} dx_j \frac{\partial \phi_i}{\partial x_j} = \sum_{j=1}^{n} dx_j \frac{\partial y_i}{\partial x_j}, \quad i = 1, 2, \ldots, m.$$

Using matrix symbols we write this as

$$dy = dx \frac{\partial(y_1, \ldots, y_n)}{\partial(x_1, \ldots, x_n)},$$

where $dx = (dx_1, \ldots, dx_n)$ is a differential vector, and

$$\frac{\partial(y_1, \ldots, y_n)}{\partial(x_1, \ldots, x_n)} = \begin{pmatrix} \frac{\partial y_1}{\partial x_1} & \cdots & \frac{\partial y_m}{\partial x_1} \\ \vdots & & \vdots \\ \frac{\partial y_1}{\partial x_n} & \cdots & \frac{\partial y_m}{\partial x_n} \end{pmatrix}$$

is called the Jacobi matrix. When $m = n$ the determinant of the Jacobi matrix is called the Jacobian of the transformation. If $z_1, \ldots, z_\ell$ are also functions of $y_1, \ldots, y_m$ then

$$dz = dy \frac{\partial(z_1, \ldots, z\ell)}{\partial(y_1, \ldots, y_m)},$$

and thus

$$dz = dx \frac{\partial(y_1, \ldots, y_n)}{\partial(x_1, \ldots, x_n)} \frac{\partial(z_1, \ldots, z\ell)}{\partial(y_1, \ldots, y_m)}.$$

In other words, we have the product of Jacobi matrices

$$\frac{\partial(z_1,\ldots,z_\ell)}{\partial(x_1,\ldots,x_n)} = \frac{\partial(y_1,\ldots,y_n)}{\partial(x_1,\ldots,x_n)}\frac{\partial(z_1,\ldots,z\ell)}{\partial(y_1,\ldots,y_m)}.$$

If $m = n = \ell$ and $z_1 = x_1, \ldots, z_n = x_n$, then

$$I = \frac{\partial(y_1,\ldots,y_n)}{\partial(x_1,\ldots,x_n)}\frac{\partial(z_1,\ldots,z\ell)}{\partial(y_1,\ldots,y_m)}.$$

In other words, if the inverse to the transformation (16.7) exists, then the Jacobi matrix for the inverse transformation is just the inverse of the square Jacobi matrix.

16.9 Implicit function theorem

Theorem 16.11 *Suppose that the system of equations*

$$F_1(x_1,\ldots,x_m,y_1,\ldots,y_n) = 0, \ \ldots, \ F_n(x_1,\ldots,x_m,y_1,\ldots,y_n) = 0 \tag{16.8}$$

has the solution

$$x_k = x_k^{(0)},\ y_\ell = y_\ell^{(0)}, \quad 1 \le k \le m,\ 1 \le \ell \le n. \tag{16.9}$$

Suppose also that F_ℓ is a continuous function in the region relevant to (16.9), and that it has a first order partial derivative in the region with

$$\frac{\partial(F_1,\ldots,F_n)}{\partial(y_1,\ldots,y_n)} \neq 0.$$

Then, when x_k is sufficiently near $x_k^{(0)}$, the system (16.8) definitely has functions $y_\ell(x_1,\ldots,x_m)$ which are continuous, have first order partial derivatives and satisfy the condition

$$y_\ell\left(x_1^{(0)},\ldots,x_m^{(0)}\right) = y_\ell^{(0)}.$$

Proof. We use induction on n; the case $n = 1$ is trivial and, for the inductive step, we take as induction hypothesis that the result holds when n is replaced by $n - 1$.

By a suitable relabelling of the functions, we may assume that, in the region relevant to (16.9),

$$\frac{\partial(F_2,\ldots,F_n)}{\partial(y_2,\ldots,y_n)} \neq 0,$$

so that there are uniquely determined functions

$$y_2 = \phi_2(x_1,\ldots,x_m,y_1), \ \ldots, \ y_n = \phi_n(x_1,\ldots,x_m,y_1). \tag{16.10}$$

Substituting these into F_1 we have

$$F_1(x_1, \ldots, x_m, y_1, \phi_2, \ldots, \phi_n) = 0. \tag{16.11}$$

This is a function of $x_1, \ldots, x_m, y_1$, and if we can show that its partial derivative with respect to y_1 does not vanish then we can solve for y_1, which is then a function of $x_1, \ldots, x_m$ satisfying the required condition of the theorem, and the inductive step is complete when it is substituted back into (16.10).

The problem then becomes one of showing that the partial derivative of (16.11) with respect to y_1 (when substituted into $\phi_2, \ldots, \phi_n$) does not vanish:

$$\left(\frac{\partial F_1}{\partial y_1}\right) = \frac{\partial F_1}{\partial y_1} + \sum_{s=2}^{n} \frac{\partial F_1}{\partial \phi_2}\frac{\partial \phi_s}{\partial y_1} \neq 0, \tag{16.12}$$

and from

$$F_\ell(x_1, \ldots, x_m, y_1, \phi_2, \ldots, \phi_n) = 0$$

we find that

$$\frac{\partial F_\ell}{\partial y_1} + \sum_{s=2}^{n} \frac{\partial F_\ell}{\partial \phi_2}\frac{\partial \phi_s}{\partial y_1} = 0, \quad \ell = 2, 3, \ldots, n. \tag{16.13}$$

Eliminating $1, \partial\phi_2/\partial y_1, \ldots, \partial\phi_n/\partial y_1$, we have

$$\begin{vmatrix} -\left(\frac{\partial F_1}{\partial y_1}\right) + \frac{\partial F_1}{\partial y_1} & \frac{\partial F_1}{\partial \phi_2} & \cdots & \frac{\partial F_1}{\partial \phi_n} \\ \vdots & \vdots & & \vdots \\ \frac{\partial F_n}{\partial y_1} & \frac{\partial F_n}{\partial \phi_2} & \cdots & \frac{\partial F_n}{\partial \phi_n} \end{vmatrix} = 0.$$

Thus

$$\left(\frac{\partial F_1}{\partial y_1}\right)\frac{\partial F_2, \ldots, F_n}{\partial y_2, \ldots, y_n} - \frac{\partial F_1, \ldots, F_n}{\partial y_1, \ldots, y_n} = 0,$$

so that

$$\left(\frac{\partial F_1}{\partial y_1}\right) \neq 0.$$

The theorem is proved. □

Setting $F_i = f_i - y_i$ in Theorem 16.11, we have the following theorem.

Theorem 16.12 *Let*

$$y_i = f_i(x_1, \ldots, x_n), \quad i = 1, 2, \ldots, n, \tag{16.14}$$

where f_i and their first order partial derivatives are continuous in the neighbourhood of $x_i = x_i^{(0)}$ $(i = 1, \ldots, n)$, at which

$$\frac{\partial(y_1, \ldots, y_n)}{\partial(x_1, \ldots, x_n)} \neq 0.$$

Then, in the neighbourhood of

$$y_i^{(0)} = f_i\big(x_1^{(0)}, \ldots, x_n^{(0)}\big),$$

there is a unique set of continuous functions $x_k(y_1, \ldots, y_n)$ of $y_1, \ldots, y_n$ for the system (16.14). Moreover, such functions also possess first order partial derivatives and they satisfy

$$x_k^{(0)} = x_k\big(y_1^{(0)}, \ldots, y_n^{(0)}\big).$$

16.10 The Jacobian of a complex transformation

If

$$f_j = \sum_{k=1}^{n} \phi_k c_{kj}, \quad x_j = \sum_{k=1}^{n} y_k c_{kj}, \quad j = 1, 2, \ldots, n,$$

then

$$\begin{aligned}\frac{\partial(\phi_1, \ldots, \phi_n)}{\partial(y_1, \ldots, y_n)} &= \frac{\partial(x_1, \ldots, x_n)}{\partial(y_1, \ldots, y_n)} \frac{\partial(f_1, \ldots, f_n)}{\partial(x_1, \ldots, x_n)} \frac{\partial(\phi_1, \ldots, \phi_n)}{\partial(f_1, \ldots, f_n)} \\ &= C \frac{\partial(f_1, \ldots, f_n)}{\partial(x_1, \ldots, x_n)} C^{-1},\end{aligned}$$

so that

$$\det\left(\frac{\partial(\phi_1, \ldots, \phi_n)}{\partial(y_1, \ldots, y_n)}\right) = \det\left(\frac{\partial(f_1, \ldots, f_n)}{\partial(x_1, \ldots, x_n)}\right).$$

Accordingly, we prove the following theorem.

Theorem 16.13 *Let*

$$f_j(z_1, \ldots, z_n), \quad j = 1, 2, \ldots, n,$$

be analytic functions of n complex variables, with their real and imaginary parts being given by

$$z_k = x_k + i y_k, \quad f_k = u_k + i v_k.$$

Then

$$\det\left(\frac{\partial(u_1, v_1, \ldots, u_n, v_n)}{\partial(x_1, y_1, \ldots, x_n, y_n)}\right) = \left|\det\left(\frac{\partial(f_1, \ldots, f_n)}{\partial(z_1, \ldots, z_n)}\right)\right|^2.$$

Proof. From

$$(z_j, \bar{z}_j) = (x_j, y_j)\begin{pmatrix} 1 & 1 \\ i & -i \end{pmatrix}, \quad (f_j, \bar{f}_j) = (u_j, v_j)\begin{pmatrix} 1 & 1 \\ i & -i \end{pmatrix},$$

we find that

$$\det\left(\frac{\partial(u_1, v_1, \ldots, u_n, v_n)}{\partial(x_1, y_1, \ldots, x_n, y_n)}\right) = \det\left(\frac{\partial(f_1, \bar{f}_1, \ldots, f_n, \bar{f}_n)}{\partial(z_1, \bar{z}_1, \ldots, z_n, \bar{z}_n)}\right)$$
$$= \det\left(\frac{\partial(f_1, \ldots, f_n; \bar{f}_1, \ldots, \bar{f}_n)}{\partial(z_1, \ldots, z_n; \bar{z}_1, \ldots, \bar{z}_n)}\right).$$

If $f(z)$ is an analytic function of z then $\partial f(z)/\partial \bar{z} = 0$ and $\partial \bar{f}/\partial z = 0$. Therefore, the determinant here is equal to

$$\det\left(\frac{\partial(f_1, \ldots, f_n)}{\partial(z_1, \ldots, z_n)}\right) \det\left(\frac{\partial(\bar{f}_1, \ldots, \bar{f}_n)}{\partial(\bar{z}_1, \ldots, \bar{z}_n)}\right),$$

as required. □

Concerning analytic functions, we have the following implicit function theorem.

Theorem 16.14 *Let*

$$F_j(w_1, \ldots, w_n, z_1, \ldots, z_n), \quad j = 1, 2, \ldots, n,$$

be analytic near the origin, with $F_j(0, \ldots, 0) = 0$ *and*

$$\det\left(\frac{\partial(F_1, \ldots, F_n)}{\partial(w_1, \ldots, w_n)}\right) \neq 0, \quad w = z = 0.$$

Then the system of equations

$$F_j(w_1, \ldots, w_n, z_1, \ldots, z_n) = 0, \quad j = 1, 2, \ldots, n, \tag{16.15}$$

has the unique solutions

$$w_j = w_j(z_1, \ldots, z_n), \quad j = 1, 2, \ldots, n, \tag{16.16}$$

with $w = 0$ *at* $z = 0$*, and they are analytic near* $z = 0$.

Proof. We separate (16.15) into real and imaginary parts to obtain $2n$ equations for $w_j = u_j + iv_j$ in terms of the $2n$ real variables u_i, v_i. By Theorem 16.13, the Jacobian of the real transformation does not vanish at the origin, so that there is a unique set of solutions with continuous partial derivatives, and they take the value 0 at the origin.

Differentiating (16.15) with respect to $\bar{z}_m$, we have

$$\sum_{p=1}^{n} \frac{\partial F_j}{\partial w_p}\frac{\partial w_p}{\partial \bar{z}_m} + \sum_{\lambda=1}^{n} \frac{\partial F_j}{\partial z_\lambda}\frac{\partial z_\lambda}{\partial \bar{z}_m} = 0,$$

and since $\partial z_\lambda/\partial \bar{z}_m = 0$ this becomes

$$\sum_{p=1}^{n} \frac{\partial F_j}{\partial w_p}\frac{\partial w_p}{\partial \bar{z}_m} = 0, \quad j, m = 1, 2, \dots, n.$$

Since

$$\det\left(\frac{\partial(F_1, \dots, F_n)}{\partial(w_1, \dots, w_n)}\right) \neq 0,$$

so that $\partial w_p/\partial \bar{z}_m = 0$, it follows that w_p is an analytic function of $z_1, \dots, z_n$. □

16.11 Functional dependence

We consider m functions

$$\begin{cases} u_1 = u_1(x_1, \dots, x_n), \\ \quad \cdots \\ u_m = u_m(x_1, \dots, x_n), \end{cases} \tag{16.17}$$

of n variables; we shall assume that these functions are continuous and that they have continuous first order partial derivatives. If there is a non-trivial function

$$F(u_1, \dots, u_m)$$

with the property that, when (16.17) is substituted into it, we have an expression for $x_1, \dots, x_n$ which is identically 0, then we say that the m functions are functionally dependent.

Consider the Jacobi matrix

$$\frac{\partial(u_1, \dots, u_m)}{\partial(x_1, \dots, x_n)} = \begin{pmatrix} \frac{\partial u_1}{\partial x_1} & \cdots & \frac{\partial u_1}{\partial x_n} \\ \vdots & & \vdots \\ \frac{\partial u_m}{\partial x_1} & \cdots & \frac{\partial u_m}{\partial x_n} \end{pmatrix}.$$

Our principal result is as follows: If there is an rth order sub-determinant of the matrix that is not identically 0, and every $(r+1)$th sub-determinant is identically 0, then there are $m - r$ functions which are representable in terms of the remaining functions u. When $n = m = r$, the functions $u_1, \dots, u_n$ are functionally independent.

If $m < n$ then it does not matter much if we insert $n - m$ functions

$$u_{m+1} = x_{m+1}, \quad \dots, \quad u_n = x_n;$$

and if $m > n$ then it will not matter if we add $m - n$ new variables $x_{n+1}, \dots, x_m$. Therefore it does not really matter if we assume that $m = n$ in the following.

The argument is rather complicated, and we divide it into several transparent steps. In the following we apply the analyst's method in which the ideas involved are revealed successively, and in Section 16.12 we tackle the problem using an algebraic approach.

To avoid confusion with the Jacobi matrix, we let the Jacobian, that is the determinant of the square Jacobi matrix $\partial(u_1, \ldots, u_n)/\partial(x_1, \ldots, x_n)$, be denoted by

$$\frac{D(u_1, \ldots, u_n)}{D(x_1, \ldots, x_n)}.$$

Theorem 16.15 *Let $m = n$. Then a necessary and sufficient condition for the functions in (16.7) to be functionally dependent is that*

$$\frac{D(u_1, \ldots, u_n)}{D(x_1, \ldots, x_n)} \equiv 0. \tag{16.18}$$

Proof. (1) Necessity. For the sake of simplicity, we take $n = 3$:

$$X = f_1(x, y, z), \quad Y = f_2(x, y, z), \quad Z = f_3(x, y, z). \tag{16.19}$$

Suppose that

$$\frac{D(X, Y, Z)}{D(x, y, z)} \not\equiv 0,$$

that is there is a point $(x, y, z) = (x_0, y_0, z_0)$ at which

$$\left(\frac{D(X, Y, Z)}{D(x, y, z)}\right)_{(x,y,z)=(x_0,y_0,z_0)} \neq 0,$$

and, by (16.19), $(X, Y, Z) = (X_0, Y_0, Z_0)$ in correspondence.

From the implicit function theorem, there exists h with the property that, for every (X, Y, Z) satisfying

$$|X - X_0| \leq h, \quad |Y - Y_0| \leq h, \quad |Z - Z_0| \leq h, \tag{16.20}$$

there are x, y, z satisfying (16.19). In other words, f_1, f_2, f_3 may take any values in the intervals (16.20), so that it is not possible to have a functional dependence

$$F(X, Y, Z) = 0.$$

The same argument can be used to show that if one of

$$\frac{D(X, Y)}{D(x, y)}, \quad \frac{D(X, Y)}{D(y, z)}, \quad \frac{D(X, Y)}{D(z, x)}$$

is not identically 0, then X, Y are functionally independent.

More generally, if $n > m$ and the m functions $u_1, \ldots, u_m$ defined in (16.17) are functionally dependent, then

$$\frac{D(u_1, \ldots, u_m)}{D(x_{\alpha_1}, \ldots, x_{\alpha_m})} \equiv 0,$$

where $\alpha_1, \ldots, \alpha_m$ are any m numbers chosen from $1, \ldots, n$, and so the 'rank' of the Jacobi matrix should be less than m.

(2) Sufficiency. We illustrate the argument by setting $n = 4$:

$$\begin{cases} X = f_1(x, y, z, t), \\ Y = f_2(x, y, z, t), \\ Z = f_3(x, y, z, t), \\ T = f_4(x, y, z, t), \end{cases} \tag{16.21}$$

and we assume that

$$\Delta = \begin{vmatrix} \dfrac{\partial f_1}{\partial x} & \dfrac{\partial f_1}{\partial y} & \dfrac{\partial f_1}{\partial z} & \dfrac{\partial f_1}{\partial t} \\ \dfrac{\partial f_2}{\partial x} & \dfrac{\partial f_2}{\partial y} & \dfrac{\partial f_2}{\partial z} & \dfrac{\partial f_2}{\partial t} \\ \dfrac{\partial f_3}{\partial x} & \dfrac{\partial f_3}{\partial y} & \dfrac{\partial f_3}{\partial z} & \dfrac{\partial f_3}{\partial t} \\ \dfrac{\partial f_4}{\partial x} & \dfrac{\partial f_4}{\partial y} & \dfrac{\partial f_4}{\partial z} & \dfrac{\partial f_4}{\partial t} \end{vmatrix} \equiv 0.$$

Suppose first that

$$\delta = \frac{D(f_1, f_2, f_3)}{D(x, y, z)} \not\equiv 0,$$

so that we can solve for

$$x = \phi_1(X, Y, Z, t), \quad y = \phi_2(X, Y, Z, t), \quad z = \phi_3(X, Y, Z, t). \tag{16.22}$$

Substituting these into the fourth equation in (16.21) we have

$$T = f_4(\phi_1, \phi_2, \phi_3, t) = F(X, Y, Z, t),$$

say, and we need only show that F does not depend on t, that is

$$\frac{\partial F}{\partial t} \equiv 0.$$

We have

$$\frac{\partial F}{\partial t} = \frac{\partial f_4}{\partial x}\frac{\partial \phi_1}{\partial t} + \frac{\partial f_4}{\partial y}\frac{\partial \phi_2}{\partial t} + \frac{\partial f_4}{\partial z}\frac{\partial \phi_3}{\partial t} + \frac{\partial f_4}{\partial t}, \tag{16.23}$$

where $\partial\phi_1/\partial t, \partial\phi_2/\partial t, \partial\phi_3/\partial t$ are the partial derivatives with respect to t of the implicit functions ϕ_1, ϕ_2, ϕ_3, which can be determined from (16.21):

$$\begin{cases} \dfrac{\partial f_1}{\partial x}\dfrac{\partial \phi_1}{\partial t} + \dfrac{\partial f_1}{\partial y}\dfrac{\partial \phi_2}{\partial t} + \dfrac{\partial f_1}{\partial z}\dfrac{\partial \phi_3}{\partial t} + \dfrac{\partial f_1}{\partial t} = 0, \\ \dfrac{\partial f_2}{\partial x}\dfrac{\partial \phi_1}{\partial t} + \dfrac{\partial f_2}{\partial y}\dfrac{\partial \phi_2}{\partial t} + \dfrac{\partial f_2}{\partial z}\dfrac{\partial \phi_3}{\partial t} + \dfrac{\partial f_2}{\partial t} = 0, \\ \dfrac{\partial f_3}{\partial x}\dfrac{\partial \phi_1}{\partial t} + \dfrac{\partial f_3}{\partial y}\dfrac{\partial \phi_2}{\partial t} + \dfrac{\partial f_3}{\partial z}\dfrac{\partial \phi_3}{\partial t} + \dfrac{\partial f_3}{\partial t} = 0. \end{cases} \tag{16.24}$$

On eliminating $\partial\phi_1/\partial t$, $\partial\phi_2/\partial t$, $\partial\phi_3/\partial t$ from (16.23) and (16.24), we have

$$\begin{vmatrix} \frac{\partial f_1}{\partial x} & \frac{\partial f_1}{\partial y} & \frac{\partial f_1}{\partial z} & \frac{\partial f_1}{\partial t} \\ \frac{\partial f_2}{\partial x} & \frac{\partial f_2}{\partial y} & \frac{\partial f_2}{\partial z} & \frac{\partial f_2}{\partial t} \\ \frac{\partial f_3}{\partial x} & \frac{\partial f_3}{\partial y} & \frac{\partial f_3}{\partial z} & \frac{\partial f_3}{\partial t} \\ \frac{\partial f_4}{\partial x} & \frac{\partial f_4}{\partial y} & \frac{\partial f_4}{\partial z} & \frac{\partial f_4}{\partial t} - \frac{\partial F}{\partial t} \end{vmatrix} = 0,$$

that is

$$0 \equiv \Delta - \delta \frac{\partial F}{\partial t}.$$

Since $\delta \not\equiv 0$, we have $\partial F/\partial t \equiv 0$. This means that the functions X, Y, Z, T are functionally dependent, so that

$$T = F(X, Y, Z). \tag{16.25}$$

□

Note. There is no other functional dependence besides this one; if there were, by (16.25) there would be a functional relationship between X, Y, Z, so that $\delta \equiv 0$, contrary to the assumption.

We consider the situation when all third order determinants in Δ are identically 0 and suppose that there is a second order determinant

$$\delta' = \frac{D(f_1, f_2)}{D(x, y)} \not\equiv 0.$$

The first two equations in (16.21) have the solution

$$x = \phi_1(X, Y, z, t), \quad y = \phi_2(X, Y, z, t),$$

so that

$$Z = f_3(x, y, z, t) = F_1(X, Y, z, t),$$
$$T = f_4(x, y, z, t) = F_2(X, Y, z, t).$$

From

$$\frac{\partial F_1}{\partial t} = \frac{\partial f_3}{\partial x}\frac{\partial \phi_1}{\partial t} + \frac{\partial f_3}{\partial y}\frac{\partial \phi_2}{\partial t} + \frac{\partial f_3}{\partial t},$$
$$0 = \frac{\partial f_1}{\partial x}\frac{\partial \phi_1}{\partial t} + \frac{\partial f_1}{\partial y}\frac{\partial \phi_2}{\partial t} + \frac{\partial f_1}{\partial t},$$
$$0 = \frac{\partial f_2}{\partial x}\frac{\partial \phi_1}{\partial t} + \frac{\partial f_2}{\partial y}\frac{\partial \phi_2}{\partial t} + \frac{\partial f_2}{\partial t},$$

we deduce that

$$\frac{D(f_1, f_2, f_3)}{D(x, y, z)} = \delta' \frac{\partial F_1}{\partial t},$$

so that $\partial F_1/\partial t \equiv 0$. The same argument shows that $\partial F_1/\partial z = 0$, so that t and z do not occur in F_1; similarly, F_2 is independent of t and z, so that we have the two relationships

$$Z = F_1(X, Y), \quad T = F_2(X, Y).$$

Finally we consider the case when all the second order determinants are identically 0. Suppose there is a first order determinant $\not\equiv 0$; then three of X, Y, Z, T can be represented as functions of the remaining one.

To summarise: If the rank of

$$\frac{\partial(F_1, \ldots, F_n)}{\partial(x_1, \ldots, x_n)}$$

is r, then $n - r$ of $F_1, \ldots, F_n$ can be represented as functions of the remaining r functions, so that there are r functionally independent functions, and no other functional dependence exists.

For this reason it is not difficult to deduce the conclusion mentioned at the beginning of this section.

Example 16.16 (A special example) The condition for the functional dependence for the two functions $F_1(x_1, \ldots, x_n)$, $F_2(x_1, \ldots, x_n)$ is that the ratio

$$\frac{\partial F_1}{\partial x_i} : \frac{\partial F_2}{\partial x_i}$$

should be independent of i.

Note 1. For the functions $F_1, F_2, \ldots, F_n$ discussed in the above, if, after the elimination of the variables $x_1, \ldots, x_n$, there are still the variables $y_1, \ldots, y_\ell$, then $D(F_1, \ldots, F_n)/D(x_1, \ldots, x_n) \equiv 0$ may indicate that $F_1, F_2, \ldots, F_n$ are functionally dependent, but such a relationship should only involve $y_1, \ldots, y_\ell$.

Note 2. If X, Y are functions of x, y, z, and x, y, z are functions of u, v, then

$$\frac{D(X, Y)}{D(u, v)} = \frac{D(X, Y)}{D(x, y} \frac{D(x, y)}{D(u, v)} + \frac{D(X, Y)}{D(y, z} \frac{D(y, z)}{D(u, v)} + \frac{D(X, Y)}{D(z, x} \frac{D(z, x)}{D(u, v)}.$$

Application. A direct proof of the intrinsic property of the logarithm function: Suppose that $f(x)$ is a function with $f(1) = 0$ and $f'(x) = 1/x$. Let

$$u = f(x) + f(y), \quad v = xy.$$

Then

$$\frac{D(u, v)}{D(x, y)} = \begin{vmatrix} 1/x & 1/y \\ y & x \end{vmatrix} = 0,$$

so that there is a relationship between u and v, say $u = \phi(v)$. Thus

$$f(x) + f(y) = \phi(xy),$$

and, on taking $y = 1$, we have $f(x) = \phi(x)$, which delivers the required property.

16.12 Algebraic considerations

Before embarking on the algebraic method used to deal with the problem, we first consider

$$y = xA, \tag{16.26}$$

where $y = y^{(1,m)}$, $x = x^{(1,n)}$, $A = A^{(n,m)}$. Equation (16.26) represents a linear system of m equations with the n variables $x_1, \ldots, x_n$, and we suppose that A has rank r.

By reordering the elements in the vectors y and x, we can ensure that the determinant of order r in the upper-left corner of the matrix A is not zero, so that

$$(y_1, \ldots, y_r, y_{r+1}, \ldots, y_m) = (x_1, \ldots, x_r, x_{r+1}, \ldots, x_n)\begin{pmatrix} A_1 & A_2 \\ A_3 & A_4 \end{pmatrix}, \tag{16.27}$$

where $A_1 = A_1^{(r)}$, $A_2 = A_2^{(r,m-r)}$, $A_3 = A_3^{(n-r,r)}$, $A_4 = A_4^{(n-r,m-r)}$. Thus

$$(y_1, \ldots, y_r, x_{r+1}, \ldots, x_n) = (x_1, \ldots, x_n)\begin{pmatrix} A_1 & 0 \\ A_3 & I^{(n-r)} \end{pmatrix}, \tag{16.28}$$

and, when substituted into the preceding equation, this yields

$$(y_1, \ldots, y_m) = (y_1, \ldots, y_r, x_{r+1}, \ldots, x_n)\begin{pmatrix} A_1 & 0 \\ A_3 & I^{(n-r)} \end{pmatrix}^{-1}\begin{pmatrix} A_1 & A_2 \\ A_3 & A_4 \end{pmatrix}. \tag{16.29}$$

From

$$\begin{pmatrix} A_1 & 0 \\ A_3 & I^{(n-r)} \end{pmatrix}^{-1} = \begin{pmatrix} A_1^{-1} & 0 \\ -A_3A_1^{-1} & I \end{pmatrix}$$

and A having rank r, we find that

$$\begin{pmatrix} A_1 & 0 \\ A_3 & I \end{pmatrix}^{-1}\begin{pmatrix} A_1 & A_2 \\ A_3 & A_4 \end{pmatrix} = \begin{pmatrix} I^{(r)} & A_1^{-1}A_2 \\ 0 & 0 \end{pmatrix},$$

so that

$$(y_1, \ldots, y_m) = (y_1, \ldots, y_r, x_{r+1}, \ldots, x_n)\begin{pmatrix} I & A_1^{-1}A_4 \\ 0 & 0 \end{pmatrix}, \tag{16.30}$$

that is

$$(y_{r+1}, \ldots, y_m) = (y_1, \ldots, y_r)A_1^{-1}A_2. \tag{16.31}$$

Thus $y_{r+1}, \ldots, y_m$ are representable as linear combinations of $y_1, \ldots, y_r$.

We can now tackle the problem in Section 16.11, starting with

$$du = dx\frac{\partial(u_1,\ldots,u_m)}{\partial(x_1,\ldots,x_n)}. \tag{16.26$'$}$$

By reordering the elements in the vectors u and x we can ensure that the determinant of order r in the upper-left corner of the Jacobi matrix is not identically zero, so that

$$(du_1,\ldots,du_r,\ldots,du_m) = (dx_1,\ldots,dx_r,\ldots,dx_n)\begin{pmatrix} A_1 & A_2 \\ A_3 & A_4 \end{pmatrix}, \tag{16.27$'$}$$

where

$$A_1 = \frac{\partial(u_1,\ldots,u_r)}{\partial(x_1,\ldots,x_r)}, \quad \cdots.$$

We then have

$$(du_1,\ldots,du_r,dx_{r+1}\ldots,dx_n) = (dx_1,\ldots,dx_r,\ldots,dx_n)\begin{pmatrix} A_1 & 0 \\ A_3 & I^{(n-r)} \end{pmatrix} \tag{16.28$'$}$$

and

$$(du_1,\ldots,du_m) = (du_1,\ldots,du_r,dx_{r+1}\ldots,dx_n)\begin{pmatrix} I & A^{-1}A_2 \\ 0 & 0 \end{pmatrix}. \tag{16.29$'$}$$

By the product of Jacobi matrices, we have

$$\begin{pmatrix} I & A^{-1}A_2 \\ 0 & 0 \end{pmatrix} = \frac{\partial(u_1,\ldots,u_r,u_{r+1},\ldots,u_m)}{\partial(u_1,\ldots,u_r,x_{r+1},\ldots,x_n)} \tag{16.30$'$}$$

so that

$$\frac{\partial u_{r+1}}{\partial x_{r+1}} = \cdots = \frac{\partial u_{r+1}}{\partial x_n} = 0. \tag{16.31$'$}$$

Thus u_{r+1} is independent of $x_{r+1},\ldots,x_n$; similarly, $u_{r+2},\ldots,u_m$ are independent of $x_{r+1},\ldots,x_n$, so that $u_{r+1},\ldots,u_m$ are functions of $u_1,\ldots,u_r$, that is

$$\begin{cases} u_{r+1} = v_{r+1}(u_1,\ldots,u_r), \\ \qquad \cdots \\ u_m = v_m(u_1,\ldots,u_r). \end{cases} \tag{16.32}$$

We now examine whether there are other functional relationships for these $m-r$ functions. Clearly the answer is yes, for example

$$u_{r+1} + u_{r+2} = v_{r+1} + v_{r+2},$$
$$u_{r+1}^3 + 7u_m^8 = v_{r+1}^3 + 7v_m^8,$$

etc. Therefore we need to be clear on what we mean by other functional relationships. We rewrite (16.32) as

$$\begin{cases} r_1(u_1, \ldots, u_m) = 0, \\ \quad \ldots \\ r_p(u_1, \ldots, u_m) = 0, \quad p = n - r; \end{cases} \tag{16.33}$$

we note that the rank of the determinant for these functions is p.

If $u_1, \ldots, u_m$ satisfy (16.33) then obviously they also satisfy

$$F(r_1, \ldots, r_p) = F(0, \ldots, 0), \tag{16.34}$$

so that if a function depends on $r_1, \ldots, r_p$ then so will $u_1, \ldots, u_m$.

Consequently, what we really demand is the following: As well as (16.33), does there still exist

$$r_{p+1}(u_1, \ldots, u_m) = 0,$$

which is functionally independent of $r_1, \ldots, r_p$, so that when substituted into (16.32) the system is not identically 0? The answer is no. For if there are functions $u_1, \ldots, u_r$ from such a substitution into (16.32) then these functions would be functionally dependent, which then contradicts

$$\frac{\partial(u_1, \ldots, u_r)}{\partial(x_1, \ldots, x_r)} \not\equiv 0.$$

We have therefore arrived at the result mentioned at the beginning of Section 16.11.

The reader is invited to compare the two methods used to tackle the problem: Examine the source before considering the advantages when the finer points are added to enhance the arguments.

Exercise 16.16 Show that if

$$u_1 = \frac{x_1}{\sqrt{1 - x_1^2 - \cdots - x_n^2}}, \quad \ldots, \quad u_n = \frac{x_n}{\sqrt{1 - x_1^2 - \cdots - x_n^2}},$$

then

$$\frac{\partial(u_1, \ldots, u_n)}{\partial(x_1, \ldots, x_n)} = \frac{1}{(1 - x_1^2 - \cdots - x_n^2)^{1+n/2}}.$$

Exercise 16.17 Show that if

$$\begin{aligned}
x_1 &= \cos\phi_1,\\
x_2 &= \sin\phi_1\cos\phi_2,\\
x_3 &= \sin\phi_1\sin\phi_2\cos\phi_3,\\
&\dots\\
x_n &= \sin\phi_1\sin\phi_2\cdots\sin\phi_{n-1}\cos\phi_n,
\end{aligned}$$

then

$$\frac{D(x_1,\dots,x_n)}{D(\phi_1,\dots,\phi_n)} = (-1)^n \sin^n\phi_1 \sin^{n-1}\phi_2 \sin^{n-2}\phi_3\cdots\sin^2\phi_{n-1}\sin\phi_n.$$

16.13 The Wronskian

A collection of n functions $\phi_i(x)$, $i = 1, 2, \dots, n$, of a single variable are said to be linearly dependent if there are constants C_i, not all zero, such that

$$\sum_{i=1}^{n} C_i\phi_i(x) \equiv 0. \tag{16.35}$$

Suppose that these functions are differentiable $n-1$ times. Then, on differentiation,

$$\sum_{i=1}^{n} C_i\phi_i^{(j-1)}(x) \equiv 0, \quad j = 1, 2, \dots, n,$$

and so, after the elimination of $C_1, \dots, C_n$,

$$W(\phi_1,\dots,\phi_n) = \begin{vmatrix} \phi_1 & \cdots & \phi_n \\ \phi_1' & \cdots & \phi_n' \\ \vdots & & \vdots \\ \phi_1^{(n-1)} & \cdots & \phi_n^{(n-1)} \end{vmatrix} \equiv 0;$$

here $W(\phi_1,\dots,\phi_n)$ is called the Wronskian. The above is a necessary condition for linear dependence, but more hypotheses are required before we can assert the sufficiency of the condition.

We now suppose that $W(\phi_1,\dots,\phi_n) \equiv 0$, and that $W(\phi_1,\dots,\phi_{n-1}) \neq 0$ in a certain region. From

$$\phi_n^{(j-1)}(x) = \sum_{i=1}^{n-1} u_i(x)\phi_i^{(j-1)}(x), \quad j = 1, 2, \dots, n, \tag{16.36}$$

where $u_i(x)$ can be uniquely determined, we differentiate to give

$$\phi_n^{(j)}(x) = \sum_{i=1}^{n-1} u_i(x)\phi_i^{(j)}(x) + \sum_{i=1}^{n-1} u_i'(x)\phi_i^{(n-1)}(x), \tag{16.37}$$

so that

$$0 = \sum_{i=1}^{n-1} u_i'(x)\phi_i^{(n-1)}(x), \quad j = 1, 2, \dots, n-1.$$

Since $W(\phi_1, \dots, \phi_{n-1}) \neq 0$ we find that $u_i'(x) \equiv 0$, so that $u_i(x) = C_i$, and hence

$$\phi_n(x) = \sum_{i=1}^{n-1} C_i\phi_i(x).$$

We have therefore the following theorem.

Theorem 16.18 *Suppose that $W(\phi_1, \dots, \phi_n) \equiv 0$ in $a \le x \le b$, and suppose further that the Wronskians of all the $n-1$ functions are not simultaneously 0. Then $\phi_1, \dots, \phi_n$ are linearly dependent.*

We need to take note of the second part of the hypothesis, because otherwise there is the following counterexample due to Peano:

$$\begin{aligned} \phi_1(x) &= 2x^2, \quad \phi_2(x) = x^2 \quad (x \ge 0); \\ \phi_1(x) &= x^2, \quad \phi_2(x) = 2x^2 \quad (x \le 0); \end{aligned}$$

thus, for $x \ge 0$, we have $\phi_1(x) = 2\phi_2(x)$, whereas, for $x \le 0$, we have $\phi_1(x) = \frac{1}{2}\phi_2(x)$. The theorem does not apply in an interval containing $x = 0$ because, at this point, $W(\phi_1) = W(\phi_2) = 0$.

However, if we impose the condition of being analytic, then we have the following theorem.

Theorem 16.19 *A necessary and sufficient condition for the analytic functions $\phi_1(x), \dots, \phi_n(x)$ to be linearly dependent is that $W(\phi_1, \dots, \phi_n) \equiv 0$.*

17

Functions, sequences and series of square matrices

17.1 Similarity of square matrices

We do not go over the concept of elementary divisors again and will only describe briefly the notion of similarity and the Jordan normal form.

Definition 17.1 Two complex square matrices A, B of order n are said to be similar if there is a square matrix P with full rank such that

$$A = PBP^{-1},$$

and we denote this by writing $A \overset{E}{=} B$.

Clearly, (i) $A \overset{E}{=} A$; (ii) if $A \overset{E}{=} B$ then $B \overset{E}{=} A$; (iii) if $A \overset{E}{=} B$ and $B \overset{E}{=} C$ then $A \overset{E}{=} C$.

Theorem 17.2 *If the elementary divisors for A are*

$$(\lambda - \lambda_1)^{q_1},\ (\lambda - \lambda_2)^{q_2},\ \ldots,\ (\lambda - \lambda_s)^{q_s},$$

then A is similar to

$$\begin{pmatrix} J_1 & 0 & \cdots & 0 \\ 0 & J_2 & \cdots & 0 \\ \vdots & \vdots & & \vdots \\ 0 & 0 & \cdots & J_s \end{pmatrix},$$

where

$$J_i (= J_i^{(q_i)}) = \begin{pmatrix} \lambda_i & 1 & 0 & \cdots & 0 \\ 0 & \lambda_i & 1 & \cdots & 0 \\ \vdots & \vdots & \vdots & & \vdots \\ 0 & 0 & 0 & \cdots & 1 \\ 0 & 0 & 0 & \cdots & \lambda_i \end{pmatrix}.$$

This is called the Jordan normal (or canonical) form in textbooks.

For convenience, when $\lambda_i \neq 0$, we sometimes replace J_i by

$$\lambda_i J^{(q_i)}, \quad J = J^{(q_i)} = \begin{pmatrix} 1 & 1 & \cdots & 0 & 0 \\ 0 & 1 & \cdots & 0 & 0 \\ \vdots & \vdots & & & \vdots \\ 0 & 0 & \cdots & 1 & 1 \\ 0 & 0 & \cdots & 0 & 1 \end{pmatrix}.$$

We can do this because $\lambda_i J^{(q_i)} \stackrel{E}{=} J_i$, and thus

$$J_i = [1, \lambda_i, \lambda_i^2, \ldots, \lambda_i^{q_i - 1}] \lambda_i J [1, \lambda_i, \lambda_i^2, \ldots, \lambda_i^{q_i - 1}]^{-1}.$$

The similarity relation was given in the context of the complex field. In some applications we may wish that complex numbers do not come into the transformations involved, so that we are dealing with real numbers only.

Definition 17.3 Two real square matrices A, B of order n are said to be real similar if there is a real square matrix P with full rank such that

$$A = PBP^{-1}.$$

Obviously we can start all over again, but now that we have Theorem 17.2 we hope that it can be applied to the new situation.

Theorem 17.4 *If the elementary divisors for real A are*

$$\begin{aligned} &(\lambda - \lambda_1)^{q_1}, \ldots, (\lambda - \lambda_s)^{q_s}, \quad \lambda, \ldots, \lambda_s \text{ are real,} \\ &(\lambda - \rho_1 e^{i\theta_1})^{p_1}, \ldots, (\lambda - \rho_r e^{i\theta_r})^{p_r}, \\ &(\lambda - \rho_1 e^{-i\theta_1})^{p_1}, \ldots, (\lambda - \rho_r e^{-i\theta_r})^{p_r}, \end{aligned}$$

with r pairs of conjugate numbers $\rho_j e^{\pm i\theta_j}$, then A is real similar to

$$T = \begin{pmatrix} \lambda_1 J^{(q_1)} & 0 & 0 & \cdots & 0 & 0 & 0 & \cdots & 0 \\ 0 & \lambda_2 J^{(q_2)} & 0 & \cdots & 0 & 0 & 0 & \cdots & 0 \\ \vdots & \vdots & \vdots & & \vdots & \vdots & \vdots & & \vdots \\ 0 & 0 & 0 & \cdots & \lambda_s J^{(q_s)} & 0 & 0 & \cdots & 0 \\ 0 & 0 & 0 & \cdots & 0 & \rho_1 K_1 & 0 & \cdots & 0 \\ \vdots & \vdots & \vdots & & \vdots & \vdots & \vdots & & \vdots \\ 0 & 0 & 0 & \cdots & 0 & 0 & 0 & \cdots & \rho_r K_r \end{pmatrix},$$

where

$$K_j = \begin{pmatrix} \cos\theta_j J^{(p_j)} & \sin\theta_j J^{(p_j)} \\ -\sin\theta_j J^{(p_j)} & \cos\theta_j J^{(p_j)} \end{pmatrix},$$

and the normal form is real.

The reader should verify that the imaginary parts of the elementary divisors do form conjugate pairs.

Proof. From

$$\rho \begin{pmatrix} I & -iI \\ I & iI \end{pmatrix} \begin{pmatrix} \cos\theta J & \sin\theta J \\ -\sin\theta J & \cos\theta J \end{pmatrix} \begin{pmatrix} I & -iI \\ I & iI \end{pmatrix}^{-1}$$

$$= \frac{\rho}{2} \begin{pmatrix} e^{i\theta} J & -ie^{i\theta} J \\ e^{-i\theta} J & ie^{-i\theta} J \end{pmatrix} \begin{pmatrix} I & I \\ iI & -iI \end{pmatrix} = \begin{pmatrix} \rho e^{i\theta} J & 0 \\ 0 & \rho e^{-i\theta} J \end{pmatrix}$$

and Theorem 17.2, it follows that A and T are (complex) similar, so that there is a matrix Q such that

$$QAQ^{-1} = T. \tag{17.1}$$

Here A and T are both real, and we proceed to show that there must be a real Q satisfying (17.1). Write

$$Q = V + iW, \quad \det(V + iW) \neq 0,$$

so that

$$VA = TV, \quad WA = TW. \tag{17.2}$$

Consider

$$f(x) = \det(V + xW),$$

a polynomial in x with real coefficients. Since $f(i) \neq 0$, the coefficients for $f(x)$ are not all zero, and so there is a real x_0 such that $f(x_0) \neq 0$, and

$$P = V + x_0 W$$

will now serve as the required real Q, because from (17.2) we have

$$(V + x_0 W)A = T(V + x_0 W),$$

that is $PAP^{-1} = T$. □

17.2 Powers of a square matrix

Let

$$J = J^{(n)} = \begin{pmatrix} 1 & 1 & \cdots & 0 & 0 \\ 0 & 1 & \cdots & 0 & 0 \\ \vdots & \vdots & & \vdots & \vdots \\ 0 & 0 & \cdots & 1 & 1 \\ 0 & 0 & \cdots & 0 & 1 \end{pmatrix};$$

that is, $a_{ii} = a_{i,i+1} = 1$ and $a_{ij} = 0$ otherwise.

Theorem 17.5 *We have*

$$J^\ell = \begin{pmatrix} 1 & \ell & \binom{\ell}{2} & \cdots & \binom{\ell}{n-1} \\ 0 & 1 & \ell & \cdots & \binom{\ell}{n-1} \\ \vdots & \vdots & \vdots & & \vdots \\ 0 & 0 & 0 & \cdots & \ell \\ 0 & 0 & 0 & \cdots & 1 \end{pmatrix};$$

that is, $a_{ii}^{(\ell)} = 1$, $a_{i,i+1}^{(\ell)} = \binom{\ell}{1}$, $a_{i,i+2}^{(\ell)} = \binom{\ell}{2}, \ldots$, *and the entries below the main diagonal are all 0.*

Proof. From

$$\binom{\ell}{i} + \binom{\ell}{i-1} = \binom{\ell+1}{i}$$

and the induction hypothesis $J^{\ell+1} = J^\ell \cdot J$, we find that

$$a_{i,i+p}^{(\ell+1)} = \sum_{k=1}^{n} a_{ik}^{(\ell)} a_{k,i+p} = a_{i,i+p}^{(\ell)} + a_{i,i+p-1}^{(\ell)} = \binom{\ell}{p} + \binom{\ell}{p-1} = \binom{\ell+1}{p},$$

which delivers the inductive step for the proof. □

Theorem 17.6 *We have*

$$\begin{pmatrix} \cos\theta J & \sin\theta J \\ -\sin\theta J & \cos\theta J \end{pmatrix}^\ell = \begin{pmatrix} \cos\ell\theta J^\ell & \sin\ell\theta J^\ell \\ -\sin\ell\theta J^\ell & \cos\ell\theta J^\ell \end{pmatrix}.$$

17.3 The limit of powers of a square matrix

We first consider the condition under which the limit

$$\lim_{\ell\to\infty} M^\ell \tag{17.3}$$

of powers of a square matrix will exist. As $\ell \to \infty$, by the matrix M^ℓ tending to M_0 we mean that the element $m_{ij}^{(\ell)}$ lying on the ith row and the jth column in M^ℓ tends to the corresponding element $m_{ij}^{(0)}$ in M_0.

The question of whether the limit (17.3) exists for M is the same as that for a matrix to which it is similar. We therefore first examine the situation when the limit is applied to a Jordan normal form.

(1) Suppose that there is an eigenvalue λ with absolute value $|\lambda| > 1$. Since $\lim_{\ell\to\infty} \lambda^\ell$ does not exist, it follows that the limit (17.3) does not exist if M has such an eigenvalue.

(2) If $|\lambda| = 1$ and $\lambda \neq 1$ then, because

$$\lim_{\ell\to\infty} e^{i\ell\alpha}, \quad 0 < \alpha < 2\pi,$$

does not exist, the limit (17.3) still does not exist if M has such an eigenvalue.

(3) If 1 is an eigenvalue, but is not the value of a simple elementary divisor, then from

$$\begin{pmatrix} 1 & 1 \\ 0 & 1 \end{pmatrix}^{\ell} = \begin{pmatrix} 1 & \ell \\ 0 & 1 \end{pmatrix} \to \begin{pmatrix} 1 & \infty \\ 0 & 1 \end{pmatrix}$$

we see that the limit (17.3) still does not exist.

Summarising, for the limit (17.3) to exist, the matrix M has to have eigenvalues with absolute values less than 1, or if the eigenvalue is 1 then it is also the value of a simple elementary divisor.

(4) Consider the situation when $|\lambda| < 1$. From

$$(\lambda J)^{\ell} = \lambda^{\ell} \begin{pmatrix} 1 & \binom{\ell}{1} & \binom{\ell}{2} & \binom{\ell}{3} & \cdots \\ & 1 & \binom{\ell}{1} & \binom{\ell}{2} & \cdots \\ & & 1 & \binom{\ell}{1} & \cdots \\ & & & 1 & \cdots \\ & & & & \ddots \end{pmatrix},$$

together with $\lambda^{\ell} \to 0$, $\ell\lambda^{\ell} \to 0$, $\binom{\ell}{2}\lambda^{\ell} \to 0, \ldots$, we see that $(\lambda J)^{\ell} \to 0$ (when $\lambda = 0$, we have $J_{\lambda}^{n} = 0$). We therefore have the following theorem.

Theorem 17.7 *Suppose that the absolute values of all the eigenvalues of A do not exceed* 1, *that if the absolute value is* 1 *then the eigenvalue itself is* 1, *and that the corresponding elementary divisor is simple. Then*

$$\lim_{\ell \to \infty} A^{\ell} = A_0$$

exists; moreover, such a condition for its existence is also necessary.

The rank of the limit A_0 is equal to the multiplicity of the eigenvalue 1 for A, and $A_0^2 = A_0$ (an idempotent matrix).

Theorem 17.8 *Under the hypothesis of Theorem 17.7, if* 1 *is a simple root 'for A' then*

$$\lim_{\ell \to \infty} A^{\ell} = u'v, \qquad vu' = 1;$$

here the vector u', or u, is the eigenvector belonging to 1, *and the representation is unique apart from a constant factor.*

Proof. By Theorem 17.7, the matrix A_0 has rank 1, so that it can be written as $u'v$, where u, v are (row) vectors. Also, by idempotency,

$$u'v = u'vu'v = (vu')(u'v),$$

so that

$$(1 - vu')u'v = 0.$$

Since $u'v \neq 0$ we deduce that $vu' = 1$. □

We proceed to show that v is a (row) eigenvector of A belonging to 1. Suppose that v_1 is an eigenvector of A belonging to 1, so that $v_1 A = v_1$, and hence $v_1 A^\ell = v_1$ when applied ℓ times. Letting $\ell \to \infty$, we find that $v_1 = v_1(u'v) = (v_1 u')v$, which shows that v differs from v_1 only by the factor $v_1 u'$, and so $vA = v$. The same method shows that $Au' = u'$.

Theorem 17.9 (Limit theorem) *Under the hypothesis of Theorem 17.8 write, for any non-zero vector x,*

$$x_\ell = xA^\ell.$$

Then the limit of x_ℓ differs from v only by a constant factor. In other words, regardless of the initial vector, the direction of the limit vector is invariant.

Proof. This follows at once from $\lim_{\ell\to\infty} xA^\ell = x(u'v) = (xu')v$. □

17.4 Power series

We consider the power series

$$\sum_{\ell=0}^{\infty} a_\ell X^\ell, \tag{17.4}$$

where a_ℓ is a complex number and X is a complex square matrix of order n. We suppose that the series

$$\sum_{\ell=0}^{\infty} a_\ell x^\ell \tag{17.5}$$

has the radius of convergence ρ, so that (17.5) converges for $|x| < \rho$.

The problem of convergence amounts to that when X is a Jordan normal form.

(1) We already know that, when $|x| < \rho$, all the series

$$\sum a_\ell x^\ell, \quad \sum a_\ell \binom{\ell}{1} x^\ell, \quad \sum a_\ell \binom{\ell}{2} x^\ell, \quad \ldots, \sum a_\ell \binom{\ell}{n} x^\ell, \quad \ldots$$

are convergent. Therefore, if all the eigenvalues of X have absolute values less than ρ, then (17.4) is convergent.

(2) If X has an eigenvalue with absolute value exceeding ρ, then (17.4) is divergent.

(3) If all the eigenvalues of X have absolute values $\le \rho$, and there is an eigenvalue $\lambda_0 = \rho e^{i\theta}$, with the corresponding elementary divisors $(\lambda - \lambda_0)^{\ell_1}, \ldots, (\lambda - \lambda_0)^{\ell_s}$,

$\ell_1 \geq \ell_2 \geq \cdots \geq \ell_s$, then we need to examine the series

$$f(\lambda_0) = \sum_{\ell=0}^{\infty} a_\ell (\rho e^{i\theta})^\ell,$$
$$f'(\lambda_0) = \sum_{\ell=1}^{\infty} \ell a_\ell (\rho e^{i\theta})^{\ell-1},$$
$$\cdots$$
$$f^{(\ell_1)}(\lambda_0) = \sum_{\ell=\ell_1}^{\infty} \ell(\ell-1)\cdots(\ell-\ell_1+1) a_\ell (\rho e^{i\theta})^{\ell-\ell_1}$$

for convergence. If they are all convergent, then (17.4) is convergent, and if one of them is divergent then (17.4) is divergent.

17.5 Examples of power series

17.5.1 Exponential series

The exponential series

$$\exp X = e^X = \sum_{\ell=0}^{\infty} \frac{1}{\ell!} X^\ell$$

is convergent for any square matrices X.

If $XY = YX$ then

$$\begin{aligned} e^X \cdot e^Y &= \sum_{\ell=0}^{\infty} \sum_{\ell=0}^{\infty} \frac{X^\ell Y^m}{\ell!\, m!} = \sum_{n=0}^{\infty} \frac{1}{n!} \sum_{\ell=0}^{n} \frac{n!}{\ell!(n-\ell)!} X^\ell Y^{n-\ell} \\ &= \sum_{n=0}^{\infty} \frac{(X+Y)^n}{n!} = e^{X+Y}. \end{aligned}$$

However, if $XY \neq YX$ then this formula cannot be established.

If X has the eigenvalues $\lambda_1, \ldots, \lambda_n$ then e^X has the eigenvalues $e^{\lambda_1}, \ldots, e^{\lambda_n}$. Moreover, from (if $\lambda \neq 0$ take $J = J^{(s)}$)

$$\begin{aligned} \exp \lambda J &= \sum_{\ell=0}^{\infty} \frac{\lambda^\ell}{\ell!} J^\ell \\ &= \sum_{\ell=0}^{\infty} \frac{\lambda^\ell}{\ell!} \begin{pmatrix} 1 & \binom{\ell}{1} & \binom{\ell}{2} & \cdots & \binom{\ell}{s-1} \\ & 1 & \binom{\ell}{1} & \cdots & \binom{\ell}{s-2} \\ & & 1 & \cdots & \binom{\ell}{s-3} \\ & & & \ddots & \vdots \\ & & & & 1 \end{pmatrix} = \begin{pmatrix} e^\lambda & \lambda e^\lambda & \frac{\lambda^2}{2!} e^\lambda & \cdots & \frac{\lambda^{s-1}}{(s-1)!} e^\lambda \\ & e^\lambda & \lambda e^\lambda & \cdots & \frac{\lambda^{s-2}}{(s-2)!} e^\lambda \\ & & e^\lambda & \cdots & \frac{\lambda^{s-3}}{(s-3)!} e^\lambda \\ & & & \ddots & \vdots \\ & & & & e^\lambda \end{pmatrix}, \end{aligned}$$

we see that the corresponding elementary divisor is equal to $(x - e^\lambda)^s$.

When $\lambda = 0$, the expression $\exp J_0$ is actually a polynomial, and the elementary divisor is $(x - 1)^s$.

17.5.2 Trigonometric series

Similarly we can define

$$\cos X = \frac{1}{2}(e^{iX} + e^{-iX}) = \sum_{\ell=0}^{\infty}(-1)^{\ell}\frac{X^{2\ell}}{(2\ell)!},$$

$$\sin X = \frac{1}{2i}(e^{iX} - e^{-iX}) = \sum_{\ell=0}^{\infty}(-1)^{\ell}\frac{X^{2\ell+1}}{(2\ell+1)!},$$

and it is not difficult to prove directly that

$$\cos^2 X + \sin^2 X = I.$$

We can also define

$$\cosh X = \frac{1}{2}(e^{X} + e^{-X}) = \sum_{\ell=0}^{\infty}\frac{X^{2\ell}}{(2\ell)!},$$

$$\sinh X = \frac{1}{2}(e^{X} - e^{-X}) = \sum_{\ell=0}^{\infty}\frac{X^{2\ell+1}}{(2\ell+1)!}.$$

17.5.3 Logarithmic series

The logarithm function is defined by

$$\log(I + X) = X - \frac{X^2}{2} + \frac{X^3}{3} - \cdots,$$

where the series is convergent when all the eigenvalues of X have absolute values less than 1. It is not difficult to prove that, under such a condition,

$$e^{\log(I+X)} = I + X.$$

Hint: First develop various identities from $e^{\log(1+x)} = 1 + x$.

17.5.4 Binomial expansion

The binomial expansion

$$(I + X)^{\alpha} = I + \alpha X + \frac{\alpha(\alpha-1)}{2}X^2 + \frac{\alpha(\alpha-1)(\alpha-2)}{3!}X^3 + \cdots$$

is also convergent when all the eigenvalues of X have absolute values less than 1.

17.6 Method of successive substitutions

The very simple formula

$$(I - A)^{-1} = I + A + A^2 + \cdots \tag{17.6}$$

is the source of the method of successive substitutions in computational mathematics. When all the eigenvalues for A have absolute values less than 1, the series (17.6) converges.

The following is the gist of the method of successive substitutions: In order to solve the system

$$x(I - A) = b, \tag{17.7}$$

we simply take $x_0 = b$ and successively evaluate

$$x_i = b + x_{i-1}A, \quad i = 1, 2, \ldots .$$

Then $\lim_{i\to\infty} x_i$ is the required solution to (17.7).

The proof is very simple – it follows easily by induction that

$$x_i = b(I + A + A^2 + \cdots + A^i),$$

so that, by (17.6),

$$\lim_{i\to\infty} x_i = b(I - A)^{-1}.$$

What criteria need to be met to determine whether all the eigenvalues for A should have absolute values less than 1? One condition that can be established easily is as follows: If

$$\sum_{i=1}^{n} |a_{ij}| \le q < 1 \tag{17.8}$$

then the series (17.6) converges.

Proof. Write $A^\ell = \left(a_{ij}^{(\ell)}\right)$, and we prove by induction that $\sum_{i=1}^{n} |a_{ij}^{(\ell)}| \le q^\ell$; in fact, the inductive step is

$$\sum_{i=1}^{n} |a_{ij}^{(\ell+1)}| = \sum_{i=1}^{n} \left| \sum_{k=1}^{n} a_{ik}^{(\ell)} a_{kj} \right| \le \sum_{k=1}^{n} \left(\sum_{i=1}^{n} |a_{ik}^{(\ell)}| \right) |a_{kj}| \le q^\ell \sum_{k=1}^{n} |a_{kj}| \le q^{\ell+1}.$$

Therefore

$$\sum_{\ell=0}^{\infty} |a_{ij}^{(\ell)}|$$

converges. □

Table 17.1.

A	$T_1 = I + A$
A^2	$T_2 = T_1 + A^2 T_1$
$A^4 = (A^2)^2$	$T_3 = T_2 + A^4 T_2$
$A^8 = (A^4)^2$	$T_4 = T_3 + A^8 T_3$
...	...

Remark 1 It is not only convergence that has been established, but also the argument delivers an error estimate, namely that, after m steps,

$$\sum_{\ell=m+1}^{\infty} \left|a_{ij}^{(\ell)}\right| \leq \sum_{\ell=m+1}^{\infty} q^{\ell} = \frac{q^{m+1}}{1-q}.$$

Remark 2 A general square matrix A with all its eigenvalues having absolute value less than 1 can, after the manipulations below, be made to satisfy condition (17.8). We can find a diagonal matrix $\Lambda = [\lambda_1, \ldots, \lambda_n]$ so that $\Lambda A \Lambda^{-1}$ satisfies condition (17.8). The existence of such $\lambda_1, \ldots, \lambda_n$ can be established as follows: Let ρ denote one of the eigenvalues with the largest modulus, and let $(\lambda_1, \ldots, \lambda_n)$ be (one of) its corresponding eigenvectors, so that

$$\sum_{i=1}^{n} \lambda_i a_{ij} = \rho \lambda_j.$$

Writing the matrix form as $[\lambda_1, \ldots, \lambda_n](a_{ij})[\lambda_1, \ldots, \lambda_n]^{-1} = (b_{ij})$, we find that

$$\sum_{i=1}^{n} b_{ij} = \rho.$$

When performing a general computation there is no need to find the actual values for $\lambda_1, \ldots, \lambda_n$ because any $\lambda_1, \ldots, \lambda_n$ which can deliver $\sum_{i=1}^{n} b_{ij} < 1$ will suffice.

Remark 3 Naturally the method of successive substitutions can also be used to compute the inverse of $I - A$ by setting

$$B_0 = I, \quad B_\ell = I + AB_{\ell-1}.$$

Indeed, the process can be speeded up by applying the following formulae, as outlined in Table 17.1.

17.7 The exponential function

Theorem 17.10 *We have*

$$\lim_{n\to\infty}\left(I+\frac{1}{n}A\right)^n = e^A.$$

Proof. From

$$n\log\left(I+\frac{1}{n}A\right) = A - \frac{1}{2n}A^2 + \frac{1}{3n^2}A^3 + \cdots$$

we deduce that

$$\lim_{n\to\infty} n\log\left(I+\frac{1}{n}A\right) = A,$$

and hence the required result follows. □

Theorem 17.11 *A square matrix with full rank must be a value taken by the exponential function.*

Proof. (1) We first consider the special case

$$J = \begin{pmatrix} 1 & 1 & \cdots & 0 & 0 \\ 0 & 1 & \cdots & 0 & 0 \\ \vdots & \vdots & & \vdots & \vdots \\ 0 & 0 & \cdots & 1 & 1 \\ 0 & 0 & \cdots & 0 & 1 \end{pmatrix} = I + L \quad (J = J^{(q)}).$$

Since $L^q = 0$ we have

$$\log J = \log(I+L) = L - \frac{1}{2}L^2 + \frac{1}{3}L^3 + \cdots + \frac{(-1)^q}{q-1}L^{q-1},$$

so that

$$\log J = \begin{pmatrix} 0 & 1 & -\frac{1}{2} & \frac{1}{3} & \cdots & \frac{(-1)^q}{q-1} \\ 0 & 0 & 1 & -\frac{1}{2} & \cdots & \frac{(-1)^{q-1}}{q-2} \\ 0 & 0 & 0 & 1 & \cdots & \frac{(-1)^{q-2}}{q-3} \\ \vdots & \vdots & \vdots & \vdots & & \vdots \\ 0 & 0 & 0 & 0 & \cdots & 1 \\ 0 & 0 & 0 & 0 & \cdots & 0 \end{pmatrix} = K,$$

that is $J = e^K$.

(2) It can then be deduced that $\lambda J = e^{\log \lambda I + K}$.

(3) The deduction of the result for any square matrix with full rank is now not difficult. □

Theorem 17.12 *A necessary and sufficient condition for all the elements of e^A to be non-negative is that*

$$A = (a_{ij}), \quad a_{ij} \geq 0, \quad i \neq j.$$

Proof. (1) Suppose first that $a_{ij} > 0$. Then from

$$e^{At} = I + At + \frac{1}{2}A^2t^2 + \cdots, \qquad t > 0,$$

we see that, for sufficiently small t, the diagonal elements have the same sign as that of the corresponding elements of I, and those off the diagonal have the same sign as that of the corresponding elements of At. Therefore, for small t, all the elements of e^{At} are positive, and from

$$e^A = \left(e^{A/n}\right)^n$$

it now follows that all the elements of e^A are positive.

(2) The required result when $a_{ij} \geq 0$ follows from a continuity argument.

(3) The same argument can be applied to the necessity part. Thus, if $a_{ij} < 0$ for some $i \neq j$, then the corresponding element b_{ij} of e^{At} is negative when t is sufficiently small. The theorem is proved. □

17.8 The derivative of a square matrix of a single variable

Suppose that $A(t) = (a_{ij}(t))$ is a square matrix of order n, depending on the variable t. We define the derivative by

$$\frac{d}{dt}A(t) = \lim_{h \to 0} \frac{1}{h}(A(t+h) - A(t)) = \left(\frac{d}{dt}a_{ij}(t)\right),$$

when the right-hand side exists. The derivative has the following properties:

(i) $\dfrac{d}{dt}(A + B) = \dfrac{d}{dt}A + \dfrac{d}{dt}B;$

(ii) $\dfrac{d}{dt}(AB) = A\dfrac{dB}{dt} + \dfrac{dA}{dt}B.$

If A has full rank, then (taking $B = A^{-1}$) we deduce at once that

(iii) $\dfrac{d}{dt}(A^{-1}) = -A^{-1}\dfrac{dA}{dt}A^{-1}.$

Note, however, that dA/dt and A may not commute, so that

$$\frac{dA^2}{dt} = A\frac{dA}{dt} + \frac{dA}{dt}A, \qquad \frac{dA^3}{dt} = A^2\frac{dA}{dt} + A\frac{dA}{dt}A + \frac{dA}{dt}A^2, \ldots .$$

Inside the region of convergence for the power series

$$F(t) = A_0 + A_1 t + A_2 t^2 + A_3 t^3 + \cdots$$

we may differentiate term by term to give

$$\frac{dF(t)}{dt} = A_1 + 2A_2 t + 3A_3 t^2 + \cdots .$$

For example, we have

(iv) $\frac{d}{dt}e^{At} = Ae^{At}$; and

(v)

$$\begin{aligned}\frac{d}{dt}\log(I + At) &= \frac{d}{dt}\Big(At - \frac{1}{2}A^2t^2 + \frac{1}{3}A^3t^3 - \cdots\Big) \\ &= A(I - At + A^2t^2 - \cdots) = A(I + At)^{-1}\end{aligned}$$

and similarly for the higher derivatives.

We define the integral of a matrix by

$$\int_c A(t)dt = \int_c (a_{ij}(t))dt = \Big(\int_c a_{ij}(t)dt\Big).$$

The integral can be that for a real variable over an interval, or it can be along a curve on the complex plane. From (ii) we deduce the formula for integration by parts:

$$\int_c A\frac{dB}{dt}\,dt + \int_c \frac{dA}{dt}B\,dt = (AB)\Big|_c .$$

17.9 Power series for the Jordan normal form

Let

$$J_\lambda = \begin{pmatrix} \lambda & 1 & 0 & \cdots & 0 \\ 0 & \lambda & 1 & \cdots & 0 \\ \vdots & \vdots & \vdots & & \vdots \\ 0 & 0 & 0 & \cdots & 1 \\ 0 & 0 & 0 & \cdots & \lambda \end{pmatrix};$$

then

$$J_\lambda^p = \begin{pmatrix} \lambda^p & p\lambda^{p-1} & \frac{p(p-1)}{2!}\lambda^{p-2} & \cdots \\ 0 & \lambda^p & p\lambda^{-1} & \cdots \\ \vdots & \vdots & \vdots & \\ 0 & 0 & 0 & \cdots \end{pmatrix} = \begin{pmatrix} \lambda^p & (\lambda^p)' & \frac{1}{2!}(\lambda^p)'' & \cdots \\ 0 & \lambda^p & (\lambda^p)' & \cdots \\ \vdots & \vdots & \vdots & \\ 0 & 0 & 0 & \cdots \end{pmatrix}.$$

Proof. We use induction, together with $(\lambda \cdot \lambda^p)^{(t)} = \lambda(\lambda^p)^{(t)} + t(\lambda^p)^{(t-1)}$, and the required result follows from

$$\frac{1}{t!}\lambda(\lambda^p)^{(t)} + \frac{1}{(t-1)!}(\lambda^p)^{(t-1)} = \frac{1}{t!}(\lambda^{p+1})^{(t)}. \qquad \square$$

Let

$$f(x) = \sum_{p=0}^{\infty} a_p x^p;$$

then

$$f(J_\lambda) = \sum_{p=0}^{\infty} a_p \begin{pmatrix} \lambda^p & (\lambda^p)' & \frac{1}{2!}(\lambda^p)'' & \cdots \\ 0 & \lambda^p & (\lambda^p)' & \cdots \\ 0 & 0 & \lambda^p & \cdots \\ \vdots & \vdots & \vdots & \end{pmatrix} = \begin{pmatrix} f(\lambda) & f'(\lambda) & \frac{1}{2!}f''(\lambda) & \cdots \\ 0 & f(\lambda) & f'(\lambda) & \cdots \\ 0 & 0 & f(\lambda) & \cdots \\ \vdots & \vdots & \vdots & \end{pmatrix}.$$

If

$$X = P(J_{\lambda_1} \dotplus J_{\lambda_s} \dotplus \cdots \dotplus J_{\lambda_s})P^{-1},$$

then

$$f(X) = P(K_{\lambda_1} \dotplus K_{\lambda_s} \dotplus \cdots \dotplus K_{\lambda_s})P^{-1},$$

where

$$K_{\lambda_i} = \begin{pmatrix} f(\lambda_i) & f'(\lambda_i) & \frac{1}{2!}f''(\lambda_i) & \cdots \\ 0 & f(\lambda_i) & f'(\lambda_i) & \cdots \\ 0 & 0 & f(\lambda_i) & \cdots \\ \vdots & \vdots & \vdots & \end{pmatrix}.$$

In particular, we have

$$e^J\lambda = \begin{pmatrix} e^\lambda & e^\lambda & \frac{1}{2}e^\lambda & \cdots \\ 0 & e^\lambda & e^\lambda & \cdots \\ 0 & 0 & e^\lambda & \cdots \\ \vdots & \vdots & \vdots & \end{pmatrix},$$

and

$$e^{tJ}\lambda = \begin{pmatrix} e^{t\lambda} & te^{t\lambda} & \frac{1}{2}t^2e^{t\lambda} & \cdots \\ 0 & e^{t\lambda} & te^{t\lambda} & \cdots \\ 0 & 0 & e^{t\lambda} & \cdots \\ \vdots & \vdots & \vdots & \end{pmatrix}.$$

17.10 A number raised to a matrix power

We define

$$x^A = e^{A \log x} = \sum_{n=0}^{\infty} \frac{(A \log x)^n}{n!};$$

we note that $x^A \cdot x^B$ may not be the same as x^{A+B}. We have the following rule for differentiation:

$$\begin{aligned} \frac{d}{dx} x^A &= \frac{d}{dx} e^{A \log x} \frac{d}{dx} \sum_{\ell=0}^{\infty} \frac{(A \log x)^\ell}{\ell!} \\ &= \frac{1}{x} \sum_{\ell=0}^{\infty} \frac{(A \log x)^{\ell-1} A}{(\ell-1)!} = A x^{A-I} (= x^{A-I} A), \end{aligned}$$

where in the derivation we have used

$$x^{-I} = e^{-I \log x} = \sum_{n=0}^{\infty} \frac{(-I \log x)^n}{n!} = \sum_{n=0}^{\infty} \frac{(-\log x)^n}{n!} I = x^{-1} I.$$

We have

$$x^{J\lambda} = e^{J\lambda \log x} = \begin{pmatrix} x^\lambda & (\log x) x^\lambda & \frac{(\log x)^2}{2} x^\lambda & \cdots \\ 0 & x^\lambda & (\log x) x^\lambda & \cdots \\ \vdots & \vdots & \vdots & \end{pmatrix},$$

and if λ has a positive real part then

$$\lim_{x \to 0} x^{J\lambda} = 0.$$

We have the following theorem.

Theorem 17.13 *If all the eigenvalues of A have positive real parts, then*

$$\lim_{x \to 0} x^A = 0.$$

Proof. If $A = PBP^{-1}$ then $x^A = P x^B P^{-1}$. The problem is thus reduced to one for a Jordan normal form, the result for which has been established above. □

17.11 The matrix e^X for some special X

Consider $n = 2$ first, and take $X = -X'$, that is

$$X = \begin{pmatrix} 0 & \lambda \\ -\lambda & 0 \end{pmatrix}.$$

We then have $X^2 = -\lambda^2 I$, so that

$$e^X = \sum_{\ell=0}^{\infty} \frac{(-\lambda^2)^\ell}{(2\ell)!} I + \sum_{\ell=0}^{\infty} \frac{(-\lambda^2)^\ell \lambda}{(2\ell+1)!} \begin{pmatrix} 0 & 1 \\ -1 & 0 \end{pmatrix} = \begin{pmatrix} \cos\lambda & \sin\lambda \\ -\sin\lambda & \cos\lambda \end{pmatrix},$$

a matrix representing a rotation on the plane.

We next take $n = 3$. A rotation in three dimensional space has the matrix representation

$$\begin{pmatrix} \cos\alpha\cos\beta\cos\gamma - \sin\alpha\sin\gamma & \cos\alpha\cos\beta\sin\gamma + \sin\alpha\cos\gamma & -\cos\alpha\sin\beta \\ -\sin\alpha\cos\beta\cos\gamma - \cos\alpha\sin\gamma & -\sin\alpha\cos\beta\sin\gamma + \sin\alpha\cos\gamma & \sin\alpha\sin\beta \\ \sin\beta\cos\gamma & \sin\beta\sin\gamma & \cos\beta \end{pmatrix}$$

$$= \begin{pmatrix} \cos\alpha & \sin\alpha & 0 \\ -\sin\alpha & \cos\alpha & 0 \\ 0 & 0 & 1 \end{pmatrix} \begin{pmatrix} \cos\beta & 0 & -\sin\beta \\ 0 & 1 & 0 \\ \sin\beta & 0 & \cos\beta \end{pmatrix} \begin{pmatrix} \cos\gamma & \sin\gamma & 0 \\ -\sin\gamma & \cos\gamma & 0 \\ 0 & 0 & 1 \end{pmatrix},$$

which is the same as

$$\exp\begin{pmatrix} 0 & \alpha & 0 \\ -\alpha & 0 & 0 \\ 0 & 0 & 0 \end{pmatrix} \exp\begin{pmatrix} 0 & 0 & -\beta \\ 0 & 0 & 0 \\ \beta & 0 & 0 \end{pmatrix} \exp\begin{pmatrix} 0 & \gamma & 0 \\ -\gamma & 0 & 0 \\ 0 & 0 & 0 \end{pmatrix}.$$

Conversely, starting with a general skew-symmetric matrix of order 3,

$$X = \theta \begin{pmatrix} 0 & -n & m \\ n & 0 & -\ell \\ -m & \ell & 0 \end{pmatrix}, \quad \ell^2 + m^2 + n^2 = 1,$$

we have

$$X^2 = -\theta^2(I - u'u),$$

where $u(\ell, m, n)$, $uu' = \ell^2 + m^2 + n^2 = 1$, and from $(I - u'u)^2 = I - u'u$ we find that

$$X^{2k} = (-\theta^2)^k (I - u'u), \quad k = 1, 2, \dots,$$

and

$$uX = 0.$$

Therefore

$$\begin{aligned} e^X &= I + \sum_{k=1}^{\infty} \frac{(-\theta^2)^k}{(2k)!}(I - u'u) + \sum_{k=0}^{\infty} \frac{(-\theta^2)^k}{(2k+1)!}(I - u'u)X \\ &= \cos\theta I + (1 - \cos\theta)u'u + \sin\theta X \\ &= \begin{pmatrix} \cos\theta + \ell^2(1-\cos\theta) & -n\sin\theta + m\ell(1-\cos\theta) & m\sin\theta + n\ell(1-\cos\theta) \\ n\sin\theta + m\ell(1-\cos\theta) & \cos\theta + m^2(1-\cos\theta) & -\ell\sin\theta + mn(1-\cos\theta) \\ -m\sin\theta + n\ell(1-\cos\theta) & \ell\sin\theta + mn(1-\cos\theta) & \cos\theta + n^2(1-\cos\theta) \end{pmatrix}. \end{aligned}$$

This is the matrix representing a rotation by an angle θ along an axis with the direction cosines (ℓ, m, n).

More generally, if $X = -X'$ then

$$e^X \cdot (e^X)' = e^X \cdot e^{X'} = e^X \cdot e^{-X} = I.$$

Thus the square matrix

$$\Gamma = e^X$$

satisfies

$$\Gamma\Gamma' = I.$$

In other words, the exponential of a skew-symmetric matrix is an orthogonal matrix. We call X the 'infinitely small' square matrix of the orthogonal matrix Γ. The infinitely small matrix of Γ^ℓ is then ℓX, and we find that

$$\lim_{t\to 0} \frac{I - \Gamma^t}{t} = -X.$$

Conversely, if we start from $\Gamma\Gamma' = I$ and set $\Gamma = e^X$, then we deduce from

$$e^X = \Gamma = (\Gamma')^{-1} = e^{-X'}$$

that $X = -X'$.

Suppose that $\operatorname{tr} X = 0$, that is the sum of the eigenvalues for X is 0. Then the product of the eigenvalues for e^X is 1. Thus, if $\operatorname{tr} X = 0$ then

$$|e^X| = 1.$$

Let $n = 2\ell$ and

$$F = \begin{pmatrix} 0^{(\ell)} & I \\ -I & 0 \end{pmatrix}.$$

Suppose that X satisfies

$$FX + X'F = 0.$$

Then $FX^m = (-1)^m (X')^m F$, so that

$$Fe^X = \sum_{m=0}^{\infty} F \frac{X^m}{m!} = \sum_{m=0}^{\infty} \frac{(-X')^m}{m!} F = e^{-X'} F.$$

Setting $P = e^X$, we then have

$$P'FP = F.$$

If

$$X = -\bar{X}',$$

then

$$e^X(e^{\bar{X}})' = e^X e^{-X} = I.$$

17.12 The relationship between e^X and X

Given a square matrix X, we may set

$$M(X) = e^X = \sum_{\ell=0}^{\infty} \frac{X^\ell}{\ell!}$$

to give e^X. On the other hand, if M is non-singular, then we can represent it by e^X, and we have the relationship

$$\lim_{t\to 0} \frac{M^t - I}{t} = \lim_{t\to 0} \frac{e^{Xt} - I}{t} = X.$$

We then have the following:

(i) M^{-1} corresponds to $-X$;
(ii) M' corresponds to X';
(iii) $\bar{M}$ corresponds to $\bar{X}$.

The property of being orthogonal for M, that is $M' = M^{-1}$, is then transferred to the property of being skew-symmetric for X, that is $X' = -X$. Being unitary for M, that is $\bar{M}' = M^{-1}$, is transferred to being Hermitian for iX.

The most important relationship is the following: For

$$e^{sX} \cdot e^{tY} \cdot (e^{sX})^{-1} \cdot (e^{ty})^{-1},$$

we find that

$$\begin{aligned}
&\lim_{t\to 0}\lim_{s\to 0} \frac{e^{sX} \cdot e^{tY} \cdot e^{-sX} \cdot e^{-tY} - I}{st} \\
&= \lim_{t\to 0}\lim_{s\to 0} \frac{e^{sX} - I + e^{sX}e^{tY}(e^{-sX} - I)e^{-tY}}{st} \\
&= \lim_{t\to 0} \frac{X - e^{tY}Xe^{-tY}}{t} \\
&= \lim_{t\to 0} \frac{X(I - e^{-tY}) + (I - e^{tY})Xe^{-tY}}{t} \\
&= XY - YX.
\end{aligned}$$

18

Difference and differential equations with constant coefficients

18.1 Difference equations

By a difference equation with a single variable we mean the equation

$$F(y(x+n\delta), y(x+(n-1)\delta), \ldots, y(x+\delta), y(x), x) = 0, \tag{18.1}$$

where δ is a fixed number, and we need to solve for the function $y = y(x)$.

When studying (18.1) we may, without loss of generality, assume that $\delta = 1$. Suppose that a solution

$$y(x+n) = \phi(y(x+(n-1)), \ldots, y(x+1), y(x), x) = 0 \tag{18.2}$$

is at our disposal. If we already know the values for

$$y(0) = c_0, \ y(1) = c_1, \ \ldots, \ y(n-1) = c_{n-1}, \tag{18.3}$$

then we can apply formula (18.2) successively to evaluate the values

$$y(n), \ y(n+1), \ y(n+2), \ \ldots.$$

We call n the order of such a difference equation, and the values in (18.3) the accompanying initial values. Note that, with the given initial values, we can determine all the values for y at the natural numbers n, but not at other places; however, if we know the values for

$$y(\tau), \ y(\tau+1), \ \ldots, \ y(\tau+n-1), \tag{18.4}$$

then we can evaluate

$$y(\tau+\ell), \quad \ell = 0, 1, 2, 3 \ldots.$$

Only when y is specified by

$$y(\tau+p) = \phi_p(\tau), \quad p = 0, 1, 2, \ldots, n-1, \tag{18.5}$$

where ϕ_p are known functions in the interval $0 \le \tau < 1$, can $y(x)$ be determined in the whole line $x \ge 0$.

If (18.2) can be used to deliver $y(x)$, then we may also determine its value for $x \le 0$.

Difference equations have become much more prevalent and useful since the development of computers. First, when tackling a differential equation, we often change it to a difference equation, which is then solved by some method. Next, we may wish to apply a difference equation method directly to a problem, without passing through the process of having to deal with a differential equation, especially when the initial data are given in a discrete form. Finally, if the answers that we seek are meant to be discrete anyway then we may deal with the problem by algebraic means. For this reason, in the following few sections, we first deal with the computational aspect of linear equations, then difference equations, and then the theory of differential equations.

We first consider the shape of the difference equation corresponding to the ordinary differential equation

$$y^{(n)}(x) = \phi(y^{(n-1)}, \dots, y'(x), y(x), x), \tag{18.6}$$

when the derivatives $y^{(i)}(x)$ are replaced by their ratios before limits are taken; thus we replace $y'(x), y''(x), \dots$ by

$$\frac{y(x+\delta)-y(x)}{\delta}, \frac{y(x+2\delta)-2y(x+\delta)+y(x)}{\delta^2}, \dots .$$

The resulting difference equation depends on δ and, speaking generally, the solution to (18.6) may then be obtained by letting $\delta \to 0$. We present a couple of examples to illustrate the idea.

Example 18.1 Solve the homogeneous linear difference equation

$$y(x+1) = ay(x), \quad y(0) = c.$$

By induction it is easy to see that the solution is

$$y(n) = a^n c, \quad n = 0, 1, 2, \dots . \tag{18.7}$$

Consider the homogeneous linear differential equation

$$y'(x) = \lambda y(x), \quad y(0) = k, \tag{18.8}$$

for which the corresponding difference equation takes the form

$$y(x+\delta) - y(x) = \delta\lambda y(x),$$

that is

$$y(x+\delta) = (1+\delta\lambda)y(x),$$

which then has the solution

$$y(n\delta) = k(1+\lambda\delta)^n.$$

Writing $x = n\delta$, we have

$$y(x) = k\left(1+\frac{x\lambda}{n}\right)^n,$$

so that, on letting $n \to \infty$,

$$y(x) = ke^{\lambda x},$$

which is the solution to the differential equation.

Example 18.2 Solve the difference equation

$$y(x+1) = ay(x) + f(x), \quad y(0) = c.$$

Let

$$y(x) = a^x c(x), \quad c(0) = c;$$

then

$$c(x+1) = c(x) + a^{-x-1} f(x).$$

This then gives

$$c(\ell) = c + \sum_{p=1}^{\ell} a^{-p} f(p-1),$$

and hence

$$y(\ell) = ca^{\ell} + \sum_{p=1}^{\ell} a^{\ell-p} f(p-1).$$

Consider the differential equation

$$\frac{dy}{dx} = \lambda y(x) + g(x), \quad y(0) = c,$$

corresponding to which we may form the difference equation

$$y(x+\delta) = (1+\delta\lambda)y(x) + \delta g(x),$$

which has the solution

$$y(\ell\delta) = c(1+\delta\lambda)^{\ell} + \delta \sum_{p=1}^{\ell} (1+\delta\lambda)^{\ell-p} g((p-1)\delta).$$

Write $x = \ell\delta$, and we have

$$y(x) = c\left(1+\frac{x\lambda}{\ell}\right)^{\ell} + \frac{x}{\ell}\sum_{p=1}^{\ell}\left(1+\frac{x\lambda}{\ell}\right)^{\ell-p} g((p-1)x/\ell)$$

$$= c\left(1+\frac{x\lambda}{\ell}\right)^{\ell} + \left(1+\frac{x\lambda}{\ell}\right)^{\ell}\cdot\frac{x}{\ell}\sum_{p=1}^{\ell} g\left(\frac{(p-1)x}{\ell}\right)e^{-p\log(1+x\lambda/\ell)}.$$

Letting $\ell \to \infty$, the expression $y(x)$ here tends to

$$e^{\lambda x}\left(c + \int_0^x g(t)e^{-\lambda t}\,dt\right),$$

which is the solution to the differential equation.

Note. Although we often use

$$\frac{y(x+\delta)-y(x)}{\delta}$$

to replace $y'(x)$, sometimes we may want to use

$$\frac{y(x+\delta)-y(x-\delta)}{2\delta}$$

instead. Indeed, generally speaking, the latter is a better choice because from (the mean-value theorem)

$$y(x+\delta) = y(x) + \delta y'(x) + \frac{\delta^2}{2}y''(x) + \frac{\delta^3}{6}y'''(x+\theta\delta),$$

the approximation

$$\frac{y(x+\delta)-y(x-\delta)}{2\delta} - y'(x) \approx \frac{\delta^2}{6}y'''(x+\theta\delta)$$

should be superior to

$$\frac{y(x+\delta)-y(x)}{\delta} - y'(x) \approx \frac{\delta}{2}y''(x+\theta'\delta).$$

18.2 Linear difference equations with constant coefficients – the method of generating functions

We try to solve the nth order difference equation

$$a_0 y(x+n) + a_1 y(x+n-1) + \cdots + a_n y(x) = g(x), \tag{18.9}$$

with the initial conditions

$$y(0) = c_0, \ \ldots, \ y(n-1) = c_{n-1}. \tag{18.10}$$

Consider first the case when $g(x) = 0$. Take the generating function

$$F(x) = \sum_{\ell=0}^{\infty} y(\ell)x^{\ell} \tag{18.11}$$

and multiply it by

$$\Phi(x) = \sum_{i=0}^{n} a_i x^i, \tag{18.12}$$

giving

$$\begin{aligned} F(x)\Phi(x) &= \sum_{\ell=0}^{\infty}\sum_{i=0}^{n} y(\ell)a_i x^{i+\ell} \\ &= \sum_{p=0}^{\infty} x^p \sum_{i\le\min(n,p)} a_i y(p-i). \end{aligned} \tag{18.13}$$

When $p \ge n$, we have

$$\sum_{i=0}^{n} a_i y(p-i) = 0,$$

and, when $p < n$,

$$\sum_{i=0}^{p} a_i y(p-i) = \sum_{i=0}^{p} a_i c_{p-i} = b_p,$$

say, so that, by (18.13),

$$F(x)\Phi(x) = B(x), \tag{18.14}$$

where

$$B(x) = \sum_{p=0}^{n-1} b_p x^p,$$

and hence

$$F(x) = \frac{B(x)}{\Phi(x)}. \tag{18.15}$$

Although the coefficients $y(\ell)$ for x^{ℓ} in the power series expansion of $F(x)$ are given by

$$y(\ell) = \frac{1}{\ell!}\frac{d^{\ell}}{dx^{\ell}}F(x)\bigg|_{x=0},$$

it is more convenient to use the following partial fraction method for their determination: Write

$$\sum_{\nu=0}^{n} a_\nu x^\nu = a_0 \prod_{\nu=1}^{s} (1 - \lambda_\nu x)^{\ell_\nu},$$

and set

$$\frac{B(x)}{\Phi(x)} = \sum_{\nu=1}^{s} \Big(\frac{a_{\nu 1}}{1 - \lambda_\nu x} + \cdots + \frac{a_{\nu \ell_\nu}}{(1 - \lambda_\nu x)^{\ell_\nu}} \Big);$$

then from

$$\frac{1}{(1 - \lambda_\nu x)^p} = \sum_{m=0}^{\infty} \frac{p(p+1)\cdots(p+m-1)}{m!} (\lambda_\nu x)^m,$$

we have

$$y(m) = \sum_{\nu=1}^{s} \sum_{p=1}^{\ell_\nu} a_{\nu p} \frac{p(p+1)\cdots(p+m-1)}{m!} \lambda_\nu^m. \tag{18.16}$$

Now let $g(x)$ be general, and not assumed to be zero. In (18.13) we have, for $p \geq n$,

$$\sum_{i=0}^{n} a_i y(p - i) = g(p - n),$$

so that

$$F(x)\Phi(x) = B(x) + \sum_{p=n}^{\infty} g(p - n) x^p = B(x) + x^n \sum_{p=0}^{\infty} g(p) x^p.$$

Setting

$$a(x) = \sum_{p=0}^{\infty} g(p) x^p$$

to be the generating function for $g(x)$, we then have

$$F(x) = (B(x) + x^n (a(x)) / \Phi(x). \tag{18.17}$$

Naturally we can also use partial fractions to deal with $1/\Phi(x)$, and then use the expansion to find the coefficients for x^ℓ.

18.3 Second method – order reduction method

Let

$$\Phi(p) = \sum_{\nu=0}^{n-1} b_\nu y(n-1-\nu+p).$$

We then have

$$\begin{aligned}\Phi(p+1) - \lambda\Phi(p) &= \sum_{\nu=0}^{n-1} b_\nu y(n-\nu+p) - \lambda \sum_{\nu=1}^{n} b_{\nu-1} y(n-\nu+p) \\ &= \sum_{\nu=0}^{n} (b_\nu - \lambda b_{\nu-1}) y(n-\nu+p) \ \ (b_{-1} = b_n = 0) \\ &= \sum_{\nu=0}^{n} a_\nu y(n-\nu+p),\end{aligned}$$

where

$$a_\nu = b_\nu - \lambda b_{\nu-1},$$

and we construct the polynomial

$$\sum_{\nu=0}^{n} a_\nu x^\nu = \sum_{\nu=0}^{n-1} b_\nu x^\nu - \lambda \sum_{\nu=1}^{n} b_{\nu-1} x^\nu = (1-\lambda x) \sum_{\nu=0}^{n-1} b_\nu x^\nu. \tag{18.18}$$

Thus, if (18.18) is established, then the difference equation

$$\sum_{\nu=0}^{n} a_\nu y(n+p-\nu) = g(p) \tag{18.19}$$

can be written as

$$\Phi(p+1) - \lambda\Phi(p) = g(p), \quad \Phi(0) = \sum_{\nu=0}^{n-1} b_\nu c_{n-1-\nu}. \tag{18.20}$$

The difference equation of the form (18.20) has been solved in Section 18.1, and with the solution $\Phi(p)$, the difference equation

$$\sum_{\nu=0}^{n-1} b_\nu y(n-1-\nu+p) = \Phi(p),$$

which has order $n-1$, can now be solved.

18.4 Third method – Laplace transform method

We solve the difference equation

$$y(t+1) = ay(t) + g(t), \tag{18.21}$$

with the accompanying condition that

$$y(t) = \phi(t), \quad 0 \le t \le 1.$$

We take the Laplace transform

$$x(s) = \int_0^\infty e^{-st} y(t)dt, \tag{18.22}$$

which gives

$$\int_0^\infty e^{-st} y(t+1)dt = e^s \int_1^\infty e^{-st} y(t)dt = e^s \int_0^\infty e^{-st} y(t)dt - e^s \int_0^1 e^{-st}\phi(t)dt$$
$$= e^s x(s) - e^s \int_0^1 e^{-st}\phi(t)dt.$$

On the other hand, by (18.21),

$$\int_0^\infty e^{-st} y(t+1)dt = a\int_0^\infty e^{-st} y(t)dt + \int_0^\infty e^{-st} g(t)dt = ax(s) + \int_0^\infty e^{-st} g(t)dt.$$

Therefore

$$e^s x(s) - e^s \int_0^1 e^{-st}\phi(t)dt = ax(s) + \int_0^\infty e^{-st} g(t)dt,$$

that is

$$x(s) = \frac{1}{e^s - a}\left[e^s \int_0^1 e^{-st}\phi(t)dt + \int_0^\infty e^{-st} g(t)dt\right],$$

and $y(t)$ can then be found from the inversion formula.

Note. The reader should note that, although we speak of three different methods (and the fourth method below), the real substance in the methods is essentially the same: we are just dancing around the difference relationship within a difference equation.

18.5 Fourth method – matrix method

Consider the system of linear difference equations

$$y_i(t+1) = \sum_{j=1}^{n} y_j(t)a_{ji} + b_i(t), \quad i = 1, 2, \ldots, n. \tag{18.23}$$

The nth order difference equation

$$a_0 y(t+n) + a_1 y(t+n-1) + \cdots + a_n y(t) = g(t)$$

can be considered to be a special case of (18.23):

$$\begin{aligned}
y_1(t+1) &= y_2(t), \quad & y_1(t) &= y(t),\\
y_2(t+1) &= y_3(t), \quad & y_2(t) &= y(t+1),\\
&\cdots & y_3(t) &= y(t+2),\\
y_{n-1}(t+1) &= y_n(t), & &\cdots\\
y_n(t+1) &= -\frac{1}{a_0}(a_1 y_n(t) + \cdots + a_n y_1(t)) + \frac{1}{a_0} g(t). & &
\end{aligned}$$

We rewrite (18.23) in the matrix form

$$y(t+1) = y(t)A + b(t), \qquad y(0) = c;$$

here $y(t)$ is a vector, $b(t)$ is a known vector, A is a square matrix and c is a constant vector.

We first consider the case $b(t) = 0$. By induction, it is easy to see (compare with Section 18.1) that

$$y(\ell) = cA^{\ell}.$$

For a general $b(t)$ we then find that

$$y(\ell) = cA^{\ell} + \sum_{p=1}^{\ell} b(p-1)A^{\ell-p}.$$

When written in matrix form, a rather complicated problem can sometimes be transformed to one that is simpler.

18.6 Linear differential equations with constant coefficients

We consider the differential equation

$$\frac{dx}{dt} = xA + b(t), \qquad x(0) = c. \tag{18.24}$$

Here x is an unknown vector, $A(= A^{(n)})$ is a square matrix with constant elements and $b(t)$ is a known vector. Let

$$X(t) = e^{At},$$

so that

$$\frac{dX}{dt} = XA, \qquad X(0) = I,$$

and, with the constant vector c, then

$$\frac{d(cX)}{dt} = (cX)A, \quad cX(0) = c.$$

This shows that $x = cX$ is the solution to (18.24) when $b(t) = 0$.

If $b(t) \neq 0$, then we let

$$x = y(t)e^{At}, \quad y(0) = c,$$

so that

$$\frac{dy}{dt}e^{At} + ye^{At}A = ye^{At}A + b(t),$$

that is

$$\frac{dy}{dt} = b(t)e^{-At},$$

giving

$$y(t) - c = \int_0^t b(\tau)e^{-A\tau}\,d\tau,$$

that is

$$x(t) = \left(c + \int_0^t b(\tau)e^{-A\tau}\,d\tau\right)e^{At}.$$

Exercise 18.3 Replace dX/dt by

$$\frac{X(t+\delta) - X(t)}{\delta},$$

and then use the results in the preceding sections to deduce the results in the present section.

18.7 The motion of a particle over the surface of the Earth

We take, as an example, the motion of a particle in space not far from the surface of the Earth, and subject to its rotation.

Let the particle have mass m and velocity v with respect to the Earth, and let ω denote the angular velocity of the Earth. The particle is then subject to the Coriolis effect, with the force being equal to $2m\omega \times v$, where $\times$ denotes the operation of vector multiplication. There is also the force mg due to the gravitational pull from the Earth, so that the equation of motion for the particle is given by the differential equation

$$\frac{dv}{dt} = g - 2\omega \times v. \tag{18.25}$$

From the definition for a vector product, we have

$$\omega \times v = (\omega_1, \omega_2, \omega_3) \times (v_1, v_2, v_3) = (\omega_2 v_3 - \omega_3 v_2, \omega_3 v_1 - \omega_1 v_3, \omega_1 v_2 - \omega_2 v_1),$$

so that

$$2\omega \times v = -vA, \tag{18.26}$$

where

$$A = -2 \begin{pmatrix} 0 & \omega_3 & -\omega_2 \\ -\omega_3 & 0 & \omega_1 \\ \omega_2 & -\omega_1 & 0 \end{pmatrix}.$$

Substituting (18.26) into (18.25) we then have

$$\frac{dv}{dt} = vA + g, \tag{18.27}$$

giving the solution

$$v = v_0 e^{At} + g \int_0^t e^{A(t-s)}\, ds = v_0 e^{At} + g \int_0^t e^{A\tau}\, d\tau, \tag{18.28}$$

where $v_0 = v|_{t=0}$. Integrating again, we have the radial vector for the motion given by

$$r = r_0 + v_0 \int_0^t e^{A\tau}\, d\tau + g \int_0^t dr \int_0^\tau e^{A\tau_1}\, d\tau_1, \tag{18.29}$$

where $r_0 = r|_{t=0}$, and we ought to point out that we cannot simply write $\int_0^t e^{A\tau}\, d\tau = (e^{At} - I)A^{-1}$ because A^{-1} does not exist.

Let us consider A^2 and A^3. First, we have

$$A^2 = -4(\omega\omega' I - \omega'\omega),$$

and from $\omega A = 0$ we also have

$$A^3 = -4\omega\omega' A + 4\omega'\omega A = -4\omega\omega' A.$$

Therefore

$$A^{2\ell+1} = (-4\omega\omega')^\ell A, \quad \ell = 0, 1, 2, \ldots$$

and

$$A^{2\ell} = (-4\omega\omega')^{\ell-1} A^2 = (-4\omega\omega')^\ell I + 4(-4\omega\omega')^{\ell-1}\omega'\omega, \quad \ell = 1, 2, \ldots .$$

We then have

$$
\begin{aligned}
e^{At} &= I + \sum_{\ell=0}^{\infty} \frac{(-4\omega\omega')^{\ell} t^{2\ell+1}}{(2\ell+1)!} A + \sum_{\ell=1}^{\infty} \frac{t^{2\ell}}{(2\ell)!} [(-4\omega\omega')^{\ell} I + 4(-4\omega\omega')^{\ell-1} \omega'\omega] \\
&= \cos(2\sqrt{\omega\omega'}t) I + \frac{\sin(2\sqrt{\omega\omega'}t)}{2\sqrt{\omega\omega'}} A - \frac{1}{\omega\omega'} (\cos(2\sqrt{\omega\omega'}t) - 1)\omega'\omega,
\end{aligned}
$$

and so

$$
\begin{aligned}
\int_0^t e^{A\tau} d\tau &= \frac{\sin(2\sqrt{\omega\omega'}t)}{2\sqrt{\omega\omega'}} I + \frac{1 - \cos 2\sqrt{\omega\omega'}t}{4\omega\omega'} A - \frac{1}{\omega\omega'} \Big(\frac{\sin 2\sqrt{\omega\omega'}t}{2\sqrt{\omega\omega'}} - t \Big) \omega'\omega, \\
\int_0^t d\tau \int_0^{\tau} e^{A\tau_1} d\tau_1 &= \frac{1 - \cos 2\sqrt{\omega\omega'}t}{4\omega\omega'} I + \frac{t - (\sin 2\sqrt{\omega\omega'}t / 2\sqrt{\omega\omega'})}{4\omega\omega'} A \\
&\quad - \frac{1}{\omega\omega'} \Big(\frac{1 - \cos 2\sqrt{\omega\omega'}t}{4\omega\omega'} - \frac{1}{2} t^2 \Big) \omega'\omega.
\end{aligned}
$$

Therefore

$$
\begin{aligned}
r = r_0 &+ v_0 \Big[\frac{\sin(2\sqrt{\omega\omega'}t)}{2\sqrt{\omega\omega'}} I + \frac{1 - \cos 2\sqrt{\omega\omega'}t}{4\omega\omega'} A - \frac{1}{\omega\omega'} \Big(\frac{\sin 2\sqrt{\omega\omega'}t}{2\sqrt{\omega\omega'}} - t \Big) \omega'\omega \Big] \\
&+ g \Big[\frac{1 - \cos 2\sqrt{\omega\omega'}t}{4\omega\omega'} I + \frac{2\sqrt{\omega\omega'}t - \sin 2\sqrt{\omega\omega'}t}{8(\sqrt{\omega\omega'})^3} A \\
&- \frac{1}{\omega\omega'} \Big(\frac{1 - \cos 2\sqrt{\omega\omega'}t}{4\omega\omega'} - \frac{1}{2} t^2 \Big) \omega'\omega \Big].
\end{aligned}
$$

From $xA = -2\omega \times x$, we have

$$
\begin{aligned}
r = r_0 &+ \frac{\sin(2\sqrt{\omega\omega'}t)}{2\sqrt{\omega\omega'}} v_0 - \frac{1 - \cos 2\sqrt{\omega\omega'}t}{2\omega\omega'} \omega \times v_0 - \frac{1}{\omega\omega'} \Big(\frac{\sin 2\sqrt{\omega\omega'}t}{2\sqrt{\omega\omega'}} - t \Big) v_0 \cdot \omega' \cdot \omega \\
&+ \frac{1 - \cos 2\sqrt{\omega\omega'}t}{4\omega\omega'} g - \frac{2\sqrt{\omega\omega'}t - \sin 2\sqrt{\omega\omega'}t}{4(\sqrt{\omega\omega'})^3} \omega \times g \\
&- \frac{1}{\omega\omega'} \Big(\frac{1 - \cos 2\sqrt{\omega\omega'}t}{4\omega\omega'} - \frac{1}{2} t^2 \Big) g\omega'\omega.
\end{aligned}
$$

When $v_0 = 0$,

$$
\begin{aligned}
r = r_0 &+ \frac{t^2}{2} g - \frac{2\sqrt{\omega\omega'}t - \sin 2\sqrt{\omega\omega'}t}{4(\omega\omega')^{3/2}} \omega \times g \\
&+ \frac{1 - \cos 2\sqrt{\omega\omega'}t - 2\omega\omega' t^2}{4(\omega\omega')^{3/2}} \Big(- \frac{g\omega\omega'}{\sqrt{\omega\omega'}} + \sqrt{\omega\omega'} g \Big).
\end{aligned}
$$

The first two terms on the right-hand side correspond to the initial position and the gravitational pull, while the third term represents the swing toward the east on the plane orthogonal to the perpendicular at the meridian; the final term represents the difference between the pull along the said perpendicular axis.

For a very small angular velocity (for example, for the rotation of the Earth we have $\sqrt{\omega\omega'} \approx 7.3\times10^{-5}$ rad s^{-1}) we may ignore the second and higher order terms in the calculation of $\sqrt{\omega\omega'}$. Thus, the difference in position due to the influence of the Earth has the following approximation formula:

$$-\omega \times \left(t^2 v_0 + \frac{1}{3} t^3 g\right).$$

18.8 Oscillations

In the study of oscillations we frequently encounter the system of second order differential equations,

$$\frac{d^2x}{dt^2} + xA = 0, \tag{18.30}$$

where $x = (x_1, \ldots, x_n)$, $A = A^{(n)}$.

Let

$$\frac{dx}{dt} = y,$$

so that (18.30) becomes

$$\left(\frac{dx}{dt}, \frac{dy}{dt}\right) = (x, y) \begin{pmatrix} 0 & -A \\ I & 0 \end{pmatrix}.$$

From

$$\begin{pmatrix} I & 0 \\ 0 & \sqrt{A} \end{pmatrix} \begin{pmatrix} 0 & -A \\ I & 0 \end{pmatrix} \begin{pmatrix} I & 0 \\ 0 & \sqrt{A} \end{pmatrix}^{-1} = \begin{pmatrix} 0 & -\sqrt{A} \\ \sqrt{A} & 0 \end{pmatrix}$$

and

$$\begin{pmatrix} I & iI \\ I & -iI \end{pmatrix} \begin{pmatrix} 0 & -I \\ I & 0 \end{pmatrix} \begin{pmatrix} I & iI \\ I & -iI \end{pmatrix}^{-1} = \begin{pmatrix} iI & -I \\ -iI & -I \end{pmatrix} \begin{pmatrix} -iI & -iI \\ -I & I \end{pmatrix} \frac{1}{-2i} = \begin{pmatrix} iI & 0 \\ 0 & -iI \end{pmatrix},$$

if we let

$$\Gamma = \begin{pmatrix} I & iI \\ I & -iI \end{pmatrix} \begin{pmatrix} I & 0 \\ 0 & \sqrt{A} \end{pmatrix} = \begin{pmatrix} I & i\sqrt{A} \\ I & -i\sqrt{A} \end{pmatrix}$$

then

$$\begin{pmatrix} 0 & -\sqrt{A} \\ I & 0 \end{pmatrix} = \Gamma^{-1} \begin{pmatrix} i\sqrt{A} & 0 \\ 0 & -i\sqrt{A} \end{pmatrix} \Gamma,$$

that is

$$\left(\frac{dx}{dt}, \frac{dy}{dt}\right)\Gamma^{-1} = (x, y)\Gamma^{-1}\begin{pmatrix} i\sqrt{A} & 0 \\ 0 & -i\sqrt{A} \end{pmatrix}.$$

Let

$$(x, y)\Gamma^{-1} = (u, v);$$

then

$$\left(\frac{du}{dt}, \frac{dv}{dt}\right)\Gamma^{-1} = (u, v)\Gamma^{-1}\begin{pmatrix} i\sqrt{A} & 0 \\ 0 & -i\sqrt{A} \end{pmatrix},$$

that is

$$u = ce^{i\sqrt{A}t}, \quad v = de^{-i\sqrt{A}t},$$

so that we have

$$(x, y) = (u, v)\Gamma = (c, d)\begin{pmatrix} e^{i\sqrt{A}t} & 0 \\ 0 & e^{-i\sqrt{A}t} \end{pmatrix}\begin{pmatrix} I & i\sqrt{A} \\ I & -i\sqrt{A} \end{pmatrix},$$

that is

$$\begin{cases} x = ce^{i\sqrt{A}t} + de^{-i\sqrt{A}t}, \\ y = i(ce^{i\sqrt{A}t} - de^{-i\sqrt{A}t})\sqrt{A}. \end{cases} \tag{18.31}$$

The initial condition for the system (18.30) when $t = 0$ is

$$x = x_0, \quad \left(\frac{dx}{dt}\right)_0 = v_0 \ (= y_0),$$

so that, from (18.31),

$$\begin{aligned} x_0 &= c + d, \\ v_0 &= i(c - d)\sqrt{A}, \quad c - d = -iv_0A^{-1/2}, \end{aligned}$$

that is

$$c = \frac{1}{2}(x_0 - iv_0A^{-1/2}), \quad d = \frac{1}{2}(x_0 + iv_0A^{-1/2}).$$

Substituting these into (18.31) we have

$$\begin{aligned} x &= \frac{1}{2}(x_0 - iv_0A^{-1/2})e^{i\sqrt{A}t} + \frac{1}{2}(x_0 + iv_0A^{-1/2})e^{-i\sqrt{A}t} \\ &= x_0 \cos\sqrt{A}t + v_0A^{-1/2}\sin\sqrt{A}t, \end{aligned} \tag{18.32}$$

which is the solution to the system (18.30).

The inhomogeneous system

$$\frac{d^2x}{dt^2} + xA = f(t)$$

can be written as

$$\left(\frac{dx}{dt}, \frac{dy}{dt}\right) = (x, y)\begin{pmatrix} 0 & -A \\ I & 0 \end{pmatrix} + (0, f(t)),$$

and it is then not difficult to derive the following:

$$x = x_0(\cos\sqrt{A}t) + v_0(\sqrt{A})^{-1}\sin\sqrt{A}t + \left[\int_0^t f(\tau)\sin(\sqrt{A}(t-\tau))d\tau\right](\sqrt{A})^{-1}.$$

If the initial time is $t = t_0$ then the solution becomes

$$x = x_0(\cos\sqrt{A}(t-t_0)) + v_0(\sqrt{A})^{-1}\sin\sqrt{A}(t-t_0) + \left[\int_{t_0}^t f(\tau)\sin(\sqrt{A}(t-\tau))d\tau\right](\sqrt{A})^{-1}.$$

Exercise 18.4 Deal with the case when $f(t) = h\sin(pt+\alpha)$.

Note. We have a glimpse of the use of the matrix function e^{At}. However, the more important aspect is that it serves as the kernel of the linear functional equations. After a suitable generalisation it becomes the foundation of the theory of semi-groups; see, for example, Functional Analysis and Semi-Groups by E. Hille and R. S. Phillips (American Mathematical Society, 1975).

18.9 Absolute values of matrices

There are several ways to introduce the absolute value of a matrix. We first introduce the one that follows: Let $A = (a_{ij})$, $1 \le i \le m$, $1 \le j \le n$, and we define the absolute value of this matrix by

$$||A|| = \sum_{i=1}^{m}\sum_{j=1}^{n}|a_{ij}|. \tag{18.33}$$

It is not difficult to show that

$$||A+B|| \le ||A|| + ||B||, \tag{18.34}$$

$$||AB|| \le ||A||\,||B||; \tag{18.35}$$

thus, the latter (when $B = (b_{ij})$ is of order $n\times\ell$) follows from

$$\sum_{i=1}^{m}\sum_{k=1}^{\ell}\left|\sum_{j=1}^{n}a_{ij}b_{jk}\right| \le \sum_{i=1}^{m}\sum_{k=1}^{\ell}\sum_{j=1}^{n}|a_{ij}b_{jk}| \le \sum_{i=1}^{m}\sum_{j=1}^{n}|a_{ij}|\cdot\sum_{j_1=1}^{n}\sum_{k=1}^{\ell}|b_{j_1k}|;$$

Table 18.1.

t	$A(t)$	$b(t)$	$x(t+1)$
0	$A(0)$	$b(0)$	$x(1) = cA(0) + b(0)$
1	$A(1)$	$b(1)$	$x(2) = x(1)A(1) + b(1)$
2	$A(2)$	$b(2)$	$x(3) = x(2)A(2) + b(2)$
$\vdots$	$\vdots$	$\vdots$	$\vdots$

and we also have

$$\left|\left| \int A(t)dt \right|\right| \leq \int ||A(t)||dt. \tag{18.36}$$

If

$$\sum_{\ell=0}^{\infty} ||A_\ell|| < \infty,$$

then $\sum_{\ell=0}^{\infty} A_\ell$ represents $m \times n$ absolutely convergent series.

Cauchy's criterion: A sequence of matrices A_ℓ must converge to a certain matrix if, to every $\epsilon > 0$, there exists N such that

$$||A_p - A_q|| < \epsilon, \quad \text{for all} \quad p, q > N.$$

18.10 Existence and uniqueness problems for a linear differential equation

We first consider the linear difference equation with variable coefficients

$$x(t+1) = x(t)A(t) + b(t), \quad x(0) = c.$$

The evaluation of the solution can be dealt with as in Table 18.1, so that there is no problem concerning the existence and the uniqueness of a solution. For this reason, a discrete problem should, if possible, not be set up as a continuous one, because there is then no need to deal with such irrelevant difficulties.

The situation for a differential equation, however, is quite different, and we have to deal with such problems.

Theorem 18.5 *Suppose that the square matrix $A(t)$ is a continuous function of $t \geq 0$. Then the vector differential equation*

$$\frac{dx}{dt} = xA(t), \quad x(0) = c, \tag{18.37}$$

has a unique solution, which can be written in the form

$$x = cX(t), \quad t \geq 0, \tag{18.38}$$

where $X(t)$ is the unique solution to the differential equation

$$\frac{dX}{dt} = XA(t), \quad X(0) = I. \tag{18.39}$$

Furthermore, the solution to

$$\frac{dx}{dt} = xA(t) + g(t), \quad x(0) = c,$$

is given by

$$x = cX(t) + \int_0^t g(\tau)X^{-1}(\tau)d\tau X(t).$$

Proof. (1) We first use the method of successive approximations to show that (18.39) does have a solution. Replace the differential equation by the integral equation

$$X(t) = I + \int_0^t X(s)A(s)ds, \tag{18.40}$$

and form a sequence of square matrices $\{X_\ell\}$ by setting $X_0 = I$, and

$$X_{\ell+1} = I + \int_0^t X_\ell A(s)ds, \quad \ell = 0, 1, \ldots, \tag{18.41}$$

so that

$$X_{\ell+1} - X_\ell = \int_0^t (X_\ell - X_{\ell-1})A(s)ds, \quad \ell = 1, 2, \ldots.$$

Let

$$m = \max_{0 \le s \le t_1} ||A(s)||;$$

then, when $0 \le t \le t_1$,

$$\begin{aligned} ||X_{\ell+1} - X_\ell|| &\le \left|\left| \int_0^t (X_\ell - X_{\ell-1})A(s)ds \right|\right| \le \int_0^t ||A(s)||\, ||X_\ell - X_{\ell-1}||ds \\ &\le m \int_0^t ||X_\ell - X_{\ell-1}||ds. \end{aligned} \tag{18.42}$$

Within this region we have

$$||X_1 - X_0|| \le \int_0^t ||A(s)||ds \le mt,$$

and an inductive argument based on (18.42) can be used to show that

$$||X_{\ell+1} - X_\ell|| \le \frac{m^{\ell+1}t^{\ell+1}}{(\ell+1)!},$$

so that the series

$$\sum_{\ell=0}^{\infty}(X_{\ell+1} - X_{\ell})$$

converges uniformly in $0 \leq t \leq t_1$. When $p \to \infty$, the sequence of square matrices

$$I + \sum_{\ell=0}^{p}(X_{\ell+1} - X_{\ell}) = X_{p+1}$$

converges uniformly to $X(t)$. Taking limits on both sides of (18.41), we find that $X(t)$ satisfies the integral equation (18.40), and therefore also satisfies the differential equation (18.39).

(2) The existence of the solution to the differential equation (18.37) now follows from (18.39) together with an appropriately chosen vector c.

(3) Uniqueness. Suppose that Y is another solution to (18.39), so that

$$X - Y = \int_0^t (X(s) - Y(s))A(s)ds, \tag{18.43}$$

and hence

$$||X - Y|| \leq \int_0^t ||X(s) - Y(s)||\,||A(s)||ds. \tag{18.44}$$

Since Y is differentiable, it is continuous, so that

$$m_1 = \max_{0 \leq t \leq t_1} ||X - Y|| \tag{18.45}$$

exists and, by (18.44),

$$||X - Y|| \leq m_1 \int_0^t ||A(s)||ds. \tag{18.46}$$

When substituted back into (18.44), we have

$$\begin{aligned} ||X - Y|| &\leq m_1 \int_0^t ||A(s)||ds \int_0^s ||A(s_1)||ds_1 \\ &= m_1 \iint_{0 \leq s_1 \leq s \leq t} ||A(s)||\,||A(s_1)||ds\,ds_1, \end{aligned}$$

and, substituting back into (18.44) again,

$$||X - Y|| \leq m_1 \iiint_{0 \leq s_2 \leq s_1 \leq s \leq t} ||A(s)||\,||A(s_1)||\,||A(s_2)||ds\,ds_1\,ds_2,$$

and so on. From

$$\int \cdots \int_{0 \le s_\ell \le s_{\ell-1} \le \cdots \le s_1 \le t} f(s_1) \cdots f(s_\ell) ds_1 \cdots ds_\ell$$
$$= \frac{1}{\ell!} \int_0^t \cdots \int_0^t f(s_1) \cdots f(s_\ell) ds_1 \cdots ds_\ell = \frac{1}{\ell!} \Big(\int_0^t f(s) ds \Big)^\ell,$$

we now have

$$||X - Y|| \le \frac{m_1}{\ell!} \Big(\int_0^t f(s) ds \Big)^\ell,$$

and hence $X = Y$, by letting $\ell \to \infty$. □

The proof of the final part of the theorem is left to the reader, who should contemplate whether the structure of the argument has been used before.

We can write explicitly the following:

$$X_0 = I,$$
$$X_1 = I + \int_0^t A(s) ds,$$
$$X_2 = I + \int_0^t A(s) ds + \int_0^t A(s) ds \int_0^s A(s_1) ds_1$$
$$= I + \int_0^t A(s) ds + \iint_{0 \le s_1 \le s \le t} A(s) A(s_1)\, ds\, ds_1,$$

and we remark that $A(s)A(a_1)$ may not be the same as $A(s_1)A(s)$, so that the double integral cannot be replaced by $\frac{1}{2}\Big(\int_0^t A(s) ds \Big)^2$;

$$X_3 = I + \int_0^t A(s) ds + \iint_{0 \le s_1 \le s \le t} A(s) A(s_1)\, ds\, ds_1$$
$$+ \iiint_{0 \le s_2 \le s_1 \le s \le t} A(s) A(s_1) A(s_2)\, ds\, ds_1 ds_2,$$

so that X has the series representation

$$X = I + \int_0^t A(s) ds + \iint_{0 \le s_1 \le s \le t} A(s) A(s_1)\, ds\, ds_1 + \cdots$$
$$+ \int \cdots \int_{0 \le s_\ell \le \cdots \le s_1 \le t} A(s_1) A(s_2) \cdots A(s_\ell)\, ds_1 \cdots ds_\ell + \cdots .$$

If the condition $X(0) = I$ is changed to $X(t_0) = I$, then

$$X = I + \int_{t_0}^t A(s) ds + \iint_{t_0 \le s_1 \le s \le t} A(s) A(s_1)\, ds\, ds_1 + \cdots$$
$$+ \int \cdots \int_{t_0 \le s_\ell \le \cdots \le s_1 \le t} A(s_1) A(s_2) \cdots A(s_\ell)\, ds_1 \cdots ds_\ell + \cdots . \tag{18.47}$$

Definition 18.6 The matrix X defined by (18.47) is called the iterated integral of the matrix function $A(t)$ from t_0 to t, and will be denoted by

$$r_{t_0}^t(A) = \widetilde{\int_{t_0}^t} (I + A(t))dt.$$

18.11 Iterated integrals

We examine the notion of an iterated integral from the approximations arising from a difference equation. Consider the difference equation

$$X(t+\delta) - X(t) = X(t)\delta A,$$

that is

$$X(t+\delta) = X(t)(I + \delta A(t)).$$

Thus

$$X(\ell\delta) = X(0)\prod_{p=0}^{\ell-1}(I + \delta A(p\delta)),$$

so that, on setting $\ell\delta = t$,

$$X(t) = X(0)\prod_{p=0}^{\ell-1}\left(I + \frac{t}{\ell}A\left(\frac{pt}{\ell}\right)\right), \quad X(0) = I.$$

Therefore we have

$$r_{t_0}^t(A) = \widetilde{\int_{t_0}^t} (I + A(t))dt = \lim_{\ell\to\infty}\prod_{p=0}^{\ell-1}\left(I + \frac{t}{\ell}A\left(\frac{pt}{\ell}\right)\right).$$

The following are some properties of $r_{t_0}^t$:

(i) $r_{t_0}^t = r_{t_0}^{t_1} r_{t_1}^t$.

Proof. By definition, $r_{t_0}^t$ and $r_{t_1}^t$ are the solutions to

$$\frac{dX}{dt} = XA, \; X(t_0) = I \quad \text{and} \quad \frac{dY}{dt} = YA, \; Y(t_1) = I,$$

respectively. Let $Z = XY^{-1}$; then

$$\frac{dZ}{dt} = \frac{dX}{dt}Y^{-1} - XY^{-1}\frac{dY}{dt}Y^{-1} = (XA - XA)Y^{-1} = 0,$$

so that Z is a constant matrix C. Setting $t = t_1$ gives $C = r_{t_0}^{t_1}$, which then delivers the required result. □

(ii) $r_{t_0}^t(A+B) = r_{t_0}^t(P)r_{t_0}^t(A)$, where $P = r_{t_0}^t(A)B(r_{t_0}^t(A))^{-1}$.

Proof. The expressions $r_{t_0}^t(A+B)$ and $r_{t_0}^t(A)$ are the solutions to

$$\frac{dX}{dt} = X(A+B),\ X(t_0) = I \quad \text{and} \quad \frac{dY}{dt} = YA,\ Y(t_0) = I,$$

respectively. Let $Z = XY^{-1}$; then

$$\begin{aligned}\frac{dZ}{dt} &= \frac{dX}{dt}Y^{-1} - XY^{-1}\frac{dY}{dt}Y^{-1}\\ &= [X(A+B) - XA]Y^{-1} = XBY^{-1} = Z(YBY^{-1}),\ Z(t_0) = I.\end{aligned}$$

□

(iii) $r_{t_0}^t = (r_t^{t_0})^{-1}$.
(iv) $r_{t_0}^t(CA(t)C^{-1}) = Cr_{t_0}^t(A(t))C^{-1}$, where C is a constant matrix.
(v) We introduce the derivative of an iteration:

$$D_t X = X^{-1}\frac{dX}{dt}.$$

From $dX/dt = XA$ so that $X^{-1}(dX/dt) = A$, we see that D_t and r have a reciprocal relationship in the sense that A is obtained from X in one, while X is obtained from A in the other. Clearly we have

$$D_t(XY) = Y^{-1}D_t(X)Y + D_t(Y),$$

and, in particular,

$$\begin{aligned}D_t(XC) &= C^{-1}D_t(X)C,\\ D_t(CY) &= D_t(Y),\\ D_t(X^{-1}) &= -XD_t(X)X^{-1}.\end{aligned}$$

(vi) $D_t(X') = X'^{-1}(D_t(X)')X'$. This is because

$$\begin{aligned}X'^{-1}\frac{dX'}{dt} &= \left(\frac{dX}{dt}X^{-1}\right)'\\ &= \left(XX^{-1}\frac{dX}{dt}X^{-1}\right)' = X'^{-1}\left(X^{-1}\frac{dX}{dt}\right)'X'.\end{aligned}$$

(vii) Similar to the formula for integration by parts, we have

$$r_{t_0}^t(Q + D_t X) = X(t_0)^{-1}r_{t_0}^t(XQX^{-1})X(t).$$

Proof. The left-hand side is the solution to

$$\frac{dZ}{dt} = Z\left(Q + X^{-1}\frac{dX}{dt}\right), \quad Z(t_0) = I.$$

Let $Y = ZX^{-1}$; then

$$\begin{aligned}\frac{dY}{dt} &= \frac{dZ}{dt}X^{-1} - ZX^{-1}\frac{dX}{dt}X^{-1} = Z\Big(Q + X^{-1}\frac{dX}{dt}\Big)X^{-1} - ZX^{-1}\frac{dX}{dt}X^{-1} \\ &= ZQX^{-1} = YXQX^{-1}, \quad Y(t_0) = X(t_0)^{-1}.\end{aligned}$$

Therefore

$$Y(t) = X(t_0)^{-1} r_{t_0}^{t}(XQX^{-1}).$$

□

18.12 Solutions with full rank

Theorem 18.7 *Suppose that $\int_0^{t_1} ||A(t)||dt$ exists. Then, in $0 \le t \le t_1$, the solution $X(t)$ to the differential equation*

$$\frac{dX}{dt} = XA(t), \quad X(0) = I, \tag{18.48}$$

has full rank.

If $A(t) = A$ is a constant matrix then, from Section 18.6, we already know that e^{At} is the solution to (18.48), and the fact that it has full rank has already been proved in Chapter 17.

The theorem here is a corollary of the Jacobi identity below.

Theorem 18.8 *The determinant of $X(t)$ is given by*

$$|X(t)| = e^{\int_0^t tr(A(s))ds}.$$

Proof. We differentiate the determinant

$$\frac{d}{dt}\begin{vmatrix} x_{11}(t) & x_{12}(t) & \cdots & x_{1n}(t) \\ x_{21}(t) & x_{22}(t) & \cdots & x_{2n}(t) \\ \vdots & \vdots & & \vdots \\ x_{n1}(t) & x_{n2}(t) & \cdots & x_{nn}(t) \end{vmatrix} = \begin{vmatrix} \frac{dx_{11}}{dt} & x_{12} & \cdots & x_{1n} \\ \frac{dx_{21}}{dt} & x_{22} & \cdots & x_{2n} \\ \vdots & \vdots & & \vdots \\ \frac{dx_{n1}}{dt} & x_{n2} & \cdots & x_{nn} \end{vmatrix}$$

$$+ \begin{vmatrix} x_{11} & \frac{dx_{12}}{dt} & \cdots & x_{1n} \\ x_{21} & \frac{dx_{22}}{dt} & \cdots & x_{2n} \\ \vdots & \vdots & & \vdots \\ x_{n1} & \frac{dx_{n2}}{dt} & \cdots & x_{nn} \end{vmatrix} + \cdots + \begin{vmatrix} x_{11} & x_{12} & \cdots & \frac{dx_{1n}}{dt} \\ x_{21} & x_{22} & \cdots & \frac{dx_{2n}}{dt} \\ \vdots & \vdots & & \vdots \\ x_{n1} & x_{n2} & \cdots & \frac{dx_{nn}}{dt} \end{vmatrix}.$$

Take any one term, say one from the first column of the first determinant on the right-hand side; because

$$\frac{dx_{i1}}{dt} = \sum_{j=1}^{n} x_{ij} a_{j1},$$

we find that

$$\begin{vmatrix} \frac{dx_{11}}{dt} & x_{12} & \cdots & x_{1n} \\ \frac{dx_{21}}{dt} & x_{22} & \cdots & x_{2n} \\ \vdots & \vdots & & \vdots \\ \frac{dx_{n1}}{dt} & x_{n2} & \cdots & x_{nn} \end{vmatrix} = \begin{vmatrix} \sum_{j=1}^{n} x_{1j} a_{j1} & x_{12} & \cdots & x_{1n} \\ \sum_{j=1}^{n} x_{2j} a_{j1} & x_{22} & \cdots & x_{2n} \\ \vdots & \vdots & & \vdots \\ \sum_{j=1}^{n} x_{nj} a_{j1} & x_{n2} & \cdots & x_{nn} \end{vmatrix} = a_{11} \begin{vmatrix} x_{11} & \cdots & x_{1n} \\ x_{21} & \cdots & x_{2n} \\ \vdots & & \vdots \\ x_{n1} & \cdots & x_{nn} \end{vmatrix},$$

so that

$$\frac{d}{dt}|X(t)| = (a_{11} + \cdots + a_{nn})|X(t)|.$$

The theorem is proved. □

For any constant matrix C, we see that $Y = CX$ satisfies

$$\frac{dY}{dt} = YA(t). \tag{18.49}$$

Therefore CX is a solution to (18.49). Conversely, if Y satisfies (18.49), and $Y(0) = C$, then we can write Y and C in the vector form

$$\begin{pmatrix} y^{(1)} \\ \vdots \\ y^{(n)} \end{pmatrix}, \begin{pmatrix} C^{(1)} \\ \vdots \\ C^{(n)} \end{pmatrix},$$

so that

$$\frac{dy^{(i)}}{dt} = y^{(i)} A(t), \quad y^{(i)}(0) = C^{(i)}.$$

Thus $y^{(i)} = C^{(i)} X$ and $Y = CX$, which shows that the rank of Y is the same as that of the initial C.

18.13 Inhomogeneous equations

Consider the equation

$$\frac{dx}{dt} = xA(t) + f(t), \quad x(0) = c, \tag{18.50}$$

which is to be dealt with by the method of 'variation of parameters'. Let

$$x = y(t)X, \tag{18.51}$$

where X is the solution to

$$\frac{dX}{dt} = XA(t), \quad X(0) = I. \tag{18.52}$$

Substituting (18.51) into (18.50) we have

$$\frac{dy}{dt}X + y\frac{dX}{dt} = yXA + f(t),$$

that is

$$\frac{dy}{dt}X = f(t).$$

Thus

$$\frac{dy}{dt} = f(t)X^{-1},$$

or

$$y = c + \int_0^t f(s)X^{-1}(s)ds.$$

It follows that (18.50) has the solution

$$x = cX(t) + \Big[\int_0^t f(s)X^{-1}(s)ds\Big]X(t). \tag{18.53}$$

Note 1. Formula (18.52) can be written as

$$X^{-1}\frac{dX}{dt} = A(t), \quad X(0) = I.$$

From $dX^{-1}/dt = -X^{-1}(dX/dt)X^{-1}$ we have

$$\frac{dX^{-1}}{dt}X = -A(t).$$

If $X^{-1} = Y$ then Y satisfies

$$\frac{dY}{dt} = -A(t)Y, \quad Y(0) = I,$$

and the solution (18.53) can be written as

$$x = \Big[c + \int_0^t f(s)Y(s)ds\Big]Y(t)^{-1}.$$

Note 2. If the initial condition for (18.52) is changed to

$$X(t_0) = I,$$

then we let X_i be the solution to

$$\frac{dX_1}{dt} = X_1 A(t), \quad X_1(t_0) = I.$$

Since $CX(t)$, where C is any square matrix with full rank, is a solution to (18.52), we find that

$$X_1(t) = CX(t), \quad X_1(t_0) = CX(t_0),$$

that is

$$X_1(t) = X^{-1}(t_0)X(t).$$

18.14 Asymptotic expansions

Another application of the inhomogeneous equation is to the theory of asymptotic expansions. Suppose that we wish to expand

$$e^{A+\epsilon B}$$

into a power series in ϵ:

$$e^{A+\epsilon B} = e^A + \sum_{n=1}^{\infty} \epsilon^n Q_n(A, B).$$

If A and B commute, that is $AB = BA$, then there is not much of a problem. However, when they do not commute, the situation can be quite complicated, and we apply the method of inhomongeneous equations to deal with it.

Consider the equation

$$\frac{dX}{dt} = X(A + \epsilon B), \quad X(0) = I, \tag{18.54}$$

of which $e^{(A+\epsilon B)t}$ is a solution. From

$$\frac{d(Xe^{-At})}{dt} = \frac{dX}{dt}e^{-At} - XAe^{-At} = \epsilon XBe^{-At},$$

we see that

$$X(t) = \left(I + \epsilon \int_0^t XBe^{-As}\, ds\right)e^{At} = e^{At} + \epsilon \int_0^t X(s)Be^{A(t-s)}\, ds. \tag{18.55}$$

This is a formula suitable for repeated substitutions – thus, on substituting $X(s)$ under the integral sign by (18.55), we have

$$X(t) = e^{At} + \epsilon \int_0^t \left[e^{As} + \epsilon \int_0^s X(s_1) B e^{A(s-s_1)} \, ds_1 \right] B e^{A(t-s)} \, ds$$

$$= e^{At} + \epsilon \int_0^t e^{As} B e^{A(t-s)} \, ds + \epsilon^2 \int_0^t ds \int_0^s X(s_1) B e^{A(s-s_1)} B e^{A(t-s)} \, ds_1,$$

and we can also obtain the terms involving $\epsilon^3, \epsilon^4, \ldots$.

Taking $t = 1$, we have

$$e^{A+\epsilon B} = e^A + \epsilon \int_0^1 e^{As} B e^{A(1-s)} \, ds + \epsilon^2 \int_0^1 ds \int_0^s X(s_1) B e^{A(s-s_1)} B e^{A(t-s)} \, ds_1.$$

Note. The reader should pause to see whether there is any similarity between the method here and that used in Section 18.13.

18.15 A functional equation

The function

$$Y(t) = e^{At}$$

satisfies the functional equation

$$Y(s+t) = Y(s)Y(t), \quad -\infty < s, t < \infty. \tag{18.56}$$

The problem is whether or not there are other functions of matrices that also satisfy (18.56). The answer is 'yes, definitely', because

$$Y(t) = \begin{pmatrix} e^{at} & 0 \\ 0 & 0 \end{pmatrix}$$

is also a solution. However, if we add the requirement that there is a value t_0 so that $Y(t_0)$ has full rank, then the answer becomes 'no'. From

$$Y(t_0) = Y(0)Y(t_0),$$

we have $Y(0) = I$, so that

$$Y(t)Y(-t) = Y(0)I,$$

and hence all $Y(t)$ need to have full rank.

If $Y(t)$ is differentiable then the problem can be disposed of easily. On differentiating (18.56) with respect to s and t, we find that

$$Y'(s)Y(t) = Y'(s+t) = Y(s)Y'(t),$$

so that, for all s, t,

$$Y^{-1}(s)Y'(s) = Y'(t)Y^{-1}(t).$$

Therefore $Y^{-1}(s)Y'(s)$ is a constant square matrix A, that is

$$\frac{d}{ds}Y(s) = Y(s)A, \quad Y(0) = I,$$

and the solution is $Y(t) = e^{At}$.

We can weaken the hypothesis as follows.

Theorem 18.9 *Suppose that (i) $Y(s+t) = Y(s)Y(t)$ in $0 \leq, s, t, s+t \leq t_0$; (ii) $Y(t)$ is continuous in $0 \leq t \leq t_0$; and (iii) there is at least one t_0 at which $Y(t_1)$ is of full rank. Then*

$$Y = e^{At}.$$

Proof. Writing the integral as the limit of a sum we have

$$\begin{aligned}\int_0^t Y(s)ds &= \lim_{\delta\to 0}\sum_{k=0}^{N-1} Y(k\delta)\delta \quad (\delta = t/N)\\ &= \lim_{\delta\to 0}\sum_{k=0}^{N-1} Y(\delta)^k\delta\\ &= \lim_{\delta\to 0}\frac{Y(N\delta)-I}{Y(\delta)-I}\delta = \lim_{\delta\to 0}\frac{\delta}{Y(\delta)-I}(Y(t)-I). \end{aligned} \tag{18.57}$$

By continuity we have

$$Y(t) = I + o(1),$$

so that

$$\int_0^t Y(s)ds = tI + o(t).$$

Thus, when t is sufficiently small,

$$\int_0^t Y(s)ds$$

has full rank, and so, for such fixed t, when δ is sufficiently small,

$$\sum_{k=0}^{N-1} Y(\delta)^k\delta$$

also has full rank.

As $\delta \to 0$,

$$\frac{Y(\delta)-I}{\delta}\sum_{k=0}^{N-1} Y(k\delta)\delta \to Y(t) - I. \tag{18.58}$$

Since the second factor on the left-hand side has full rank, it follows from (18.56) and (18.57) that

$$A = \lim_{\delta \to 0} \frac{Y(\delta) - I}{\delta} = (Y(t) - I)\left(\int_0^t Y(s)ds\right)^{-1},$$

that is

$$A \int_0^t Y(s)ds = Y(t) - I.$$

Since the left-hand side is differentiable, so is $Y(t)$, and, when t is sufficiently small,

$$AY(t) = Y'(t), \quad Y(0) = I.$$

Thus, for such t, we have $Y(t) = e^{At}$, and from $Y(nt) = (Y(t))^n$, we see that the same is true even for values of t which are not small. □

18.16 Solutions to the differential equation $dX/dt = AX + XB$

We consider the more complicated differential equation

$$\frac{dX}{dt} = A(t)X + XB(t), \quad X(0) = C. \tag{18.59}$$

The equation can indeed be complicated, but if we write X in terms of its elements $x_{rs}(t)$, and consider a system of n^2 equations in n^2 unknowns, then we have dealt with the problem concerning the existence and the uniqueness of the solution. Our problem here is the search for the actual solution.

We start with

$$\frac{dY}{dt} = A(t)Y, \quad Y(0) = I,$$

and

$$\frac{dZ}{dt} = ZB(t), \quad Z(0) = I.$$

From

$$\frac{d(YCZ)}{dt} = \frac{dY}{dt}CZ + YC\frac{dZ}{dt} = A(t)YCZ + YCZB(t)$$

and $(YCZ)_{t=0} = C$, we see that

$$X = YCZ$$

is a solution to (18.59), and the unique one.

In particular, we have the following theorem.

Theorem 18.10 *For constant square matrices A and B, the solution to the differential equation*

$$\frac{dX}{dt}AX + XB, \quad X(0) = C, \tag{18.60}$$

is

$$X = e^{At}Ce^{Bt}. \tag{18.61}$$

This simple differential equation is rather useful in the study of vector mechanics, in which the solution has a very natural meaning. See, for example, p. 149 of *Elements of Statistical Mechanics*, by D. Ter Haar (Holt, Rinehart and Winston, 1954).

The theorem has the following application.

Theorem 18.11 *If*

$$X = -\int_0^\infty e^{At}Ce^{Bt}\,dt \tag{18.62}$$

exists for all C, then the matrix equation

$$AX + XB = C \tag{18.63}$$

has a unique solution.

Proof. We first consider the equation

$$\frac{dZ}{dt} = AZ + ZB, \quad Z(0) = C.$$

Integrating both sides with respect to t from 0 to ∞, and assuming

$$\lim_{t\to\infty} Z(t) = 0, \tag{18.64}$$

we have

$$-C = -Z(0) = A\int_0^\infty Z(t)dt + \int_0^\infty Z(t)dt\,B.$$

Therefore

$$X = -\int_0^\infty Z(t)dt = -\int_0^\infty e^{At}Ce^{Bt}\,dt$$

is a solution to (18.63) – that condition (18.64) holds follows from (18.62). For uniqueness we consider (18.63) to be a linear system of n^2 equations with n^2 unknowns, which is soluble for all C, so that the determinant of the system does not vanish, and therefore there is a unique solution for each C. □

Note. Under what condition will

$$\int_0^\infty e^{At} C e^{Bt} \, dt$$

exist for all C? We shall prove later that the following is a necessary and sufficient condition for (18.63) to have a solution: Let λ_i and μ_j denote the eigenvalues for A and B. None of the sums $\lambda_i + \mu_j$ is equal to 0.

19
Asymptotic properties of solutions

19.1 Difference equations with constant coefficients

Let us start again with difference equations. We already know that the difference equation with constant coefficients

$$x(t+1) = x(t)A, \quad x(0) = c, \tag{19.1}$$

has the solution

$$x(t) = cA^t, \quad t = 0, 1, 2, \ldots . \tag{19.2}$$

Our current problem is on the nature of the solution $x(t)$ when $t \to \infty$. As t takes the values $0, 1, 2, \ldots,$

$$x(t) = (x_1(t), \ldots, x_n(t))$$

represents a sequence of points in an n-dimensional space. We may ask: Is the sequence bounded? Is there a limit? Will it return to its initial point? Will it be 'dense' within a certain curve? Will there be a certain property corresponding to 'any c', and a certain other property corresponding to 'some c'? Much of these depend, of course, on what we can find out from the sequence of powers of square matrices

$$A^t, \quad t = 0, 1, 2, \ldots .$$

We have, to some extent, dealt with such problems in Chapter 17. As part of the discussion we need to investigate the behaviour of

$$\xi^t = e^{2\pi i \alpha t}, \quad t = 0, 1, 2, \ldots ,$$

when α is a real number. These are points of the unit circle in the complex plane. First, if $\alpha = p/q$ (p, q coprime and $q > 0$) is a rational number, then ξ^t can take only finitely many values corresponding to $t = 0, 1, \ldots, q-1$, after which it repeats itself periodically. If α is an irrational number, then ξ^t forms a set which is everywhere dense on the unit circle. By this we mean that, corresponding to any point $\xi_0 = e^{2\pi i \alpha_0}$ $(0 \le \alpha_0 < 1)$ and any $\epsilon > 0$, there are infinitely many t such that

$$|\xi_0 - \xi^t| < \epsilon.$$

The proof of this fact requires some knowledge from number theory – more precisely, Kronecker's theorem states that, for any $\epsilon > 0$, there are infinitely many natural numbers q and integers p such that

$$|\alpha p - \alpha_0 - q| < \epsilon,$$

from which it follows that $|\xi^p - \xi_0| < 2\pi\epsilon$. We describe such a property by saying that ξ^t is dense on the unit circle.

Returning to our investigation of

$$A^t, \quad t = 0, 1, 2, \dots,$$

there is a P such that PAP^{-1} is equal to a Jordan piece

$$J_\lambda = \begin{cases} \lambda J, & \text{when } \lambda \neq 0, \ J = \begin{pmatrix} 1 & 1 & & & \\ & 1 & 1 & & \\ & & 1 & \ddots & \\ & & & \ddots & 1 \\ & & & & 1 \end{pmatrix}, \\ J_0, & \text{when } \lambda = 0, \ J_0 = \begin{pmatrix} 0 & 1 & & \\ & \ddots & \ddots & \\ & & 0 & 1 \end{pmatrix}. \end{cases}$$

If $|\lambda| < 1$, then

$$\lim_{t\to\infty} J_\lambda^t = 0;$$

if $|\lambda| > 1$, then

$$\lim_{t\to\infty} J_\lambda^t \text{ does not exist.}$$

If $|\lambda| = 1$, $\lambda = e^{2\pi i\alpha}$, and J_λ is not simple, then the limit clearly does not exist. If $|\lambda| = 1$ and J_λ is simple, then the conclusion is that given in the problem discussed in the above.

We decompose the square matrix PAP^{-1} into three parts:

(i) Keep the various pieces corresponding to the eigenvalues with absolute values less than 1, and replace the remaining parts by 0; use B_- to denote such a square matrix.
(ii) Keep the various pieces corresponding to the eigenvalues with absolute values equal to 1, and simple, and replace the remaining parts by 0; use B_0 to denote such a square matrix.
(iii) Take the various pieces that have not been kept before, and replace the remaining parts by 0; use B_+ to denote such a square matrix.

We then have the decomposition

$$PAP^{-1} = B_- + B_0 + B_+,$$

in which the product of any two terms of the decomposition is 0; also, $\lim B_-^t$ exists, B_0^t is bounded, and B_+^t is unbounded. We now have our decomposition of the matrix A:

$$\begin{aligned} A &= P^{-1}(B_- + B_0 + B_+)P \\ &= A_- + A_0 + A_+. \end{aligned}$$

Returning to our original investigation, we may write

$$x = cA^t = c(A_-^t + A_0^t + A_+^t).$$

If $cA_+ = 0$ then $x(t)$ is bounded, and if $cA_0 = 0$ then

$$\lim_{t\to\infty} x(t) = 0.$$

We can now deduce several simple conclusions.

Theorem 19.1 *If all the eigenvalues of A have absolute values less than* 1, *then, regardless of the initial values,*

$$\lim_{t\to\infty} x(t) = 0.$$

In this case we say that the solution is 'asymptotically stable'.

Theorem 19.2 *If all the eigenvalues of A have absolute values not exceeding* 1, *and those with absolute values equal to* 1 *are simple, then $x(t)$ is bounded. Moreover, given $\epsilon > 0$, we can find $\delta > 0$ such that*

$$||x(t) - x_1(t)|| < \epsilon$$

when $||c - c_1|| < \delta$, with $x_1(t) = c_1 A^t$.

Theorem 19.3 *If one of the eigenvalues of A has the value* 1, *and the remaining ones have absolute values less than* 1, *then $\lim_{t\to\infty} x(t)$ exists and its direction is fixed.*

Proof. The rank of $\lim_{t\to\infty} A^t$ is 1, that is

$$\lim_{t\to\infty} A^t = u'v, \quad uv' = 1,$$

where u, v are vectors. It follows that

$$\lim_{t\to\infty} x(t) = \lim_{t\to\infty} cA^t = (cu')v,$$

which is proportional to v, as required. □

We leave it to the reader to investigate the situation when all the eigenvalues have absolute values equal to 1 and are simple.

It may appear that there are many possibilities, but actually, once we are familiar with the handling of the situation for $n = 1$, normal forms can then be used to deduce the more general case.

19.2 Generalised similarity

Let us consider a more general set of difference equations:

$$x(t+1) = x(t)A(t). \tag{19.3}$$

Making the substitution

$$y(t) = x(t)L(t), \tag{19.4}$$

we have

$$\begin{aligned} y(t+1) &= x(t+1)L(t+1) \\ &= x(t)A(t)L(t+1) \\ &= y(t)L^{-1}(t)A(t)L(t+1). \end{aligned} \tag{19.5}$$

Thus, the system of difference equations with $A(t)$ as the coefficient square matrix has been transformed into one with

$$B(t) = L^{-1}(t)A(t)L(t+1) \tag{19.6}$$

as the coefficient square matrix.

Definition 19.4 If $L(t)$ and $L^{-1}(t)$ are bounded, then $L(t)$ is called a Lyapunov square matrix.

If $L(t)$ is a Lyapunov matrix then from $\lim_{t\to\infty} x(t) = 0$ we deduce that $\lim_{t\to\infty} y(t) = 0$; the converse also holds.

Definition 19.5 If $A(t)$ and $B(t)$ satisfy (19.6), and $L(t)$ is a Lyapunov square matrix, then the two square matrices are said to be L-similar, and we write

$$A \stackrel{L}{=} B.$$

It is clear that being L-similar is an equivalence relation in the sense that (i) $A \stackrel{L}{=} A$, (ii) if $A \stackrel{L}{=} B$ then $B \stackrel{L}{=} A$, (iii) if $A \stackrel{L}{=} B$ and $B \stackrel{L}{=} C$ then $A \stackrel{L}{=} C$.

Also, if there is a constant square matrix P such that $PAP^{-1} = B$ then $A \stackrel{L}{=} B$.

Definition 19.6 Square matrices which are L-similar to a constant square matrix are said to be reducible.

Theorem 19.7 *A reducible square matrix must be L-similar to a square matrix with real eigenvalues.*

Proof. Since reducible matrices must be L-similar to a constant square matrix, we need only deal with such a matrix. Take the case $n = 1$ and consider

$$x(t+1) = x(t)\rho e^{i\theta}.$$

Let $y(t) = x(t)\ell^{-1}(t)$; then

$$y(t+1)y(t)\ell(t)\rho e^{i\theta}/\ell(t+1).$$

Taking $\ell(t) = e^{i\theta}$ we find that $y(t+1) = y(t)\rho$. □

The general case can be deduced from normal forms.

19.3 Ordinary linear differential equations with constant coefficients

We study the asymptotic properties of solutions to ordinary differential equations. The system of equations

$$\frac{dx}{dt} = xA, \quad x(0) = c, \tag{19.7}$$

has the solution

$$x(t) = ce^{At}. \tag{19.8}$$

As before, we need only study the property of e^{At} as $t \to \infty$, and we again reduce the discussion to the situation when A is in Jordan normal form.

(i) If λ has a negative real part, then $\lim_{t\to\infty} e^{\lambda Jt} = 0$.
(ii) If λ is purely imaginary, and is simple, then $\lim_{t\to\infty} e^{\lambda t}$ does not exist (unless $\lambda = 0$), and $z = e^{\lambda t}$ describes a circle on the plane.
(iii) Otherwise

$$x(t), \quad 0 \le t \le \infty,$$

describes a curve passing through $x = c$ tending toward ∞.

Consequently we have the following.

Theorem 19.8 *If the real parts of the eigenvalues of A are all negative, then, regardless of the initial values, the integral curve $(x_1(t), \dots, x_n(t))$ tends toward the origin.*

Theorem 19.9 *If all the eigenvalues of A are purely imaginary and are simple, then the integral curve describes a closed curve in space.*

Proof. There exists P such that

$$PAP^{-1} = \begin{pmatrix} 0 & \lambda_1 \\ -\lambda_1 & 0 \end{pmatrix} \dotplus \cdots \dotplus \begin{pmatrix} 0 & \lambda_\nu \\ -\lambda_\nu & 0 \end{pmatrix} \dotplus 0 \dotplus \cdots \dotplus 0.$$

Let

$$y = xP^{-1};$$

then

$$\frac{dy}{dt} = y(PAP^{-1}).$$

From

$$\begin{aligned} e^{\begin{pmatrix} 0 & \lambda \\ -\lambda & 0 \end{pmatrix} t} &= \sum_{n=0}^{\infty} \frac{1}{n!} \begin{pmatrix} 0 & \lambda \\ -\lambda & 0 \end{pmatrix}^n t^n \\ &= \sum_{\ell=0}^{\infty} \frac{1}{(2\ell)!} \begin{pmatrix} (-\lambda^2 t^2)^\ell & 0 \\ 0 & (-\lambda^2 t^2)^\ell \end{pmatrix} + \sum_{\ell=0}^{\infty} \frac{(-\lambda^2 t^2)^\ell}{(2\ell+1)!} \begin{pmatrix} 0 & \lambda \\ -\lambda & 0 \end{pmatrix} t \\ &= \begin{pmatrix} \cos \lambda t & \sin \lambda t \\ -\sin \lambda t & \cos \lambda t \end{pmatrix}, \end{aligned}$$

we find that

$$\begin{aligned} y &= c\left(\begin{pmatrix} \cos \lambda_1 t & \sin \lambda_1 t \\ -\sin \lambda_1 t & \cos \lambda_1 t \end{pmatrix} \dotplus \cdots \right), \\ y_1 &= c_1 \cos \lambda_1 t - c_2 \sin \lambda_1 t, \\ y_2 &= c_1 \sin \lambda_1 t + c_2 \cos \lambda_1 t, \\ &\ldots, \end{aligned}$$

and

$$y_1^2 + y_2^2 = c_1^2 + c_2^2, \quad y_3^2 + y_4^2 = c_3^2 + c_4^2, \quad \ldots,$$

which shows clearly that the integral curve is a closed one. □

Note 1. From $yy' = cc'$ we find that

$$xP^{-1}P'^{-1}x' = cc',$$

which shows that the integral curve lies on a similar ellipsoid; indeed much more can be said beyond such a property.

Note 2. The integral curves that we considered all start from $x(0) = c(\neq 0)$. If $c = 0$ then, from the above, there may be more than one (or none) integral curves passing through the origin – the point concerned is then said to be singular.

Exercise 19.10 Examine the situation when there are two variables in the system of equations in the neighbourhood of the origin under the various types of eigenvalues (already dealt with).

Exercise 19.11 Examine the situation when there are three variables in a system of three equations.

Exercise 19.12 Solve the system

$$\frac{dx}{dt} = xA + b, \quad x(0) = c,$$

where b is a constant vector.

19.4 Introduction to Lyapunov's method

The introduction to Lyapunov's method can be extended to the study of the stability of solutions to non-linear functional equations. However, the real worth of the method is much clearer in what we consider here.

We consider

$$\frac{dx}{dt} = xA, \quad x(0) = c, \tag{19.9}$$

where c and A are real. Let

$$u = xYx',$$

where Y is a fixed constant symmetric matrix. On differentiation we have

$$\frac{du}{dt} = \frac{dx}{dt}Yx' + xY\left(\frac{dx}{dt}\right)' = x(Ay + YA')x'.$$

Suppose that we can select a positive definite Y such that

$$AY + YA' = -I.$$

Denoting by λ_0 the largest eigenvalue of Y, we then have

$$\frac{du}{dt} = -xx' \leq -\lambda_0^{-1}u,$$

and, from $u(0)$ being positive, we deduce that

$$d\log u(t) = \frac{du}{u} \leq -\lambda_0^{-1}dt,$$

$$\int_0^t d\log u(t) \leq -\lambda_0^{-1}t,$$

so that

$$u(t) \leq u(0)e^{-\lambda_0^{-1}t}.$$

It follows that

$$u = xYx' \to 0 \quad \text{as} \quad t \to \infty,$$

and hence $x \to 0$, because Y is positive definite.

Thus, the problem of whether

$$\lim_{t\to\infty} x(t) = 0$$

is transformed to the problem of whether such a matrix Y exists.

If all the real parts of the eigenvalues of A are negative, then Y definitely exists. Such a matrix can be

$$Y = \int_0^\infty e^{At} e^{A't}\, dt.$$

That the integral converges to a symmetric matrix is clear, and we show that it is positive definite. Suppose then that

$$0 = xYx' = \int_0^\infty (xe^{At})(xe^{At})'\, dt,$$

so that $xe^{At} = 0$. Since the exponential matrix e^{At} has full rank, we find that $x = 0$, so that Y is positive definite.

Next, from integration by parts, we have

$$\begin{aligned} AY + YA' &= \int_0^\infty (Ae^{At} \cdot e^{A't} + e^{At} \cdot e^{A't} A')dt \\ &= \int_0^\infty \frac{d}{dt}(e^{At} \cdot e^{A't})dt = -I. \end{aligned}$$

If there is a positive definite symmetric matrix Y such that

$$AY + YA' = -I,$$

then all the eigenvalues of A have negative real parts.

Proof. Let ρ be an eigenvalue of A, with v being the corresponding eigenvector, so that $vA = \rho v$. Since A is real we have $\bar{v}A = \overline{\rho v}$.

On multiplying $AY + YA' = -I$ from the left and the right by v and $\bar{v}'$, respectively, we have $v(AY + YA')\bar{v}' = -v\bar{v}'$, that is $(\rho + \bar{\rho})vY\bar{v}' = -v\bar{v}'$. Thus $2\Re\rho = \rho + \bar{\rho} < 0$, as required. □

We use Lyapunov's method to deal with a more general problem:

$$\frac{dx}{dt} = x(A + B(t)),$$

where A still has the same property, but

$$\lim_{t\to\infty} ||B(t)|| = 0.$$

Using the same Y as in the above, we then have

$$\begin{aligned}\frac{d}{dt}(xYx') &= \frac{dx}{dt}Yx' + xY\left(\frac{dx}{dt}\right)' \\ &= x(A + B(t))Yx' + xY(A' + B'(t))x' \\ &= -xx' + x(B(t)Y + YB'(t))x'.\end{aligned}$$

When t is sufficiently large, say $t \geq t_0$,

$$\frac{d}{dt}(xYx') \leq -\lambda_1 xYx',$$

where λ_1 is a positive constant, and, on integrating,

$$xYx' \leq (xYx')_{t=t_0} e^{-\lambda_1 t}.$$

Therefore $xYx' \to 0$, and so $x \to 0$, as $t \to \infty$.

Note. If A is not a constant square matrix we may not have the same conclusion, even when $A(t)$ exists. Thus, the solution to

$$\frac{dy}{dt} = yA(t)$$

satisfies $\lim_{t\to\infty} y(t) = 0$, whereas the solution to

$$\frac{dx}{dt} = x(A(t) + B(t))$$

may not have such a property, even under the hypothesis

$$\lim_{t\to\infty} ||B(t)|| = 0.$$

For a counterexample, we note that

$$\frac{dy_1}{dt} = -ay_1, \quad \frac{dy_2}{dt} = (\sin\log t + \cos\log t - 2a)y_2$$

has the solution

$$y_1 = c_1 e^{-at}, \quad y_2 = c_2 e^{t\sin\log t - 2at}.$$

If $a > \frac{1}{2}$ then $y_1, y_2 \to 0$ as $t \to \infty$. If we insert

$$B(t) = \begin{pmatrix} 0 & e^{-at} \\ 0 & 0 \end{pmatrix}$$

into the system then the new system

$$\frac{dx_1}{dt} = -ax_1,$$
$$\frac{dx_2}{dt} = (\sin\log t + \cos\log t - 2a)x_2 + x_1 e^{-at}$$

has the solutions

$$x_1 = c_1 e^{-at},$$
$$x_2 = e^{t\sin\log t - 2at}\left(c_2 + c_1\int_0^t e^{-\tau\sin\log\tau}\,d\tau\right).$$

If we take the points $t = e^{(2n+1/2)\pi}$, $n = 1, 2, \ldots,$ then

$$\begin{aligned}\int_0^t e^{-\tau\sin\log\tau}\,d\tau &> \int_{te^{-\pi}}^{te^{-2\pi/3}} e^{-\tau\sin\log\tau}\,d\tau\\ &= \pi\int_{2/3}^{1}\exp(-te^{-\alpha\pi}\cos\alpha\pi)te^{-\alpha\pi}\,d\alpha \quad (\text{taking } \tau = te^{-\alpha\pi})\\ &> \pi t\exp(te^{-\pi}/2)\int_{2/3}^{1} e^{-\alpha\pi}\,d\alpha\\ &= t(e^{-2\pi/3} - e^{-\pi})\exp(te^{-\pi}/2),\end{aligned}$$

so that the coefficient c_1 of the solution x_2 exceeds

$$e^{t(1-2a)}t(e^{-2\pi/3} - e^{-\pi})e^{(1/2)te^{-\pi}}.$$

When $1 - 2a > -\frac{1}{2}e^{-\pi}$, we find that $x_2(t)$ tends to ∞, or 0, or $-\infty$, depending on whether $c_1 > 0$, or $= 0$, or < 0. Therefore we cannot possibly have the expected conclusion.

Exercise 19.13 How do we apply Lyapunov's method when A is a complex square matrix?

19.5 Stability

A physical system can sometimes be described by a set of differential equations:

$$\frac{dx_i}{dt} = f_i(x_1, \ldots, x_n, t), \quad i = 1, 2, \ldots, n, \tag{19.10}$$

where $x_1, \ldots, x_n$ are parameters and t denotes time. A very important problem concerning such a physical system is whether equilibrium exists. Thus we want to know if, under a small disturbance, the system will return to its original state – that is, whether there is stability in the system or not. Naturally, whether there is stability in a physical system can be determined, or tested, by a physical experiment. However, such experiments are usually costly or time-consuming, and it is often much better to use mathematical methods to find out if there is stability.

We need to fix our notion on stability.

(1) Will there be a substantial change in the solution corresponding to a tiny alteration of the initial values?

Let $(c_1, \ldots, c_n)$ be a set of initial values at time $t = t_0$, and for which the corresponding solution is

$$x_i = g_i(t, c_1, \ldots, c_n).$$

Let $(c_1^0, \ldots, c_n^0)$ be a given set of initial values. If, given any $\epsilon > 0$, we can find $\delta > 0$ so that, when

$$|c_1 - c_1^0| < \delta, \ \cdots, \ |c_n - c_n^0| < \delta$$

we have

$$|g_i(t, c_1, \ldots, c_n) - g_i(t, c_1^0, \ldots, c_n^0)| < \epsilon, \quad i = 1, 2 \ldots, n,$$

for all $t \geq t_0$, then we say that the system is stable.

With no loss of generality, we may assume that $c_1^0 = \cdots = c_n^0 = 0$ and that $g_i(t, c_1, {}^0, \ldots, c_n^0) = 0$, so that our definition is simplified as follows: When

$$|c_i| < \delta, \quad i = 1, 2 \ldots, n,$$

we have

$$|x_i(t)| < \epsilon, \quad t \geq t_0.$$

(2) Strengthening the condition: If there exists $\delta > 0$ such that, when $|c_i| < \delta$,

$$\lim_{t \to \infty} x_i(t) = 0,$$

then we say the system concerned is asymptotically stable. If necessary we may add the phrase 'in the neighbourhood of 0'.

We already know that the linear system of equations

$$\frac{dx}{dt} = xA(t)$$

has the solution

$$x = cX(t),$$

where c is an initial value, that is $x = c$ when $t = t_0$, and $X(t)$ is a square matrix satisfying

$$\frac{dX}{dt} = XA(t), \quad X(t_0) = T.$$

(i) If $X(t)$ is a bounded square matrix in (t_0, ∞), then there is an M such that

$$|x_{ij}(t)| \leq M, \quad t \geq t_0, \quad i, j = 1, 2, \ldots, n;$$

we then have

$$|x_i(t)| \le Mn \max_{i \le j \le n} |c_j|.$$

Thus, on taking $\delta < \epsilon/(nM)$, we find that $|x_i(t)| < \epsilon$ when $|c_j| \le \delta$; that is, the trivial solution $x_1 = \cdots = x_n = 0$ to the physical system being described is stable.

(ii) $\lim_{t\to\infty} X(t) = 0$. If so, then $X(t)$ is obviously bounded, and the system is stable. Moreover, from $\lim_{t\to\infty} x(t) = 0$, we know that the system is asymptotically stable.

(iii) $X(t)$ is unbounded. Then, there is at least one $x_{ij}(t)$ which is unbounded in (t_0, ∞). We let $c_i \ne 0$, and $c_j = 0$ for $j \ne i$, so that $x_j = c_i x_{ij}(t)$ is also unbounded, and the system is unstable.

Note. A system may be stable corresponding to a certain set of initial values, but unstable corresponding to another set.

19.6 Lyapunov transformation

We consider the general linear equation

$$\frac{dx}{dt} = xA(t).$$

Definition 19.14 We call

$$x = yL(t)$$

a Lyapunov transformation if the square matrix $L(t)$ satisfies the following:

(i) $L(t)$ has a continuous derivative dL/dt in (t_0, ∞);
(ii) $L(t)$ and dL/dt are bounded in (t_0, ∞);
(iii) there is a positive constant which is smaller than the absolute value of the determinant of $L(t)$.

Such a matrix $L(t)$ is called a Lyapunov square matrix.

Example 19.15 A constant square matrix with full rank is a Lyapunov square matrix.

Example 19.16 If the eigenvalues of D are simple and purely imaginary then

$$L(t) = e^{Dt}$$

is a Lyabpunov square matrix.

We omit the proofs of the following two theorems; they are not difficult.

Theorem 19.17 *The inverse of a Lyapunov square matrix is also a Lyapunov square matrix.*

Theorem 19.18 *The properties of being stable, asymptotically stable and unstable are invariant under a Lyapunov transformation.*

We consider y, which satisfies the differential equation

$$yL(t)A(t) = xA(t) = \frac{dx}{dt} = \frac{dy}{dt}L(t) + y\frac{dL}{dt},$$

that is

$$\frac{dy}{dt} = yL(t)A(t)L^{-1}(t) - y\frac{dL}{dt}L^{-1}(t).$$

Definition 19.19 If there is a Lyapunov square matrix $L(t)$ such that

$$B(t) = L(t)A(t)L^{-1}(t) - \frac{dL}{dt}L^{-1}(t), \tag{19.11}$$

then we say that $A(t)$ and $B(t)$ are Lyapunov equivalent, and we write

$$A \stackrel{L}{=} B.$$

Let us verify that being Lyapunov equivalent is an equivalence relation: It is clear that (i) $A \stackrel{L}{=} A$. Next, (ii) if $A \stackrel{L}{=} B$ then $B \stackrel{L}{=} A$; thus from (19.11) we have

$$A(t) = L^{-1}B(t)L + L^{-1}\frac{dL}{dt} = L^{-1}B(t)L - \frac{dL^{-1}}{dt}L.$$

Also, (iii) if $A \stackrel{L}{=} B$ and $B \stackrel{L}{=} C$ then $A \stackrel{L}{=} C$; for we have

$$\begin{aligned} C &= MBM^{-1} - \frac{dM}{dt}M^{-1} = M\Big(LAL^{-1} - \frac{dL}{dt}L^{-1}\Big)M^{-1} - \frac{dM}{dt}M^{-1} \\ &= MLA(ML)^{-1} - \Big(M\frac{dL}{dt}L^{-1}M^{-1} + \frac{dM}{dt}M^{-1}\Big) \\ &= MLA(ML)^{-1} - \frac{d(ML)}{dt}(ML)^{-1}. \end{aligned}$$

Definition 19.20 A system of equations with a coefficient matrix being Lyapunov equivalent to a constant square matrix is called a reducible system.

Thus, the solution to a reduced system of equations must have the form

$$X(t) = L(t)e^{At},$$

where $L(t)$ is a Lyapunov square matrix and A is a constant square matrix. We need say no more about solutions to a reducible system.

19.7 Differential equations with periodic coefficients

Theorem 19.21 *A system of differential equations with a periodic coefficient is reducible.*

Proof. Let us suppose that

$$A(t + \tau) = A(t), \quad -\infty \leq t \leq \infty;$$

then

$$\frac{dX(t+\tau)}{dt} = X(t+\tau)A(t),$$

so that $X(t+\tau)$ is also a solution, that is

$$X(t+\tau) = VX(t),$$

where V is a constant square matrix with full rank. Since $|V| \neq 0$, we can define

$$V^{t/\tau} = e^{(t/\tau)\log V}$$

(noting the branch taken by $\log V$) and take

$$L(t) = V^{-t/\tau}X(t);$$

this is a periodic function,

$$L(t+\tau) = L(t),$$

with $|L(t)| \neq 0$, so that $L(t)$ is a Lyapunov square matrix. We may then define

$$X(t) = Y(t)L(t), \quad Y(t) = V^{t/\tau},$$

to give

$$\frac{dY}{dt} = Y\Big(\frac{1}{\tau}\log V\Big).$$

□

The system

$$\frac{dx}{dt} = x(A(t) + B(t)), \tag{19.12}$$

where $A(t)$ is periodic and $\lim_{t\to\infty}||B(t)|| = 0$, can now be dealt with by means of a Lyapunov transformation, so that it becomes one with

$$A = (a_{ij})_{1\leq i,j\leq n} = \frac{1}{\tau}\log V,$$

to which the method given in Section 19.4 can be applied.

If all the eigenvalues of the square matrix V have absolute values less than 1, then the system (19.12) is asymptotically stable. If there is one with an absolute value exceeding 1, then the system is unstable.

Remark 1 It follows that the solution to

$$\frac{dX}{dt} = XA(t), \quad X(0) = I,$$

where $A(t+1) = A(t)$, has the form

$$X(t) = e^{Ct} Q(t),$$

where $Q(t+1) = Q(t)$, and C is a constant square matrix.

Remark 2 Problems involving systems of differential equations with periodic coefficients form an important and difficult subject.

For example, in mathematical physics, the Mathieu equation

$$\frac{d^2u}{dt^2} + (a + b\cos 2t)u = 0$$

is heavily involved in some important problems, and the general equation

$$\frac{d^2u}{dt^2} + \Big(\sum_{n=0}^{\infty} a_n \cos nt + b_n \sin nt\Big) = 0$$

often crops up in the study of the motion of the moon. Indeed whole treatises are devoted to such topics.

19.8 Lyapunov equivalence

We consider the problem of constant square matrices being Lyapunov equivalent. We already know that if

$$A \overset{E}{=} B, \quad \text{that is} \quad A = PBP^{-1},$$

then

$$A \overset{L}{=} B.$$

Consequently any square matrix A must be Lyapunov equivalent to a Jordan normal form. We therefore consider the problem of equivalence applied to a Jordan piece.

For $n = 1$, the condition of being Lyapunov equivalent becomes

$$b = a - \frac{d\ell}{dt}\ell^{-1} = a - \frac{d}{dt}\log \ell,$$

so that

$$\ell(t) = e^{(a-b)(t-t_0)}.$$

From $\ell(t)$ and $\ell^{-1}(t)$ being bounded, we find that

$$\ell(t) = ce^{i\lambda t}, \quad \lambda \text{ real}.$$

Thus, if two numbers are Lyapunov equivalent then they have the same real parts. Conversely, let $a = \alpha + i\beta$; taking $\ell(t) = e^{i\beta t}$ we have

$$b = a - \frac{d}{dt}\log \ell = \alpha,$$

so that a number must be Lyapunov equivalent to its own real part.

In general, for

$$aJ, \quad a = \alpha + \beta i,$$

we take $L = e^{i\beta t} I$, which then transforms aJ to

$$\alpha J.$$

We therefore have the following theorem.

Theorem 19.22 (Erugin) *Every reducible system can be transformed to*

$$\frac{dY}{dt} = YJ$$

by a Lyapunov transformation; here J is a square matrix with real eigenvalues.

We do not prove the fact that the normal form concerned is unique, apart from the ordering of the various pieces along the main diagonal for J.

19.9 Difference and differential equations with coefficients which are asymptotically constant

Concerning the difference equation

$$x(t+1) = x(t)A(t), \quad \lim_{t\to\infty} A(t) = A,$$

we leave it to the reader to consult the Tom relevant literature.

Concerning the differential equation

$$\frac{dx}{dt} = xA(t), \quad \lim_{t\to\infty} A(t) = A,$$

we leave it to the reader to consult pp. 52–60 of *Stability Theory of Differential Equations* by Richard Bellman (McGraw-Hill, 1953).

20
Quadratic forms

20.1 Completing squares

The method of 'completing squares', which we learned in the study of quadratic equations at high school, has many useful applications, and becoming familiar with the technique will enable us to tackle a variety of problems. We have seen already how the method can be applied to problems involving maxima and minima in elementary calculus, and it is even more important in the study of quadratic forms.

A quadratic form in n variables is an expression of the form

$$Q = \sum_{i=1}^{n}\sum_{j=1}^{n} a_{ij}x_i x_j. \tag{20.1}$$

Since $a_{ij}x_i x_j + a_{ji}x_j x_i = (a_{ij} + a_{ji})x_i x_j$, we may as well assume that

$$a_{ij} = a_{ji}.$$

Given a quadratic form there is a square matrix

$$A = (a_{ij}), \tag{20.2}$$

where $a_{ij} = a_{ji}$, that is $A = A'$, a symmetric matrix. Conversely, given such a matrix, we have the quadratic form

$$xAx', \tag{20.3}$$

where $x = (x_1, \dots, x_n)$.

If $x_1, \dots, x_n$ are given the linear substitution

$$x_i = \sum_{j=1}^{n} y_j p_{ji}, \quad x = yP, \tag{20.4}$$

for the new variables $y_1, \dots, y_n$, the new quadratic form becomes

$$\sum_{i=1}^{n}\sum_{j=1}^{n} b_{ij}y_i y_j, \quad b_{ij} = b_{ji}, \quad B = (b_{ij}).$$

What can be said about the relationship between B and A? From (20.1) and (20.3) we have

$$\begin{aligned}\sum_{i=1}^{n}\sum_{j=1}^{n} a_{ij}x_i x_j &= \sum_{i=1}^{n}\sum_{j=1}^{n} a_{ij}\sum_{s=1}^{n} y_s p_{si}\sum_{t=1}^{n} y_t p_{tj}\\ &= \sum_{s=1}^{n}\sum_{t=1}^{n}\Big(\sum_{i=1}^{n}\sum_{j=1}^{n} a_{ij}p_{si}p_{tj}\Big)y_s y_t,\end{aligned}$$

that is

$$b_{st} = \sum_{i=1}^{n}\sum_{j=1}^{n} p_{si}a_{ij}p_{tj},$$

and we may write

$$B = PAP'. \tag{20.5}$$

Definition 20.1 If there is a matrix P with full rank such that (20.5) holds, then we say that A and B are congruent to each other.

Being congruent is an equivalence relation. Our problem is as follows: What sort of quadratic form can (20.1) be 'reduced or simplified' to under the linear substitution (20.4)?

Suppose that $a_{11} \neq 0$, and write (20.1) as

$$\begin{aligned}Q = a_{11}x_1^2 + 2a_{12}x_1x_2 + 2a_{13}x_1x_3 + \cdots + 2a_{1n}x_1x_n + a_{22}x_2^2\\ + 2a_{23}x_2x_3 + \cdots + 2a_{2n}x_2x_n + \cdots + a_{nn}x_n^2.\end{aligned}$$

By completing the square we have

$$\begin{aligned}Q &= a_{11}\Big(x_1 + \frac{a_{12}}{a_{11}}x_2 + \cdots + \frac{a_{1n}}{a_{11}}x_n\Big)^2 - a_{11}\Big(\frac{a_{12}}{a_{11}}x_2 + \cdots + \frac{a_{1n}}{a_{11}}x_n\Big)^2\\ &\quad + a_{22}x_2^2 + 2a_{23}x_2x_3 + \cdots + 2a_{2n}x_2x_n + \cdots + a_{nn}x_n^2\\ &= a_{11}\Big(x_1 + \frac{a_{12}}{a_{11}}x_2 + \cdots + \frac{a_{1n}}{a_{11}}x_n\Big)^2 + \frac{a_{22}a_{11} - a_{12}^2}{a_{11}}x_2^2\\ &\quad + 2\frac{a_{11}a_{23} - a_{12}a_{13}}{a_{11}}x_2x_3 + \cdots + 2\frac{a_{11}a_{2n} - a_{12}a_{1n}}{a_{11}}x_2x_n\\ &\quad + 2\frac{a_{33}a_{11} - a_{13}^2}{a_{11}}x_3^2 + \cdots + 2\frac{a_{11}a_{3n} - a_{13}a_{1n}}{a_{11}}x_3x_n + \cdots + \frac{a_{nn}a_{11} - a_{1n}^2}{a_{11}}x_n^2.\end{aligned}$$

If $a_{22}a_{11} - a_{12}^2 \neq 0$, then we can complete a square again, and it is not difficult to obtain

$$\begin{aligned}Q &= a_{11}(x_1 + \beta_{12}x_2 + \cdots + \beta_{1n}x_n)^2 + \frac{D\binom{1,2}{1,2}}{a_{11}}(x_2 + \beta_{23}x_3 + \cdots + \beta_{2n}x_n)^2\\ &\quad + \frac{D\binom{1,2,3}{1,2,3}}{D\binom{1,2}{1,2}}(x_3 + \beta_{34}x_4 + \cdots + \beta_{3n}x_n)^2 + \cdots + \frac{D\binom{1,\ldots,n}{1,\ldots,n}}{D\binom{1,\ldots,n-1}{1,\ldots,n-1}}x_n^2,\end{aligned} \tag{20.6}$$

where

$$\beta_{ij} = \frac{D\binom{1,2,\ldots,i-1,i}{1,2,\ldots,i-1,j}}{D\binom{1,2,\ldots,i}{1,2,\ldots,i}}, \quad i < j.$$

Let

$$\begin{aligned} y_1 &= x_1 + \beta_{12}x_2 + \beta_{13}x_3 + \cdots + \beta_{1n}x_n, \\ y_2 &= \qquad x_2 + \beta_{23}x_3 + \cdots + \beta_{2n}x_n, \\ &\cdots \\ y_n &= \qquad\qquad\qquad\qquad x_n; \end{aligned}$$

if none of the principal minors of A,

$$D\binom{1}{1}, \quad D\binom{1,2}{1,2}, \quad \ldots, \quad D\binom{1,\ldots,n}{1,\ldots,n}$$

vanishes, then there is a triangular matrix

$$P = \begin{pmatrix} 1 & 0 & 0 & \cdots & 0 \\ \beta_{12} & 1 & 0 & \cdots & 0 \\ \beta_{13} & \beta_{23} & 1 & \cdots & 0 \\ \vdots & \vdots & \vdots & & \vdots \\ \beta_{1n} & \beta_{2n} & \beta_{2n} & \cdots & 1 \end{pmatrix}$$

such that

$$PAP'$$

becomes a diagonal matrix, and the determination of P involves only the usual arithmetic operations (and not involving any expansion in the computation).

If the quadratic form is identically 0, then all the coefficients must be 0. Why? Because it follows from

$$e_iAe_i' = a_{ii}, \quad (e_i + e_j)A(e_i + e_j)' - e_iAe_i' - e_jAe_j' = 2a_{ij},$$

where e_i is the vector $(0, \ldots, 0, 1, 0 \ldots, 0)$ in which the ith component is 1 and the others are 0.

Therefore a non-trivial quadratic form must have a non-zero coefficient. If $a_{ii} \neq 0$, we can swap x_i with x_1 and proceed as before. If $a_{11} = \cdots = a_{nn} = 0$ and $a_{ij} \neq 0$, then we may suppose without loss that $a_{12} \neq 0$. In this case we make the substitution $x_1 = y_1$, $x_2 = y_1 + y_2$, $x_i = y_i$ for $3 \leq i \leq n$, so that the form

$$0x_1^2 + a_{12}x_1x_2 + 0x_2^2 + \cdots = a_{12}y_1^2 + \cdots$$

now has a non-zero coefficient for y_1^2. We therefore have the following.

Theorem 20.2 *Any symmetric matrix must be congruent to a diagonal matrix; alternatively, any quadratic form can be reduced to*

$$\lambda_1 x_1^2 + \cdots + \lambda_r x_r^2, \quad \lambda_1, \ldots, \lambda_r \neq 0,$$

where r is the rank of the original matrix.

The reduction process involves only the usual arithmetic operations.

Over the complex field, we let

$$y_i = \sqrt{\lambda_i} x_i, \tag{20.7}$$

so that

$$Q = y_1^2 + \cdots + y_r^2. \tag{20.8}$$

Therefore we have Theorem 20.3.

Theorem 20.3 *Over the complex field, a symmetric matrix must be congruent to*

$$\begin{pmatrix} I^{(r)} & 0 \\ 0 & 0 \end{pmatrix}.$$

Here r is the rank of the given matrix, and it follows that any two symmetric matrices of the same order and the same rank must be congruent to each other.

Over the real field, $\sqrt{\lambda_i}$ may not be real, and (20.8) may not apply. Instead, we let s denote the number of positive λ_i, and label them so that

$$\lambda_1 > 0, \cdots, \lambda_s > 0, \quad \lambda_{s+1} < 0, \cdots, \lambda_r < 0.$$

The substitution $y_i = \sqrt{|\lambda_i|} x_i$ then transforms Q to

$$Q = y_1^2 + \cdots + y_s^2 - y_{s+1}^2 - \cdots - y_r^2.$$

Thus, over the real field, the matrix A is congruent to

$$\begin{pmatrix} I^{(s)} & 0 & 0 \\ 0 & -I^{(r-s)} & 0 \\ 0 & 0 & 0 \end{pmatrix}. \tag{20.9}$$

Definition 20.4 We call the difference between the number of positive and the number of negative terms (that is $s - (r - s) = 2s - r$) the signature of the quadratic form.

Theorem 20.5 *Over the real field, a necessary and sufficient condition for two quadratic forms to be congruent is that they have the same rank and the same signature; in this case we call (20.9) their standard form.*

We have already shown that a real quadratic form must be congruent to one with a coefficient matrix in the standard form (20.9), so that it remains to show that if

$$x_1^2 + \cdots + x_s^2 - x_{s+1}^2 - \cdots - x_r^2 = y_1^2 + \cdots + y_{s_1}^2 - y_{s_1+1}^2 - \cdots - y_{r_1}^2,$$

then $s = s_1, r = r_1$. Indeed $r = r_1$ follows at once from the fact that x and y are related to each other by a linear transformation with full rank.

Suppose, if possible, that $s < s_1$. We set

$$x_1 = \cdots = x_s = 0, \quad y_{s_1+1} = \cdots = y_r = 0. \tag{20.10}$$

By considering $x_1, \ldots, x_s$ as linear forms in y, there are $s + r - s_1 < r$ equations with $x_{s+1}, \ldots, x_r, y_1, \ldots, y_{s_1}$ unknowns. The number $s + r - s_1 < r$ of equations being smaller than the number of unknowns, there is a non-trivial set $x_{s+1}, \ldots, x_r, y_1, \ldots, y_{s_1}$, together with (20.10), such that

$$-(x_{s+1}^2 + \cdots + x_r^2) = y_1^2 + \cdots + y_{s_1}^2,$$

which is impossible. Similarly, we cannot have $s_1 < s$, so that $s = s_1$.

20.2 Completing the square on pieces

We are still applying the basic method of completing squares. Let the matrix A be partitioned as

$$A = \begin{pmatrix} A_1^{(s)} & L \\ L' & A_2^{(n-s)} \end{pmatrix}.$$

If $|A_1| \neq 0$ then

$$\begin{pmatrix} I & 0 \\ -L'A_1^{-1} & I \end{pmatrix} \begin{pmatrix} A_1 & L \\ L' & A_2 \end{pmatrix} \begin{pmatrix} I & 0 \\ -L'A_1^{-1} & I \end{pmatrix}'$$

$$= \begin{pmatrix} A_1 & L \\ 0 & A_2 - L'A_1^{-1}L \end{pmatrix} \begin{pmatrix} I & -A_1^{-1}L \\ 0 & I \end{pmatrix} = \begin{pmatrix} A_1 & 0 \\ 0 & A_2 - L'A_1^{-1}L \end{pmatrix}.$$

From completing the square on pieces, we see at once that

$$|A| = |A_1||A_2 - L'A_1^{-1}L|.$$

Definition 20.6 If $A' = -A$ then we say that A is skew-symmetric; two skew-symmetric matrices A, B are said to be congruent if there is a matrix P with full rank such that

$$PAP' = B.$$

Theorem 20.7 *A skew-symmetric matrix A must be congruent to*

$$\begin{pmatrix} F & 0 & \\ 0 & F & \\ & & \ddots \end{pmatrix}, \quad F = \begin{pmatrix} 0 & 1 \\ -1 & 0 \end{pmatrix}.$$

A necessary and sufficient condition for two skew-symmetric matrices to be congruent is that they should have the same rank.

Proof. (1) The elements in the main diagonal of a skew-symmetric matrix are all 0.

(2) $n = 2$. The required result follows from

$$\begin{pmatrix} a^{-1} & 0 \\ 0 & 1 \end{pmatrix} \begin{pmatrix} 0 & a \\ -a & 0 \end{pmatrix} \begin{pmatrix} a^{-1} & 0 \\ 0 & 1 \end{pmatrix} = \begin{pmatrix} 0 & 1 \\ -1 & 0 \end{pmatrix}.$$

(3) Assume, without loss, that $a_{12} \neq 0$, and partition A as

$$\begin{pmatrix} A_1 & L \\ -L' & A_2 \end{pmatrix}, \quad A_1 = \begin{pmatrix} 0 & a \\ -a & 0 \end{pmatrix}, \quad A_2 = A_2^{(n-2)}.$$

Then

$$\begin{pmatrix} I & 0 \\ L'A_1^{-1} & I \end{pmatrix} \begin{pmatrix} A_1 & L \\ -L' & A_2 \end{pmatrix} \begin{pmatrix} I & 0 \\ L'A_1^{-1} & I \end{pmatrix}' = \begin{pmatrix} A_1 & 0 \\ 0 & A_2 + L'A_1^{-1}L \end{pmatrix},$$

and $A_2 + L'A_1^{-1}L$ is a skew-symmetric matrix of order $n - 2$, so that the theorem is proved. □

It is then easy to deduce the following theorems.

Theorem 20.8 *A skew-symmetric matrix with odd order does not have full rank.*

Theorem 20.9 *The determinant of a skew-symmetric matrix with full rank is a perfect square.*

Remark We have only applied the usual arithmetic operations, without any expansion.

20.3 Affine types of the affine geometry of quadratic surfaces

Definition 20.10 Let $x = (x_1, \ldots, x_n)$ and $y = (y_1, \ldots, y_n)$. The substitution

$$x = yA + c, \quad |A| \neq 0, \tag{20.11}$$

where $A = A^{(n)}$ and c is an n-dimensional vector, is called an affine transformation. The space formed by $x = (x_1, \ldots, x_n)$ is then called an affine space.

Affine geometry is the study of the properties of geometric figures under affine transformations. Two figures which can be mapped into each other under an affine transformation are said to be affine equivalent. Being affine equivalent is an equivalence relation.

Example 20.11 Any point is affine equivalent to any other point – we say that affine space is reachable; under such a transformation, a straight line is mapped into a straight line.

Example 20.12 Under an affine transformation a plane

$$xa' = \lambda$$

is mapped into another plane

$$yAa' = \lambda - ca'.$$

Also, any plane can be mapped to $x_1 = 0$.

Example 20.13 Any $n+1$ points not on the same plane can be mapped to

$$0, e_1, \ldots, e_n.$$

Proof. We first map a point to 0. Next, from the points $x^{(i)} = (x_1^{(i)}, \ldots, x_n^{(i)})$, $i = 1, 2, \ldots, n$, we form the matrix

$$A = \begin{pmatrix} x_1^{(1)} & \cdots & x_n^{(1)} \\ \vdots & & \vdots \\ x_1^{(n)} & \cdots & x_n^{(n)} \end{pmatrix},$$

which has full rank (because the $n+1$ points do not lie on the same plane). The affine transformation

$$x = yA$$

now maps $x = x^{(i)}$ to $y = e_i$. □

Example 20.14 Take three points $x^{(1)}, x^{(2)}, x^{(3)} = tx^{(1)} + (1-t)x^{(2)}$ on a line, and let them be mapped to $y^{(1)}, y^{(2)}, y^{(3)}$ under the transformation (20.11). Then, from

$$\begin{aligned} x^{(3)} &= tx^{(1)} + (1-t)x^{(2)} = t(y^{(1)}A + c) + (1-t)(y^{(2)}A + c) \\ &= (ty^{(1)} + (1-t)y^{(2)})A + c, \end{aligned}$$

we find that

$$y^{(3)} = ty^{(1)} + (1-t)y^{(2)}.$$

We deduce the following: The ratio between the two distances $x^{(1)}, x^{(3)}$ and $x^{(2)}, x^{(3)}$ is the same as the ratio between the two distances $y^{(1)}, y^{(3)}$ and $y^{(2)}, y^{(3)}$. Note, however, that the distance between points is not an invariant under an affine transformation.

The general second degree (quadratic) surface takes the form

$$xSx' + 2bx' + \gamma = 0; \tag{20.12}$$

here $S = S' = S^{(n)}$, b is a $1 \times n$ vector and γ is a constant vector. Such a quadratic surface can be assigned the matrix

$$G = \begin{pmatrix} S & b' \\ b & \gamma \end{pmatrix} \tag{20.13}$$

of order $n+1$.

When (20.12) has been transformed by (20.11) to

$$\begin{aligned}&(yA+c)S(yA+c)'+2b(yA+c)'+\gamma\\&\quad=yASAy'+2(cSA'+bA')y'+cSc'+2bc'+\gamma,\end{aligned}$$

the corresponding matrix is

$$\begin{aligned}F&=\begin{pmatrix}T&a'\\a&B\end{pmatrix}=\begin{pmatrix}ASA'&A(Sc'+b')\\(cS+b)A'&cSc'+2bc'+\gamma\end{pmatrix}=\begin{pmatrix}A&0\\c&1\end{pmatrix}\begin{pmatrix}S&b'\\b&\gamma\end{pmatrix}\begin{pmatrix}A&0\\c&1\end{pmatrix}'\\&=\begin{pmatrix}A&0\\c&1\end{pmatrix}G\begin{pmatrix}A&0\\c&1\end{pmatrix}'.\end{aligned}$$

Thus, if the quadratic surfaces represented by G and F are affine equivalent, then S and T are congruent. Over the complex field, S and T have the same rank, and G and F also have the same rank; over the real field, S and T have the same signature and rank, and G and F also have the same signature and rank.

More explicitly, over the complex field, there is an A such that

$$ASA'=\begin{pmatrix}I^{(r)}&0\\0&0\end{pmatrix},$$

that is

$$x_1^2+\cdots+x_r^2+2\sum_{i=1}^{n}b_ix_i+\gamma=0.$$

Completing squares then yields

$$(x_1+b_1)^2+\cdots+(x_r+b_r)^2+2\sum_{i=r+1}^{n}b_ix_i+\gamma'=0,\quad \gamma'=\gamma-\sum_{i=1}^{r}b_i^2.$$

There are three cases to be considered.

(i) All the b_i $(i=r+1,\ldots,n)$ are 0 and $\gamma'\neq 0$. Then

$$x_1+b_1=\sqrt{\gamma'}y_1,\quad\cdots,\quad x_r+b_r=\sqrt{\gamma'}y_r$$

transform (20.12) to

$$y_1^2+\cdots+y_r^2+1=0. \tag{20.14}$$

(ii) All the b_i $(i=r+1,\ldots,n)$ and γ' are 0. Then (20.12) is affine equivalent to

$$y_1^2+\cdots+y_r^2=0. \tag{20.15}$$

(iii) Not all the b_i $(i=r+1,\ldots,n)$ are 0. Then from

$$x_1+b_1=y_1,\quad\cdots,\quad x_r+b_r=y_r,\quad \sum_{i=r+1}^{n}b_ix_i+\frac{1}{2}\gamma=y_{r+1},$$

we have

$$y_1^2 + \cdots + y_r^2 + 2y_{r+1} = 0. \tag{20.16}$$

Thus, over the complex field, any quadratic surface must be affine equivalent to one of (20.14), (20.15) or (20.16), and we have also established in passing that the rank of G differs from that of S by at most 2.

Theorem 20.15 *If the ranks of S and G have the same value r, then the quadratic surface is affine equivalent to (20.15); if the rank of S is 1 less than that of G, then the quadratic surface is affine equivalent to (20.14); if the rank of S is 2 less than that of G, then the quadratic surface is affine equivalent to (20.16).*

Consider next the situation over the real field, and suppose first that there is an A such that

$$ASA' = \begin{pmatrix} I^{(r)} & 0 & 0 \\ 0 & -I^{(r-s)} & 0 \\ 0 & 0 & 0 \end{pmatrix}.$$

Then (20.12) is affine equivalent to

$$x_1^2 + \cdots + x_s^2 - x_{s+1}^2 - \cdots - x_r^2 + 2\sum_{i=1}^{n} b_i x_i + \gamma = 0. \tag{20.17}$$

On completing squares we have

$$(x_1 + b_1)^2 + \cdots + (x_s + b_s)^2 - (x_{s+1} - b_{s+1})^2 - \cdots - (x_r - b_r)^2$$
$$+ 2\sum_{i=r+1}^{n} b_i x_i + \gamma' = 0, \quad \gamma' = \gamma - \sum_{i=1}^{s} b_i^2 + \sum_{i=s+1}^{n} b_i^2.$$

(i) If b_i $(i = r+1, \ldots, n)$ and γ' are all 0, then (20.12) is equivalent to

$$x_1^2 + \cdots + x_s^2 - x_{s+1}^2 - \cdots - x_r^2 = 0. \tag{20.18}$$

(ii) If b_i $(i = r+1, \ldots, n)$ are all 0 and $\gamma' \neq 0$, then, from

$$x_1 + b_1 \mapsto \sqrt{|\gamma'|}x_1, \quad \cdots, \quad x_r - b_r \mapsto \sqrt{|\gamma'|}x_r,$$

we find that (20.12) is equivalent to

$$x_1^2 + \cdots + x_s^2 - x_{s+1}^2 - \cdots - x_r^2 = \pm 1. \tag{20.19}$$

(iii) If there is one $b_i \neq 0$ $(i = r+1, \ldots, n)$ then, from

$$x_1 + b_1 \mapsto x_1, \quad \cdots, \quad x_r - b_r \mapsto x_r, \quad \sum_{i=r+1}^{n} b_i x_i + \frac{1}{2}\gamma' \mapsto x_{r+1},$$

we find that (20.12) is equivalent to

$$x_1^2 + \cdots + x_s^2 - x_{s+1}^2 - \cdots - x_r^2 + 2x_{r+1} = 0. \tag{20.20}$$

We therefore have the following theorem.

Theorem 20.16 *Suppose that S has the signature $2s - r$. If the rank of G is the same as that of S, then the quadratic surface is affine equivalent to (20.18); if the rank of G exceeds that of S by 1, then the surface is affine equivalent to (20.19); if the rank of G exceeds that of S by 2, then the surface is affine equivalent to (20.20).*

Special examples. On the real plane, that is $n = 2$, a quadratic curve is affine equivalent to one of the following curves:

(i)	$x_1^2 + x_2^2 = 0,$	a point circle,
(ii)	$x_1^2 - x_2^2 = 0,$	two intersecting straight lines,
(iii)	$x_1^2 = 0,$	a straight line,
(iv)	$x_1^2 + x_2^2 = 1,$	a circle,
(v)	$x_1^2 + x_2^2 = -1,$	an imaginary circle,
(vi)	$x_1^2 - x_2^2 = 1,$	a hyperbola,
(vii)	$x_1^2 = 1,$	two parallel straight lines,
(viii)	$x_1^2 = -1,$	two imaginary straight lines,
(ix)	$x_1^2 = 2x_2,$	a parabola.

In three dimensional space, a quadratic surface is affine equivalent to (considering only the situation when G has full rank):

(i)	$x_1^2 + x_2^2 + x_3^2 = 0,$	a point sphere,
(ii)	$x_1^2 + x_2^2 - x_3^2 = 0,$	a circular bowl,
(iii)	$x_1^2 + x_2^2 + x_3^2 = 1,$	a sphere,
(iv)	$x_1^2 + x_2^2 + x_3^2 = -1,$	an imaginary sphere,
(v)	$x_1^2 + x_2^2 - x_3^2 = 1,$	a single-leaf hyperbolic surface,
(vi)	$x_1^2 - x_2^2 - x_3^2 = 1,$	a double-leaf hyperbolic surface,
(vii)	$x_1^2 + x_2^2 + x_3 = 0,$	an elliptic-parabolic surface,
(viii)	$x_1^2 - x_2^2 + x_3 = 0,$	a hyperbolic-parabolic surface.

Exercise 20.17 Consider the situation when F does not have full rank, and identify the geometric surfaces concerned.

20.4 Projective geometry

Definition 20.18 A projective transformation is a mapping of the form

$$y = \frac{xA + b}{xc + d}, \tag{20.21}$$

where $A = A^{(n,n)}$, $b = b^{(1,n)}$, $c = c^{(n,1)}$, $d = d^{(1)}$, with

$$P = \begin{pmatrix} A & c \\ b & d \end{pmatrix}, \qquad \det P \neq 0. \tag{20.22}$$

If

$$z = \frac{yA^* + b^*}{yc^* + d^*}, \qquad P^* = \begin{pmatrix} A^* & c^* \\ b^* & d^* \end{pmatrix},$$

then

$$z = \frac{(xA + b)A^* + (xc + d)b^*}{(xA + b)c^* + (xc + d)d^*} = \frac{x(AA^* + cb^*) + bA^* + db^*}{x(Ac^* + cd^*) + bc^* + dd^*}$$

is also a projective transformation, with the matrix

$$\begin{pmatrix} AA^* + cb^* & Ac^* + cd^* \\ bA^* + db^* & bc^* + dd^* \end{pmatrix} = \begin{pmatrix} A & c \\ b & d \end{pmatrix} \begin{pmatrix} A^* & c^* \\ b^* & d^* \end{pmatrix}.$$

Thus, the successive application of two projective transformations is also a projective transformation, and the corresponding matrix is just the product of the two matrices of the original transformations. Consequently, corresponding to the inverse of the matrix, we have the inverse transformation to (20.21).

Note. Corresponding to the transformation (20.21), we have the matrix P in (20.22). However, different matrices may correspond to the same transformation – for example, the matrix ρP, where ρ is any non-zero number, also corresponds to the transformation (20.21).

Let us consider those matrices which correspond to (20.21) and are different from P. We need only examine those P which correspond to the identity transformation, by taking

$$x = 0, \qquad e_1, \ldots, e_n$$

in (20.21) and asking for

$$y = 0, \qquad e_1, \ldots, e_n;$$

this yields $b = 0$ and

$$e_i(e_i c + d) = e_i A.$$

With $e_i c + d$ being a number, we find that $a_{ji} = 0$ for $i \neq j$; taking $x = y = \lambda e_i$, we find that, for any $\lambda \neq 0$,

$$(e_i c)\lambda^2 e_i + \lambda d e_i = \lambda a_{ii} e_i, \quad (e_i c)\lambda + d = a_{ii}.$$

Thus $e_i c = 0$, so that $c = 0$, and $a_{ii} = d$, giving

$$P = [d, d, \ldots, d].$$

In other words, only $P = \rho I$ can represent the identity transformation in (20.21). It is not difficult to show that if P and Q represent the same transformation then they differ only by a constant factor. A projective transformation also preserves linear relationships.

An affine transformation is a projective transformation, so that invariants of the latter transformations are also invariants of the former ones. In the following we present some properties of a projective transformation.

Any $n+2$ points, no $n+1$ of which lie on an n-dimensional hyperplane, can be mapped to

$$0,\ e_1,\ \ldots\ ,\ e_n,\ e_1 + \cdots + e_n.$$

Proof. An affine transformation can already map $n+1$ points to

$$0,\ e_1,\ \ldots\ ,\ e_n;$$

let the remaining point be $(a_1, \ldots, a_n)$. By hypothesis, $a_1 \cdots a_n \neq 0$ (because if $a_1 = 0$, say, then the point would go into the hyperplane defined by $0, e_2, \ldots, e_n$). The projective transformation

$$x_i = a_i y_1$$

maps $x_1 = a_i$ to $y_i = 1$, and we are done. □

An alternative form for a projective transformation is as follows: Let

$$\begin{pmatrix} A & c \\ b & d \end{pmatrix} \begin{pmatrix} A^* & c^* \\ b^* & d^* \end{pmatrix} = I;$$

then

$$(xc + d)^{-1}(xA + b) = (d^*x - b^*)(-c^*x + A^*)^{-1},$$

which is easily verified because

$$\begin{aligned}
&(xA + b)(-c^*x + A^*) - (xc + d)(d^*x - b^*) \\
&\quad = -x(Ac^* + cd^*)x + x(AA^* + cb^*) - (bc^* + dd^*)x + bA^* + db^* \\
&\quad = x - x = 0.
\end{aligned}$$

Suppose that $x^{(1)}, x^{(2)}, x^{(3)}, x^{(4)}$ are mapped to $y^{(1)}, y^{(2)}, y^{(3)}, y^{(4)}$ by (20.21). Then

$$\begin{aligned}
y^{(i)} - y^{(j)} &= (x^{(i)}c + d)^{-1}(x^{(i)}A + b) - (d^*x^{(j)} - b^*)(-c^*x^{(j)} + A^*)^{-1} \\
&= (x^{(i)}c + d)^{-1}[(x^{(i)}A + b)(-c^*x^{(j)} + A^*) \\
&\quad - (x^{(i)}c + d)(d^*x^{(j)} - b^*)](-c^*x^{(j)} + A^*)^{-1} \\
&= (x^{(i)}c + d)^{-1}(x^{(i)} - x^{(j)})(-c^*x^{(j)} + A^*)^{-1} \\
&= p^{(i)}(x^{(i)} - x^{(j)})Q^{(j)}.
\end{aligned}$$

Suppose that these points lie on the line joining x^* and x^{**}, and that

$$x^{(i)} = t^{(i)}x^* + (1 - t^{(i)})x^{**},$$

and also that x^*, x^{**} are mapped to y^*, y^{**}, and that $x^{(i)}$ is mapped to $y^{(i)} = \lambda^{(i)}y^* + (1 - \lambda^{(i)})y^{**}$. Then from

$$x^{(i)} - x^{(j)} = (t^{(i)} - t^{(j)})(x^* - x^{**})$$

we find that

$$(\lambda^{(i)} - \lambda^{(j)})(y^* - y^{**}) = p^{(i)}(t^{(i)} - t^{(j)})(x^* - x^{**})Q^{(j)}.$$

The square of the distance between x^i and x^j is $(x^i - x^j)(x^i - x^j)^{-1}$, so that the relationship between the squares of the distances between x^i, x^j and y^i, y^j is

$$\begin{aligned}(\lambda^{(i)} - \lambda^{(j)})^2&(y^* - y^{**})(y^* - y^{**})' \\ &= (p^{(i)})^2(t^{(i)} - t^{(j)})^2(x^* - x^{**})Q^{(i)}Q^{(j)\prime}(x^* - x^{**})'.\end{aligned}$$

The ratio of the distances between $y^{(1)}$, $y^{(2)}$ and between $y^{(3)}$, $y^{(2)}$ is

$$(\lambda^{(1)} - \lambda^{(2)})(\lambda^{(3)} - \lambda^{(2)})^{-1} = p^{(1)}(t^{(1)} - t^{(2)})(t^{(3)} - t^{(2)})^{-1}/p^{(3)}.$$

Therefore

$$\frac{\lambda^{(1)} - \lambda^{(2)}}{\lambda^{(3)} - \lambda^{(2)}} \Bigg/ \frac{\lambda^{(1)} - \lambda^{(4)}}{\lambda^{(3)} - \lambda^{(4)}} = \frac{t^{(1)} - t^{(2)}}{t^{(3)} - t^{(2)}} \Bigg/ \frac{t^{(1)} - t^{(4)}}{t^{(3)} - t^{(4)}}.$$

The number

$$\frac{\lambda^{(1)} - \lambda^{(2)}}{\lambda^{(3)} - \lambda^{(2)}} \Bigg/ \frac{\lambda^{(1)} - \lambda^{(4)}}{\lambda^{(3)} - \lambda^{(4)}}$$

is called the cross-ratio of the four points, and we see that a projective transformation preserves the cross-ratio of four points on a straight line.

20.5 Projective types of quadratic surfaces

Taking the quadratic surface

$$xSx' + 2bx' + \gamma = 0, \quad F = \begin{pmatrix} S & b' \\ b & \gamma \end{pmatrix}$$

through a projective transformation, it becomes

$$\begin{aligned}
&(xA+b)S(xA+b)' + 2(xc+d)b(xA+b)' + \gamma(xc+d)^2 \\
&\quad = (xA+b)S(xA+b)' + (xc+d)b(xA+b)' \\
&\qquad + (xA+b)b'(xc+d)' + \gamma(xc+d)(xc+d)' \\
&\quad = x(ASA' + cbA' + Ab'c' + \gamma cc')x' + 2(bSA' + dbA' + bb'c' + \gamma dc')x' \\
&\qquad + bSb' + 2dbb' + \gamma d^2,
\end{aligned}$$

with the matrix

$$\begin{aligned}
G &= \begin{pmatrix} ASA' + cbA' + Ab'c' + \gamma cc' & bSA' + dbA' + bb'c' + \gamma dc' \\ bSA' + dbA' + bb'c' + \gamma dc' & bSb' + 2dbb' + \gamma d^2 \end{pmatrix} \\
&= \begin{pmatrix} A & c \\ b & d \end{pmatrix} \begin{pmatrix} S & b' \\ b & \gamma \end{pmatrix} \begin{pmatrix} A & c \\ b & d \end{pmatrix}' = PFP'.
\end{aligned}$$

It is therefore rather easy to separate quadratic surfaces into their projective types.

Over the complex field, the rank of F is the only invariant. For any F with rank r, there is a P such that

$$PFP' = \begin{pmatrix} I^{(r)} & 0 \\ 0 & 0 \end{pmatrix}.$$

This means that any quadratic surface must be projectively equivalent to

$$\begin{aligned}
x_1^2 + \cdots + x_r^2 &= 0, \quad \text{if } r < n, \\
x_1^2 + \cdots + x_r^2 &= 1, \quad \text{if } r = n;
\end{aligned}$$

the former one is said to be degenerate. Thus a non-degenerate quadratic surface must be projectively equivalent to the sphere.

Over the real field, if F is non-degenerate, that is F has full rank, then the surface must be equivalent to

$$x_1^2 + \cdots + x_s^2 - x_{s+1}^2 - \cdots - x_n^2 = 1;$$

on the plane, a non-degenrate quadratic surface must be projectively equivalent to a real circle, an imaginary circle or a hyperbola.

20.6 Positive definite forms

We consider quadratic forms over the real field.

Definition 20.19 Suppose that, for any non-trivial vector x,

$$xSx' > 0, \quad (S' = S).$$

Then S, and all its equivalent quadratic forms, are said to be positive definite. A positive semi-definite form is one in which $xSx' \geq 0$ for any vector x.

A positive definite form S must have full rank, since otherwise there would be a non-trivial vector x such that $xS = 0$, and hence $xSx' = 0$.

Being positive definite (or positive semi-definite) is invariant under congruence; that is, if S is positive definite, and $T = PSP'$ (P being a real square matrix with $|P| \neq 0$), then T is also positive definite.

Since any symmetric matrix must be congruent to

$$\begin{pmatrix} I^{(s)} & 0 & 0 \\ 0 & -I^{(r-s)} & 0 \\ 0 & 0 & 0 \end{pmatrix},$$

the corresponding form is

$$(x_1^2 + \cdots + x_s^2 - x_{s+1}^2 - \cdots - x_r^2).$$

Therefore, the condition for being positive semi-definite is $r = s$, and for being positive definite is $r = s = n$. Thus, if S is positive semi-definite then

$$S = P \begin{pmatrix} I^{(r)} & 0 \\ 0 & 0 \end{pmatrix} P'.$$

Here the columns $r + 1, \ldots, n$ of P are not relevant, and if we write the first r columns as

$$Q = \begin{pmatrix} p_{11} & \cdots & p_{1r} \\ \vdots & & \vdots \\ p_{n1} & \cdots & p_{nr} \end{pmatrix}, \qquad Q = Q^{(n,r)},$$

then

$$S = QQ'.$$

There are two special cases:

(1) If $r = 1$ then

$$S = u'u,$$

where u is an n-dimensional vector.

(2) If $r = n$ then any positive definite matrix can be represented as

$$S = PP',$$

with P having full rank.

The converse is also true, because if $y = xP$ then

$$xSx' = yy' = \sum_{i=1}^{n} y_i^2 = 0,$$

so that $y_1 = \cdots = y_n = 0$, and hence $x_1 = \cdots = x_n = 0$.

Theorem 20.20 *The main sub-square matrices of a positive semi-definite matrix are also positive semi-definite.*

To see this, we assume without loss that the sub-matrix is made up of the first ℓ rows and columns of the matrix, that is

$$S = \begin{pmatrix} S_1 & L \\ L' & S_2 \end{pmatrix}, \quad S_1 = S_1^{(\ell)}.$$

Now, for $x = (x_1, \ldots, x_\ell, 0, \ldots, 0)$, we have

$$(x_1, \ldots, x_\ell)S_1(x_1, \ldots, x_\ell)' = xSx' \geq 0,$$

so that S_1 is positive semi-definite.

Theorem 20.21 *The determinant of a positive semi-definite matrix is non-negative.*

This follows at once from

$$\det(PP') = (\det P)^2 \geq 0.$$

We can also deduce the Schwarz inequality:

$$\left| \begin{pmatrix} x \\ y \end{pmatrix} \begin{pmatrix} x \\ y \end{pmatrix}' \right| = \begin{vmatrix} xx' & xy' \\ yx' & yy' \end{vmatrix} = xx'yy' - (xy')^2 \geq 0.$$

Indeed, more generally we have

$$\left| \begin{pmatrix} x \\ y \\ z \end{pmatrix} \begin{pmatrix} x \\ y \\ z \end{pmatrix}' \right| = \begin{vmatrix} xx' & xy' & xz' \\ yx' & yy' & yz' \\ zx' & zy' & zz' \end{vmatrix} \geq 0,$$

and so on.

20.7 Finding the least value from completion of squares

20.7.1 Least value

Let S be a positive definite matrix, and let C be a vector. We proceed to determine the least value for the function

$$F(x_1, \ldots, x_n) = xSx' + 2xC' + \gamma. \tag{20.23}$$

From

$$xSx' + 2xC' + \gamma = (x + CS^{-1})S(x + CS^{-1})' + \gamma - CS^{-1}C' \geq \gamma - CS^{-1}C',$$

we see that the sought-after least value is

$$\gamma - CS^{-1}C', \tag{20.24}$$

which is taken by F when and only when $x = -CS^{-1}$.

If S is negative definite, then (20.23) has a greatest value. If the signature of S is not equal to $\pm n$, then (20.23) has neither a least, nor a greatest, value. Thus, with $S = P'[1, \ldots, 1, -1, \ldots, -1]P'$, we let $x = Py$, which yields

$$F(x_1, \ldots, x_n) = \sum_{i=1}^{s} y_i^2 - \sum_{i=s+1}^{n} y_i^2 + \cdots .$$

It then follows that $F \to \infty$ as $y_1 \to \infty$, and $F \to -\infty$ as $y_n \to \infty$ (keeping the remaining variables fixed).

20.7.2 Conditional maxima and minima

We seek the least value for

$$xSx', \quad \text{subject to the constraint} \quad xC' = \alpha.$$

Consider the least value for

$$xSx' + 2\lambda(xC' - \alpha),$$

where λ is a real number. From Section 20.7.1 we see that

$$\begin{aligned} xSx' + 2\lambda xC' - 2\lambda\gamma &\geq -2\lambda\alpha - \lambda^2 CS^{-1}C' \\ &= \frac{\alpha^2}{CS^{-1}C'} - \left(\lambda + \frac{\alpha}{CS^{-1}C'}\right)^2 CS^{-1}C', \end{aligned}$$

with equality when and only when $x = -\lambda CS^{-1}$. The constraint condition now gives $-\lambda CS^{-1}C' = \alpha$, so that $xSx' \geq \alpha^2/(CS^{-1}C')$, with equality if and only if $x = \alpha CS^{-1}/(CS^{-1}C')$.

It now follows that, for any positive definite matrix, we have the inequality

$$(xC')^2 (= \alpha^2) \leq (xSx')(CS^{-1}C'),$$

which is also a generalised version of the Schwarz inequality. More specifically, the ordinary Schwarz inequality corresponds to $S = I$, and the reader may care to use this to deduce the more general inequality here.

20.7.3 General constraint condition

Suppose next that the constraint condition is more general, say

$$xC = \alpha,$$

where $C = C^{(n,\ell)}$, and α is an ℓ-dimensional vector. We consider the function

$$xSx' + 2(xC - \alpha)\lambda' = xSx' + 2xC\lambda' - 2\alpha\lambda',$$

where λ is an ℓ-dimensional vector. By Section 20.7.1, we have

$$xSx' + 2(xC - \alpha)\lambda' \geq -2\alpha\lambda' - \lambda C' S^{-1} C\lambda',$$

with equality only when $x = -\lambda C'S^{-1}$. From the constraint condition

$$xC = -\lambda C'S^{-1}C = \alpha, \quad \lambda = -\alpha(C'S^{-1}C)^{-1},$$

we arrive at

$$xSx' \geq -\alpha\lambda' = \alpha(C'S^{-1}C)^{-1}\alpha'.$$

20.8 The Hessian

The Hessian of a function $F(x_1, \ldots, x_n)$ is the square matrix

$$H(x_1, \ldots, x_n, F) = \left(\frac{\partial^2 F}{\partial x_i \partial x_j}\right)_{1 \leq i, j \leq n}.$$

With the substitution $x_i = f_i(y_1, \ldots, y_n)$, we have

$$\frac{\partial F}{\partial x_i} = \sum_{k=1}^{n} \frac{\partial F}{\partial y_k} \frac{\partial y_k}{\partial x_i},$$

$$\frac{\partial^2 F}{\partial x_i \partial x_j} = \sum_{k=1}^{n} \sum_{\ell=1}^{n} \frac{\partial^2 F}{\partial y_k \partial y_\ell} \frac{\partial y_k}{\partial x_i} \frac{\partial y_\ell}{\partial x_j} + \sum_{k=1}^{n} \frac{\partial F}{\partial y_k} \frac{\partial^2 y_k}{\partial x_i \partial x_j},$$

that is

$$H(x_1, \ldots, x_n, F) = \left(\frac{\partial(y_1, \ldots, y_n)}{\partial(x_1, \ldots, x_n)}\right)' H(y_1, \ldots, y_n, F) \left(\frac{\partial(y_1, \ldots, y_n)}{\partial(x_1, \ldots, x_n)}\right) + \sum_{k=1}^{n} \frac{\partial F}{\partial y_k} H(x_1, \ldots, x_n, y_k).$$

If the substitution is linear,

$$x_i = \sum_{i=1}^{n} y_j a_{ji}, \quad x = yA,$$

then

$$H(x_1, \ldots, x_n) = (A^{-1})' H(y_1, \ldots, y_n) A^{-1}.$$

Definition 20.22 If, within a region D, the Hessian $H(x_1, \ldots, x_n)$ is positive, then the function is said to be concave downward ('cup'), and if $H(x_1, \ldots, x_n)$ is negative, then it is convex upward ('cap').

From the above, we see that being concave or convex is an invariant property with respect to a linear substitution.

20.9 Types of second order partial differential equations with constant coefficients

We consider the second order partial differential equation with constant coefficients

$$\sum_{i,j=1}^{n} a_{ij} \frac{\partial^2 u}{\partial x_i \partial x_j} + 2 \sum_{i=1}^{n} b_i \frac{\partial u}{\partial x_i} + cu = 0, \quad a_{ij} = a_{ji}. \tag{20.25}$$

With the substitution

$$x_s = \sum_{t=1}^{n} y_t p_{ts}, \quad x = yP, \tag{20.26}$$

equation (20.25) becomes

$$\sum_{s,t=1}^{n} a_{st}^* \frac{\partial^2 u}{\partial y_s \partial y_t} + 2 \sum_{s=1}^{n} b_s^* \frac{\partial u}{\partial y_s} + c^* u = 0. \tag{20.27}$$

We consider the relationship between the coefficients:

$$\frac{\partial u}{\partial y_s} = \sum_{i=1}^{n} \frac{\partial u}{\partial x_i} \frac{\partial x_i}{\partial y_s} = \sum_{i=1}^{n} \frac{\partial u}{\partial x_i} p_{si}$$

and

$$\frac{\partial^2 u}{\partial y_s \partial y_t} = \sum_{i,j=1}^{n} \frac{\partial^2 u}{\partial x_i \partial x_j} p_{ti} p_{sj},$$

so that

$$\begin{aligned} &\sum_{s,t=1}^{n} a_{st}^* \frac{\partial^2 u}{\partial y_s \partial y_t} + 2 \sum_{s=1}^{n} b_s^* \frac{\partial u}{\partial y_s} + c^* u \\ &= \sum_{i,j=1}^{n} \frac{\partial^2 u}{\partial x_i \partial x_j} \sum_{s,t=1}^{n} p_{ti} a_{st}^* p_{sj} + 2 \sum_{i=1}^{n} \frac{\partial u}{\partial x_i} \sum_{s=1}^{n} b_s^* p_{si} + c^* u, \end{aligned}$$

that is

$$a_{ij} = \sum_{s,t=1}^{n} p_{ti} a_{st}^* p_{sj}, \quad b_i = \sum_{s=1}^{n} b_s^* p_{si}, \quad c = c^*;$$

or

$$A = P' A^* P, \quad b = b^* P,$$

that is

$$A^* = (P')^{-1} A P^{-1}, \quad b^* = b P^{-1}.$$

Thus A^* and Λ are congruent, so that, over the real field, there is a matrix P such that

$$(P')^{-1}AP^{-1} = \begin{pmatrix} I^{(p)} & 0 & 0 \\ 0 & -I^{(r-p)} & 0 \\ 0 & 0 & 0 \end{pmatrix}.$$

Therefore (20.25) is reduced to

$$\sum_{i=1}^{p} \frac{\partial^2 u}{\partial x_i^2} - \sum_{i=p+1}^{r} \frac{\partial^2 u}{\partial x_i^2} + 2\sum_{i=1}^{n} b_i \frac{\partial u}{\partial x_i} + cu = 0. \tag{20.28}$$

Let

$$u = ve^{\alpha_1 x_1 + \cdots + \alpha_r x_r},$$

and substitute

$$\frac{\partial u}{\partial x_i} = \left(\frac{\partial v}{\partial x_i} + \alpha_i v\right)e^{\alpha_1 x_1 + \cdots + \alpha_r x_r},$$
$$\frac{\partial^2 u}{\partial x_i^2} = \left(\frac{\partial^2 v}{\partial x_i^2} + 2\alpha_i \frac{\partial v}{\partial x_i} + \alpha^2 v\right)e^{\alpha_1 x_1 + \cdots + \alpha_r x_r}$$

into (20.28) to give

$$\sum_{i=1}^{p} \frac{\partial^2 v}{\partial x_i^2} - \sum_{i=p+1}^{r} \frac{\partial^2 v}{\partial x_i^2} + 2\sum_{i=1}^{p}(\alpha_i + b_i)\frac{\partial v}{\partial x_i}$$
$$+ 2\sum_{i=p+1}^{r}(-\alpha_i + b_i)\frac{\partial v}{\partial x_i} + 2\sum_{r+1}^{n} b_i \frac{\partial v}{\partial x_i} + c_0 v = 0,$$

and we can take $\alpha_i = -b_i$ for $i = 1, 2, \ldots, p$, and $\alpha_i = b_i$ for $i = p+1, \ldots, n$.

Moreover, on taking P so that

$$(b_{r+1}, \ldots, b_n)P^{-1} = (1, 0, \ldots, 0),$$

we arrive at the standard form

$$\sum_{i=1}^{p} \frac{\partial^2 v}{\partial x_i^2} - \sum_{i=p+1}^{r} \frac{\partial^2 v}{\partial x_i^2} + \epsilon \frac{\partial v}{\partial x_{r+1}} + cv = 0, \quad \epsilon = 0 \text{ or } 1. \tag{20.29}$$

If $\epsilon = 1$, then we take $v = we^{-cx_{r+1}}$, so that c may be reduced to 0. If $\epsilon = 0$, then we may use $x_i \to |c|^{-1/2} y_i$ and change c to ± 1. Thus we have the following types of standard forms:

$$\text{(i)} \qquad \sum_{i=1}^{p} \frac{\partial^2 v}{\partial x_i^2} - \sum_{i=p+1}^{r} \frac{\partial^2 v}{\partial x_i^2} = 0,$$

$$\text{(ii)} \qquad \sum_{i=1}^{p} \frac{\partial^2 v}{\partial x_i^2} - \sum_{i=p+1}^{r} \frac{\partial^2 v}{\partial x_i^2} \pm v = 0,$$

$$\text{(iii)} \qquad \sum_{i=1}^{p} \frac{\partial^2 v}{\partial x_i^2} - \sum_{i=p+1}^{r} \frac{\partial^2 v}{\partial x_i^2} + \frac{\partial v}{\partial x_{r+1}} = 0.$$

20.10 Hermitian forms

Take the complex field, and write $\bar{a}$ for the complex conjugate of a. By a Hermitian form we mean an algebraic expression:

$$\sum_{i,j=1}^{n} a_{ij} z_i \bar{z}_j, \qquad a_{ij} = \bar{a}_{ji}.$$

The corresponding square matrix

$$H = (a_{ij}), \quad \text{which satisfies} \quad H = \bar{H}',$$

is called a Hermitian matrix.

If $K = -\bar{K}'$, then K is said to be skew-Hermitian; it is clear that, in this case, iK is Hermitian. Thus, although the notions for symmetric and skew-symmetric are somewhat different, there is nothing new in the study of skew-Hermitian matrices.

A Hermitian form can be written as

$$zH\bar{z}'.$$

After the substitution

$$z = wP,$$

the form becomes

$$wPH\bar{P}'\bar{w},$$

with the matrix

$$H_1 = PH\bar{P}'.$$

We say that H_1 and H are conjunctive, or H-congruent, and we write

$$H_1 \overset{H}{=} H.$$

It is clear that if A is a Hermitian matrix, and $B \overset{H}{=} A$, then B is also Hermitian. Being conjunctive is an equivalence relation: (i) $A \overset{H}{=} A$; (ii) $A \overset{H}{=} B$ implies $B \overset{H}{=} A$; (iii) $A \overset{H}{=} B$ and $B \overset{H}{=} C$ implies $A \overset{H}{=} C$.

Corresponding to symmetric matrices over the real field, we can prove the following theorem.

Theorem 20.23 *Any Hermitian matrix must be H-congruent to*

$$[1, \ldots, 1, -1, \ldots, -1, 0 \ldots, 0].$$

Suppose that there are p entries with the value 1 and q entries with the value -1 in the diagonal matrix. Then $r = p + q$ is the rank, and we also call $p - q$ the signature of the Hermitian matrix. We deduce the following.

Theorem 20.24 *A necessary and sufficient condition for two Hermitian matrices to be H-congruent is that they should have the same rank and have the same signature.*

Definition 20.25 Suppose that $zH\bar{z}' > 0$, except when $z = 0$. Then we say that H is positive definite. If we always have $zH\bar{z}' \geq 0$ then H is positive semi-definite.

A necessary and sufficient condition for positive definiteness is that the order, the rank and the signature are the same integer.

20.11 The real shape of a Hermitian form

Write $z = x + iy$ and

$$H = S + iK,$$

where the real matrices S and K are symmetric and skew-symmetric, respectively. Then

$$\begin{aligned} zH\bar{z}' &= (x + iy)(S + iK)(x' - iy') = [xS - yK + i(yS + xK)](x' - iy') \\ &= (xS - yK)x' + (yS + xK)y' - [(xS - yK)y' - (yS + xK)x']i. \end{aligned}$$

From $yKy' = 0$, $xKx' = 0$ and $xSy' = ySx'$, we find that the imaginary part vanishes, so that

$$zH\bar{z}' = xS\bar{x}' + 2xKy' + ySy' = (x, y)\begin{pmatrix} S & K \\ -K & S \end{pmatrix}(x, y)'.$$

Therefore, a Hermitian matrix of order n can be considered as a special symmetric matrix of order $2n$:

$$\begin{pmatrix} S & K \\ -K & S \end{pmatrix}.$$

Also, the substitution $z = wP$ can be written as $w = u + iv$, $P = A + iB$, so that

$$x + iy = (u + iv)(A + iB) = uA - vB + i(vA + uB),$$

$$(x, y) = (u, v)\begin{pmatrix} A & B \\ -B & A \end{pmatrix}.$$

The square matrix

$$\begin{pmatrix} A & B \\ -B & A \end{pmatrix} \tag{20.30}$$

can be extracted from the following: Any matrix which commutes with

$$\begin{pmatrix} 0 & I \\ -I & 0 \end{pmatrix} \tag{20.31}$$

must have the form (20.30), that is, from

$$\begin{pmatrix} 0 & I \\ -I & 0 \end{pmatrix}\begin{pmatrix} A & B \\ C & D \end{pmatrix} = \begin{pmatrix} A & B \\ C & D \end{pmatrix}\begin{pmatrix} 0 & I \\ -I & 0 \end{pmatrix},$$

we have $B = -C$, $A = D$.

Therefore, one may say that the theory of Hermitian matrices can be considered as the study of some special symmetric matrices under specific substitutions, and by special we mean matrices of the form (20.30).

21

Orthogonal groups corresponding to quadratic forms

21.1 Orthogonal groups

We consider the orthogonal transformations, which are linear transformations in a real vector space that preserve the lengths of vectors. In other words, we have

$$y = x\Gamma, \tag{21.1}$$

with

$$xx' = yy'. \tag{21.2}$$

Substituting (21.1) into (21.2) we have the identity

$$x\Gamma\Gamma'x' = xx',$$

so that

$$\Gamma\Gamma' = I. \tag{21.3}$$

A matrix satisfying (21.3) is said to be orthogonal. It is clear that the result of two successive orthogonal transformations is an orthogonal transformation, that is the product of two orthogonal matrices is orthogonal. The collection of orthogonal transformations (or matrices) of order n is called the orthogonal group. From (21.3) we see that $|\Gamma| = \pm 1$ for members of the group.

We proceed to investigate certain problems associated with this orthogonal group. An orthogonal transformation not only preserves the lengths of vectors, but also the cosine of the angle subtended between two vectors. That is, if two vectors $x^{(1)}$, $x^{(2)}$ are mapped to $y^{(1)}$, $y^{(2)}$, respectively, then

$$L(x^{(1)}, x^{(2)}) = \frac{x^{(1)}x^{(2)'}}{\sqrt{x^{(1)}x^{(1)'}x^{(2)}x^{(2)'}}}$$

becomes

$$\frac{y^{(1)}y^{(2)\prime}}{\sqrt{y^{(1)}y^{(1)\prime}y^{(2)}y^{(2)\prime}}}.$$

The condition for a matrix to be orthogonal can also be written as folllows: Let $\Gamma = (c_{ij})$; then

$$\sum_{i=1}^{n} c_{ij}c_{jk} = \delta_{jk}, \tag{21.4}$$

where $\delta_{jk} = 0$ if $j \neq k$ and $\delta_{ii} = 1$. From $\Gamma'\Gamma = I$ we see that Γ' is also orthogonal, so that there is also a similar set of equations corresponding to the columns. From (21.4) we also deduce that an orthogonal matrix is made up of n mutually orthogonal unit vectors

$$c^{(j)} = (c_{1j}, \ldots, c_{nj}).$$

Consider the situation when $n = 2$. From

$$c_{11}^2 + c_{12}^2 = 1,$$

we may take $c_{11} = \cos\theta$, giving $c_{12} = \pm\sin\theta$, and we may assume without loss that $c_{12} = \sin\theta$ (otherwise, we replace θ by $-\theta$). From

$$c_{11}c_{21} + c_{12}c_{22} = 0,$$

we see that $c_{21} = -\rho\sin\theta$, $c_{22} = \rho\cos\theta$, and again, from $c_{21}^2 + c_{22}^2 = 1$, we deduce that $\rho = \pm 1$. Therefore a two dimensional orthogonal matrix must have one of the forms

$$\begin{pmatrix} \cos\theta & \sin\theta \\ -\sin\theta & \cos\theta \end{pmatrix}, \quad \begin{pmatrix} \cos\theta & \sin\theta \\ \sin\theta & -\cos\theta \end{pmatrix}.$$

The former represents a rotation, and the value of the determinant is 1; the latter represents a reflection, so that its square is the identity matrix:

$$\begin{pmatrix} \cos\theta & \sin\theta \\ \sin\theta & -\cos\theta \end{pmatrix}^2 = \begin{pmatrix} 1 & 0 \\ 0 & 1 \end{pmatrix}.$$

The determinant of a reflection has the value -1; the transformation is the product of a rotation and the special reflection

$$\begin{pmatrix} 1 & 0 \\ 0 & -1 \end{pmatrix}.$$

In general, if we let

$$a_{ii} = \cos\theta, \quad a_{ij} = \sin\theta, \quad a_{ji} = -\sin\theta, \quad a_{jj} = \cos\theta,$$

and set the remaining elements

$$a_{st} = \delta_{st},$$

then the matrix so defined represents a rotation on the ij-plane.

Consider

$$\begin{pmatrix} c_{11} & \cdots & c_{1n} \\ c_{21} & \cdots & c_{2n} \\ \vdots & & \vdots \\ c_{n1} & \cdots & c_{nn} \end{pmatrix} \begin{pmatrix} \cos\theta & \sin\theta & 0 & \cdots & 0 \\ -\sin\theta & \cos\theta & 0 & \cdots & 0 \\ \vdots & \vdots & \vdots & & \vdots \\ 0 & 0 & 0 & \cdots & 1 \end{pmatrix}$$

and take θ so that the element in position (2, 1) assumes the value 0 (that is, the solution of $c_{21}\cos\theta - c_{22}\sin\theta = 0$). Then, on multiplication by

$$\begin{pmatrix} \cos\theta & 0 & \sin\theta & 0 & \cdots & 0 \\ 0 & 1 & 0 & 0 & \cdots & 0 \\ -\sin\theta & 0 & \cos\theta & 0 & \cdots & 0 \\ \vdots & \vdots & \vdots & \vdots & & \vdots \\ 0 & 0 & 0 & 0 & \cdots & 1 \end{pmatrix},$$

the element in position (3, 1) takes the value 0; continuing in this way, we arrive at a matrix with $c_{21} = \cdots = c_{n1} = 0$, so that $c_{11} = \pm 1$. From the property of being orthogonal, we have $c_{12} = \cdots = c_{1n} = 0$.

We next multiply the resulting matrix by

$$\begin{pmatrix} 1 & 0 & 0 & 0 & \cdots & 0 \\ 0 & \cos\theta & \sin\theta & 0 & \cdots & 0 \\ 0 & -\sin\theta & \cos\theta & 0 & \cdots & 0 \\ \vdots & \vdots & \vdots & \vdots & & \vdots \\ 0 & 0 & 0 & 0 & \cdots & 1 \end{pmatrix}$$

to deliver another 0 for the element at position (3, 2). Proceeding in this way, we see that any orthogonal matrix can be turned into the diagonal matrix

$$[\pm 1, \pm 1, \ldots, \pm 1] \tag{21.5}$$

by the multiplication of a series of matrices representing rotations on planes.

Taking $\theta = \pi$, so that

$$\begin{pmatrix} \cos\theta & \sin\theta \\ -\sin\theta & \cos\theta \end{pmatrix} = \begin{pmatrix} -1 & 0 \\ 0 & -1 \end{pmatrix},$$

the matrix (21.5) can be given rotations so that it becomes

$$\text{either } [1, 1, \ldots, 1, 1] \quad \text{or } [1, 1, \ldots, 1, -1],$$

depending on whether the value of its determinant is $+1$ or -1.

Summarising, any orthogonal matrix with its determinant having the value 1 can be represented as a product of matrices representing rotations. If the determinant has the value -1, then it is represented as a product of such matrices, together with the reflection represented by $[1, 1, \ldots, 1, -1]$.

Theorem 21.1 *Correpsonding to any m mutually orthogonal unit vectors $u^{(1)}, \ldots, u^{(m)}$, we can always find $n - m$ vectors $u^{(m+1)}, \ldots, u^{(n)}$ such that*

$$\begin{pmatrix} u^{(1)} \\ \vdots \\ u^{(n)} \end{pmatrix}$$

is an orthogonal matrix.

Proof. (1) We first show that, for any unit vector u, there is an orthogonal matrix Γ such that

$$u\Gamma = e_1.$$

We can choose θ so that the second component in

$$(u_1, u_2, \ldots, u_n) \begin{pmatrix} \cos\theta & \sin\theta & 0 & \cdots & 0 \\ -\sin\theta & \cos\theta & 0 & \cdots & 0 \\ 0 & 0 & 1 & \cdots & 0 \\ \vdots & \vdots & \vdots & & \vdots \\ 0 & 0 & 0 & \cdots & 1 \end{pmatrix}$$

vanishes, and the required result follows by continuing in such a manner.

(2) Suppose that we already have $u^{(1)}\Gamma_1 = e_1$. Then

$$\begin{pmatrix} u^{(1)} \\ \vdots \\ u^{(m)} \end{pmatrix} \Gamma_1 = \begin{pmatrix} e_1 \\ v^{(2)} \\ \vdots \\ v^{(m)} \end{pmatrix};$$

from $v^{(2)}, \ldots, v^{(m)}$ being orthogonal to e_1, we see that all their first components vanish. Proceeding in this way, we find that

$$\begin{pmatrix} u^{(1)} \\ \vdots \\ u^{(m)} \end{pmatrix} \Gamma_1 \Gamma_2 \cdots \Gamma_m = \begin{pmatrix} e_1 \\ e_2 \\ \vdots \\ e_m \end{pmatrix},$$

and

$$\begin{pmatrix} e_1 \\ \vdots \\ e_n \end{pmatrix} \Gamma'_m \cdots \Gamma'_2 \Gamma'_1 = \begin{pmatrix} u^{(1)} \\ \vdots \\ u^{(n)} \end{pmatrix}$$

then delivers the sought-after vectors. □

We extend the definition for an orthogonal group.

Definition 21.2 Let S be a positive definite symmetric matrix. Any square matrix T satisfying

$$TST' = S$$

is said to be S-orthogonal.

It is clear that being S-orthogonal is multiplicatively closed – that is, the product of any two S-orthogonal matriccs is also S-orthogonal.

Since any positive definite matrix has the representation $S = PP'$, we have

$$TPP'T' = PP',$$

that is

$$(P^{-1}TP)(P^{-1}TP)' = I,$$

so that $P^{-1}TP$ is an orthogonal matrix.

There is nothing special in the study of S-orthogonal matrices – their results can be deduced from ordinary orthogonality. However, if S is not positive definite, then it is a different matter. For example, if S has only one negative term, then we have the Lorentz group, about which we shall say nothing about here.

21.2 Square roots of positive definite forms for distance functions

In ordinary geometry, the distance between two points

$$x = (x_1, \dots, x_n), \quad y = (y_1, \dots, y_n)$$

is given by

$$d(x, y) = \sqrt{(x - y)(x - y)'} = \sqrt{\sum_{i=1}^{n} (x_i - y_i)^2}.$$

The function $d(x, y)$ satisfies the following four conditions.

(i) Positive definiteness: $d(x, y) \geq 0$, with equality if and only if $x = y$.
(ii) Symmetry: $d(x, y) = d(y, x)$.

(iii) Proportionality: $d(\lambda x, \lambda y) = |\lambda| d(x, y)$.
(iv) Triangle inequality:

$$d(x, y) + d(y, z) \geq d(x, z), \tag{21.6}$$

with equality if and only if the three points are collinear.

The following is the algebraic proof of the triangle inequality: Let $x_i - y_i = a_i$ and $y_i - z_i = b_i$, so that (21.6) becomes

$$\sqrt{aa'} + \sqrt{bb'} \geq \sqrt{(a+b)(a+b)'}.$$

On squaring, this is equivalent to

$$aa'bb' \geq (ab')^2,$$

which is the Schwarz inequality.

The cosine rule gives

$$d(x, z)^2 = d(x, y)^2 + d(y, z)^2 + 2d(x, y)d(y, z)\cos\theta,$$

that is

$$\cos\theta = \frac{(x-y)(y-z)'}{\sqrt{(x-y)(x-y)'(y-z)(y-z)'}}, \tag{21.7}$$

where θ is the angle subtended by the sides (x, y) and (y, z).

Problem 21.3 Are there other functions of the two points x, y which also possess the properties (i), (ii), (iii) and (iv)?

The answer is that there are indeed plenty of others, and the most obvious one is the following: Let S be a positive definite symmetric matrix. Then

$$d(x, y) = \sqrt{(x-y)S(x-y)'}$$

satisfies (i), (ii), (iii) and (iv). The proof is not difficult – we need only write $S = PP'$ and let $\xi = xP$, $\eta = yP$, $\zeta = zP$, and the required results then follow from the above.

For another example, we may take $p \geq 1$ and set

$$d(x, y) = \sqrt[p]{\sum_{i=1}^{n} |x_i - y_i|^p}.$$

21.3 Normed spaces

We see from the preceding sections that xSy' can be very important in applications, and this leads us to the following notion.

Let R denote the n-dimensional vector space over the real field. We define the inner product (or scalar product) between two vectors x, y to be a function $\langle x, y\rangle$ having the following properties:

(1) $\langle x, y\rangle = \langle y, x\rangle$;
(2) for any real λ, we have $\langle \lambda x, y\rangle = \lambda\langle x, y\rangle$;
(3) $\langle x+y, z\rangle = \langle x, z\rangle + \langle y, z\rangle$.

We deduce at once that

(2′) $\langle x, \lambda y\rangle = \lambda\langle x, y\rangle$;
(3′) $\langle x, y+z\rangle = \langle x, y\rangle + \langle x, z\rangle$.

We also write $N(x)$ for $\langle x, x\rangle$. If

(4) $N(x) \geq 0$, with equality when and only when $x = 0$, then we say that the function is a Euclidean norm in the vector space.

By a normed space we mean a real vector space R in which the vectors have norms. In this case, $\sqrt{\langle x, x\rangle}$ is called the length of the vector x, and vectors having unit length are called unit vectors, or normal vectors. Any non-zero vector x can be scaled to give the unit vector $x/\sqrt{\langle x, x\rangle}$.

If $\langle x, y\rangle = 0$ then the two vectors x and y are said to be orthogonal to each other, and we write $x \perp y$. In this case, we deduce at once from (1), (2) and (3) that

$$N(x+y) = \langle x+y, x+y\rangle = \langle x, x\rangle + \langle y, y\rangle = N(x) + N(y),$$

which is a generalised form of Pythagoras' theorem.

Let $a^{(1)}, \ldots, a^{(n)}$ be linearly independent vectors in the n-dimensional vector space over the real field. Then, from

$$x = \sum_{i=1}^{n} x_i a^{(i)}, \quad y = \sum_{i=1}^{n} y_i a^{(i)},$$

along with (2), (3), (2′) and (3′), we have

$$\begin{aligned}\langle x, y\rangle &= \left\langle \sum_{i=1}^{n} x_i a^{(i)}, \sum_{j=1}^{n} y_j a^{(j)} \right\rangle = \sum_{i=1}^{n} x_i \left\langle a^{(i)}, \sum_{j=1}^{n} y_j a^{(j)} \right\rangle \\ &= \sum_{i=1}^{n}\sum_{j=1}^{n} x_i y_j \left\langle a^{(i)}, a^{(j)}\right\rangle = \sum_{i,j=1}^{n} S_{ij} x_i y_j,\end{aligned}$$

where $S_{ij} = \left\langle a^{(i)}, a^{(j)}\right\rangle$, giving

$$N(x) = \sum_{i=1}^{n}\sum_{j=1}^{n} S_{ij} x_i x_j.$$

Property (4) then amounts to $S = (S_{ij})$ being positive definite, so that the abstract way of dealing with a norm includes the explicit way, and there is nothing new to the definition

$$\langle x, y\rangle = xSy'.$$

Nevertheless, the abstract definition leads us to various other notions.

Definition 21.4 Let $e_1, \ldots, e_m$ satisfy

$$\langle e_i, e_j \rangle = \delta_{ij} = \begin{cases} 0, & \text{if } i \neq j, \\ 1, & \text{if } i = j, \end{cases} \qquad i, j = 1, \ldots, m.$$

These are then m mutually orthogonal unit vectors, and we say that they form an orthonormal set; when $m = n$, we say they form an orthonormal basis.

From $S = PP'$, we set

$$\begin{pmatrix} e_1 \\ \vdots \\ e_n \end{pmatrix} = P^{-1} \begin{pmatrix} a^{(1)} \\ \vdots \\ a^{(n)} \end{pmatrix},$$

so that

$$\begin{pmatrix} e_1 \\ \vdots \\ e_n \end{pmatrix} \begin{pmatrix} e_1 \\ \vdots \\ e_n \end{pmatrix}' = I,$$

that is $\langle e_i, e_j \rangle = \delta_{ij}$, and $e_1, \ldots, e_n$ form an orthonormal basis.

21.4 The Gram–Schmidt process

Let $x^{(1)}, \ldots, x^{(m)}$ be m linearly independent vectors. The Gram–Schmidt process is the method of obtaining an orthonormal set from these m vectors.

We first form the unit vector

$$u^{(1)} = \frac{1}{\sqrt{\langle x^{(1)}, x^{(1)} \rangle}} x^{(1)}.$$

We next take

$$u^{(2)} = \alpha u^{(1)} + \beta x^{(2)},$$

where α, β are chosen so that $u^{(2)}$ is a unit vector orthogonal to $u^{(1)}$. The orthogonality requirement dictates that

$$0 = \alpha + \beta \langle x^{(2)}, u^{(1)} \rangle = \alpha + \frac{\beta}{\sqrt{\langle x^{(1)}, x^{(1)} \rangle}} \langle x^{(2)}, x^{(1)} \rangle,$$

that is

$$\alpha = -\beta \frac{\langle x^{(2)}, x^{(1)}\rangle}{\sqrt{\langle x^{(1)}, x^{(1)}\rangle}};$$

$u^{(2)}$ being a unit vector implies that

$$\begin{aligned}
1 = \langle u^{(2)}, u^{(2)}\rangle &= \langle \alpha u^{(1)} + \beta x^{(2)}, \alpha u^{(1)} + \beta x^{(2)}\rangle \\
&= \alpha^2 + 2\alpha\beta\langle u^{(1)}, x^{(2)}\rangle + \beta^2\langle x^{(2)}, x^{(2)}\rangle \\
&= \beta^2\Big(-\frac{\langle x^{(1)}, x^{(2)}\rangle^2}{\langle x^{(1)}, x^{(1)}\rangle} + \langle x^{(2)}, x^{(2)}\rangle\Big) \\
&= \beta^2 \frac{\langle x^{(1)}, x^{(1)}\rangle\langle x^{(2)}, x^{(2)}\rangle - \langle x^{(1)}, x^{(2)}\rangle^2}{\langle x^{(1)}, x^{(1)}\rangle},
\end{aligned}$$

giving

$$\begin{aligned}
\beta &= \frac{\sqrt{\langle x^{(1)}, x^{(1)}\rangle}}{\sqrt{\langle x^{(1)}, x^{(1)}\rangle\langle x^{(2)}, x^{(2)}\rangle - \langle x^{(1)}, x^{(2)}\rangle^2}}, \\
\alpha &= -\frac{\langle x^{(2)}, x^{(1)}\rangle}{\sqrt{\langle x^{(1)}, x^{(1)}\rangle\langle x^{(2)}, x^{(2)}\rangle - \langle x^{(1)}, x^{(2)}\rangle^2}}.
\end{aligned}$$

We next take

$$u^{(3)} = \alpha u^{(1)} + \beta u^{(2)} + \gamma x^{(3)},$$

where the symbols α and β are used again as new unknowns, and together with the unknown γ they can be solved from the three equations $u^{(1)}u^{(3)\prime} = 0$, $u^{(2)}u^{(3)\prime} = 0$ and $u^{(3)}u^{(3)\prime} = 1$.

Proceeding in this way, we finally arrive at a set of m unit vectors,

$$\begin{aligned}
u^{(1)} &= \alpha_{11}x^{(1)}, \\
u^{(2)} &= \alpha_{21}x^{(1)} + \alpha_{22}x^{(2)}, \\
&\cdots \\
u^{(m)} &= \alpha_{m1}x^{(1)} + \alpha_{m2}x^{(2)} + \cdots + \alpha_{mm}x^{(m)},
\end{aligned}$$

which are mutually orthogonal to each other, that is

$$\begin{pmatrix} u^{(1)} \\ \vdots \\ u^{(m)} \end{pmatrix} = \begin{pmatrix} \alpha_{11} & 0 & \cdots & 0 \\ \alpha_{21} & \alpha_{22} & \cdots & 0 \\ \vdots & \vdots & & \vdots \\ \alpha_{m1} & \alpha_{m2} & \cdots & \alpha_{mm} \end{pmatrix} \begin{pmatrix} x^{(1)} \\ \vdots \\ x^{(m)} \end{pmatrix}.$$

From

$$I^{(m)} = \begin{pmatrix} u^{(1)} \\ \vdots \\ u^{(m)} \end{pmatrix} S \begin{pmatrix} u^{(1)} \\ \vdots \\ u^{(m)} \end{pmatrix}'$$

$$= \begin{pmatrix} \alpha_{11} & 0 & \cdots & 0 \\ \alpha_{21} & \alpha_{22} & \cdots & 0 \\ \vdots & \vdots & & \vdots \\ \alpha_{m1} & \alpha_{m2} & \cdots & \alpha_{mm} \end{pmatrix} \begin{pmatrix} x^{(1)} \\ \vdots \\ x^{(m)} \end{pmatrix} S \begin{pmatrix} x^{(1)} \\ \vdots \\ x^{(m)} \end{pmatrix}' \begin{pmatrix} \alpha_{11} & 0 & \cdots & 0 \\ \alpha_{21} & \alpha_{22} & \cdots & 0 \\ \vdots & \vdots & & \vdots \\ \alpha_{m1} & \alpha_{m2} & \cdots & \alpha_{mm} \end{pmatrix}',$$

we see that the above problem has been transformed to one on the positive definite matrix

$$\begin{pmatrix} x^{(1)} \\ \vdots \\ x^{(m)} \end{pmatrix} S \begin{pmatrix} x^{(1)} \\ \vdots \\ x^{(m)} \end{pmatrix}'.$$

By 'completing the squares', this can be reduced to the unit matrix, and the values for α_{ij} have already been computed in Chapter 20.

We still need a supplement as follows.

Theorem 21.5 (Gram) *A necessary and sufficient condition for m vectors* $x^{(1)}, \ldots, x^{(m)}$ *to be linearly independent is that*

$$\begin{pmatrix} x^{(1)} \\ \vdots \\ x^{(m)} \end{pmatrix} S \begin{pmatrix} x^{(1)} \\ \vdots \\ x^{(m)} \end{pmatrix}'$$

should be positive definite.

Proof. Such a matrix is at least positive semi-definite, because

$$(u_1, \ldots, u_m) \begin{pmatrix} x^{(1)} \\ \vdots \\ x^{(m)} \end{pmatrix} S \begin{pmatrix} x^{(1)} \\ \vdots \\ x^{(m)} \end{pmatrix}' (u_1, \ldots, u_m)'$$
$$= (u_1 x^{(1)} + \cdots + u_m x^{(m)}) S (u_1 x^{(1)} + \cdots + u_m x^{(m)})' \geq 0.$$

If the m vectors $x^{(1)}, \ldots, x^{(m)}$ are linearly dependent, then there is a non-trivial set $(u_1, \ldots, u_m)$ such that $u_1 x^{(1)} + \cdots + u_m x^{(m)} = 0$, so that the matrix is only positive semi-definite, and not positive definite. Conversely, if the matrix is only positive semi-definite and not positive definite, then we have a non-trivial $(u_1, \ldots, u_m)$ such that the displayed expression has the value 0. Since S is positive definite, we deduce that

$$u_1 x^{(1)} + \cdots + u_m x^{(m)} = 0,$$

so that $x^{(1)}, \ldots, x^{(m)}$ are linearly dependent. The theorem is proved. □

21.5 Orthogonal projections

Let S be a subspace of a Euclidean space R with the basis $x^{(1)}, \ldots, x^{(m)}$. We prove that any vector x has the representation

$$x = x_S + x_N, \quad x_S \in S, \quad x_N \perp S. \tag{21.8}$$

We call x_S the orthogonal image of x onto S, and x_N the projection vector.

Take R, for example, to be the usual three dimensional space, and let S be a plane through the origin. Every vector x has the initial point at the origin, x_S is the orthogonal image on S, while x_N is a vector with its endpoint forming the line perpendicular to S, so that $|x_N|$ is the distance of the endpoint from the plane S.

Let x_S be given the representation

$$x_S = c_1 x^{(1)} + \cdots + c_m x^{(m)},$$

where $c_1, \ldots, c_m$ are real numbers. By orthogonality we have

$$\langle x - x_S, x^{(k)} \rangle = 0 \quad (k = 1, 2, \ldots, m),$$

that is

$$\begin{gathered}
\langle x^{(1)}, x^{(1)} \rangle c_1 + \cdots + \langle x^{(1)}, x^{(m)} \rangle c_m + \langle x, x^{(1)} \rangle (-1) = 0, \\
\cdots \\
\langle x^{(m)}, x^{(1)} \rangle c_1 + \cdots + \langle x^{(m)}, x^{(m)} \rangle c_m + \langle x, x^{(m)} \rangle (-1) = 0, \\
x^{(1)} c_1 + \cdots + x^{(m)} c_m + x_S(-1) = 0.
\end{gathered}$$

We then deduce that

$$\begin{vmatrix}
\langle x^{(1)}, x^{(1)} \rangle & \cdots & \langle x^{(1)}, x^{(m)} \rangle & x^{(1)} \\
\vdots & & \vdots & \vdots \\
\langle x^{(m)}, x^{(1)} \rangle & \cdots & \langle x^{(m)}, x^{(m)} \rangle & x^{(m)} \\
\langle x, x^{(1)} \rangle & \cdots & \langle x, x^{(m)} \rangle & x_S
\end{vmatrix} = 0,$$

from which we solve for x_S to give

$$x_S = \frac{\begin{vmatrix}
 & & & x^{(1)} \\
 & \Gamma & & \vdots \\
 & & & x^{(m)} \\
\langle x, x^{(1)} \rangle & \cdots & \langle x, x^{(m)} \rangle & 0
\end{vmatrix}}{\Gamma},$$

where $\Gamma(x^{(1)}, \dots, x^{(m)})$ is the Gram matrix, or determinant, according to the context, and

$$x_N = x - x_S = \frac{\begin{vmatrix} & \Gamma & & \begin{matrix} x^{(1)} \\ \vdots \\ x^{(m)} \end{matrix} \\ \langle x, x^{(1)}\rangle & \cdots & \langle x, x^{(m)}\rangle & x \end{vmatrix}}{\Gamma}.$$

Let h denote the length of x_N. Then

$$h^2 = \langle x_N, x_N\rangle = \langle x_N, x\rangle = \frac{\begin{vmatrix} & \Gamma & & \begin{matrix} \langle x^{(1)}, x\rangle \\ \vdots \\ \langle x^{(m)}, x\rangle \end{matrix} \\ \langle x, x^{(1)}\rangle & \cdots & \langle x, x^{(m)}\rangle & \langle x, x\rangle \end{vmatrix}}{\Gamma}$$

$$= \frac{\Gamma(x^{(1)}, \dots, x^{(m)}, x)}{\Gamma(x^{(1)}, \dots, x^{(m)})}.$$

Here h has the following geometric interpretation: The vectors $x^{(1)}, \dots, x^{(m)}$ together with x form an $(m+1)$ dimensional simplex, and if we consider its base to be the m-dimensional simplex formed from $x^{(1)}, \dots, x^{(m)}$, then h is the height of the tip of the vector x from the base.

Suppose that y is any vector in S, and x is any vector in R, both having the origin as the initial point. From

$$N(x-y) = N(x_N + x_S - y) = N(x_N) + N(x_S - y) \geq N(x_N) = h^2,$$

we deduce that

$$|x - y| \geq |x - x_S| = h,$$

that is the height cannot exceed the slant height. Therefore, among all the vectors y in S, the one which is closest to x is x_S. Also, $h = \sqrt{N(x - x_S)}$ can be used as the error in the least square approximation $x \approx x_S$.

Introducing $\Gamma_p = \Gamma(x^{(1)}, \dots, x^{(p)})$, $p = 1, 2, \dots, m$, we have

$$\sqrt{\Gamma_1} = |x^{(1)}| = V_1,$$

the length of the vector $x^{(1)}$;

$$\sqrt{\Gamma_2} = V_1 h_1 = V_2,$$

the area of the parallelogram formed by $x^{(1)}, x^{(2)}$; and

$$\sqrt{\Gamma_3} = V_2 h_2 = V_3,$$

the volume of the parallelepiped formed by $x^{(1)}, x^{(2)}, x^{(3)}$. Continuing in this way, we have

$$\sqrt{\Gamma_4} = V_3 h_3 = V_4,$$

and, in general,

$$\sqrt{\Gamma_m} = V_{m-1} h_{m-1} = V_m.$$

Naturally we call V_m the volume of the m-dimensional parallelepiped formed by $x^{(1)}, \ldots, x^{(m)}$.

Take a normal basis for R, and write

$$x^{(i)} = (x_{1i}, \ldots, x_{ni}), \qquad X = \begin{pmatrix} x^{(1)} \\ \vdots \\ x^{(m)} \end{pmatrix};$$

then

$$\Gamma_m = |XX'|,$$

so that

$$V_m^2 = \Gamma_m = \sum_{1 \le i_1 < i_2 < \cdots < i_m \le n} \begin{vmatrix} x_{i_1 1} & x_{i_1 2} & \cdots & x_{i_1 m} \\ x_{i_2 1} & x_{i_2 2} & \cdots & x_{i_2 m} \\ \vdots & \vdots & & \vdots \\ x_{i_m 1} & x_{i_m 2} & \cdots & x_{i_m m} \end{vmatrix}^2.$$

We therefore have the following geometric property.

Theorem 21.6 *The square of the volume of a parallelepiped is the sum of all the squares of the volumes of all its m-dimensional projected images in the coordinate system.*

In particular, when $m = n$, we have

$$V_n = \begin{vmatrix} x_{11} & x_{12} & \cdots & x_{1n} \\ x_{21} & x_{22} & \cdots & x_{2n} \\ \vdots & \vdots & & \vdots \\ x_{n1} & x_{n2} & \cdots & x_{nn} \end{vmatrix} \text{ absolute value thereof.}$$

Returning to the analysis on the projection of a vector, we deduce from

$$\langle x, x\rangle = \langle x_S + x_N, x_S + x_N\rangle = \langle x_S, x_S\rangle + \langle x_N, x_N\rangle \ge \langle x_N, x_N\rangle = h^2$$

that

$$\frac{\Gamma(x^{(1)}, \ldots, x^{(m)}, x)}{\Gamma(x^{(1)}, \ldots, x^{(m)})} = h_m^2 \le \langle x, x\rangle = \Gamma(x),$$

with equality if and only if x is orthogonal to $x^{(1)}, \ldots, x^{(m)}$. We thus have Hadamard's inequality

$$\Gamma(x^{(1)}, \ldots, x^{(m)}, x) \le \Gamma(x^{(1)})\Gamma(x^{(2)}) \cdots \Gamma(x^{(m)})$$

in which there is equality only when $x^{(1)}, \ldots, x^{(m)}$ are mutually orthogonal to each other. The inequality has the following geometric interpretation: The volume of a parallelepiped does not exceed the product of its side lengths, and there is equality only for a rectangular parallelepiped.

The usual form of Hadamard's inequality is

$$\begin{vmatrix} x_{11} & \cdots & x_{1n} \\ \vdots & & \vdots \\ x_{n1} & \cdots & x_{nn} \end{vmatrix}^2 \le \sum_{i=1}^{n} x_{i1}^2 \sum_{i=1}^{n} x_{i2}^2 \cdots \sum_{i=1}^{n} x_{in}^2,$$

and the following is an alternative proof:

(1) If S, T are positive definite, then so is $S + T$, and $|S + T| \ge |S|$ (reduce to the case $S = I, T = [\lambda_1, \ldots, \lambda_n]$).

(2) Hadamard's inequality is then a special case of

$$\begin{vmatrix} S_1 & L \\ L' & S_2 \end{vmatrix} = |S_1||S_2 - LS_1^{-1}L'| \le |S_1||S_2|.$$

21.6 Unitary spaces

In a vector space over the complex field we define the inner product of two vectors u, v by

$$u\bar{v}'.$$

A linear transformation which leaves $u\bar{u}'$ invariant is said to be unitary; such a transformation has the form

$$v = uU, \quad U\bar{U}' = I,$$

and we call U a unitary matrix.

For $n = 1$, the unitary transformation is simply $e^{i\theta}$. In n-dimensional space, the diagonal matrix

$$[e^{i\theta_1}, \ldots, e^{i\theta_n}]$$

is unitary. A real orthogonal matrix is obviously unitary.

The unitary matrix $U = (u_{ij})$ has the following property:

$$\sum_{j=1}^{n} u_{ij}\bar{u}_{kj} = \delta_{ik}.$$

From $\bar{U}'U = I$ we see that $\bar{U}', U', \bar{U}$ are all unitary.

Let

$$u_{11} = \rho_{11} e^{i\theta_1}, \quad \cdots, \quad u_{1n} = \rho_{1n} e^{i\theta_n}.$$

Then the elements of the first row in

$$U[e^{-i\theta_1}, \ldots, e^{-i\theta_n}]$$

are real; that is, the matrix has the form

$$\begin{pmatrix} u_{11} & \cdots & u_{1n} \\ & \cdots & \end{pmatrix}, \quad u_{11}^2 + \cdots + u_{1n}^2 = 1,$$

with $u_{11}, \ldots, u_{1n}$ real. From the preceding chapter, we know that there is an orthogonal matrix Γ such that the first row of the matrix

$$\begin{pmatrix} u_{11} & \cdots & u_{1n} \\ & \cdots & \end{pmatrix} \Gamma$$

is e_1. Continuing with such an argument, we find that a unitary matrix can be represented as the product of a unitary matrix having the form $[e^{i\theta_1}, \ldots, e^{i\theta_n}]$ together with orthogonal matrices. By referring to $[e^{i\theta_1}, 1, \ldots, 1]$, $[1, \ldots, e^{i\theta_i}, 1, \ldots, 1], \ldots$ as one dimensional unitary matrices, we have the following.

Theorem 21.7 *Any unitary matrix can be represented as a product of rotations on a plane and one dimensional unitary matrices.*

Notes.

(1) The matrix $[-1, 1]$ is a one dimensional unitary matrix.
(2) The determinant of a unitary matrix has absolute value 1. This follows from $|U\bar{U}'| = 1$, so that $|U||\bar{U}| = 1$.

Theorem 21.8 (Gram) *A necessary and sufficient condition for the m complex vectors*

$$u^{(1)}, \ldots, u^{(m)}$$

to be linearly independent is that the Hermitian matrix

$$(u^{(i)}\bar{u}^{(j)\prime})_{i,j=1,2,\ldots,m}$$

should be positive definite.

We omit the proof, which is the same as that given before. We also leave it to the reader to generalise the Gram–Schmidt orthogonalisation process.

Similarly it is not difficult to prove that if the m vectors satisfy

$$u^{(i)}\bar{u}^{(j)\prime} = \delta_{ij}, \quad i, j = 1, 2, \ldots, m,$$

then we can supplement $n - m$ vectors so that

$$\begin{pmatrix} u^{(1)} \\ \vdots \\ u^{(n)} \end{pmatrix}$$

is unitary.

We write $\langle u, v\rangle$ again for the inner product; that it has the following generalised properties follows from the above results:

(1) $\langle x, y\rangle = \overline{\langle y, x\rangle}$;
(2) $\langle \alpha x, y\rangle = \alpha\langle x, y\rangle$;
(3) $\langle x + y, z\rangle = \langle x, z\rangle + \langle y, z\rangle$;
(2′) $\langle x, \alpha y\rangle = \bar{\alpha}\langle x, y\rangle$;
(3′) $\langle x, y + z\rangle = \langle x, y\rangle + \langle x, z\rangle$;
(4) $N(x) = \langle x, x\rangle \geq 0$, with equality if and only if $x = 0$.

Such a space is called a unitary space, and all the generalised results can be established accordingly.

21.7 Introduction to inner product function spaces

We consider the collection of square-integrable functions $f(x)$ over the interval (a, b), which we call the L^2-space, in which we define the inner product for two such functions $f(x)$, $g(x)$ by

$$\langle f, g\rangle = \int_a^b f(x)g(x)dx.$$

If the inner product is 0 then we say that the two functions are orthogonal to each other. If

$$\int_a^b f^2(x)dx = 1$$

then we say that $f(x)$ is normal; in any case we call

$$N(f)^{1/2} = \left(\int_a^b f^2(x)dx\right)^{1/2}$$

the norm, or the length, of the function in the space.

If there is a non-trivial set of real numbers $c_1, \ldots, c_n$ such that

$$c_1 f_1(x) + \cdots + c_n f_n(x) \equiv 0,$$

that is 0 ‘almost everywhere’, then $f_1(x), \ldots, f_n(x)$ are said to be linearly dependent; otherwise they are linearly independent.

Theorem 21.9 (Gram) *A necessary and sufficient condition for the n functions $f_1(x), \ldots, f_n(x)$ to be linearly independent is that the symmetric matrix*

$$\big(\langle f_i, f_j\rangle\big)_{1\le i,j\le n}$$

should be positive definite.

Proof. Consider

$$(u_1, \ldots, u_n)\begin{pmatrix} \int_a^b f_1 f_1\,dx & \cdots & \int_a^b f_1 f_n\,dx \\ \vdots & & \vdots \\ \int_a^b f_n f_1\,dx & \cdots & \int_a^b f_n f_n\,dx \end{pmatrix}(u_1, \ldots, u_n)'$$

$$= \left(\int_a^b (u_1 f_1 + \cdots + u_n f_n) f_1\,dx, \ldots, \int_a^b (u_1 f_1 + \cdots + u_n f_n) f_n\,dx\right)\begin{pmatrix} u_1 \\ \vdots \\ u_n \end{pmatrix}$$

$$= \int_a^b (u_1 f_1 + \cdots + u_n f_n)^2\,dx.$$

If $f_1(x), \ldots, f_n(x)$ are linearly dependent then there is a non-trivial set $(u_1, \ldots, u_n)$ such that $\int_a^b (u_1 f_1 + \cdots + u_n f_n)^2\,dx = 0$. Conversely, if this equation holds then $u_1 f_1 + \cdots + u_n f_n$ has to be 0 almost everywhere. □

The matrix in the statement of the theorem is called the Gramian.

By completing squares there is a triangular matrix such that

$$\begin{pmatrix} \alpha_{11} & 0 & \cdots & 0 \\ \alpha_{21} & \alpha_{22} & \cdots & 0 \\ \vdots & \vdots & & \vdots \\ \alpha_{n1} & \alpha_{n2} & \cdots & \alpha_{nn} \end{pmatrix}\big(\langle f_i, f_j\rangle\big)_{1\le i,j\le n}\begin{pmatrix} \alpha_{11} & 0 & \cdots & 0 \\ \alpha_{21} & \alpha_{22} & \cdots & 0 \\ \vdots & \vdots & & \vdots \\ \alpha_{n1} & \alpha_{n2} & \cdots & \alpha_{nn} \end{pmatrix}' = 1,$$

so that there exists a set of functions

$$\begin{aligned} g_1 &= \alpha_{11} f_1, \\ g_2 &= \alpha_{21} f_1 + \alpha_{22} f_2, \\ &\cdots \\ g_m &= \alpha_{a1} f_1 + \alpha_{m2} f_2 + \cdots + \alpha_{mm} f_m, \\ &\cdots \end{aligned}$$

which are orthonormal, that is they are normal functions which are mutually orthogonal.

Example 21.10 We take $a = -1$ and $b = 1$. The Gramian of the functions $1, x, x^2, x^3, \dots$ is given by

$$\int_{-1}^{1} (1, x, x^2, x^3, \dots)'(1, x, x^2, x^3, \dots)dx = 2\begin{pmatrix} 1 & 0 & \frac{1}{3} & 0 & \frac{1}{5} & 0 & \cdots \\ 0 & \frac{1}{3} & 0 & \frac{1}{5} & 0 & \frac{1}{7} & \cdots \\ \frac{1}{3} & 0 & \frac{1}{5} & 0 & \frac{1}{7} & 0 & \cdots \\ & & & \cdots & & & \end{pmatrix},$$

and from

$$1 = 1 \cdot 1', \qquad \begin{pmatrix} 1 & 0 \\ 0 & \frac{1}{3} \end{pmatrix} = \begin{pmatrix} 1 & 0 \\ 0 & \frac{1}{\sqrt{3}} \end{pmatrix}\begin{pmatrix} 1 & 0 \\ 0 & \frac{1}{\sqrt{3}} \end{pmatrix}',$$

$$\begin{pmatrix} 1 & 0 & \frac{1}{3} \\ 0 & \frac{1}{3} & 0 \\ \frac{1}{3} & 0 & \frac{1}{5} \end{pmatrix} = \begin{pmatrix} 1 & 0 & 0 \\ 0 & \frac{1}{3} & 0 \\ \frac{1}{3} & 0 & \frac{2}{3\sqrt{5}} \end{pmatrix}\begin{pmatrix} 1 & 0 & 0 \\ 0 & \frac{1}{3} & 0 \\ \frac{1}{3} & 0 & \frac{2}{3\sqrt{5}} \end{pmatrix}',$$

$$\begin{pmatrix} 1 & 0 & \frac{1}{3} & 0 \\ 0 & \frac{1}{3} & 0 & \frac{1}{5} \\ \frac{1}{3} & 0 & \frac{1}{5} & 0 \\ 0 & \frac{1}{5} & 0 & \frac{1}{7} \end{pmatrix} = \begin{pmatrix} 1 & 0 & 0 & 0 \\ 0 & \frac{1}{3} & 0 & 0 \\ \frac{1}{3} & 0 & \frac{2}{3\sqrt{5}} & 0 \\ 0 & \frac{\sqrt{3}}{5} & 0 & \frac{2}{5\sqrt{7}} \end{pmatrix}\begin{pmatrix} 1 & 0 & 0 & 0 \\ 0 & \frac{1}{3} & 0 & 0 \\ \frac{1}{3} & 0 & \frac{2}{3\sqrt{5}} & 0 \\ 0 & \frac{\sqrt{3}}{5} & 0 & \frac{2}{5\sqrt{7}} \end{pmatrix}',$$

the Gram–Schmidt method delivers the orthonormal set of functions:

$$\frac{1}{\sqrt{2}}(1, x, x^2, x^3, \dots)\begin{pmatrix} 1 & 0 & 0 & 0 & \cdots \\ 0 & \frac{1}{3} & 0 & 0 & \cdots \\ \frac{1}{3} & 0 & \frac{2}{3\sqrt{5}} & 0 & \cdots \\ 0 & \frac{\sqrt{3}}{5} & 0 & \frac{2}{5\sqrt{7}} & \cdots \\ & & \cdots & & \end{pmatrix}'^{-1}.$$

There are various obvious ways to extend the notion:

(1) Let $\alpha(x)$ be a real increasing function which is not a constant. Then, applying the notion of Lebesgue–Stieltjes integration, we can define the inner product by

$$\langle f, g \rangle = \int_a^b f(x)g(x)d\alpha(x).$$

(2) For complex functions we define

$$\langle f, g \rangle = \int_a^b f(x)\overline{g(x)}d\alpha(x).$$

(3) For multiple integrals we define

$$\langle f, g \rangle = \int \cdots \int f(x_1, \dots, x_n)g(x_1, \dots, x_n)d\alpha(x_1, \dots, x_n).$$

(4) For an m-dimensional body S in an n-dimensional space we let

$$\langle f, g\rangle = \int \underset{\xi}{\cdots} \int f(x_1, \ldots, x_n) g(x_1, \ldots, x_n) d\sigma,$$

where $d\sigma$ is the flux over ξ.

(5) Abstractly, we may take any 'inner-product space'.

Generally speaking, however, the source of the basic method and the fundamentals involved come from what we have given here, after which we selectively introduce the others.

21.8 Eigenvalues

Theorem 21.11 *Any square matrix A of order n has the following unique representation:*

$$A = H + K, \quad H = \bar{H}', \quad K = -\bar{K}'.$$

Let a, h, k denote the members of A, H, K with their largest absolute values, respectively. If $\sigma = \alpha + \beta i$ is an eigenvalue of A, then

$$|\sigma| \le na, \quad |\alpha| \le nh, \quad |\beta| \le nk.$$

Proof. With σ being an eigenvalue of A, there is a non-trivial vector z such that $zA = \sigma z$, and hence

$$\sigma z\bar{z}' = zA\bar{z}' \quad \text{and} \quad \bar{\sigma} z\bar{z}' = z\bar{A}'\bar{z}'.$$

Adding and subtracting we find that

$$\alpha z\bar{z}' = \frac{1}{2} z(A + \bar{A}')\bar{z}' = \sum_{i,j} \frac{1}{2}(a_{ij} + \bar{a}_{ji}) z_i \bar{z}_j,$$

$$\beta z\bar{z}' = z\Big(\frac{A - \bar{A}'}{2i}\Big)\bar{z}' = \sum_{i,j} \frac{1}{2i}(a_{ij} - \bar{a}_{ji}) z_i \bar{z}_j.$$

Thus

$$|\sigma| z\bar{z}' \le \sum_{i,j} |a_{ij}||z_i||z_j| \le a \sum_{i,j} |z_i||z_j| = a\Big(\sum_i |z_i|\Big)^2$$
$$\le an \sum_i |z_i|^2 = anz\bar{z}',$$

and hence $|\sigma| \le na$. The same method shows that

$$|\alpha| z\bar{z}' \le \sum_{i,j} |h_{ij}||z_i||z_j| \le h\Big(\sum_i |z_i|\Big)^2 \le nhz\bar{z}',$$

and that $|\beta| z\bar{z}' \le nkz\bar{z}'$. The theorem is proved. □

We have the following corollaries.

Theorem 21.12 *The eigenvalues of a Hermitian matrix are real.*

Proof. We have $K = 0$, so that $k = 0$, and hence $\beta = 0$. □

Theorem 21.13 *The eigenvalues of a real symmetric matrix are real.*

Theorem 21.14 *The eigenvalues of a skew-symmetric matrix are purely imaginary.*

Theorem 21.15 *If the matrix K in Theorem 21.11 is real, then*

$$|\beta| \le \sqrt{\frac{1}{2}n(n-1)}k.$$

Proof. We have

$$\beta z\bar{z}' = z\frac{A - \bar{A}'}{2i}\bar{z}' = \frac{1}{i}zK\bar{z}'.$$

If K is real then it is a real skew-symmetric matrix, so that

$$\beta z\bar{z}' = \sum_{i<j} k_{ij}\frac{\bar{z}_i z_j - z_i\bar{z}_j}{i},$$

and hence

$$|\beta| z\bar{z}' \le k\sum_{i<j}\left|\frac{\bar{z}_i z_j - z_i\bar{z}_j}{i}\right|.$$

By Schwarz's inequality we have

$$\begin{aligned}
\Bigl(\sum_{i<j}\left|\frac{\bar{z}_i z_j - z_i\bar{z}_j}{i}\right|\Bigr)^2 &\le \frac{1}{2}n(n-1)\sum_{i<j}\left|\frac{\bar{z}_i z_j - z_i\bar{z}_j}{i}\right|^2 \\
&= -\frac{1}{2}n(n-1)\sum_{i,j}(\bar{z}_i z_j - z_i\bar{z}_j)^2 \\
&= \frac{1}{2}n(n-1)\Bigl[\Bigl(\sum_i z_i\bar{z}_i\Bigr)^2 - \sum_i z_i^2\sum_i \bar{z}_i^2\Bigr] \\
&\le \frac{1}{2}n(n-1)(z\bar{z}')^2,
\end{aligned}$$

so that the required result follows. □

Note. Theorem 21.15 improves slightly on the upper bound for the eigenvalues when the matrix is skew-symmetric.

Theorem 21.16 *Let $\sigma = \alpha + i\beta$ be an eigenvalue of $A = H + K$. Then*

$$m \le \alpha \le M,$$

where m and M are the least and greatest eigenvalues of H, respectively.

Proof. Let z be an eigenvector corresponding to σ, and let $H = (h_{ij})$. Then $\alpha z\bar{z}' = zH\bar{z}'$, so that

$$\alpha = \frac{zH\bar{z}'}{z\bar{z}'}.$$

Denote the right-hand side here by $f(z)$, so that

$$\min f(z) \leq \alpha \leq \max f(z).$$

It suffices to show that $\min f(z)$ and $\max f(z)$ are actual eigenvalues of H. Let $z + j = x_j + iy_j$. From

$$\frac{\partial f}{\partial x_j} = \frac{\partial f}{\partial y_j} = 0$$

we deduce that

$$\frac{\partial f}{\partial z_j} = 0, \quad \frac{\partial f}{\partial \bar{z}_j} = 0$$

and

$$\frac{\partial f}{\partial z_j} = \frac{e_j H\bar{z}'}{z\bar{z}'} - \frac{e_j\bar{z}'}{(z\bar{z}')^2} zH\bar{z}' = 0.$$

The extreme values λ $(zH\bar{z}'/z\bar{z}')$ satisfy

$$e_j(H - \lambda I)\bar{z}' = 0,$$

so that $(H - \lambda I)\bar{z}' = 0$, and hence $|H - \lambda I| = 0$. The theorem is proved. □

We have also proved, en passant, the following theorem.

Theorem 21.17 *Let S be a real symmetric matrix. Then its least (greatest) eigenvalue is equal to the least (greatest) value assumed by*

$$\frac{xSx'}{xx'}.$$

This value is the least (greatest) value taken by xSx' on the sphere $xx' = 1$.

Let us use a more descriptive language: The equation $xSx' = \lambda$ represents an ellipsoid, and when λ is an eigenvalue of S, the ellipsoid is tangential to the sphere $xx' = 1$. If λ is less than the least eigenvalue, then the whole sphere lies inside the ellipsoid. If λ exceeds the greatest eigenvalue, then the whole ellipsoid lies inside the sphere.

Theorem 21.18 *The eigenvalues of a unitary matrix U have absolute values equal to* 1.

Proof. Let x be an eigenvalue of U. Then $x + 1/x$ is an eigenvalue of the Hermitian matrix $U + U^{-1} = U + \bar{U}'$, and $(1/i)\big(x - 1/x)$ is an eigenvalue of $(1/i)(U - \bar{U}')$. Such eigenvalues are real, so that

$$x + \frac{1}{x} = 2r, \quad x - \frac{1}{x} = 2is \quad (r, s \text{ real}).$$

This gives $r + is = x$ and $r - is = 1/x$, and hence $r^2 + s^2 = 1$. The theorem is proved. □

Finally, we deduce Theorem 21.19.

Theorem 21.19 *The complex eigenvalues of a real orthogonal matrix form conjugate pairs.*

21.9 Eigenvalues of integral equations

Let us clarify the situation concerning a real symmetric matrix: If α is an eigenvalue of a real symmetric matrix, then there is a vector z such that $zS = \alpha z$, so that

$$zS\bar{z}' = \alpha z\bar{z}'.$$

From $\overline{zS\bar{z}'}' = z\bar{S}'\bar{z}' = zS\bar{z}'$, so that $zS\bar{z}'$ is real, we find that α is also real. We apply such a reasoning to the eigenvalue problem associated with integral equations.

Let $K(x, y) = K(y, x)$ be a symmetric continuous function in the square $a \le x \le b, a \le y \le b$. We wish to find the eigenvalues λ for which

$$f(x) = \lambda \int_a^b K(x, y) f(y) dy$$

is soluble in $f(x)$.

Theorem 21.20 *The eigenvalues λ are real.*

Proof. Multiply the equation by $\overline{f(x)}$, and then integrate over $a \le x \le b$; we have

$$\int_a^b |f(x)|^2 dx = \lambda \int_a^b \overline{f(x)} dx \int_a^b K(x, y) f(y) dy.$$

From

$$\int_a^b \int_a^b \overline{K(x, y)\overline{f(x)} f(y)} dx\, dy = \int_a^b \int_a^b K(x, y) f(x) \overline{f(y)} dx\, dy$$
$$= \int_a^b \int_a^b K(y, x) \overline{f(x)} f(y) dx\, dy = \int_a^b \int_a^b K(x, y) \overline{f(x)} f(y) dx\, dy,$$

we see that the expressions are real, and therefore λ is also real.

Exercise 21.21 Generalise the theorem for complex functions.

Exercise 21.22 Use the method in Section 21.5 to tackle the case when $K(x, y)$ is not symmetric.

21.10 Orthogonal types of symmetric matrices

Theorem 21.23 *Under an orthogonal mapping, a quadratic form can be transformed to*

$$\lambda_1 x_1^2 + \cdots + \lambda_n x_n^2.$$

Thus, corresponding to any symmetric matrix S, there is an orthogonal matrix Γ such that

$$\Gamma S \Gamma' = [\lambda_1, \ldots, \lambda_n],$$

where $\lambda_1, \ldots, \lambda_n$ are real and are the eigenvalues of S.

Proof. Corresponding to an eigenvalue λ_1, there is an eigenvector x such that

$$xS = \lambda_1 x,$$

and we may assume that x has a unit length, that is $xx' = 1$. There is an orthogonal matrix Γ which has x as the first row, that is $e_1\Gamma = x$, and so

$$e_1 \Gamma S \Gamma' = \lambda_1 e_1,$$

which shows that $\Gamma S \Gamma'$ has the form

$$\begin{pmatrix} \lambda_1 & 0 \\ 0 & S_1 \end{pmatrix}, \qquad S_1 = S_1^{(n-1)}.$$

The proof of the theorem is now clear. □

It follows from Theorem 21.23 that

$$xSx' = \sum_{i=1}^{n} \lambda_i (\alpha_{i1} x_1 + \cdots + \alpha_{in} x_n)^2,$$

where $\Gamma = (\alpha_{ij})$, so that

$$S = \Gamma[\lambda_1, \ldots, \lambda_n]\Gamma'$$

and $e_1\Gamma'$ is an eigenvector for S, that is $(\alpha_{11}, \ldots, \alpha_{n1})$ is the eigenvector belonging to the eigenvalue λ_1 of S.

Theorem 21.24 *Eigenvectors belonging to different eigenvalues are orthogonal to each other.*

Proof. If

$$x^{(1)} S = \lambda_1 x^{(1)}, \quad x^{(2)} S = \lambda_2 x^{(2)}, \quad \lambda_1 \neq \lambda_2,$$

then

$$\lambda_1 x^{(1)} x^{(2)\prime} = x^{(1)} S x^{(2)\prime} = \lambda_2 x^{(1)} x^{(2)\prime},$$

which implies $x^{(1)} x^{(2)\prime} = 0$. □

We can also deduce the following from Theorem 21.23: Let λ_i be ordered so that

$$\lambda_1 \le \lambda_2 \le \cdots \le \lambda_n;$$

it then follows that

$$\frac{\lambda_1 x_1^2 + \cdots + \lambda_n x_n^2}{x_1^2 + \cdots + x_n^2} \ge \lambda_1.$$

Thus

$$\min \frac{\lambda_1 x_1^2 + \cdots + \lambda_n x_n^2}{x_1^2 + \cdots + x_n^2} = \lambda_1,$$
$$\min \frac{\lambda_2 x_2^2 + \cdots + \lambda_n x_n^2}{x_2^2 + \cdots + x_n^2} = \lambda_2.$$

With a change in labelling, we have the following theorem in which $z^{(i)}$ is an eigenvector belonging to λ_i.

Theorem 21.25 *We have*

$$\min \frac{xSx'}{xx'} = \lambda_1,$$

with equality when $x = z^{(1)}$*; also*

$$\min_{x \perp z^{(1)}} \frac{xSx'}{xx'} = \lambda_2,$$

with equality when $x = z^{(2)}$*, and*

$$\min_{x \perp z^{(1)}, z^{(2)}} \frac{xSx'}{xx'} = \lambda_3,$$
$$\cdots.$$

21.11 Euclidean types of quadratic surfaces

The mapping

$$y = x\Gamma + c,$$

where Γ is a real orthogonal matrix and c is a real vector, is called a Euclidean transformation. Besides rotation and reflection, there is also the translation map $y = x + c$. The set of all Euclidean transformations form the Euclidean group, and Euclidean geometry is the study of geometric properties which are invariant under transformations of the group.

First, the distance between two points $x^{(1)}, x^{(2)}$ is an invariant, as can be seen from

$$\begin{aligned}(y^{(1)} - y^{(2)})(y^{(1)} - y^{(2)})' &= (x^{(1)} - x^{(2)})\Gamma\Gamma'(x^{(1)} - x^{(2)})' \\ &= (x^{(1)} - x^{(2)})(x^{(1)} - x^{(2)})'.\end{aligned}$$

Next, any hyperplane $ax' = \lambda$ can be mapped into $ax' = 0$ by a Euclidean transformation.

Let us examine the different types of quadratic surfaces:

$$xAx' + 2xb' + \gamma = 0.$$

We can take Γ so that $\Gamma A\Gamma' = [\lambda_1, \ldots, \lambda_n]$. Thus, with the transformation $x = y\Gamma$ the surface becomes

$$\sum_{i=1}^{n} \lambda_i y_i^2 + 2\sum_{i=1}^{n} b_i y_i + \gamma = 0.$$

Suppose first that no λ_i is 0. Then from

$$\sum_{i=1}^{n} \lambda_i \left(y_i + \frac{b_i}{\lambda_i}\right)^2 + \gamma - \sum_{i=1}^{n} \frac{b_i^2}{\lambda_i} = 0,$$

we see that the general quadratic surface can be transformed to

$$\sum_{i=1}^{n} \lambda_i x_i^2 = \lambda.$$

If $\lambda \neq 0$, then this can be written in the form

$$\sum_{i=1}^{p} \left(\frac{x_i}{a_i}\right)^2 - \sum_{i=p+1}^{n} \left(\frac{x_i}{a_i}\right)^2 = 1;$$

if $\lambda = 0$ then we have the degenerate quadratic surface

$$\sum_{i=1}^{p} \left(\frac{x_i}{a_i}\right)^2 - \sum_{i=p+1}^{n} \left(\frac{x_i}{a_i}\right)^2 = 0.$$

If there are some $\lambda_i = 0$, then we may suppose that

$$\sum_{i=1}^{r} \lambda_i y_i^2 + 2\sum_{i=r+1}^{n} b_i y_i + \gamma = 0, \quad \lambda_1, \ldots, \lambda_r \neq 0.$$

If the b_i are not all 0, then we can further transform it to

$$\sum_{i=1}^{r} \lambda_i y_i^2 + y_{r+1} = 0.$$

Summarising, we first consider

$$\begin{pmatrix} A & b' \\ b & \gamma \end{pmatrix};$$

if this is singular, then the original quadratic surface is a degenerate one. A non-degenerate quadratic surface must, via Euclidean transformations, be equivalent to either of the following standard forms:

$$\sum_{i=1}^{p} \left(\frac{x_i}{a_i}\right)^2 - \sum_{i=p+1}^{n} \left(\frac{x_i}{a_i}\right)^2 = 1, \tag{21.9}$$

$$\sum_{i=1}^{p} \left(\frac{x_i}{a_i}\right)^2 - \sum_{i=p+1}^{n-1} \left(\frac{x_i}{a_i}\right)^2 + x_n = 0. \tag{21.10}$$

When $p = n$, the surface is an ellipsoid, with $a_1, a_2, \dots, a_n$ as the major axes.

We leave it to the reader to write out the standard forms corresponding to the degenerate surfaces.

21.12 Pairs of square matrices

We first recall some relationships between pairs of square matrices. Two pairs of square matrices A_1, A_2 and B_1, B_2 are said to be equivalent if there are two square matrices P, Q with full ranks such that

$$A_1 = PB_1Q, \quad A_2 = PB_2Q.$$

There is the following important result.

Theorem 21.26 *If A_1, A_2 have full ranks, then a necessary and sufficient condition for equivalence is that $A_1\lambda + A_2$ and $B_1\lambda + B_2$ should have the same invariant factors (or elementary factors).*

The theorem has already been proved.

We consider the case when A_1, A_2 may not have full ranks, starting with

$$\lambda A_1 + \mu A_2, \quad -\infty < \lambda, \mu < \infty.$$

If there are λ, μ such that $|\lambda A_1 + \mu A_2| \neq 0$, then such matrices are said to form a non-singular family of square matrices. It is clear that Theorem 21.26 can be generalised to any such families.

Let us return to consider the families of equivalent matrices.

Theorem 21.27 *If*

$$A = PBQ,$$

with A and B symmetric (or skew-symmetric), and P and Q have full ranks, then there is a square matrix R with full rank, depending on P, Q, but not on A, B, such that $A = R'BR$.

Proof. (1) From $A = PBQ$ we have

$$A = Q'BP' = PBQ.$$

Let $U = Q'^{-1}P$; then

$$UB = BU',$$

and indeed, for any polynomial $f(x)$, we always have

$$f(U)B = B(f(U))'.$$

(2) The square root of a square matrix U with full rank. We prove that there is a polynomial $f(x)$ such that

$$(f(U))^2 = U.$$

Suppose first that this has been done, and we let $X = f(U)$, with

$$XB = BX'.$$

Taking $R = X'Q$, we then have

$$R'BR = Q'XBX'Q = Q'X^2BQ = Q'UBQ = PBQ = A,$$

as required.

Let $g(\lambda)$ be the characteristic polynomial of U. If we can prove that there exists $f(\lambda)$ such that

$$f(\lambda)^2 \equiv \lambda \pmod{g(\lambda)},$$

then the conclusion for (2) is established.

(3) Factorise $g(x)$ into

$$g(x) = (x-a)^r(x-b)^s(x-c)^t \cdots,$$

where $a, b, c, \dots$ are distinct non-zero roots.

Expand $\sqrt{x}$ into a power series in $(x-a)$:

$$\sqrt{x} = \sqrt{a} + \frac{\sqrt{a}}{2a}(x-a) - \frac{(x-a)^2}{8a^2} + \cdots,$$

and let $F(x)$ denote the sum of the first r terms in the expansion. Then $\sqrt{x} - F(x)$ is divisible by $(x-a)^r$. Similarly $\sqrt{x}$ can be expanded about b, and we let $G(x)$ denote the first s terms of the expansion, so that $\sqrt{x} - G(x)$ is divisible by $(x-b)^s$. We proceed in this way for all the other roots.

Next, we take the partial fractions expansion for $F(x)/g(x)$:

$$\frac{F(x)}{g(x)} = \frac{A_0}{(x-a)^r} + \cdots + \frac{A_{r-1}}{x-a} + R(x) = \frac{A(x)}{(x-a)^r} + R(x).$$

Similarly we can define $B(x)$ corresponding to $G(x)$, etc. Let

$$f(x) = g(x)\Big(\frac{A(x)}{(x-a)^r} + \frac{B(x)}{(x-b)^s} + \cdots\Big),$$

and we see that $\sqrt{x} - f(x)$ is divisible by $g(x)$, that is

$$x \equiv f(x)^2 \pmod{g(x)}.$$

□

From the above we can deduce the following.

Theorem 21.28. *If*

$$PAQ = A_1, \quad PBQ = B_1,$$

where A and A_1 are both symmetric or skew-symmetric, and B and B_1 are both symmetric or skew-symmetric, then there is a matrix R with full rank such that

$$R'AR = A_1, \quad R'BR = B_1.$$

Theorem 21.29. *Suppose that $|A| \neq 0$. Then a necessary and sufficient condition for the existence of a matrix P with full rank such that*

$$A = PBP'$$

is that

$$\lambda A + \mu A' \quad \textit{and} \quad \lambda B + \mu B'$$

should have the same invariant divisors.

Proof. (1) If $A = PBP'$ then $A' = PB'P'$, so that

$$\lambda A + \mu A' = P(\lambda B + \mu B')P',$$

and the condition is satisfied.

(2) If the condition holds, then we have $\lambda A + \mu A' = P(\lambda B + \mu B')Q$, and

$$A + A' = P(B + B')Q, \quad A - A' = P(B - B')Q,$$

where $A + A'$ and $B + B'$ are symmetric, and $A - A'$ and $B - B'$ are skew-symmetric, so that there exists R such that

$$A + A' = R(B + B')R', \quad A - A' = R(B - B')R',$$

and hence we obtain the required result. □

Remark Theorem 21.29 still holds when $|A| = 0$, but the lengthier proof is omitted here.

21.13 Orthogonal types of skew-symmetric matrices

Over the real field, the problem of the equivalence between pairs of symmetric matrices is not that simple. We have already discussed the particular example of the types of quadratic surfaces under an orthogonal group. This amounts to the problem of the equivalence of the pair of matrices

$$(I, S), \quad (I, T).$$

We can generalise a little here: If one of the matrices is positive definite, then it can be reduced to the standard form

$$(I, [\lambda_1, \dots, \lambda_n]).$$

If there is a positive definite matrix in the family, then it is possible to have a simultaneous reduction to diagonal forms. Such generalisations are not that difficult. However, without the assumption of one of the matrices being positive definite, the problem becomes much more complicated, and it will not be considered here.

A similar situation exists for pairs of Hermitian matrices, although such a problem is tackled erroneously in some books.

If A_1 is symmetric and A_2 is skew-symmetric, we can prepare ourselves for some discussion under the situation when A_1 is positive definite. We may as well assume that $A_1 = I$, and this then reduces to the problem of the type of skew-symmetric matrix $A_2 = K$ under an orthogonal group.

When $n = 2$, we have

$$\begin{pmatrix} a & b \\ c & d \end{pmatrix} \begin{pmatrix} 0 & 1 \\ -1 & 0 \end{pmatrix} \begin{pmatrix} a & b \\ c & d \end{pmatrix}' = (ad - bc) \begin{pmatrix} 0 & 1 \\ -1 & 0 \end{pmatrix},$$

so that any skew-symmetric matrix of order 2 must be orthogonally equivalent to

$$\begin{pmatrix} 0 & k \\ -k & 0 \end{pmatrix}, \qquad k > 0.$$

Theorem 21.30. *Any skew-symmetric matrix must be orthogonally equivalent to*

$$\begin{pmatrix} 0 & k_1 \\ -k_1 & 0 \end{pmatrix} \dotplus \begin{pmatrix} 0 & k_2 \\ -k_2 & 0 \end{pmatrix} \dotplus \cdots \dotplus \begin{pmatrix} 0 & k_\nu \\ -k_\nu & 0 \end{pmatrix} \dotplus 0 \dotplus \cdots \dotplus 0,$$

with $k_1 \geq k_2 \geq \cdots \geq k_\nu > 0$.

The following is a sketch for the proof.

We have already seen that an eigenvalue ik of a skew-symmetric matrix K is purely imaginary. Let w be an eigenvector, so that $wK = ikw$. Write the complex vector in the form $w = u + iv$, so that

$$uK = -kv, \quad vK = ku.$$

From $uKu' = 0$ we deduce that $uv' = 0$, so that u and v are orthogonal.

If $k < 0$ then we find that

$$\begin{pmatrix} u \\ v \end{pmatrix} K = \begin{pmatrix} 0 & -k \\ k & 0 \end{pmatrix} \begin{pmatrix} u \\ v \end{pmatrix}$$

and

$$kuu' = vKu' = -uKv' = kvv',$$

so that $uu' = vv'$. Let $u_1 = u/\sqrt{uu'}$, $v_1 = v/\sqrt{vv'}$. Then

$$\begin{pmatrix} u_1 \\ v_1 \end{pmatrix} K = \begin{pmatrix} 0 & -k \\ k & 0 \end{pmatrix} \begin{pmatrix} u_1 \\ v_1 \end{pmatrix},$$

and $u_1u_1' = v_1v_1' = 1$, $u_1v_1' = 0$. Supplementing $\begin{pmatrix} u_1 \\ v_1 \end{pmatrix}$ to form an orthogonal matrix Γ, we then have

$$\Gamma K \Gamma' = \begin{pmatrix} 0 & -k & \\ k & 0 & \\ & & K_1 \end{pmatrix}.$$

The same method can be used to deal with the case when $k > 0$, and continuing in this way the theorem is proved.

21.14 Symplectic groups and symplectic types

Definition 21.31. A matrix which leaves a (fixed) non-singular skew-symmetric matrix invariant is called a symplectic matrix. More precisely, for a fixed $K = -K'$, $|K| \neq 0$, a matrix P such that

$$PKP' = K$$

is said to be symplectic.

To simplify matters we let

$$K = \begin{pmatrix} O & I \\ -I & O \end{pmatrix}, \quad I = I^{(\nu)}, \quad O = O^{(\nu)}.$$

Writing

$$P = \begin{pmatrix} A & B \\ C & D \end{pmatrix},$$

it follows from

$$\begin{pmatrix} A & B \\ C & D \end{pmatrix}\begin{pmatrix} O & I \\ -I & O \end{pmatrix}\begin{pmatrix} A & B \\ C & D \end{pmatrix}' = \begin{pmatrix} O & I \\ -I & O \end{pmatrix}$$

that

$$AB' = BA', \quad CD' = DC', \quad AD' - BC' = I.$$

It follows at once that

$$\begin{pmatrix} I & S \\ O & I \end{pmatrix}, \quad \begin{pmatrix} A & O \\ O & A'^{-1} \end{pmatrix}, \quad S' = S, \quad |A| \neq 0$$

are symplectic matrices. Let J be a diagonal matrix satisfying $J^2 = J$, so that the elements of J are 1 or 0. Then

$$\begin{pmatrix} J & I - J \\ -(I - J) & J \end{pmatrix}$$

is also symplectic.

It is not difficult to prove that any symplectic matrix is representable as a product of these matrices.

21.15 Various types

We have a collection of groups: the orthogonal group, the symplectic group, the unitary group. We also have a collection of different types of objects: symmetric matrices, skew-symmetric matrices, Hermitian matrices, orthogonal matrices, symplectic matrices, unitary matrices.

The following problem then arises: Given a certain group, we consider the classification of the matrices under the group. For example, the symplectic type, or the unitary type, of symmetric matrices. If we add further the 'encirclement of mathematics' then we have various sorts of associated problems, to which specialist schools are attached and for which examples will not be given here.

Nevertheless, we ought to point out the following: A good number of such problems do have practical associated notions, and there are reciprocal relationships between the different problems.

For example, the orthogonal types of skew-symmetric matrices can be used to deal with the orthogonal types of orthogonal matrices, say in the determination of the standard form of the orthogonal matrix. Thus, given an orthogonal matrix Γ, there must be an orthogonal matrix T such that

$$T\Gamma T' = \begin{pmatrix} \cos\theta_1 & \sin\theta_1 \\ -\sin\theta_1 & \cos\theta_1 \end{pmatrix} \dotplus \cdots \dotplus \begin{pmatrix} \cos\theta_\nu & \sin\theta_\nu \\ -\sin\theta_\nu & \cos\theta_\nu \end{pmatrix} \dotplus I.$$

Proof. Let $\Gamma = (I - K)(I + K)^{-1}$. From $\Gamma\Gamma' = I$ we have $K = -K'$. We saw earlier that there is an orthogonal matrix T such that

$$TKT' = \begin{pmatrix} 0 & k_1 \\ -k_1 & 0 \end{pmatrix} \dotplus \cdots \dotplus \begin{pmatrix} 0 & k_\nu \\ -k_\nu & 0 \end{pmatrix} \dotplus 0 \dotplus \cdots,$$

and we deduce that

$$T\Gamma T' = M_1^{(2)} \dotplus M_2^{(2)} \dotplus \cdots,$$

where

$$M^{(2)} = \frac{I - \begin{pmatrix} 0 & k \\ -k & 0 \end{pmatrix}}{I + \begin{pmatrix} 0 & k \\ -k & 0 \end{pmatrix}} = \begin{pmatrix} 1 & -k \\ k & 1 \end{pmatrix} \begin{pmatrix} 1 & k \\ -k & 1 \end{pmatrix}^{-1}$$

$$= \frac{1}{1+k^2} \begin{pmatrix} 1 & -k \\ k & 1 \end{pmatrix}^2 = \frac{1}{1+k^2} \begin{pmatrix} 1-k^2 & -2k \\ 2k & 1-k^2 \end{pmatrix}.$$

The required result follows. □

21.16 Particle oscillations

The motion of particles in classical mechanics is governed by the kinetic energy equation

$$2T = \sum_{i,j=1}^{n} \dot{q}_i a_{ij} \dot{q}_j, \quad a_{ij} = a_{ji},$$

with the position given by

$$2V = \sum_{i,j=1}^{n} q_i b_{ij} q_j, \quad b_{ij} = b_{ji};$$

here q is the generalised coordinate, and a and b are independent of q and the time t; we are dealing with real numbers, and $\dot{q}_i = dq_i/dt$.

Let

$$A = (a_{ij}), \quad B = (b_{ij}), \quad q = (q_1, \ldots, q_n),$$

so that we may write

$$2T = \dot{q} A \dot{q}', \tag{21.11}$$

$$2V = q B q'. \tag{21.12}$$

Since the kinetic energy cannot be negative, and equals 0 only when there is no motion, that is when $\dot{q} = 0$, we conclude that A must be positive definite. We already know that there is a matrix P such that

$$PAP' = I, \quad PBP' = \Lambda. \tag{21.13}$$

We examine the equation of motion for the particle. The Lagrangian for the particle is

$$2L \equiv 2T - 2V = \dot{q}A\dot{q}' - qBq',$$

so that the equation of motion is

$$\frac{d}{dt}\Big(\frac{\partial L}{\partial \dot{q}_i}\Big) - \frac{\partial L}{\partial q_i} = 0,$$

which can be written as

$$\ddot{q}A + qB = 0.$$

Let $q = \theta P$, so that

$$\ddot{\theta}PAP' + \theta PBP' = 0,$$

that is

$$\ddot{\theta} + \theta\Lambda = 0.$$

In component form, this may be written as

$$\frac{d^2\theta_i}{dt^2} + \lambda_i\theta_i = 0, \quad i = 1, 2, \ldots, n.$$

If the motion can be described as a periodical oscillation (or the position of the centre of the system of particles is in a fixed position), then $\lambda_i > 0$, so that B is also positive definite.

Let $\lambda_i = p_i^2$, so that

$$\theta_i = A_i \sin(p_i t + \alpha_i);$$

from $q = \theta P$ we deduce that

$$q_i = \sum_{k=1}^{n} A_k \sin(p_k t + \alpha_k) v_{ki}.$$

Example 21.32 Two connected periodic oscillations: Let x_1 denote the displacement to the right of the material m_1, and let x_2 be the displacement of the material m_2. By Hooke's law, we have

$$\begin{aligned} 2T &= m_1 x_1^2 + m_2 x_2^2, \\ 2V &= k_1' x_1^2 + k_2' x_2^2 + k_3(x_1 - x_2)^2 \\ &= k_1 x^2 + k_2 x_2^2 - 2k_3 x_1 x_3. \end{aligned}$$

Now

$$A = \begin{pmatrix} m_1 & 0 \\ 0 & m_2 \end{pmatrix}, \quad B = \begin{pmatrix} k_1 & -k_3 \\ -k_3 & k_2 \end{pmatrix},$$

and the equations of motion are given by

$$m_1 \frac{d^2 x_1}{dt^2} + k_1 x_1 - k_3 x_2 = 0, \quad m_2 \frac{d^2 x_2}{dt^2} - k_3 x_1 + k_2 x_2 = 0.$$

Now

$$P = \begin{pmatrix} \frac{1}{\sqrt{m_1}} \cos\theta & \frac{1}{\sqrt{m_2}} \sin\theta \\ -\frac{1}{\sqrt{m_1}} \sin\theta & \frac{1}{\sqrt{m_2}} \cos\theta \end{pmatrix},$$

where

$$\tan 2\theta = \frac{2k_3/\sqrt{m_1 m_2}}{k_2/m_2 - k_1/m_1}$$

so that

$$x_1 = \frac{\theta_1}{\sqrt{m_1}} \cos\theta - \frac{\theta_2}{\sqrt{m_2}} \sin\theta,$$
$$x_2 = \frac{\theta_1}{\sqrt{m_2}} \cos\theta + \frac{\theta_2}{\sqrt{m_1}} \sin\theta$$

and

$$\frac{d^2\theta_1}{dt^2} + \lambda_1 \theta_1 = 0, \quad \frac{d^2\theta_2}{dt^2} + \lambda_2 \theta_2 = 0,$$

where

$$\lambda_1 = \omega_1^2 = \frac{k_1}{m_1} \cos^2\theta - \frac{2k_3}{\sqrt{m_1 m_2}} \cos\theta \sin\theta + \frac{k_2}{m_2} \sin^2\theta,$$
$$\lambda_2 = \omega_2^2 = \frac{k_1}{m_1} \sin^2\theta + \frac{2k_3}{\sqrt{m_1 m_2}} \cos\theta \sin\theta + \frac{k_2}{m_2} \cos^2\theta.$$

22
Volumes

22.1 Volume elements in m-dimensional manifolds

Let

$$a^{(1)}, \ldots, a^{(m)}$$

be vectors with the origin as the initial point in n-dimensional space. These m vectors form a parallelepiped with m pairs of parallel faces, and we saw earlier that such a body has a volume

$$|G|^{1/2},$$

where G is the value of the determinant (Gramian)

$$G = \left| \begin{pmatrix} a^{(1)} \\ \vdots \\ a^{(m)} \end{pmatrix} \begin{pmatrix} a^{(1)} \\ \vdots \\ a^{(m)} \end{pmatrix}' \right|. \tag{22.1}$$

More precisely, this parallelepiped with $2m$ faces can be described by the parametric equation

$$x = \lambda_1 a^{(1)} + \cdots + \lambda_m a^{(m)}, \quad 0 \leq \lambda_\nu \leq 1, \tag{22.2}$$

and G is the determinant of the quadratic form derived from the norm

$$xx' = (\lambda_1, \ldots, \lambda_m) \begin{pmatrix} a^{(1)} \\ \vdots \\ a^{(m)} \end{pmatrix} \begin{pmatrix} a^{(1)} \\ \vdots \\ a^{(m)} \end{pmatrix}' (\lambda_1, \ldots, \lambda_m)'.$$

More generally, an m-dimensional manifold in n-dimensional space can be given by

$$x_i = x_i(\lambda_1, \ldots, \lambda_m), \quad i = 1, 2, \ldots, n. \tag{22.3}$$

We shall assume that the functions have continuous partial derivatives and that, at a given point, there is a vector derivative:

$$dx_i = \sum_{j=1}^{m} \frac{\partial x_j}{\partial \lambda_j} d\lambda_j, \quad \text{that is} \quad dx = d\lambda \frac{\partial(x)}{\partial(\lambda)}.$$

Thus, in the neighbourhood of a point, the volume of an element in such an m-dimensional manifold is given by the square root of the determinant of the quadratic form

$$dx\, dx' = d\lambda \frac{\partial(x)}{\partial(\lambda)} \left(\frac{\partial(x)}{\partial(\lambda)}\right)' d\lambda',$$

that is

$$\left| \frac{\partial(x)}{\partial(\lambda)} \left(\frac{\partial(x)}{\partial(\lambda)}\right)' \right|^{1/2}.$$

Using $d\sigma$ to denote the volume of such an element, we then have

$$d\sigma = \left| \frac{\partial(x)}{\partial(\lambda)} \left(\frac{\partial(x)}{\partial(\lambda)}\right)' \right|^{1/2} d\lambda_1 \cdots d\lambda_m.$$

Thus if an m-dimensional manifold D^* can be described by pieces of such elements for $x \in D^*$, then the volume of the manifold is given by

$$\int \cdots \int_{D^*} \left| \frac{\partial(x)}{\partial(\lambda)} \left(\frac{\partial(x)}{\partial(\lambda)}\right)' \right|^{1/2} d\lambda_1 \cdots d\lambda_m. \tag{22.4}$$

With the substitution

$$\lambda_j = \lambda_j(t_1, \ldots, t_m),$$

so that

$$\frac{\partial(x)}{\partial(\lambda)} \frac{\partial(\lambda)}{\partial(t)} = \frac{\partial(x)}{\partial(t)},$$

we then have

$$\left| \frac{\partial(x)}{\partial(\lambda)} \left(\frac{\partial(x)}{\partial(\lambda)}\right)' \right|^{1/2} d\lambda_1 \cdots d\lambda_m = \left| \frac{\partial(x)}{\partial(t)} \left(\frac{\partial(x)}{\partial(t)}\right)' \right|^{1/2} dt_1 \cdots dt_m.$$

Naturally, a more rigorous consideration is required at several places in what we have described here. First, we need to be a little more specific in our notion of an m-dimensional manifold, for which we may need to introduce a coordinate system in the region. This is because such a manifold may take on a certain shape in some context and another shape in a different context (there may not be a fixed universal formula for its representation). Next, we may need to specify the 'directional volume': If we use λ to be fundamental to the derivative vector then, on changing it to t, we need to take note of the sign for $\partial(\lambda)/\partial(t)$; if it is positive then the volume has the same direction, otherwise it is in the opposite direction. If we fix $e_1, \ldots, e_m$ so that the volume of the manifold with $2m$ faces is positive, then

$$a_i^{(i)} = \sum_{j=1}^{m} p_{ij} e_j,$$

and the manifold formed from the m vectors now has a volume which is positive if $|p_{ij}| > 0$ and negative if $|p_{ij}| < 0$.

We consider some examples.

Example 22.1 Let us determine the surface area of an n-dimensional sphere, that is the surface area of

$$xx' = r^2.$$

On differentiation, we have $x dx' = 0$, which can be 'solved' to give

$$x_1\, dx_1 - -x_2\, dx_2 - \cdots \quad x_n\, dx_n = -X\, dX',$$

where $X = (x_2, \ldots, x_n)$. Therefore

$$dx\, dx' = \Big(\frac{X\, dX'}{x_1}\Big)^2 + dX\, dX' = dX\Big(I + \frac{X'X}{x_1^2}\Big)dX'.$$

The determinant of the quadratic form is

$$\Big|I + \frac{1}{x_1^2} X'X\Big| = 1 + \frac{XX'}{x_1^2} = \frac{xx'}{x_1^2} = \frac{r^2}{x_1^2};$$

here we have used $|I + u'u| = 1 + uu'$. It follows that the surface area is given by

$$r \int_{r^2 - XX' > 0} \cdots \int \frac{dx_2 \cdots dx_n}{|x_1|} = 2^n r \int_{\substack{r^2 - x_2^2 - \cdots - x_n^2 \geq 0 \\ x_\nu \geq 0}} \cdots \int \frac{dx_2 \cdots dx_n}{\sqrt{r^2 - x_2^2 - \cdots - x_n^2}}.$$

This is a Dirichlet integral, the value for which we show in Section 22.2 to be given by

$$\frac{2\pi^{n/2} r^{n-1}}{\Gamma(n/2)}.$$

The expression is equal to the derivative, with respect to the radius r, of the volume of the n-dimensional sphere, which in turn is given by

$$\frac{\pi^{n/2} r^n}{\Gamma(1 + n/2)};$$

we leave it to the reader to ponder over the property being manifested in the geometric configuration.

Example 22.2 The length of the curve

$$x_i = x_i(t), \quad a \leq t \leq b, \quad i = 1, \ldots, n,$$

in n-dimensional space is given by

$$\int_a^b \sqrt{dx\,dx'} = \int_a^b \sqrt{\sum_{i=1}^{n} \left(\frac{dx_i}{dt}\right)^2}\,dt.$$

Example 22.3 In the spherical coordinates system we have

$$x = \rho u, \qquad uu' = 1,$$
$$dx = d\rho\, u + \rho\, du, \qquad du\, u' + u\, du' = 0,$$

so that

$$dx\,dx' = (d\rho)^2 + \rho\, d\rho(du\, u' + u\, du') + \rho^2\, du\, du' = (d\rho)^2 + \rho^2\, du\, du'.$$

The spherical coordinates are

$$\begin{aligned} u_1 &= \cos\theta_1, \\ u_2 &= \sin\theta_1 \cos\theta_2, \\ &\cdots \\ u_{n-1} &= \sin\theta_1 \cdots \sin\theta_{n-2} \cos\theta_{n-1}, \\ u_n &= \sin\theta_1 \cdots \sin\theta_{n-2} \sin\theta_{n-1}, \end{aligned}$$
$$0 \le \theta_i \le \pi \quad (i = 1, \dots, n-2), \quad 0 \le \theta_{n-1} \le 2\pi.$$

Thus

$$\begin{aligned} dx\,dx' = (d\rho)^2 + \rho^2[(d\theta_1)^2 + \sin^2\theta_1(d\theta_2)^2 + \sin^2\theta_1 \sin^2\theta_2(d\theta_3)^2 \\ + \cdots + \sin^2\theta_1 \cdots \sin^2\theta_{n-2}(d\theta_{n-1})^2]. \end{aligned}$$

Therefore the volume of an element in spherical coordinates is given by

$$\rho^{n-1} \sin^{n-2}\theta_1 \sin^{n-3}\theta_2 \cdots \sin\theta_{n-2}\, d\rho\, d\theta_1 \cdots d\theta_{n-1};$$

the area of an element on the surface of a sphere (ρ being constant) is given by

$$\rho^{n-1} \sin^{n-2}\theta_1 \sin^{n-3}\theta_2 \cdots \sin\theta_{n-2}\, d\theta_1 \cdots d\theta_{n-1},$$

which gives the answer to the problem posed at the end of Example 22.1.

Let us determine the geodesic between two points on the sphere. We may assume the points concerned to be

$$(1, 0, \dots, 0),\ (\cos\alpha, \sin\alpha, 0, \dots, 0), \quad 0 \le \alpha \le \pi,$$

since any two points can, by a suitable rotation, be moved there. Any curve on the surface of the sphere can be written as $\theta_\nu = \theta_\nu(t)$, $0 \le t \le 1$. If such a curve passes through the two points then

$$\theta_1(0) = \theta_2(0) = \cdots = \theta_{n-1}(0) = 0,$$
$$\theta_1(1) = \alpha, \quad \theta_2(1) = \cdots = \theta_{n-1}(1) = 0.$$

The length of such a curve is given by

$$\int_0^1 \sqrt{\left(\frac{d\theta_1}{dt}\right)^2 + \sin^2\theta_1\left(\frac{d\theta_2}{dt}\right)^2 + \cdots + \sin^2\theta_1 \cdots \sin^2\theta_{n-2}\left(\frac{d\theta_{n-1}}{dt}\right)^2}\, dt$$
$$\geq \int_0^1 \frac{d\theta_1}{dt}\, dt = \alpha,$$

with equality only when $\theta_2 = 0$, so that the curve has to lie on the plane $(\cos\theta_1, \sin\theta_1, 0, \ldots, 0)$. This shows that the geodesic has to lie on a plane through the centre of the sphere, so that it has to lie on a 'great circle'.

22.2 Dirichlet's integral

We consider the n-fold multiple integral

$$I = \int \cdots \int f(t_1 + \cdots + t_n) t_1^{\alpha_1 - 1} \cdots t_n^{\alpha_n - 1}\, dt_1 \cdots dt_n$$

over the region defined by

$$t_\nu \geq 0, \quad t_1 + \cdots + t_n \leq 1.$$

If f is continuous and $\alpha_\nu > 0\,(\nu = 1, 2, \ldots, n)$, then I can be reduced to a single integral.

We first reduce the double integral

$$F(\lambda) = \int_0^{1-\lambda} \int_0^{1-\lambda-T} f(t + T + \lambda) t^{\alpha-1} T^{\beta-1}\, dt\, dT$$

by setting $t = T(1-v)/v$, which then gives

$$F(\lambda) = \int_0^{1-\lambda} \int_{T/(1-\lambda)}^{1} f\left(\lambda + \frac{T}{v}\right)(1-v)^{\alpha-1} v^{-\alpha-1} T^{\alpha+\beta-1}\, dv\, dT$$
$$= \int_0^1 \int_0^{(1-\lambda)v} f\left(\lambda + \frac{T}{v}\right)(1-v)^{\alpha-1} v^{-\alpha-1} T^{\alpha+\beta-1}\, dT\, dv,$$

on interchanging the order of integration. Setting $T = v\tau_2$ now gives

$$F(\lambda) = \int_0^1 \int_0^{1-\lambda} f(\lambda + \tau_2)(1-v)^{\alpha-1} \tau_2^{\alpha+\beta-1} v^{-\alpha-1+(\alpha+\beta-1)+1}\, d\tau_2\, dv$$
$$= \int_0^{1-\lambda} f(\lambda + \tau_2)\tau_2^{\alpha+\beta-1}\, d\tau_2 \int_0^1 (1-v)^{\alpha-1} v^{\beta-1}\, dv$$
$$= \frac{\Gamma(\alpha)\Gamma(\beta)}{\Gamma(\alpha+\beta)} \int_0^{1-\lambda} f(\lambda + \tau_2)\tau_2^{\alpha+\beta-1}\, d\tau_2.$$

Therefore

$$I = \underset{\substack{t_3+\cdots+t_n\le 1\\ t_\nu\ge 0}}{\int\cdots\int} t_3^{\alpha_3-1}\cdots t_n^{\alpha_n-1}\,dt_3\cdots dt_n$$
$$\times \int_0^{1-t_3-\cdots-t_n}\int_0^{1-t_2-\cdots-t_n} f(t_1+\cdots+t_n)t_1^{\alpha_1-1}t_2^{\alpha_2-1}\,dt_1dt_2$$
$$= \underset{\substack{t_3+\cdots+t_n\le 1\\ t_\nu\ge 0}}{\int\cdots\int} t_3^{\alpha_3-1}\cdots t_n^{\alpha_n-1}\,dt_3\cdots dt_n$$
$$\times \frac{\Gamma(\alpha_1)\Gamma(\alpha_2)}{\Gamma(\alpha_1+\alpha_2)}\int_0^{1-t_3-\cdots-t_n} f(t_3+\cdots+t_n+\tau_2)\tau_2^{\alpha_1+\alpha_2-1}\,d\tau_2$$
$$= \frac{\Gamma(\alpha_1)\Gamma(\alpha_2)}{\Gamma(\alpha_1+\alpha_2)}\times \underset{\substack{\tau_2+t_3+\cdots+t_n\le 1\\ t_\nu\ge 0,\tau_2\ge 0}}{\int\cdots\int} f(\tau_2+t_3+\cdots+t_n)\tau_2^{\alpha_1+\alpha_2-1}t_3^{\alpha_3-1}\cdots t_n^{\alpha_n-1}\,d\tau_2\,dt_3\cdots dt_n.$$

Continuing in this way, we finally arrive at

$$I = \frac{\Gamma(\alpha_1)\Gamma(\alpha_2)\cdots\Gamma(\alpha_n)}{\Gamma(\alpha_1+\cdots+\alpha_n)}\int_0^1 f(\tau)\tau^{\alpha_1+\cdots+\alpha_n-1}\,d\tau,$$

which is Dirichlet's integral.

Example 22.4 Determine the volume of the simplex

$$x = \lambda_1 a^{(1)}+\cdots+\lambda_m a^{(m)},\quad \lambda_1+\cdots+\lambda_m\le 1,\ \lambda_\nu\ge 0.$$

From

$$dx\,dx' = d\lambda\begin{pmatrix}a^{(1)}\\ \vdots\\ a^{(m)}\end{pmatrix}\begin{pmatrix}a^{(1)}\\ \vdots\\ a^{(m)}\end{pmatrix}'d\lambda'$$

we have

$$\underset{\substack{\lambda_1+\cdots+\lambda_m\le 1\\ \lambda_\nu\ge 0}}{\int\cdots\int}\left|\begin{pmatrix}a^{(1)}\\ \vdots\\ a^{(m)}\end{pmatrix}\begin{pmatrix}a^{(1)}\\ \vdots\\ a^{(m)}\end{pmatrix}'\right|^{1/2} d\lambda_1\cdots d\lambda_m$$
$$= |G|^{1/2}\frac{\Gamma(1)^m}{\Gamma(m)}\int_0^1\tau^{m-1}\,d\tau = |G|^{1/2}\frac{1}{\Gamma(m+1)} = \frac{1}{m!}|G|^{1/2};$$

in other words, the volume of the m-dimensional simplex is the square root of the Gramian of the m vectors, divided by $m!$.

Remark We note that

$$\begin{aligned}1 &= \int_0^1 \cdots \int_0^1 dx_1 \cdots dx_m \\ &= m! \int_{0 \le x_1 \le \cdots \le x_m \le 1} \cdots \int dx_1 \cdots dx_m = m! \int_{\substack{t_1+\cdots+t_m \le 1 \\ t_\nu \ge 0}} \cdots \int dt_1 \cdots dt_m \\ &\quad (x_1 = t_1,\ x_2 = x_1 + t_2,\ x_3 = x_2 + t_3, \ldots).\end{aligned}$$

Example 22.5 We have

$$J = \int_{\sum_{\nu=1}^n x_\nu^p \le r^p} \cdots \int dx_1 \cdots dx_n = \frac{(2r)^n \Gamma(1+1/p)^n}{\Gamma(1+n/p)}.$$

Divide the integral into 2^n pieces, the value of each piece being given by

$$\int_{\substack{\sum_{\nu=1}^n x_\nu^p \le r^p \\ x_\nu \ge 0}} \cdots \int dx_1 \cdots dx_n = r^n \int_{\substack{\sum_{\nu=1}^n x_\nu^p \le 1 \\ x_\nu \ge 0}} \cdots \int dx_1 \cdots dx_n.$$

The substitution $x_\nu^p = y_\nu$ then gives

$$\begin{aligned}\frac{1}{2^n} J &= r^n \int_{\substack{\sum_{\nu=1}^n y_\nu \le 1 \\ y_\nu \ge 0}} \cdots \int \left(\frac{1}{p}\right)^n y_1^{1/p-1} \cdots y_n^{1/p-1}\, dy_1 \cdots dy_n \\ &= \frac{r^n \Gamma(1/p)^n}{p^n \Gamma(n/p)} \int_0^1 \tau^{n/p-1}\, d\tau = \frac{r^n \Gamma(1+1/p)^n}{\Gamma(1+n/p)};\end{aligned}$$

in particular, for $p = 2$, the volume of an n-dimensional sphere with radius r is equal to $\theta_n r^n$, with

$$\theta_n = \frac{\pi^{\pi/2}}{\Gamma(1+n/2)}.$$

Example 22.6 We have

$$\int_{xx' \le r^2} \cdots \int x'x\, dx_1 \cdots dx_n = \frac{\theta_n}{n+2} r^{n+2} I.$$

On replacing x_i by $-x_i$, it is easy to see that

$$\int_{xx' \le r^2} \cdots \int x_i x_j\, dx_1 \cdots dx_n = 0, \quad i \ne j.$$

On the other hand, we have

$$\underset{xx' \le r^2}{\int \cdots \int} x_1^2 \, dx_1 \cdots dx_n = r^{n+2} 2^{-n} \underset{\substack{y_1 + \cdots + y_n \le 1 \\ y_\nu \ge 0}}{\int \cdots \int} y_1^{1/2} y_2^{-1/2} \cdots y_n^{-1/2} \, dy_1 \cdots dy_n$$

$$= r^{n+2} \frac{\Gamma(1/2)^{n-1} \Gamma(3/2)}{2^n \Gamma(1 + n/2)} \int_0^1 \tau^{n/2} \, d\tau$$

$$= r^{n+2} \frac{\pi^{n/2}}{\Gamma(1 + n/2)} \frac{1}{n+2},$$

as required.

22.3 Normal distribution integrals

Theorem 22.7 *Suppose that S is positive definite. Then*

$$\underset{x \in \mathbf{R}^n}{\int \cdots \int} e^{itx' - (1/2)xSx'} \, dx_1 \cdots dx_n = \frac{(2\pi)^{n/2}}{\sqrt{|S|}} e^{-(1/2)tS^{-1}t'}.$$

Proof. When $S = I$, we have

$$\underset{x \in \mathbf{R}^n}{\int \cdots \int} e^{itx' - (1/2)xx'} \, dx_1 \cdots dx_n = \prod_{\nu=1}^{n} \int_{-\infty}^{\infty} e^{it_\nu x_\nu - (1/2)x_\nu^2} \, dx_\nu$$

$$= \prod_{\nu=1}^{n} \left(\sqrt{2\pi} e^{-(1/2)t_\nu^2} \right) = (\sqrt{2\pi})^{n/2} e^{-(1/2)tt'}.$$

In general, we let $S = TT'$, $y = xT$, $t = uT'$, so that

$$\underset{x \in \mathbf{R}^n}{\int \cdots \int} e^{itx' - (1/2)xSx'} \, dx_1 \cdots dx_n = \frac{1}{|T|} \underset{y \in \mathbf{R}^n}{\int \cdots \int} e^{iuy' - (1/2)yy'} \, dy_1 \cdots dy_n$$

$$= \frac{(2\pi)^{n/2}}{|S|^{1/2}} e^{-(1/2)uu'} = \frac{(2\pi)^{n/2}}{\sqrt{|S|}} e^{-(1/2)tS^{-1}t'}.$$

□

Theorem 22.8 *Let S be positive definite and suppose that u and v are two linearly independent vectors. Then*

$$\underset{\substack{x \in \mathbf{R}^n \\ ux' \ge 0, vx' \ge 0}}{\int \cdots \int} e^{-(1/2)xSx'} \, dx_1 \cdots dx_n = \frac{(2\pi)^{n/2-1}}{\sqrt{|S|}} \cos^{-1} \frac{-uS^{-1}v'}{\sqrt{uS^{-1}u' vS^{-1}v'}}.$$

Proof. Write S as $S = TT'$, so that $xT \to x$, $uT'^{-1} \to u$, $vT'^{-1} \to v$. Then thc required equation becomes

$$\int \cdots \int_{\substack{x \in \mathbf{R}^n \\ ux' \geq 0, vx' \geq 0}} e^{-(1/2)xx'}\, dx_1 \cdots dx_n = (2\pi)^{n/2-1} u^\wedge v, \tag{22.5}$$

where $u^\wedge v$ denotes the angle subtended by the two vectors u, v.

It is clear that we may assume that u, v are unit vectors, so that there exists an orthogonal matrix Γ such that

$$y\Gamma = e_1, \qquad v = \Gamma = (\cos\alpha, \sin\alpha, 0, \ldots, 0),$$

where $\alpha = u^\wedge v$. Let $x = y\Gamma'$; then

$$\int \cdots \int_{\substack{x \in \mathbf{R}^n \\ ux' \geq 0, vx' \geq 0}} e^{-(1/2)xx'}\, dx_1 \cdots dx_n = \int \cdots \int_{\substack{y_1 > 0 \\ y_1 \cos\alpha + y_2 \sin\alpha \geq 0}} e^{-(1/2)yy'}\, dy_1 \cdots dy_n$$

$$= \int \cdots \int_{\substack{y_1 \geq 0 \\ y_1 \cos\alpha + y_2 \sin\alpha \geq 0}} e^{-(1/2)(y_1^2 + y_2^2)}\, dy_1 dy_2 (2\pi)^{(1/2)(n-2)}.$$

Let $y_1 = \rho \sin\theta$, $y_2 = \rho\cos\theta$; then

$$\int \cdots \int_{\substack{y_1 \geq 0 \\ y_1 \cos\alpha + y_2 \sin\alpha \geq 0}} e^{-(1/2)(y_1^2 + y_2^2)}\, dy_1\, dy_2 = \int_0^\infty e^{-(1/2)\rho^2} \rho\, d\rho \int_0^\alpha d\theta = \alpha,$$

as required, that is (22.5). □

The geometry associated with (22.5) tells us that the integral of $e^{-(1/2)xx'}$ is proportional to the angle between the two planes through the origin. This suggests the following generalisation: We may conjecture that the value of the integral over three planes through the origin should be proportional to the 'solid angle' formed by the three planes. This is actually the case, and we have

$$\int \cdots \int_{\substack{x \in \mathbf{R}^n \\ ux' \geq 0, vx' \geq 0, wx' \geq 0}} e^{-(1/2)xx'} dx_1 \cdots dx_n = \frac{1}{2}(2\pi)^{n/2-1}(\alpha + \beta + \gamma - \pi),$$

where α, β, γ are the angles subtended by the three vectors. More generally,

$$\int \cdots \int_{\substack{x \in \mathbf{R}^n \\ ux' \geq 0, vx' \geq 0, wx' \geq 0}} e^{-(1/2)xSx'}\, dx_1 \cdots dx_n = \frac{1}{2}(2\pi)^{n/2-1}|S|^{-1/2}\Big(\cos^{-1}\frac{-uS^{-1}v'}{\sqrt{uS^{-1}u'vS^{-1}v'}}$$

$$+\cos^{-1}\frac{-vS^{-1}w'}{\sqrt{vS^{-1}v'wS^{-1}w'}} + \cos^{-1}\frac{-wS^{-1}u'}{\sqrt{wS^{-1}w'uS^{-1}u'}} - \pi\Big).$$

22.4 Normal Parent's distribution

In the investigation of the normal Parent's distribution, we can apply the integral below, where the symmetric matrix $X = (x_{ij})$ can be considered to be a point in a space of dimension $\frac{1}{2}n(n+1)$. The region where the matrix X is positive definite is denoted by $X > 0$.

Theorem 22.9 *Let $k > n$. For any $A > 0$, we have*

$$\int \cdots \int_{X>0} |X|^{(1/2)(k-n-2)} e^{-\sigma(AX)} \dot{X}$$
$$= \pi^{(1/4)n(n+1)} \Gamma\left(\tfrac{1}{2}(k-1)\right) \Gamma\left(\tfrac{1}{2}(k-2)\right) \cdots \Gamma\left(\tfrac{1}{2}(k-n)\right) |A|^{-(1/2)(k-1)},$$

where $\dot{X} = \prod_{1 \le i \le j \le n} dx_{ij}$.

Proof. Let $A = TT'$ and $T'AT = Y$. Then from

$$|T|^{n+1} \dot{X} = \dot{Y}$$

(see Section 22.5 for the proof) we find that

$$\int \cdots \int_{X>0} |X|^{(1/2)(k-n-2)} e^{-\sigma(AX)} \dot{X}$$
$$= \int \cdots \int_{Y>0} |Y|^{(1/2)(k-n-2)} |T|^{-(k-n-2)} \times e^{-\sigma(Y)} |T|^{-(n+1)} \dot{Y}$$
$$= |A|^{-(1/2)(k-1)} \int \cdots \int_{Y>0} |Y|^{(1/2)(k-n-2)} e^{-\sigma(Y)} \dot{Y}.$$

We may as well assume that $A = I$, so that we need to show that if

$$C_{n,k} = \int \cdots \int_{X>0} |X|^{(1/2)(k-n-2)} e^{-\sigma(X)} \dot{X},$$

then

$$C_{n,k} = \pi^{(1/2)n(n+1)} \Gamma\left(\tfrac{1}{2}(k-1)\right) \Gamma\left(\tfrac{1}{2}(k-2)\right) \cdots \Gamma\left(\tfrac{1}{2}(k-n)\right).$$

With the substitution $x_{ij} = y_{ij}\sqrt{x_{ii}x_{jj}}$, we let

$$Y = \begin{pmatrix} 1 & y_{12} & \cdots & y_{1n} \\ y_{21} & 1 & \cdots & y_{2n} \\ \vdots & \vdots & & \vdots \\ y_{n1} & y_{n2} & \cdots & 1 \end{pmatrix}, \quad y_{ij} = y_{ji},$$

so that

$$X = DYD, \quad D = [\sqrt{x_{11}}, \sqrt{x_{22}}, \dots, \sqrt{x_{nn}}].$$

We then have $|X| = |Y|x_{11} \cdots x_{nn}$, and the Jacobian is

$$(x_{11} \cdots x_{nn})^{(1/2)(n-1)}.$$

Therefore

$$\begin{aligned} C_{k,n} &= \int_0^\infty \cdots \int_0^\infty (x_{11} \cdots x_{nn})^{(1/2)(k-3)} e^{-\sum_{i=1}^n x_{ii}}\, dx_{11} \cdots d_{nn} \\ &\quad \times \int \cdots \int_{Y>0} |Y|^{(1/2)(n-k-2)}\, dy_{12} \cdots dy_{m-1,n} \\ &= \Gamma\big(\tfrac{1}{2}(k-1)\big)^n J_n, \end{aligned}$$

where

$$J_n = \int \cdots \int_{Y>0} |Y|^{(1/2)(n-k-2)}\, dy_{12} \cdots dy_{m-1,n}, \quad 2 \le n < k.$$

We proceed to evaluate J_n by induction. First, we have

$$J_2 = \int_{-1}^1 (1-y^2)^{(1/2)(k-4)}\, dy = \sqrt{\pi}\frac{\Gamma\big(\frac{1}{2}(k-2)\big)}{\Gamma\big(\frac{1}{2}(k-1)\big)}.$$

By the expansion of a determinant,

$$\begin{vmatrix} 1 & y_{12} & \cdots & y_{1,n+1} \\ \vdots & \vdots & & \vdots \\ y_{n+1,1} & y_{n+1,2} & \cdots & 1 \end{vmatrix} = |Y| - \sum_{i,j=1}^n Y_{ij} y_{i,n+1} y_{j,n+1},$$

where Y_{ij} is the cofactor of Y corresponding to y_{ij}, so that

$$\begin{aligned} J_{n+1} &= \int \cdots \int dy_{12} \cdots dy_{n-1,n} \\ &\quad \times \int \left(|Y| - \sum_{i,j=1}^n Y_{ij} y_{i,n+1} y_{j,n+1}\right)^{(1/2)(k-n-3)} dy_{1,n+1} \cdots dy_{n,n+1}, \end{aligned}$$

where the inner most integral is over all $y_{1,n+1}, \dots, y_{n,n+1}$ satisfying

$$|Y| - \sum_{i,j=1}^n Y_{ij} y_{i,n+1} y_{j,n+1} > 0.$$

The integral can now be evaluated as a Dirichlet's integral to give

$$J_{n+1} = J_n \frac{\Gamma\big(\frac{1}{2}(k-n-1)\big)}{\Gamma\big(\frac{1}{2}(k-1)\big)} \pi^{n/2},$$

and the theorem follows. □

The expression in Theorem 22.9 is an integral function of A, so that it still holds when a_{ij} is replaced by $a_{ij} - i\epsilon_{ij}t_{ij}$; this then gives the following theorem in which

$$\epsilon_{ij} = \begin{cases} 1, & \text{if } i = j, \\ \frac{1}{2}, & \text{if } i \neq j. \end{cases}$$

Theorem 22.10 *Let*

$$f_k(X) = \begin{cases} C_{n,k}^{-1}|X|^{(1/2)(k-n-1)}e^{-\sigma(AX)}, & X > 0, \\ 0, & \text{otherwise}. \end{cases}$$

Then the Fourier transform of $f_k(X)$ *is*

$$\phi_n(T) = \int \cdots \int e^{i\sum_{i\leq j} x_{ij}t_{ij}} f_k(X)dx_{11}dx_{12}\cdots dx_{nn} = \left(\frac{1}{|A - iT|}\right)^{(1/2)(k-1)},$$

where $T = (\epsilon_{ij}t_{ij})$.

22.5 Determinants of matrix transformations

An $m \times n$ matrix can be considered as a point in space with dimension mn. Two such matrices X, Y related by

$$X = PYQ, \quad P = P^{(m)}, \quad Q = Q^{(n)}, \tag{22.6}$$

can be considered as a linear transformation in space of dimension mn, and we proceed to compute the determinant of the transformation, which is the product of the following two transformations:

$$X = ZQ, \tag{22.7}$$

$$Z = PY. \tag{22.8}$$

Writing (22.7) in rows, we have

$$(x_{11}, \ldots, x_{1n}) = (z_{11}, \ldots, z_{1n})Q,$$
$$\cdots$$
$$(x_{m1}, \ldots, x_{mn}) = (z_{m1}, \ldots, z_{mn})Q,$$

so that the determinant of the linear transformation (22.7) is $|Q|^m$. Similarly, writing (22.8) in rows, we see that its determinant is $|P|^n$. Therefore the determinant of the transformation defined by (22.6) is

$$|P|^n|Q|^m. \tag{22.9}$$

In particular, when $m = n$, the determinant of the transformation $X = PY\bar{P}'$ is $|P\bar{P}'|^n = \text{abs}|P|^{2n}$.

Consider a Hermitian matrix

$$H = (h_{ij}), \quad h_{ii} = k_{ii}, \quad h_{ij} = k_{ij} + ik_{ji}, \quad i < j, \tag{22.10}$$

which can be considered as a point in space of dimension $2n$. The determinant of the transformation

$$h_{ii} = k_{ii},$$
$$h_{ij} = k_{ij} + ik_{ji}, \quad h_{ji} = k_{ij} - ik_{ji} \quad i < j,$$

which maps h_{ij} to k_{ij}, is equal to

$$\begin{vmatrix} 1 & i \\ 1 & -i \end{vmatrix}^{(1/2)n(n-1)} = (-2i)^{(1/2)n(n-1)}.$$

Therefore the value of the determinant of the linear transformation defined by the Hermitian relationship

$$H_1 = PH\bar{P}' \tag{22.11}$$

is

$$\mathrm{abs}|P|^{2n}. \tag{22.12}$$

Also, a symmetric matrix S of order n can be considered as a point in space of dimension $\frac{1}{2}n(n+1)$, and the relationship between the elements of S and T is also a linear transformation when

$$S = PTP'; \tag{22.13}$$

the determinant of such a transformation has the value

$$|P|^{n+1}. \tag{22.14}$$

We give a sketch of the proof: If P is a transvection (shear) then it is not difficult to show that the value of the determinant is 1; if

$$P = [\lambda, 1, \ldots, 1],$$

then, from

$$s_{11} = \lambda^2 t_{11}, \quad s_{12} = \lambda t_{12}, \quad \ldots, \quad s_{1n} = \lambda t_{1n}$$

(with no change elsewhere), the value of the determinant is λ^{n+1}. The required result then follows.

For a skew-symmetric matrix, the space has dimension $\frac{1}{2}n(n-1)$, and from

$$K = PQP', \tag{22.15}$$

the corresponding linear transformation has the determinant

$$|P|^{n-1}. \tag{22.16}$$

We leave it to the reader to establish this from the following remark: From $X = PYP'$ one can deduce that

$$X + X' = P(Y + Y')P', \quad X - X' = P(Y - Y')P';$$

also the product of (22.14) and (22.16) is $|P|^{2n}$.

Remarks (1) List the elements of X in (22.6) as

$$x_{11}, \ldots, x_{1m}, x_{21}, \ldots, x_{2m}, \ldots, x_{m1}, \ldots, x_{mn},$$

and similarly for the elements of Y, so as to form a matrix, which is then denoted by

$$P' \cdot\!\times Q;$$

the square matrix of order mn

$$\begin{pmatrix} r_{11}Q & \cdots & r_{1m}Q \\ \vdots & & \vdots \\ r_{m1}Q & \cdots & r_{mm}Q \end{pmatrix}$$

is denoted by $R \cdot\!\times Q$, and we call it the direct product of R and Q. Formula (22.9) can then be written as $|P' \cdot\!\times Q| = |P|^n |Q|^m$.

(2) List the elements of S in (22.13) as

$$s_{11}, s_{12}, \ldots, s_{1n}, s_{22}, s_{23}, \ldots, s_{22}, \ldots, s_{n-1,n}, s_{nn},$$

and similarly for those of T. Denoting the matrix defined in (22.13) by $P^{[2]}$, the value in (22.14) can be written as

$$|P^{[2]}| = |P|^{n+1}.$$

Applying the notion when the matrix is skew-symmetric, and using $P^{(n)}$ for its representation, the value in (22.16) can be written as

$$|P^{(2)}| = |P|^{n-1}.$$

(3) The matrices $P' \cdot\!\times P$, $P[2]$, $P^{(2)}$ have the following properties:

$$P[2]Q[2] = (PQ)[2], \quad P(2)Q(2) = (PQ)(2)$$

and

$$(P' \cdot\!\times P)(Q' \cdot\!\times Q) = (PQ)' \cdot\!\times PQ.$$

These properties are the most basic examples in group representation theory.

22.6 Integration elements in a unitary group

The elements of a complex matrix of order n can be considered as a point in $2n^2$ dimensional real space, with the Euclidean norm

$$\sum_{i=1}^{n}\sum_{j=1}^{n}(dx_{ij}^2+dy_{ij}^2), \quad z_{ij}=x_{ij}+iy_{ij}. \tag{22.17}$$

Such a norm can also be written as

$$\sigma\big(dZ\overline{(dZ)'}\big), \tag{22.18}$$

where $\sigma(A)$ denotes the trace of the matrix A.

The collection of all unitary matrices form a manifold in this space, and we first prove that the manifold has dimension n^2. We make the Cayley transform

$$U=(I+iH)^{-1}(I-iH), \tag{22.19}$$

which has the solution

$$iH=(I-U)(I+U)^{-1}. \tag{22.20}$$

The condition of being unitary yields

$$I=U\bar{U}'=(I+iH)^{-1}(I-iH)(I+i\bar{H}')(I-i\bar{H}')^{-1},$$

that is

$$(I-iH)(I+i\bar{H}')=(I+iH)(I-i\bar{H}'),$$

giving

$$H=\bar{H}'. \tag{22.21}$$

Thus, corresponding to a unitary matrix with $|I+U|\neq 0$, there is a Hermitian matrix H, and so, there being n^2 elements in H, the said dimension is n^2. Note, however, that unitary matrices with $|I+U|=0$ form a manifold with a lower dimension.

The matrix H can be considered as a parameter for the manifold formed from the unitary matrices. Conversely, the unitary group manifold can be considered as the extension of the space formed by Hermitian matrices, by the addition of a point at infinity to the manifold in space.

Naturally the norm in the unitary group manifold is

$$\sigma\big(dU\overline{(dU)'}\big).$$

Differentiating

$$U\bar{U}'=I$$

gives

$$dU\bar{U}' + u + d\bar{U}' = 0.$$

Let $\delta U = U^{-1}\,dU$; then $\delta U = -\overline{(\delta U)'}$, and hence

$$\sigma\big(dU\overline{(dU)'}\big) = -\sigma\big(\delta U\overline{(\delta U)'}\big). \tag{22.22}$$

On the other hand, from Cayley's formula,

$$\begin{aligned} dU &= -(I+iH)^{-1}i\,dH - i(I+iH)^{-1}\,dH(I+iH)^{-1}(I-iH) \\ &= -i(I+iH)^{-1}\,dH[I+(I+iH)^{-1}(I-iH)] \\ &= -2i(I+iH)^{-1}\,dH(I+iH)^{-1}, \end{aligned} \tag{22.23}$$

which then gives

$$\begin{aligned} \sigma\big(dU\overline{(dU)'}\big) &= 4\sigma\big((I+iH)^{-1}\,dH(I+iH)^{-1}\times(I-iH)^{-1}\,dH(I-iH)^{-1}\big) \\ &= 4\sigma\big(dH(I+H^2)^{-1}\,dH(I+H^2)^{-1}\big). \end{aligned} \tag{22.24}$$

Letting $(I+H^2)^{-1} = P\bar{P}'$ and $X = \bar{P}'dHP$, we then have

$$\sigma\big(dU\overline{(dU)'}\big) = 4\sigma(X^2) = 4\sum_{i,j=1}^{n}|x_{ij}|^2.$$

Since $\bar{x}_{ij} = x_{ji}$, setting $x_{ij} = y_{ij} + iz_{ij}$ so that $|x_{ij}|^2 + |x_{ji}|^2 = 2(y_{ij}^2 + z_{ij}^2)$, the determinant of the quadratic form displayed above is given by

$$4^{n^2}\cdot 2^{n(n-1)} = 2^{n(3n-1)}.$$

From $X = \bar{P}'dHP$ together with (22.22) and (22.23), we find that the determinant of the linear transformation from X to dH is

$$|P\bar{P}'|^n = |I+H^2|^{-n}.$$

Thus, the value of the determinant of (22.24) is

$$2^{n(3n-1)}\cdot|I+H^2|^{-2n}.$$

Therefore the volume element on the unitary group is given by

$$\dot{U} = 2^{n^2}\cdot 2^{(1/2)n(n-1)}|I+H^2|^{-n}\prod_{j=1}^{n}dh_j\prod_{j<k}dh'_{jk}\,dh''_{jk};$$

here $H = (h_{jk})$, $h_{jj} = h_j$, $h_{jk} = h'_{jk} + ih''_{jk}$, and, for the whole unitary group, the region of integration is over

$$-\infty < h_j < \infty, \quad -\infty < h'_{jk}, h''_{jk} < \infty.$$

Theorem 22.11 *The (manifold) volume of the unitary group is equal to*

$$\theta_n = \frac{(2\pi)^{(1/2)n(n+1)}}{1!\,2!\cdots(n-1)!}.$$

From the above we deduce that

$$\theta_n = \int_U \cdots \int \dot{U} = 2^{n^2} \cdot 2^{(1/2)n(n-1)} \int \cdots \int |I + H^2|^{-n} \prod_{j=1}^{n} dh_j \prod_{j<k} dh'_{jk}\, dh''_{jk}.$$

The integral here is evaluated in the following section.

22.7 Integration elements in a unitary group (continuation)

Before we evaluate the integral in the preceding section, we deal with the following integral.

Theorem 22.12 *Let $a > 0$, $b^2 - ac < 0$ and $\alpha > \frac{1}{2}$. Then*

$$\int_{-\infty}^{\infty} \frac{dx}{(ax^2 + 2bx + c)^{\alpha}} = a^{\alpha-1}(ac - b^2)^{1/2-\alpha}\sqrt{\pi}\,\frac{\Gamma(\alpha - \frac{1}{2})}{\Gamma(\alpha)}.$$

Proof. Let

$$y = \frac{a}{\sqrt{ac - b^2}}\left(x + \frac{b}{a}\right);$$

then

$$dx = \frac{\sqrt{ac - b^2}}{a}\, dy$$

and

$$ax^2 + 2bx + c = \frac{ac - b^2}{a}(y^2 + 1).$$

Therefore

$$\begin{aligned}\int_{-\infty}^{\infty} \frac{dx}{(ax^2 + 2bx + c)^{\alpha}} &= a^{\alpha-1}(ac - b^2)^{1/2-\alpha} \int_{-\infty}^{\infty} \frac{dy}{(1 + y^2)^{\alpha}} \\ &= a^{\alpha-1}(ac - b^2)^{1/2-\alpha}\sqrt{\pi}\,\frac{\Gamma(\alpha - \frac{1}{2})}{\Gamma(\alpha)}.\end{aligned}$$

□

Theorem 22.13 *Let $\alpha > n - \frac{1}{2}$, and let H be a Hermitian matrix of order n. Then*

$$H_n(\alpha) = \int \cdots \int_H \frac{\dot{H}}{(\det(I + H^2))^\alpha}$$
$$= 2^{(1/2)n(n-1)} \pi^{(1/2)n^2} \prod_{i=0}^{n-1} \frac{\Gamma(\alpha - j - \frac{1}{2})}{\Gamma(\alpha - j} \prod_{k=0}^{n-2} \frac{\Gamma(2\alpha - n - k)}{\Gamma(2\alpha - 2k - 1)},$$

where $H = (h_{jk}), h_{jj} = h_j, h_{jk} = h'_{jk} + ih''_{jk}\ (j < k)$ and

$$\dot{H} = 2^{(1/2)n(n-1)} \prod_{j=1}^{n} dh_j \prod_{j<k} dh'_{jk}\, dh''_{jk}.$$

Proof. Let

$$\begin{pmatrix} H_1 & \bar{v}' \\ v & h \end{pmatrix}, \quad (h = h_n),$$

where H_1 is a Hermitian matrix of order $n - 1$, v is a vector with dimension $n - 1$ and h is a real number. Then

$$I + H^2 = \begin{pmatrix} I + H_1^2 + \bar{v}'v & H_1\bar{v}' + \bar{v}'h \\ vH_1 + hv & 1 + h^2 + v\bar{v}' \end{pmatrix}.$$

Applying the identity

$$\begin{pmatrix} I & 0 \\ -pA^{-1} & 1 \end{pmatrix} \begin{pmatrix} A & p' \\ p & \ell \end{pmatrix} \overline{\begin{pmatrix} I & 0 \\ -pA^{-1} & 1 \end{pmatrix}}' = \begin{pmatrix} A & 0 \\ 0 & \ell - pA^{-1}\bar{p}' \end{pmatrix} \quad (A = \bar{A}'),$$

we find that

$$\det(I + H^2) = (ah^2 + 2bh + c) \det(I + H_1^2 + \bar{v}'v),$$

where

$$a = 1 - v(I + H_1^2 + \bar{v}'v)^{-1}\bar{v}',$$
$$2b = -vH_1(I + H_1^2 + \bar{v}'v)^{-1}\bar{v}' - v(I + H_1^2 + \bar{v}'v)^{-1}H_1\bar{v}',$$
$$c = 1 + v\bar{v}' - vH_1(I + H_1^2 + \bar{v}'v)^{-1}H_1\bar{v}'.$$

Since H_1 is Hermitian there is a unitary matrix U such that

$$H_1 = U[\lambda_1, \lambda_2, \ldots, \lambda_n]\bar{U}'.$$

Let

$$T = U\left[\sqrt{1 + \lambda_1^2}, \sqrt{1 + \lambda_2^2}, \ldots, \sqrt{1 + \lambda_n^2}\right]\bar{U}';$$

then

$$T = \bar{T}', \quad TH_1 = H_1 T, \quad I + H_1^2 = T^2.$$

Making the substitution

$$v = uT,$$

we have

$$\dot{v} = |\det T|^2 \dot{u} = \det(I + H_1^2)\dot{u}, \quad I + H_1^2 + \bar{v}'v = T(I + \bar{u}'u)T,$$

and because

$$(I + \bar{u}'u)^{-1}\bar{u}' = \frac{\bar{u}'}{1 + u\bar{u}'}, \quad w(I + \bar{u}'u)^{-1}\bar{w}' = w\bar{w}' - \frac{|w\bar{u}'|^2}{1 + u\bar{u}'},$$

where w is an n-dimensional vector, we find that

$$\begin{aligned}
a &= 1 - u(I + \bar{u}'u)^{-1}\bar{u}' = \frac{1}{1 + u\bar{u}'} \ (> 0), \\
b &= -uH_1(I + \bar{u}'u)^{-1}\bar{u}' = -\frac{uH_1\bar{u}'}{1 + u\bar{u}'}, \\
c &= 1 - uT^2\bar{u}' - uH_1(I + \bar{u}'u)^{-1}H_1\bar{u}' = 1 + u\bar{u}' + \frac{|uH_1\bar{u}'|^2}{1 + u\bar{u}'}.
\end{aligned}$$

Since $uH_1\bar{u}'$ is a real number, it follows that $ac - b^2 = 1$.

From Theorem 22.12, we have

$$\begin{aligned}
H_n(\alpha) &= \int_H \cdots \int \frac{\dot{H}}{(\det(I + H^2))^\alpha} \\
&= 2^{n-1} \int_{u, H_1} \cdots \int (\det(I + H^2))^{1-\alpha} (1 + u\bar{u}')^{-\alpha} \dot{u} \dot{H}_1 \int_{-\infty}^{\infty} (ah^2 + 2bh + c)^{-\alpha} \, dh \\
&= 2^{n-1} \frac{\pi^{1/2}\Gamma(\alpha - \frac{1}{2})}{\Gamma(\alpha)} \int_u \cdots \int (1 + u\bar{u}')^{1-2\alpha} \dot{u} \int_{H_1} \cdots \int (\det(I + H^2))^{1-\alpha} \dot{H}_1 \\
&= 2^{n-1} \pi^{n-1/2} \frac{\Gamma(\alpha - \frac{1}{2})\Gamma(2\alpha - n)}{\Gamma(\alpha)\Gamma(2\alpha - 1)} H_{n-1}(\alpha - 1).
\end{aligned}$$

The theorem follows on the repeated application of this reduction formula, together with the direct evaluation of

$$H_1(\alpha - n - 1) = \int_{-\infty}^{\infty} \frac{dx}{(1 + x^2)^{\alpha - n + 1}} = \frac{\pi^{1/2}\Gamma(\alpha - n + \frac{1}{2})}{\Gamma(\alpha - n + 1)} \quad \left(\alpha > n - \tfrac{1}{2}\right).$$

□

Remark The differential element $\dot{U}$ in the unitary integral defined in Section 22.6 has the following invariant property: If

$$U_1 = VUW, \quad V, W \text{ being unitary matrices,}$$

then

$$\sigma\left(dU_1\overline{(dU_1)'}\right) = \sigma\left(dU\overline{(dU)'}\right),$$

so that $\dot{U}_1 = \dot{U}$. This is an example of an invariant associated with the integral.

22.8 Volume elements for real orthogonal matrices

We give a brief description of an element in the integral associated with the real orthogonal group and its volume.

We first consider O_n, the real orthogonal group of order n, the collection of real matrices T of order n satisfying

$$T'T = I. \tag{22.25}$$

Clearly $\det T = \pm 1$, and we denote by O_n^+ those orthogonal matrices with determinants taking the value $+1$. Corresponding to a matrix T we can form the matrix

$$K = (I - T)(I + T)^{-1}, \tag{22.26}$$

although we need to exclude the case $\det(I + T) = 0$. We cannot simply dismiss it as a 'special case' because a matrix T with determinant -1 must have $\det(I + T) = 0$ (this follows from $\det(I + T) = \det(TT' + T) = \det T \det(T' + I) = -\det(I + T)$); so we need to restrict ourselves to those T belonging to O_n^+. From (22.25) we have

$$K = -K'; \tag{22.27}$$

solving (22.26) gives

$$T = (I - K)(I + K)^{-1}; \tag{22.28}$$

and from $\det(I - K) = \det(I + K') = \det(I + K)$, we see that there is confirmation that $\det T = +1$.

Differentiating (22.28) we have

$$dT = -2(I + K)^{-1}\, dK(I + K)^{-1},$$

so that

$$\sigma(dT\, dT') = -4\sigma(dK(I - K^2)^{-1}\, dK(I - K^2)^{-1}). \tag{22.29}$$

Writing dK as (dk_{ij}), where $dk_{ij} = -dk_{ji}$, and $(I - K^2)^{-1}$ as (u_{st}), where $u_{st} = u_{ts}$, the trace (22.29) becomes

$$-8 \sum_{i<j} \sum_{s<t} (u_{js}u_{it} - u_{is}u_{jt}) dk_{ij}\, dk_{st},$$

so that the differential element for the integral is

$$\dot{T} = 2^{(1/2)n(n-1)} \det(I - K^2)^{-(1/2)(n-1)} \dot{K}, \tag{22.30}$$

where $\dot{K} = 2^{n(n-1)/4} \prod_{i<j} dk_{ij}$.

22.9 The total volume of the real orthogonal group

Theorem 22.14 *Let $n \geq 2$ and $\alpha > \frac{1}{4}(2n-3)$. Then*

$$J_n(\alpha) = \int_K \cdots \int \frac{\dot{K}}{(\det(I + KK'))^{\alpha}}$$
$$= (2\pi)^{(1/4)n(n-1)} \prod_{\nu=2}^{n} \frac{\Gamma\left(2\alpha - n + \frac{1}{2}(\nu+1)\right)}{\Gamma(2\alpha - n + \nu)},$$

where $K = (k_{ij})$ runs over all real skew-symmetric matrices of order n and $\dot{K} = 2^{n(n-1)/4} \prod_{i<j} dk_{ij}$.

Proof. Write K in the form

$$K = \begin{pmatrix} K_1 & -t' \\ t & 0 \end{pmatrix},$$

where K_1 is a real skew-symmetric matrix of order $n-1$ and t is a vector with dimension $n-1$, so that

$$I + KK' = \begin{pmatrix} I + K_1K_1' + t't & K_1t' \\ tK_1' & 1 + tt' \end{pmatrix},$$

it is not difficult to show that

$$\det(I + KK') = (1 + tt' - tK_1'(I + K_1K_1' + t't)^{-1}K_1t') \det(I + K_1K_1' + t't).$$

There is an orthogonal matrix Γ such that

$$K_1 = \Gamma\left(\begin{pmatrix} 0 & \lambda_1 \\ -\lambda_1 & 0 \end{pmatrix} \dot{+} \begin{pmatrix} 0 & \lambda_2 \\ -\lambda_2 & 0 \end{pmatrix} \dot{+} \cdots\right)\Gamma',$$

where the bracket ends with either

$$\begin{pmatrix} 0 & \lambda_{\frac{n}{2}} \\ -\lambda_{\frac{n}{2}} & 0 \end{pmatrix}, \text{ or } \begin{pmatrix} 0 & \lambda_{\frac{n-1}{2}} \\ -\lambda_{\frac{n-1}{2}} & 0 \end{pmatrix} \dot{+} 0,$$

depending on whether n is even or odd.

Let

$$T = \Gamma[\sqrt{1+\lambda_1^2}, \sqrt{1+\lambda_1^2}, \sqrt{1+\lambda_2^2}, \sqrt{1+\lambda_2^2}, \dots]\Gamma';$$

then

$$T = T', \quad K_1 T = T K_1, \quad I + K_1 K_1' = T^2.$$

Now set $t = wT$, so that

$$\dot{t} = (\det T)\dot{w} = (\det(I + K_1 K_1'))^{1/2}\dot{w},$$
$$I + K_1 K_1' + t't = T^2 + Tw'wT = T(I + w'w)T$$

and

$$\begin{aligned} 1 + tt' - tK_1'(I + K_1 K_1' + t't)^{-1}K_1 t' &= 1 + wT^2 w' - wTK_1'T^{-1}(I + w'w)^{-1}T^{-1}K_1 T w' \\ &= 1 + wT^2 w' - wK_1'(I + w'w)^{-1}K_1 w' \\ &= 1 + ww' - \frac{(wK_1 w')^2}{1 + ww'} = 1 + ww'. \end{aligned}$$

Therefore

$$\begin{aligned} J_n(\alpha) &= \int \cdots \int_K \frac{\dot{K}}{(\det(I + KK'))^\alpha} \\ &= 2^{(1/2)(n-1)} \int \cdots \int_{K_1} \frac{\dot{K}_1}{(\det(I + K_1 K_1'))^{\alpha - 1/2}} \int \cdots \int_w (1 + ww')^{-2\alpha}\,\dot{w} \\ &= (2\pi)^{(1/2)(n-1)} \frac{\Gamma(2\alpha - \frac{1}{2}(n-1))}{\Gamma(2\alpha)} J_{n-1}(\alpha - \tfrac{1}{2}). \end{aligned}$$

The theorem follows on the repeated application of this reduction formula, together with the direct evaluation of

$$\begin{aligned} J_2(\alpha - \tfrac{n-2}{2}) &= \sqrt{2}\int_{-\infty}^{\infty} \frac{dt}{(1 + t^2)^{2\alpha - n + 2}} \\ &= \sqrt{2\pi}\,\frac{\Gamma(2\alpha - n + \frac{3}{2})}{\Gamma(2\alpha - n + 2)} \quad \left(\alpha > \tfrac{1}{4}(2n - 3)\right). \end{aligned}$$

□

We therefore have the following.

Theorem 22.15 *The total volume of the orthogonal group O_n^+ is given by*

$$\int \dot{T} = 2^{(1/2)n(n-1)} \underset{K}{\int \cdots \int} \det(I - K^2)^{-(1/2)(n-1)} \dot{K}$$

$$= (8\pi)^{(1/4)n(n-1)} \prod_{\nu=2}^{n} \frac{\Gamma(\frac{1}{2}(\nu - 1))}{\Gamma(\nu - 1)}.$$

Exercise 22.16 Determine the volume of the symplectic group.

Exercise 22.17 Determine the volume of the manifold associated with the symmetric unitary matrices.

Exercise 22.18 Determine the volume of the manifold associated with the skew-symmetric unitary matrices.

23

Non-negative square matrices

23.1 Similarity of non-negative matrices

Definition 23.1 A square matrix in which there is exactly one positive entry in each row and in each column and with all the remaining entries 0 is called a generalised exchange matrix.

For example, a diagonal matrix $\Lambda = [\lambda_1, \ldots, \lambda_n]$ with $\lambda_i > 0$ is a generalised exchange matrix. An ordinary exchange matrix P, in which there is exactly one entry 1 in each row and each column, and with all the remaining entries 0, is obviously a generalised exchange matrix, and indeed so is $P\Lambda$. It is not difficult to see that there are no other kinds of such matrices.

Definition 23.2 A non-negative matrix is one in which all its elements are non-negative real numbers. Two non-negative square matrices A, B of the same order are said to be similar to each other if there is a generalised exchange matrix Q such that

$$QAQ^{-1} = B,$$

and this will be denoted by $A \sim B$.

Clearly, being similar is an equivalence relation:

(1) $A \sim A$;
(2) if $A \sim B$ then $B \sim A$;
(3) if $A \sim B$ and $B \sim C$ then $A \sim C$.

We write $A \geq 0$ to mean that A is non-negative, and $A > 0$ means that all its elements are positive. The same applies to vectors when we write $x \geq 0$ and $x > 0$. Moreover, we shall write $A \geq B$ to mean that $A - B \geq 0$, etc. It should be clear that

(4) if $A \geq B$ and $B \geq C$ then $A \geq C$;
(5) if $A \geq 0$ and $B \geq C$ then $AB \geq AC$ and $BA \geq CA$;
(6) if $A \sim B$ and $A \geq 0$ (or > 0) then $B \geq 0$ (or > 0).

Their proofs rely on the fact that generalised exchange matrices, and their inverses, are non-negative. Conversely, we have the following.

Theorem 23.3 *If the inverse of an invertible non-negative matrix is also non-negative, then it must be a generalised exchange matrix.*

Proof. Let

$$B = (b_{ij}), \quad b_{ij} \geq 0,$$

be the inverse of

$$A = (a_{ij}), \quad a_{ij} \geq 0.$$

Then, for $i \neq k$,

$$\sum_{j=1}^{n} a_{ij} b_{jk} = 0,$$

so that, for any j,

$$a_{ij} b_{jk} = 0, \quad i \neq k.$$

If there were two elements of A in the ith row with $a_{ij_1} \neq 0$, $a_{ij_2} \neq 0$, then, for all $k \neq i$, we have

$$b_{j_1 k} = b_{j_2 k} = 0,$$

which means that all the elements in rows j_1, j_2 in B, apart from $b_{j_1 i}$, $b_{j_2 i}$, are 0, so that B would be singular, which is a contradiction. Therefore there is exactly one non-zero element in each row of A; the same argument applies to the columns of A. The theorem is proved. □

Exercise 23.4 Suppose that, for any non-negative square matrix A, the matrix TAT^{-1} is non-negative. Show that T must be a generalised exchange matrix.

Definition 23.5 If

$$A \sim \begin{pmatrix} A_1^{(h)} & A_2 \\ 0 & A_3^{(n-h)} \end{pmatrix},$$

then we say that A is decomposable; otherwise A is indecomposable.

23.2 Standard forms

We consider non-negative indecomposable square matrices.

Definition 23.6 A non-negative indecomposable square matrix A satisfying

$$A = (a_{ij}), \quad \sum_{j=1}^{n} a_{ij} = q,$$

is said to be in standard form, and q is called its *peak*.

Theorem 23.7 (Fundamental Theorem) *Any non-negative indecomposable square matrix must be similar to one in standard form, which is unique apart from the ordering of the rows.*

Before we give the proof of the theorem, we first introduce a method which is the source of the proof – it is also a good method for actual computations.

Suppose that

$$\begin{cases} a_{11} + \cdots + a_{1n} = q_1, \\ \qquad \cdots \\ a_{n1} + \cdots + a_{nn} = q_n, \end{cases} \tag{23.1}$$

with $q_1 \geq q_2 \geq \cdots \geq q_n$ and $q_1 > q_n$.

Take λ so that

$$a_{11} + \cdots + a_{1,n-1} + \frac{a_1 n}{\lambda} = \lambda(a_{n1} + \cdots + a_{n,n-1}) + a_n n, \tag{23.2}$$

that is

$$q_1 - a_{1n} + \frac{a_{1n}}{\lambda} = \lambda(q_n - a_{nn}) + a_{nn}$$

or

$$\lambda^2(q_n - a_{nn}) + (a_{nn} + a_{1n} - q_1)\lambda - a_{1n} = 0,$$

and we use $f(\lambda)$ to denote the left-hand side. If $q_n = a_{nn}$ then $a_{n1} = \cdots = a_{n,n-1} = 0$, and A is indecomposable, so that we need to have $q_n > a_{nn}$. From $f(\pm\infty) > 0$ and $f(1) = q_n - q_1 < 0$, we see that there is a $\lambda > 1$ satisfying (23.2), and therefore the sums of the rows of the matrix

$$[1, \ldots, 1, \lambda]A[1, \ldots, 1, \lambda]^{-1} = \begin{pmatrix} a_{11} & a_{12} & \cdots & a_{1,n-1} & \frac{a_{1n}}{\lambda} \\ a_{21} & a_{22} & \cdots & a_{2,n-1} & \frac{a_{2n}}{\lambda} \\ \vdots & \vdots & & \vdots & \vdots \\ \lambda a_{n1} & \lambda a_{n2} & \cdots & \lambda a_{n,n-1} & a_{nn} \end{pmatrix}$$

are all less than q_1 (except for the case when $a_{2n} = 0$, $q_2 = q_1$).

Therefore, generally speaking, we can use this method to reduce $\max(q_1, \ldots, q_n)$ successively until all the sums of the rows are the same.

23.3 Proof of the fundamental theorem

As we remarked, the method introduced in the preceding section is the source of the proof of the fundamental theorem, and we now introduce

$$Q(A) = \max(q_1, \ldots, q_n). \tag{23.3}$$

We first show that if $q_1, \dots, q_n$ are not the same, then we can take Λ so that

$$Q(\Lambda A \Lambda^{-1}) < Q(A). \tag{23.4}$$

Suppose that, on reordering the columns, the sums of the rows of A can be arranged so that

$$q_1 = \dots = q_s > q_{s+1} \geq q_{s+2} \geq \dots \geq q_n.$$

We proceed to show that Λ can be chosen so that rows of $\Lambda A \Lambda^{-1}$ with sums equal to q_1 have indices less than s, and the remaining sums are less than q_1. Decompose A as

$$A = \begin{pmatrix} A_1 & A_2 \\ A_3 & A_4 \end{pmatrix}, \quad A_1 = A_1^{(s)}, \dots .$$

Take

$$\Lambda = \begin{pmatrix} I^{(s)} & 0 \\ 0 & \Lambda_1 \end{pmatrix}, \quad \Lambda_1 = [\lambda_{s+1}, \dots, \lambda_n],$$

so that

$$\Lambda A \Lambda^{-1} = \begin{pmatrix} A_1 & A_2 A_1^{-1} \\ A_1 A_3 & A_1 A_4 A_1^{-1} \end{pmatrix}.$$

Because $A_2 \neq 0$, there are $\lambda_{s+1}, \dots, \lambda_n$ such that

$$A_2 \Lambda_1^{-1} \leq A_2, \quad \text{but } A_2 \Lambda_1^{-1} \neq A_2.$$

At the same time it also ensures that the sums of the last $n - s$ rows of $\Lambda A \Lambda^{-1}$ are less than q_1. From $A_2 \Lambda_1^{-1} \leq A_2$ and $A_2 \Lambda_1^{-1} \neq A_2$, we know that there is at least one sum among the first s rows of $\Lambda A \Lambda^{-1}$ which is less than q_1, so that the number of sums of rows in $\Lambda A \Lambda^{-1}$ being equal to q_1 is less than s. We continue in this way until there is no row with a sum $\geq q_1$. This then proves (23.4).

Next, we consider the set S:

$$\Lambda = [\lambda_1, \dots, \lambda_n], \quad \lambda_1 + \dots + \lambda_n = 1, \quad \lambda_\nu \geq 0. \tag{23.5}$$

(Because $\Lambda A \Lambda^{-1} = ((1/\lambda)\Lambda) A ((1/\lambda)\Lambda)^{-1}$, we have effectively discussed all such Λ already.) Corresponding to each $\Lambda (\in S)$, there is the value

$$Q(\Lambda A \Lambda^{-1}).$$

Let q denote the infimum of this set of numbers. If q is attained by some Λ then, from (23.4), we know that all the rows of $\Lambda A \Lambda^{-1}$ have the sum q, so that A is similar to a standard form.

From the definition of q, there is a sequence

$$\Lambda_1, \Lambda_2, \dots, \Lambda_i, \dots \tag{23.6}$$

such that

$$Q(\Lambda_i A \Lambda_i^{-1}) < q + \frac{1}{i}. \tag{23.7}$$

Since S is a closed set, the collection $\{\Lambda_i\}$ has a 'limit point' Λ_0, and we may therefore select a subsequence of (23.6) which tends to Λ_0; we may assume that

$$\lim_{i\to\infty} \Lambda_i = \Lambda_0.$$

We first show that $\Lambda_0[\lambda_0^{(0)}, \ldots, \lambda_n^{(0)}] > 0$. From $\lambda_1^{(0)} + \cdots + \lambda_n^{(0)} = 1$ we know that not all λ_j can be 0. If there were some $\lambda_j = 0$ then we suppose that $\lambda_1^{(0)} = \cdots = \lambda_s^{(0)} = 0$, and that $\lambda_{s+1}^{(0)}, \ldots, \lambda_n^{(0)}$ are not 0. From

$$\lambda_j^{(i)} a_{j1} \lambda_1^{(i)-1} + \cdots + \lambda_j^{(i)} a_{js} \lambda_s^{(i)-1} + \cdots + \lambda_j^{(i)} a_{jn} \lambda_n^{(i)-1} < q + \frac{1}{i}$$
$$(s+1 \le j \le n,\ i = 1, 2, \ldots),$$

it follows that $a_{j1} = \cdots = a_{js} = 0\,(s+1 \le j \le n)$, showing that A is decomposable. This is a contradiction, and therefore $\Lambda_0 > 0$.

Let $\Lambda_i = [\lambda_1^{(i)}, \ldots, \lambda_n^{(i)}]$. From (23.7) we see that

$$\max[\lambda_1^{(i)}(a_{11}\lambda_1^{(i)-1} + \cdots + a_{1n}\lambda^{(i)-1}), \ldots, \lambda_n^{(i)}(a_{n1}\lambda_1^{(i)-1} + \cdots + a_{nn}\lambda^{(i)-1})]$$
$$< q + \frac{1}{i},$$

that is, for any j, we have

$$\lambda_j^{(i)}(a_{j1}\lambda_1^{(i)-1} + \cdots + a_{jn}\lambda^{(i)-1}) < q + \frac{1}{i},$$

and so, letting $i \to \infty$,

$$\lambda_j^{(0)}(a_{j1}\lambda_1^{(0)-1} + \cdots + a_{jn}\lambda^{(0)-1}) \le q.$$

Thus

$$Q(\Lambda_0 A \Lambda_0^{-1}) = q.$$

The first part of the theorem is proved.

For the uniqueness part, let us suppose that A is in standard form, with its rows having sums equal to q, and that $\Lambda A \Lambda^{-1}$ is also a standard form, with its rows having sums equal to q_1. Then

$$a_{i1}\lambda_1^{-1} + \cdots + a_{in}\lambda_n^{-1} = q_1 \lambda_i^{-1}. \tag{23.8}$$

This means that, for all i, we have

$$q \min(\lambda_1^{-1}, \ldots, \lambda_n^{-1}) \le q_1 \lambda_i^{-1} \le q \max(\lambda_1^{-1}, \ldots, \lambda_n^{-1}). \tag{23.9}$$

Taking $\lambda_i^{-1} = \max(\lambda_1^{-1}, \ldots, \lambda_n^{-1})$ and then $(\lambda_1^{-1}, \ldots, \lambda_n^{-1})$, we see at once that $q = q_1$, and substituting this into (23.8) we have

$$a_{i1}\lambda_1^{-1} + \cdots + a_{in}\lambda_n^{-1} = (a_{i1} + \cdots + a_{in})\lambda_n^{-1}. \tag{23.10}$$

If we rearrange the terms so that

$$\lambda_1^{-1} \leq \cdots \leq \lambda_s^{-1} < \lambda_{s+1} = \cdots = \lambda_n^{-1},$$

then from (23.8) we find that, for $i = s + 1, \ldots, n$,

$$a_{i1} = \cdots = a_{is} = 0.$$

This means that A is decomposable. Therefore there is uniqueness.

23.4 Another form of the fundamental theorem

From

$$\Lambda A \Lambda^{-1} = (b_{ij}), \quad \sum_{j=1}^{n} b_{ij} = q,$$

we have

$$\sum_{j=1}^{n} a_{ij}\frac{\lambda_i}{\lambda_j} = \sum_{j=1}^{n} b_{ij} = q,$$

that is

$$\sum_{j=1}^{n} \frac{a_{ij}}{\lambda_j} = \frac{q}{\lambda_i}.$$

This means that

$$A \begin{pmatrix} \lambda_1^{-1} \\ \vdots \\ \lambda_n^{-1} \end{pmatrix} = q \begin{pmatrix} \lambda_1^{-1} \\ \vdots \\ \lambda_n^{-1} \end{pmatrix}.$$

We deduce at once the following theorem.

Theorem 23.7 *A non-negative indecomposable square matrix has a positive eigenvalue q, which has belonging to it a positive (column) eigenvector.*

Related to positive eigenvalues and positive eigenvectors, we have the following property.

Theorem 23.8 *A non-negative indecomposable square matrix has only one non-negative (column) eigenvector (disregarding multiples thereof). It is a positive vector belonging to* *the eigenvalue which takes the value of the peak, and the absolute values of the remaining eigenvalues do not exceed the peak.*

Proof. We may assume that the given matrix is in standard form, that is

$$A = (a_{ij}), \quad \sum_{j=1}^{n} a_{ij} = q.$$

Clearly $(1, 1, \ldots, 1)'$ is an eigenvector, and it is positive. If there is another non-negative eigenvector $(x_1, \ldots, x_n)'(\neq 0')$ then

$$a_{i1}x_1 + \cdots + a_{in}x_n = q_1 x_i, \quad i = 1, 2, \ldots, n, \tag{23.11}$$

from which it follows that $q_1 > 0$. Consequently,

$$q \min(x_1, \ldots, x_n) \leq q_1 x_i \leq q \max(x_1, \ldots, x_n). \tag{23.12}$$

If there are some x_i which are 0, say $x_1 = \cdots = x_s = 0$, $x_{s+1} > 0, \ldots, x_n > 0$, then from

$$a_i, s + 1x_{s+1} + \cdots + a_{in}x_n = q_1 x_i = 0, \quad i = 1, 2, \ldots, s,$$

we find that $a_{i,s+1} = \cdots = a_{in} = 0$ for $1 \leq i \leq s$, so that A would then be decomposable. Therefore all x_i are positive.

From (23.12) we deduce that $q_1 = q$, and from (23.11) we have

$$a_{i1}x_1 + \cdots + a_{in}x_n = (a_{i1} + \cdots + a_{in})x_i;$$

we further deduce that $x_1 = \cdots = x_n$. We therefore have the uniqueness of the non-negative eigenvector, that is there is only one non-negative eigenvector and the eigenvalue to which it belongs is the peak q.

Let q_1 be any eigenvalue of A, with the corresponding eigenvector $x' = (x_1, \ldots, x_n)'$ (which may possibly be complex). Then

$$\sum_{j=1}^{n} a_{ij}x_j = q_1 x_i,$$

that is

$$|q_1||x_1| \leq \sum_{j=1}^{n} a_{ij}|x_j| \leq q \max(|x_1|, \ldots, |x_n|).$$

Taking $|x_i| = \max(|x_1|, \ldots, |x_n|)$, we then have

$$|q_1| \leq q.$$

The theorem is proved. □

Theorem 23.9 *There is only one (column) eigenvector (disregarding multiples thereof) belonging to the peak.*

Proof. Suppose that, besides $e' = (1, 1, \ldots, 1)'$, there is another eigenvector x' belonging to the peak; then x' has to be real. Now, for real α,

$$e' + \alpha x'$$

is also an eigenvector belonging to the peak, and it is non-negative when α is sufficiently small. Thus $e' + \alpha x' = \beta e'$, which shows that x' is a multiple of e'. The theorem is proved. □

23.5 Arithmetic operations on standard forms

Theorem 23.10 *The sum of two non-negative matrices in standard form is also in standard form, and its peak is the sum of the two peaks; the product is also in standard form, and its peak is the product of the two peaks.*

Proof. If

$$\sum_{j=1}^{n} a_{ij} = q, \quad \sum_{ij} b_{ij} = r,$$

then

$$\sum_{j=1}^{n} (a_{ij} + b_{ij}) = q + r,$$

$$\sum_{k=1}^{n} \left(\sum_{j=1}^{n} a_{ij} b_{jk} \right) = \sum_{j=1}^{n} \left(\sum_{k=1}^{n} b_{jk} \right) a_{ij} = qr.$$

If we discard non-negativity for the moment then the sum of the rows of A^{-1} are equal to q^{-1}. The proof of this point is not difficult, because

$$A(1, 1, \ldots, 1)' = q(1, 1, \ldots, 1)',$$

so that

$$A^{-1}(1, 1, \ldots, 1)' = q^{-1}(1, 1, \ldots, 1)'.$$

Therefore the rows of A^{-1} have sums equal to q^{-1}. □

Theorem 23.11 *If $A \geq 0$ and $(I - A)^{-1} \geq 0$, then the peak of A is less than 1; also $(I - A)^{-1}$ and A have the same standard form, and the peak of the former has the value $(1 - q)^{-1}$.*

Proof. From

$$A(1, 1, \ldots, 1)' = q(1, 1, \ldots, 1)',$$

we have

$$(I - A)^{-1}(1, 1, \ldots, 1)' = (1 - q)^{-1}(1, 1, \ldots, 1)'.$$

If $(I - A)^{-1} \geq 0$ then $q < 1$. The theorem follows. □

It is not difficult to use the same method to show that if $A \geq 0$ and it has the peak q, then e^A has the peak e^q; moreover, A and e^A have the same positive eigenvector.

What we have said in the above applies to column eigenvectors, but the same applies to row vectors.

Now take row and column eigenvectors x and y belonging to the same eigenvalue of a non-negative indecomposable square matrix A, so that

$$xA = qx, \qquad Ay = yq.$$

Then

$$z_1 = x_1 y_1, \ \ldots \ , \ z_n = x_n y_n$$

stay the same after the operation $\Lambda A \Lambda^{-1}$.

A standard form in which the rows have the same sums has the eigenvector $(1, 1, \ldots, 1)'$, and the row eigenvector is $(z_1, \ldots, z_n)$. Similarly, if we consider standard forms in terms of column sums then the row eigenvector is $(1, 1, \ldots, 1)$, while the column eigenvector is $(z_1, \ldots, z_n)'$.

23.6 Large and small matrices

Theorem 23.12 *Let*

$$C = (c_{ij}) \tag{23.13}$$

be a complex matrix. If

$$|c_{ij}| \leq a_{ij}, \tag{23.14}$$

where $A = (a_{ij})$ is a non-negative indecomposable square matrix with peak q, then the absolute value of any eigenvalue γ of C cannot exceed q, that is

$$|\gamma| \leq q. \tag{23.15}$$

Moreover, if $\gamma = e^{i\theta}q$ then

$$C = e^{i\theta}[e^{i\theta_1}, \ldots, e^{i\theta_n}]A[e^{i\theta_1}, \ldots, e^{i\theta_n}]^{-1}.$$

Proof. (1) We may assume, without loss of generality, that A is in standard form. Let $x(\neq 0)$ be a (column) eigenvector belonging to an eigenvalue γ of C, so that

$$\gamma x_i = \sum_{j=1}^{n} c_{ij} x_j.$$

We then have, by hypothesis,

$$|\gamma||x_i| \leq \sum_{j=1}^{n} |c_{ij}||x_j| \leq \sum_{j=1}^{n} a_{ij}|x_j| \leq q \max(|x_1|, \ldots, |x_n|), \tag{23.16}$$

so that $|\gamma| \leq q$ follows from taking $|x_i| = \max(|x_1|, \ldots, |x_n|)$.

(2) Suppose that $\gamma = e^{i\theta}q$. We examine the various inequality signs in (23.16), and we assume, without loss of generality, that

$$|x_1| = |x_2| = \cdots = |x_s| > |x_{s+1}| \geq \cdots \geq |x_n|.$$

Thus, for $i = 1, 2, \ldots, s$, there is equality between the left-hand and the right-hand sides of (23.16), so that

$$q|x_i| = \sum_{j=1}^{n} |c_{ij}||x_j| = \sum_{j=1}^{n} a_{ij}|x_j| = \left(\sum_{j=1}^{n} a_{ij}\right)|x_i|.$$

The last equality sign holds only when

$$a_{i,s+1} = a_{i,s+2} = \cdots = a_{in} = 0, \quad 1 \leq i \leq s,$$

which means that A can be decomposed, contradicting the hypothesis. Therefore

$$|x_1| = |x_2| = \cdots = |x_n|,$$

and we may as well suppose that $|x_1| = 1$, so that

$$x_1 = e^{i\theta_1}, \; \ldots, \; x_n = e^{i\theta_n}.$$

The square matrix

$$C_1 = [e^{i\theta_1}, \ldots, e^{i\theta_n}]^{-1} C [e^{i\theta_1}, \ldots, e^{i\theta_n}]$$

has $(1, 1, \ldots, 1)'$ as a (column) eigenvector; we may assume without loss that C_1 is C itself, that is C has the (column) eigenvector $(1, 1, \ldots, 1)'$ belonging to the eigenvalue $\gamma = qe^{i\theta}$, so that

$$\gamma = qe^{i\theta} = \sum_{j=1}^{n} c_{ij}, \quad |c_{ij}| = a_{ij}, \quad \sum_{j=1}^{n} a_{ij} = q,$$

that is

$$q = \sum_{j=1}^{n} c_{ij}e^{-i\theta} = \sum_{j=1}^{n} a_{ij},$$

giving

$$c_{ij} = a_{ij}e^{i\theta}.$$

The theorem is proved. □

Theorem 23.13 *Let A be a non-negative indecomposable square matrix. Then the peak of any main sub-matrix of A must be less than the peak of A.*

Proof. Let

$$A_1 = \begin{pmatrix} a_{11} & \cdots & a_{1m} & 0 & \cdots & 0 \\ \vdots & & \vdots & \vdots & & \vdots \\ a_{m1} & \cdots & a_{mm} & 0 & \cdots & 0 \\ 0 & \cdots & 0 & 0 & \cdots & 0 \\ & & \cdots & & & \end{pmatrix};$$

then $A_1 \leq A$ and $A_1 \neq A$, so that, by Theorem 23.12, the peak of A_1 must be smaller than that of A. □

Theorem 23.14 *Let A be a non-negative indecomposable square matrix. Then its peak is not a repeated root of its characteristic equation.*

Proof. The derivative of the characteristic polynomial

$$f(\lambda) = \begin{vmatrix} \lambda - a_{11} & -a_{12} & \cdots & -a_{1n} \\ a_{21} & \lambda - a_{22} & \cdots & -a_{2n} \\ \vdots & \vdots & & \vdots \\ -a_{n1} & -a_{n2} & \cdots & \lambda - a_{nn} \end{vmatrix}$$

is

$$f'(\lambda) = \begin{vmatrix} 1 & 0 & \cdots & 0 \\ a_{21} & \lambda - a_{22} & \cdots & -a_{2n} \\ \vdots & \vdots & & \vdots \\ -a_{n1} & -a_{n2} & \cdots & \lambda - a_{nn} \end{vmatrix}$$

$$+ \begin{vmatrix} \lambda - a_{11} & -a_{12} & \cdots & -a_{1n} \\ 0 & 1 & \cdots & 0 \\ \vdots & \vdots & & \vdots \\ -a_{n1} & -a_{n2} & \cdots & \lambda - a_{nn} \end{vmatrix} + \cdots,$$

and the first determinant here is the characteristic polynomial of the matrix

$$\begin{pmatrix} a_{22} & \cdots & a_{2n} \\ \vdots & & \vdots \\ a_{n2} & \cdots & a_{nn} \end{pmatrix}.$$

By Theorem 23.13, its peak is less than the peak q of A so that, when $\lambda \geq q$, the value of the characteristic polynomial is positive. The same argument applies to the other determinants, so that $f'(\lambda) > 0$ when $\lambda \geq q$. This shows that q is a simple root of $f(\lambda)$. □

Theorem 23.15 *Suppose that A has other eigenvalues with absolute values equal to q; these values are*

$$qe^{2\pi i\ell/h}, \quad \ell = 0, 1, \ldots, h-1,$$

where $h \geq 2$ is an integer.

Proof. Take $\gamma = qe^{i\theta}$; then, by the second part of Theorem 23.13 (taking $C = A$),

$$A = e^{i\theta}[e^{i\theta_1}, \ldots, e^{i\theta_n}]A[e^{i\theta_1}, \ldots, e^{i\theta_n}]'. \tag{23.17}$$

It follows that if x_0 is an eigenvalue of A then so is $e^{i\theta}x_0$, and thus so are

$$x_0, e^{i\theta}x_0, e^{2i\theta}x_0, \ldots .$$

There can only be a finite number of eigenvalues, so that there exists h such that $h\theta$ is a multiple of 2π. Let h denote the smallest such positive integer. There being no repeated roots, the theorem follows. □

Since there are repeated roots for A^h, we have the following.

Theorem 23.16 *If, apart from q, there are other eigenvalues of A having absolute values equal to q, then there is a positive integer h such that A^h is decomposable.*

Returning to (23.17) we examine

$$A = e^{2\pi i/h}[e^{i\theta_1}, \ldots, e^{i\theta_n}]A[e^{i\theta_1}, \ldots, e^{i\theta_n}]', \tag{23.18}$$

which, on repeated substitutions, yields

$$A = [e^{ih\theta_1}, \ldots, e^{ih\theta_n}]A[e^{ih\theta_1}, \ldots, e^{ih\theta_n}]'.$$

If $e^{ih\theta_j} \neq 1$ for some j, then A is decomposable; thus each $e^{i\theta_j}$ is an hth root of unity. Suppose that, among $e^{i\theta_1}, \ldots, e^{i\theta_n}$, there are n_1 lots of 1, n_2 lots of $e^{2\pi i/h}$, n_3 lots of $e^{4\pi i/h}$, etc. After such a listing, we have

$$[e^{i\theta_1}, \ldots, e^{i\theta_n}] = \begin{pmatrix} I^{(n_1)} & & & \\ & e^{\frac{2\pi i}{h}} I^{(n_2)} & & \\ & & \ddots & \\ & & & e^{\frac{2\pi i(h-1)}{h}} \quad I^{(n_h)} \end{pmatrix};$$

on decomposing A as

$$\begin{pmatrix} A_{11} & A_{12} & \cdots & A_{1h} \\ \vdots & \vdots & & \vdots \\ A_{h1} & A_{h2} & \cdots & A_{hh} \end{pmatrix},$$

and substituting this into (23.18), we have

$$A_{st} = e^{2\pi i/h} \cdot e^{(2\pi i/h)(s-1)} \cdot e^{-(2\pi i/h)(t-1)} A_{st} = e^{(2\pi i/h)(1+s-t)} A_{st}.$$

When $t \neq 1 + s$, we need to have $A_{st} = 0$, so that A has the form

$$\begin{pmatrix} 0 & A_{12} & 0 & \cdots & 0 \\ 0 & 0 & A_{23} & \cdots & 0 \\ \vdots & \vdots & \vdots & & \vdots \\ A_{h1} & 0 & 0 & \cdots & 0 \end{pmatrix}.$$

23.7 Strongly indecomposable matrices

Definition 23.17 If no power of a square matrix is decomposable then we say that the matrix is strongly indecomposable.

We have the following property associated with strongly indecomposable matrices.

Theorem 23.18 *Let A be a strongly indecomposable non-negative square matrix with peak q. Then*

$$\lim_{\ell \to \infty} \left(\frac{A}{q}\right)^{\ell} = u'v, \quad vu' = 1, \tag{23.19}$$

where u' and v are the positive column eigenvector and positive row eigenvector of A, respectively.

Proof. Being strongly indecomposable, the eigenvalues of A distinct from q have absolute values less than q. Thus A/q has an eigenvalue equal to 1, and the remaining ones are less than 1 in modulus. It follows that

$$\lim_{\ell \to \infty} \left(\frac{A}{q}\right)^{\ell}$$

is a square matrix having eigenvalue 1 (which is a simple root) and the remaining eigenvalues are equal to 0. This shows that (23.19) holds. Let c be any vector, and set

$$v_\ell = c\frac{A^\ell}{q^\ell}; \tag{23.20}$$

then

$$\lim_{\ell \to \infty} v_\ell = (cu')v. \tag{23.21}$$

This means that, regardless of any c, the limit of v_ℓ is a certain multiple of v. Also, from letting $\ell \to \infty$ in

$$qv_{\ell+1} = c\frac{A^{\ell+1}}{q^\ell} = v_\ell A,$$

we have

$$qv = vA,$$

which shows that v is a row eigenvector. The same method shows that $qu' = Au'$, so that u, v are positive vectors. □

We deduce at once the following theorem.

Theorem 23.19 *Corresponding to any strongly indecomposable square matrix A, there is a positive integer ℓ such that*

$$A^\ell > 0,$$

meaning that all the elements of A^ℓ are positive.

23.8 Markov chains

Definition 23.20 A vector $x = (x_1, \ldots, x_n)$ satisfying

$$x_1 + \cdots + x_n = 1, \quad x_i \geq 0,$$

is called a probability vector, and a matrix

$$P = (p_{ij})_{1 \leq i,j \leq n}, \quad p_{ij} \geq 0, \quad \sum_{i=1}^{n} p_{ij} = 1,$$

is called a Markov matrix.

A Markov matrix maps a probability vector into another such vector, that is Px' is still a probability vector. This is because from

$$y_i = \sum_{j=1}^{n} p_{ij} x_j$$

we find that

$$\sum_{i=1}^{n} y_i = \sum_{j=1}^{n} \sum_{i=1}^{n} p_{ij} x_j = \sum_{j=1}^{n} x_j = 1.$$

We consider a certain physical system, in which there are only finitely many 'states', which can only change at certain times.

We use the numbers $1, 2, \ldots, n$ to represent the different states, which change at time $t = 0, 1, 2, \ldots$. We also suppose that the probability of the system having state j at time t, changing to state i at $t + 1$, is given by

$$p_{ij} \quad \text{with} \quad \sum_{j=1}^{n} p_{ij} = 1.$$

Let $x_i(t)$ denote the probability of the appearance of the state i at time t, so that

$$x_i(t+1) = \sum_{j=1}^{n} p_{ij} x_j(t).$$

Using matrix symbols, this can be written as

$$x(t+1)' = Px(t)',$$

and we assume that the initial state is

$$x(0) = (c_1, \ldots, c_n), \quad c_1 + \cdots + c_n = 1.$$

The fundamental theorem on Markov chains is as follows.

Theorem 23.21 *Let $P > 0$ be a Markov matrix. Then, corresponding to a probability vector x, the limit*

$$y = \lim_{t\to\infty} x(t)$$

is still a probability vector; it is the unique positive eigenvector of P, and it is independent of the initial state $x(0)$.

Proof. This is an obvious corollary of Theorem 23.18, because P is positive so that it is strongly indecomposable. Therefore

$$\lim_{\ell\to\infty} P^\ell = u'v, \quad vu' = 1. \tag{23.22}$$

Since P is a Markov matrix, $(1, 1, \ldots, 1)$ is its (row) eigenvector, so that $v = (1, 1, \ldots, 1)$. From

$$\lim_{t\to\infty} x(t)' = \lim_{t\to\infty} P^t x(0)' = u'vx(0)' = u',$$

we have

$$\lim_{t\to\infty} x(t) = u.$$

□

The theorem holds not only for positive P, but also for any strongly indecomposable square matrices. However, for P being merely indecomposable, because there may be eigenvalues $e^{2\pi ik/h}$, $k = 0, 1, 2, \ldots, h-1$, we cannot draw the conclusion (23.22). Nevertheless, we can consider the arithmetic mean

$$\frac{1}{L}\sum_{\ell=1}^{L} A^\ell. \tag{23.23}$$

From

$$\lim_{L\to\infty}\frac{1}{L}\sum_{\ell=1}^{L}e^{2\pi ik/h}=\begin{cases}0, & \text{if } h \text{ does not divide } k,\\ 1, & \text{if } h \text{ divides } k,\end{cases}$$

we see that, as $L\to\infty$, the expression (23.23) is also a matrix with rank 1 and having 1 as an eigenvector. Therefore

$$\lim_{L\to\infty}\frac{1}{L}\sum_{\ell=1}^{L}P^{\ell}=u'v,\quad vu'=1,$$

which then delivers

$$\lim_{L\to\infty}\frac{1}{L}\sum_{\ell=1}^{L}vP^{\ell}=v.$$

Thus $vP=v$, that is $v=(1,1,\ldots,1)$, and therefore

$$\lim_{L\to\infty}\frac{1}{L}\sum_{\ell=1}^{L}P^{\ell}x(0)'=u'(vx(0)')=u',$$

that is

$$\lim_{L\to\infty}\frac{1}{L}\sum_{\ell=1}^{L}x(\ell)=u.$$

Theorem 23.22 *Corresponding to a probability vector x, if P is an indecomposable Markov matrix, then*

$$y=\lim_{L\to\infty}\frac{1}{L}\sum_{\ell=1}^{L}x(\ell)$$

is still a probability vector; it is the unique positive eigenvector of P, and it is independent of the initial state $x(0)$.

23.9 Continuous stochastic transition processes

We change the discrete transition process in Section 23.8 to one which is continuous, but taking the time

$$t=0,\ \Delta,\ 2\Delta,\ \ldots.$$

In order to sustain the notion of being continuous, we assume that, as the time interval becomes smaller, we are approaching the situation when there is no change; that is, we suppose that, when Δ is sufficiently small, $a_{ij}\Delta$ is the probability of the state j of the system

S at time t, changing to the state i $(i \neq j)$ at time $t + \Delta$, and $1 - a_{jj}\Delta$ is the probability of the state j of the system S at time t, changing to the state j $(i \neq j)$ at time $t + \Delta$. Here

$$a_{jj} = \sum_{\substack{i=1 \\ i \neq j}}^{n} a_{ij}, \quad a_{ij} \geq 0, \quad i \neq j. \tag{23.24}$$

We then have

$$x_i(t + \Delta) = (1 - a_{ii}\Delta)x_i(t) + \Delta \sum_{j \neq i} a_{ij} x_j(t), \quad i = 1, 2, \ldots, n,$$

and $t = 0, \Delta, 2\Delta, \ldots$.

Suppose that

$$x_i(t + \Delta) = x_i(t) + \Delta \frac{dx_i(t)}{dt} + O(\Delta^2);$$

then, as $\Delta \to 0$, we have

$$\frac{dx_i}{dt} = -a_{ii}x_i + \sum_{\substack{j=1 \\ j \neq i}}^{n} a_{ij}x_j, \quad x_i(0) = c_i, \quad i = 1, 2, \ldots, n. \tag{23.25}$$

Here $c = (c_1, \ldots, c_n)$ is a probability vector

$$c_i \geq 0, \quad \sum_{i=1}^{n} c_i = 1.$$

The differential equation is called a transition differential equation. Let

$$M = \begin{pmatrix} -a_{11} & a_{12} & \cdots & a_{1n} \\ a_{21} & -a_{22} & \cdots & a_{2n} \\ \vdots & \vdots & & \vdots \\ a_{n1} & a_{n2} & \cdots & -a_{nn} \end{pmatrix}, \; a_{jj} = \sum_{\substack{i=1 \\ i \neq j}}^{n} a_{ij}, \; a_{ij} \geq 0, \; i \neq j. \tag{23.26}$$

Then equation (23.25) can be written as

$$\frac{dx}{dt} = xM', \quad x(0) = c, \tag{23.27}$$

and it is obvious that it has the solution

$$x = ce^{M't}. \tag{23.28}$$

We prove that, corresponding to any initial probability vector c, the vector x must be a probability vector. This is really easy: Let $u = (1, 1, \ldots, 1)$. From (23.26) we find that $uM = 0$, so that

$$ue^{Mt} = u\left(I + Mt + \frac{M^2}{2}t^2 + \cdots\right) = u$$

and

$$xu' = ce^{M't}u' = cu' = 1.$$

By Theorem 17.12, we know that $e^{M't}$ is a non-negative square matrix. Therefore, we have the following.

Theorem 23.23 *The solution to a transition differential equation is a probability vector.*

Finally, we consider a property of the solution to a transition equation when $t \to \infty$.

Theorem 23.24 *If $a_{ij} > 0$ $(i \neq j)$, then, as $t \to \infty$, the limit vector of the solution to the transition equation is independent of the initial vector.*

Proof. It is not difficult to show that, under the additional condition $a_{ij} > 0$ $(i \neq j)$, we have $e^M > 0$, so that e^M only has one eigenvalue equal to 1, with the others having absolute values less than 1. Therefore,

$$\lim_{t \to \infty} (e^M)^t = p'q, \quad pq' = 1.$$

From $ue^{Mt} = u$, we find that $up'q = u$, that is $(up')q = u$. Let $v = p/(up')$; then

$$\lim_{t \to \infty} (e^M)^t = v'u, \quad vu' = 1.$$

It then follows that

$$\lim_{t \to \infty} x(t) = \lim_{t \to \infty} ce^{M't} = \lim_{t \to \infty} cu'v = v.$$

The theorem is proved. □

Problem 23.25 Can the conditions in Theorem 23.24 be weakened? For example: If we introduce various types of square matrices M under some equivalence relation (such as similarity via a generalised exchange matrix) together with the notion of being indecomposable, can Theorem 23.24 still be valid for indecomposable M?

Problem 23.26 Can there be a relationship $P = e^M$ for a probability matrix P and a matrix M satisfying (23.26)? If not, then find some counterexamples and investigate the condition required for it to be possible. If M is decomposable then clearly P is also decomposable, but is the converse true?

Index

Abel summability, 242
absolute values of matrices, 463
affine equivalence, 500
affine space, 500
affine transformation, 500
analytic continuation, 131
analytic function, 55
Argand plane, 2

Bernoulli number, 191
Bieberbach conjecture, 198
bilinear transformation, 5
Borel polygon, 266
Borel summability, 244
Borel–Carathéodory theorem, 162
boundary point, 92
branch point, 133

(C, α) summability, 240
Cauchy principal value, 150
Cauchy type integral, 70
Cauchy–Riemann equations, 49, 52
Cauchy's integral formula, 70, 113
Cauchy's theorem, 108
Cesàro summability, 232
closed set, 92
closure, 93
cofactor, 383
compact set, 90
complex plane, *see* Argand plane
conformal transformation, 48
congruence, 496
conjugate functions, 50
  harmonic, 50
conjunctive, 515
connectivity index, 102
convergence on the sphere, 193
Cramer's rule, 383
cross-ratio, 12, 507

decomposable matrix, 576
dense set, 92
determinant, 382
difference equation, 449
Dirichlet problem, 72
Dirichlet's integral, 556
doubly periodic function, 305

eigenvalues, 536
elementary (matrix) operations, 406
elementary symmetric functions, 390
elliptic function, 306
essential singularity, 126
Euclidean norm, 524
exterior point, 92

Fejér's theorem, 264
function, uniformly differentiable, 119
fundamental region, 308
fundamental theorem of algebra, 154
fundamental theorem of projective geometry, 30

Γ function, 147
Gauss plane, *see* Argand plane, 27
general linear group, 27
generalised exchange matrix, 575
generating function, 453
generating function method, 452
genus, 180
Gram–Schmidt process, 525
Gramian, 534
Green's function, 71, 87

H-congruence, 515
Hadamard's factorisation theorem, 181
Hadamard's inequality, 531
Hadamard's three-circle theorem, 164
Hardy–Littlewood theorem, 247
harmonic function, 50
harmonic sequence, 12
Hermitian form, 515
Hermitian matrix, 3, 515
Hessian, 512

Hilbert–Privalov problem, 286
Hölder summability, 235

indecomposable (non-negative square) matrix, standard form, 576
inner/scalar product, 523
integral function
entire function, 175
order of, 178
integral transformation, 3
isolated singularity, 125
iterated integrals, 468

Jacobi matrix, 416
Jacobian, 416
Jacobian elliptic function, 360
Jacobi's θ function, 346
Jordan arc, 97
Jordan (closed) curve, 97
Jordan normal form, 431
Jordan's theorem, 99

Knopp–Schnee theorem, 236

L-similarity, 482
Laplace transform, 456
Laplace's equation, 50
Laplacian operator, 50
Laurent series, 121
left-equivalence, 412
limit points, 92
linear regression analysis, 380
Liouville's theorem, 83
Littlewood's Tauberian theorem, 259
Lobachevsky group, *see* non-Euclidean motion
Lyapunov equivalence, 493
Lyapunov matrix, 482
Lyapunov transformation, 490
Lyapunov's method, 485

Markov matrix, 588
matrix equivalence, 409
maximum modulus theorem, 162
mean-square deviation, 380
meromorphic function, 175
Mittag–Leffler's theorem, 182
Möbius transformation, *see* bilinear transformation
modular function, 327, 335
modular substitution, 327
module, 303
Montel's theorem, 173

natural boundary, 134
Neumann problem, 77
non-Euclidean motion, 8
norm, 524
normal distribution integral, 559
normed space (Euclidean), 524
normal family, 174

open kernel, 93
open set, 92
order, of an integral function, 178
ordinary differential equation, 450
orthogonal group, 518
orthogonal matrix, 519
orthogonal projection, 528
orthogonal transformation, 518
orthonormal base, 525
orthonormal vectors, 525

peak, 576
perfect set, 92
Phragmen–Lindelöf theorem, 163
Picard's theorem, 196
Poisson kernel, 68
Poisson–Jensen formula, 160
Poisson's formula, 67
poles, 124
Pólya–Szegö theorem, 167
positive definite form, 508
positive (semi-)definite form, 508
probability vector, 588
projection vector, 528
projective transformation, 504

rank, of a matrix, 405
reducible matrix, 482
regular point, 134
relaxation, 382
residue, 138
Riemann sphere, 10
Riemann surface, 60
Riemann ζ function, 189
Riemann–Hilbert problem, 290
Riemann's mapping theorem, 64
Riesz's theorem, 263
Rouché's theorem, 153

scalar product, 523
Schottky's theorem, 196
Schwarz reflection principle, 220
shear, *see* transvection
signature, of a quadratic form, 498
similarity, of (square) matrices, 431
Sokhotsky's formula, 282
strongly indecomposable matrix, 587
symmetric functions, 390
symplectic matrix, 547

Tauber's theorem, 251
topological mappings, 96

transformations, groups of, 5
transvection, 407

unitary matrix, 531
unitary space, 533
unitary transformation, 531
univalent function, 96, 199

Vandermonde determinant, 385
vector space, 411
Vitali's theorem, 169

Weierstrass' double series theorem, 115
Weierstrass' essential singularity theorem, 126
Weierstrass' factorisation theorem, 177
Weierstrass' function, 314
Weierstrass' γ function, 313
Weierstrass' ρ function, *see* Weierstrass' γ function
Wronskian, 429

zeros, 124